D0597652

Essentials of Statistical Analysis

for the Behavioral Sciences

Douglas R. Glasnapp
John P. Poggio

The University of Kansas

Essentials of Statistical Analysis

for the Behavioral Sciences

Charles E. Merrill Publishing Company
A Bell & Howell Company
Columbus Toronto London Sydney

153104

To Mary, James, and Jason
To Ann, Mark, Jean, and Andrew
for their support and understanding

Published by Charles E. Merrill Publishing Co.
A Bell & Howell Company
Columbus, Ohio 43216

This book was set in Garamond and Univers
Production Editor: Megan Rowe
Text Designer: Cynthia Brunk
Cover Designer: Cathy Watterson
Cover Art: Jasper Johns *Numbers in Color,* 1959 encaustic & collage on canvas, 66½ × 49½. Albright-Knox Art Gallery, Buffalo, New York. Gift of Seymour H. Knox, 1959.

Copyright © 1985 by Bell & Howell Company. All rights reserved. No part of this book may be reproduced in any form, electronic or mechanical, including photocopy, recording, or any information storage and retrieval system, without permission in writing from the publisher.

Library of Congress Catalog Card Number: 84–62227
International Standard Book Number: 0–675–20369–4
Printed in the United States of America
1 2 3 4 5 6 7 8 9 10—89 88 87 86 85

519.5
G548e
GIFT
SR. MONA SMILEY

PREFACE

As in other areas of study, advances in statistical methodology continue to broaden the horizons of our knowledge. Complex statistical manipulations that only a few years ago presented monumental challenges are now routinely and efficiently accomplished by computers. Prerequisites to learning both new and old methods are the fundamentals of statistical reasoning. This text focuses on these necessary foundational concepts.

The scope of the text includes an introduction to the role of statistics in behavioral science research, and comprehensive discussions of descriptive statistics, the fundamentals of inferential statistics, and tests of significance through the one-way analysis of variance design. These topics are appropriate for a one-semester introductory course and may supplement a second-semester course. The intended audience includes undergraduate and graduate students in education, psychology, and related areas.

We realize that no single text can fulfill each individual's needs. We believe, however, that a text's organization and strategy in presenting material can accommodate individual differences. This text incorporates several instructional design principles intended to facilitate learning for a diverse group of students. The writing style, immediate feedback, opportunities to apply statistical concepts to real data, and a variety of task-oriented exercises support the text's student orientation. Lacking previous exposure to statistics, many students do not trust their ability to acquire such knowledge and thus will appreciate the variety of opportunities for feedback on their knowledge level and understanding. Exercises use relevant problem situations and emphasize the role of statistical analysis in the total research process.

The level of presentation focuses on students with limited formal mathematical background. Where appropriate, mathematical derivations are presented and explained; they supplement information presented rather than directly convey a statistical concept. The format each chapter follows is to introduce the major concepts, then elaborate on their meaning. Examples and illustrations of the statistical concepts and procedures follow. When important concepts are introduced, questions or directions request that a specific task be performed. Feedback and answers appear in a table or in a footnote as appropriate to the situation. Students can thus practice and apply their understanding of a concept at critical points.

Examples and problem exercises are many and varied. Our experience in teaching statistics indicates that students desire relevant applications for statistical concepts, which the book's empirically based data sets provide. Many of the examples and problems throughout the text use data from the first-year evaluation of the children's television program "Sesame Street."

We are indebted to a number of individuals and organizations who provided critical reviews, assistance, and permission to reprint material. We appreciate the many helpful comments and suggestions made by Robert Abbott, University of Washington; Ken Brewer, Florida State University; Clarke Burnham, University of Texas-Austin; Carl Huberty, University of Georgia; James Treloar, Ball State University; Thomas Stecklein, University of Minnesota; William Schmidt; Michigan State University; Barbara Dodd, University of Texas-Austin; and Dawn Eros for her detailed attention to the accuracy of the many computations in the text. We also thank the students who provided feedback on initial drafts.

Production of drafts of the manuscript involved several individuals to whom we are indebted. We also acknowledge the patience and contributions of the staff at Charles E. Merrill. Megan Rowe, our production editor, deserves special recognition. We thank the authors and publishers who granted permission to use tables or figures in the text. Included in this list are the Biometrika Trustees, the Psychological Corporation, the Rand Corporation, National Assessment of Educational Progress, and S. Karger.

We are particularly grateful to the Children's Television Workshop (CTW) for granting permission to use a subset of the data from the first-year evaluation of "Sesame Street." We thank especially Drs. Keith W. Mielke and Edward L. Palmer at CTW and Gerry Ann Bogatz at Educational Testing Service (ETS) for their special efforts on our behalf. The data reported in Appendix A are not the complete set of evaluation data. Rather, it is a sample selected to meet our own requirements. The text presents these data for illustrative purposes only; they should not be used to replicate findings from the ETS evaluation, perform secondary analyses, or make specific research generalizations about the variables or subgroups involved. Any text illustration or use of the "Sesame Street" data in a manner contrary to the intent of the first-year evaluation and the goals of CTW is unintentional. For any such instances we sincerely apologize and assume all responsibility. Statements and conclusions drawn from our use of "Sesame Street" data are ours and do not imply the endorsement of either CTW or ETS.

Finally, but most importantly, we recognize the contributions of our families for their patience, understanding, and support.

Douglas R. Glasnapp and John P. Poggio

CONTENTS

3 SAMPLING 31

4 DATA REPRESENTATION: FREQUENCY DISTRIBUTIONS AND GRAPHS 51

5 MEASURES OF CENTRAL TENDENCY 87

6 MEASURES OF VARIABILITY 109

7 STANDARDIZED SCORES AND THE NORMAL CURVE 131

8 MEASURES OF RELATIONSHIP 165

12 HYPOTHESIS TESTING 301

13 HYPOTHESES ABOUT POPULATION MEANS 341

14 HYPOTHESES ABOUT OTHER STATISTICS 387

15 THE χ^2 DISTRIBUTION—ANALYSIS OF FREQUENCIES 421

16 THE ONE-WAY ANALYSIS OF VARIANCE 451

Essentials of Statistical Analysis

for the Behavioral Sciences

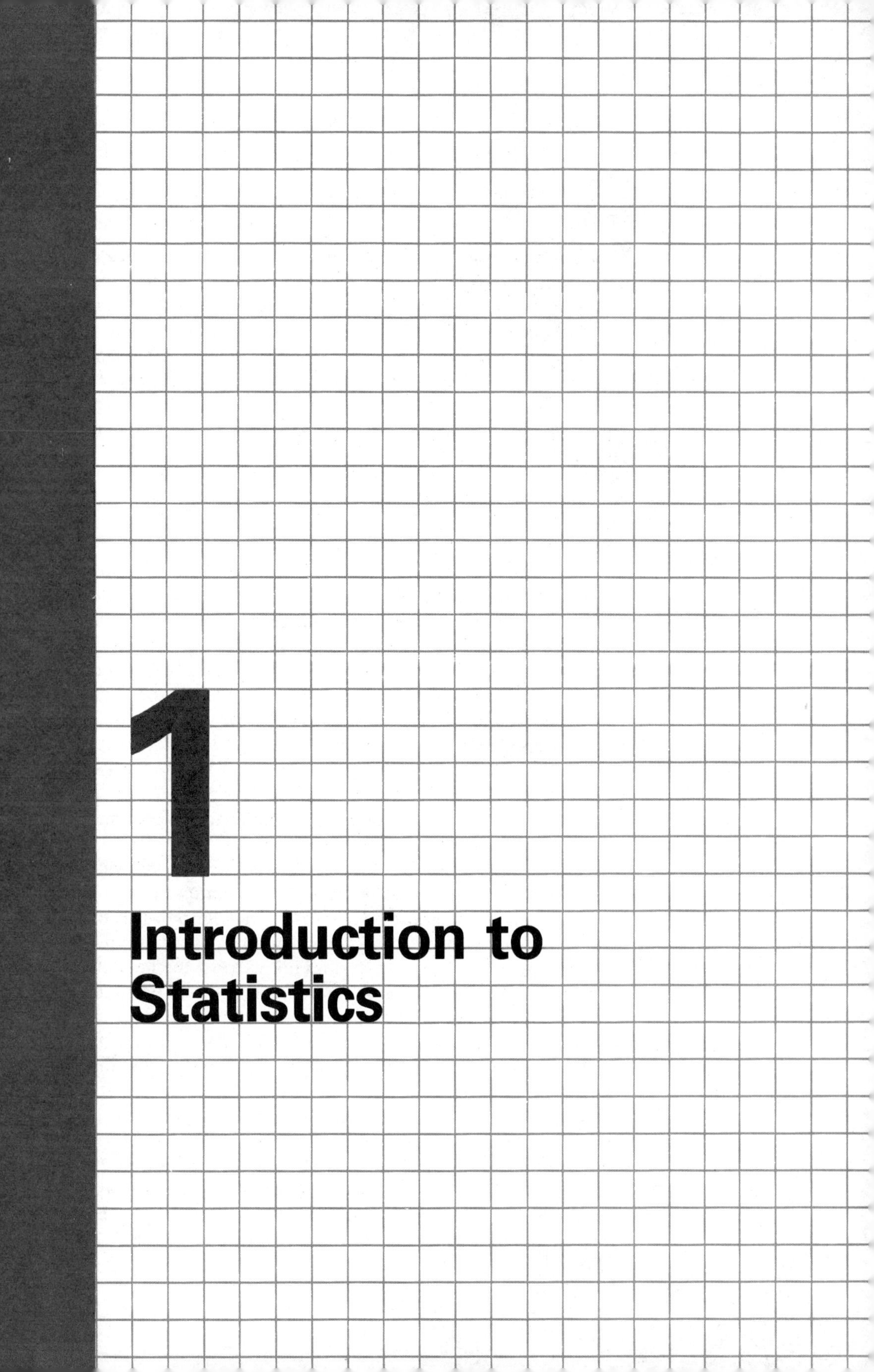

1

Introduction to Statistics

Each of us enters an introductory course in statistics with a different background and skills, personal goals, aspirations, values, and attitudes that magnify our own uniqueness. Research into the teaching/learning process suggests that the instructional system and the extent of student motivation and involvement will directly affect what can be achieved. The goal of *Essentials of Statistical Analysis for the Behavioral Sciences* is to present introductory statistics in an understandable, meaningful form. This text represents an instructional system whose effectiveness increases with student involvement, interest, and motivation.

1.1 THE TERM STATISTICS

The word *statistics* stimulates a variety of reactions from students entering an introductory course. Interestingly, initial reactions are often attitudinal or emotionally based, possibly because many in the behavioral sciences consider statistics only a required course or series of courses in their program of study. They attach expectations to the course(s) *and* the word. Fortunately, for those who teach statistics, many responses are positive. Students in programs emphasizing research and evaluation realize the importance of statistics. They know that without statistical knowledge and skills they will not be able to reach their potential as professionals. At the minimum, a professional needs familiarity with the terminology of statistics to be able to interact and communicate effectively with other behavioral scientists.

One way to think of statistics is as a body of knowledge—a content domain of knowledge and skills. This text *introduces* you to this content domain. You will note that *introduces* has been italicized. This text is just that—an introduction—aimed at building a foundation of basic terminology, concepts, and generalizable skills. It does not exhaust the field of statistics. Many institutions offer a sequence of four or more courses on statistics and do not completely cover the field. A comment made by many students illustrates the expansive domain of statistical content: "The more statistics I learn, the more I realize how limited my knowledge level is. Each new concept makes me aware of other indices and procedures I need to know about." Of the several possible reasons for this statement, the primary one is that as one's knowledge of statistics grows, so does one's awareness of the research process and the complexity of most researchable problems. Introductory statistical concepts can be applied to basic, less complex research situations; students soon recognize that limitation. These introductory concepts form a necessary foundation for learning more complex statistical methods, however.

We have discussed statistics as a body of knowledge. Consider another definition of a statistic—as a numerical index calculated on a sample set of data. Examples of statistical indices are the average salary for a sample of teachers in a school, the proportion of children in a region who live in single-parent homes, the percentage of employed individuals in a particular state, or the number of automobiles sold during a given month. Averages, counts, percentages, and proportions are the statistics commonly found in lay publications. This text will expand your knowledge about these and other statistics, but more importantly, it will stress the role of these indices in research and evaluation. By the conclusion of this course, you will be familiar with a variety of statistical measures and generalizable procedures that will form a foundation for statistical reasoning.

You probably still don't have a "good feel" for what *statistics* implies. It will probably take the better part of the course for you to have a comfortable grasp of the concept. As you progress through the text, your awareness of both its meaning as a sample index and as a domain of content should increase.

1.2 ROLE OF STATISTICS IN RESEARCH AND EVALUATION

While this is not a text on research or evaluation methods, the importance of statistics to these processes cannot be overemphasized. Statistical concepts and skills should be viewed as tools and not ends in themselves. They are essential for communicating and making decisions in scientific inquiry. The following overview of research and evaluation processes illustrates the role of statistics.

The Research Process

As with any scientific endeavor, research in the behavioral sciences aims to describe, explain, predict, or understand phenomena. Part of this process involves acquiring knowledge and developing and testing theories that allow us to understand the typically complex relationships among phenomena. Today, most research follows the scientific method, using a systematic, controlled investigation that produces objective conclusions relevant to the problem. The steps in this process are

- Step 1 Identify problem area.
- Step 2 Clarify problem by identifying researchable hypotheses or questions.
- Step 3 Develop research plan for data collection.
- Step 4 Analyze data by applying statistics.
- Step 5 Make conclusions/interpretations from findings.

As students in search of thesis and dissertation topics know, the first two steps are often the most difficult in the process. Delimiting the many factors in a problem area to a specific definition of characteristics to be included, relationships to be investigated, or treatment to be assessed requires considerable induction, deduction, and inference. When these steps are accomplished well, however, the remaining steps will flow naturally. Sources for identifying problem areas include direct experience, professional literature, and postulated theories. For problem clarification, a thorough review of the literature is an absolute necessity under any circumstance.

The adequacy of the procedures entailed in step 3 determines the validity of a researcher's conclusions. These design and data collection procedures include identification of

1. a participant sampling plan that will allow correct interpretation of effects and generalization of results
2. data collection instruments that measure what they purport to measure (validity) consistently and precisely (reliability)
3. data collection procedures that do not obscure the meaning of the data
4. an overall research design that produces appropriate data for answering the research question(s) or testing the hypotheses generated in step 2, including identification of the appropriate statistic or statistical technique(s) to be used when analyzing the data

Without proper sampling, measurement techniques, research design, and data collection procedures, the representativeness and generalizability of the results are questionable. Often overlooked or deemphasized are the measurement techniques employed. It is a mistake to believe that statistical methods provide the cure for any inadequacies in the procedures at step 3. Nothing can be further from the truth! Bad data beget invalid results and conclusions, no matter how sophisticated the statistical technique. The data collection plan should be a major concern in any research endeavor.

The procedures for steps 4 and 5 depend solely on the outcomes of steps 2 and 3. The statistical procedures employed must be compatible with the research hypotheses or questions and the research design. The task is one of choosing the appropriate statistical measure and procedure that fit precisely the question and design used. It should be evident from the sequence of activities that applying appropriate statistical techniques is a necessary component of the research process but not the only factor in the validity of results.

In the final step, the conclusions are drawn based on interpretation of evidence provided by the statistical indices and tests. Too often, statistical results are interpreted only at a rote level. It is important to remember that the numbers were derived from data and that these data represent measurements of particular variables (i.e., trait, attribute, construct, or characteristic).

The important task is interpretation of statistical results as they relate to the research questions and hypotheses in terms of the real variables under investigation.

The Evaluation Process

Researchers commonly differentiate between evaluation and research. Distinctions can be made between the goals, initial stages, and design for data collection and type and use of data gathered. Statistical knowledge and skills are as important to evaluation as they are to research, however. The same statistics serve as tools in both processes.

The goal of evaluation generally concerns the immediate usefulness of the results in making decisions about a specific practice. Evaluation is a process that provides information concerning the adequacy or effectiveness of programs to achieve goals. In evaluation, systematically collected information guides decisions to assist program efforts and possibly to judge the merits of these efforts. The steps in the evaluation process typically include

- Step 1 Assess needs/clarify goals/identify program objectives.
- Step 2 Select, develop, implement initial program.
- Step 3 Implement evaluation plan/collect data.
- Step 4 Analyze data/apply statistics.
- Step 5 Interpret results/make decisions.

While the research process is geared toward reaching conclusions about hypotheses or questions tied to specific theories, the evaluation process usually results in an interpretation of statistical indices addressing the attainment of specific program objectives. Evaluation activities are often more varied, directing attention at both process and product, and ongoing. Yet, steps 3 and 4 in the research process need to be followed in evaluation, too. Again, care must be taken in the design for data collection, in the quantification (measurement) of variables, and in the actual collection of data. The adequacy of these procedures determines the validity of the evaluation results. Statistical analysis once again serves as a tool; the results are only as good as the data on which they are based.

1.3 PREREQUISITES AND STUDY HABITS

This book has been prepared for use in an introductory statistic class for students in the behavioral sciences at the graduate or undergraduate level. The only prerequisite math skills are the abilities to add, subtract, multiply, and divide. This book stresses the skills and knowledge necessary to understand and use statistics.

Memorizing formulas or studying specific facts will not necessarily ensure success. Formulas should be considered only as symbols, abbreviated methods for communication. To memorize them does not guarantee an understanding of them. Read and study for a broader level of understanding. As an illustration, imagine that you are driving down a road and see in the distance an eight-sided red sign. Your only additional knowledge about this object is that written across the face of the sign are the letters S-T-O-P. This situation is comparable to the individual who tries to study statistics by memorizing formulas. In either instance, not knowing the intent—of the stop sign or the statistical formula—can produce disastrous results. Suppose on the other hand you are driving down a road in a foreign country in which you do not know the language and you see a red, eight-sided sign with the inscription [illegible]-[illegible]-Y-I. You stop at this sign and do not proceed until it is safe. This behavior indicates a general, conceptual understanding of the symbol representing this book's philosophy of studying statistics. Formulas represent relationships that by themselves have little meaning or value. When coupled with conceptual understandings of the relations they represent, they are a convenient, efficient method for communication. The moral is: Should you find yourself studying a formula but have no idea of its meaning, make sure your seatbelt is fastened!

A second rule of good study habits in statistics is to complete problem exercises. This text contains a wide range of task-oriented experiences and hands-on exercises. At each opportunity try your hand at these problems. As earlier, if you find yourself simply going through the motions, stop and reevaluate what you are doing. Ask questions. Work toward a level of understanding that will allow you to transfer and apply your knowledge to other situations.

A third suggestion for the study of statistics is to read material *before* going to class. Instructors will present information clearly and precisely. If a class presentation is the *first* time a student has encountered the information, the student can easily develop a *false* sense of understanding. An attitude of "Well, that sure makes sense. I guess it won't be hard to learn that" can be misleading. Realize that your instructor has studied and prepared extensively before the presentation to put it in a meaningful form. If instructor presentations were 100% foolproof, guaranteeing success, you would not need to be reading this text now. If it takes that long for the instructor to organize the material, it is likely to take you as long to learn it for yourself. Reading before class allows you to ask questions about material causing you trouble and to confirm and reinforce what you've already learned.

1.4 THE TEXT AS AN INSTRUCTIONAL SYSTEM

As should be obvious by now, this text assumes that statistics need not be studied as a separate domain of knowledge. Few people are trained to be solely statisticians. For most students in introductory statistics courses, the immediate goal is acquiring skills necessary to participate effectively in the overall research/evaluation process. As such, this book deemphasizes the idea that statistical procedures are merely manipulation of numbers and formulas. Manipulation is a necessary step, but numbers do not exist in isolation. Rather, they represent data, a quantification of some attribute, construct, trait, or characteristic of an individual or group. The statistical indices derived from manipulating these numbers help define the characteristics under investigation. Interpretation and decision making within the context of the total investigation represent the most relevant aspects of statistics for most students.

Each chapter in *Essentials of Statistical Analysis for the Behavioral Sciences* will first present the major concepts, elaborate on their meaning, and then provide situations in which these concepts become important considerations. Examples and illustrations of the chapter's statistical concepts and procedures follow. The step-by-step sequence depicts, interprets, and elaborates on the result.

For important concepts, questions or directions in the text will request that you perform a specific task. At these junctures a critical skill or concept is being considered; immediate practice may assist and then reinforce learning. Feedback and answers will be included in tables or footnotes. This format will ask you to practice and apply your understanding of a concept while you are reading. You should perform the requested tasks before seeking feedback and answers.

To provide realistic examples in the text, we have used data from the first-year evaluation of the children's television program "Sesame Street." This data base, provided in Appendix A, is the source for many of the examples and problems throughout the text. Advantages of using the "Sesame Street" data as a central focus include the following:

1. The number of variables, groups, and situations offer opportunities to apply a wide variety of statistical techniques.
2. The variables studied involve several disciplines within the behavioral sciences.
3. The data and questions that can be answered should be of interest to many students in the behavioral sciences.

Because of its key role in illustrating concepts and problems, the "Sesame Street" evaluation study is summarized in the next section.

In addition to the illustrations used to present concepts within a chapter, the problem exercises at the end of each chapter provide both conceptual and experiential applications to real research situations. These exercises are divided into three types. Problem Set A provides straightforward applications of chapter concepts and techniques. The problems in this set are included primarily for practice, repetition, and feedback, reinforcing an understanding of individual chapter topics. Problem Sets B and C integrate chapter statistical concepts and procedures into the more generalized research process framework. Problem Set B presents a brief abstract of at least one research study reported in the literature. While the study reported is always authentic, the data are usually simulated. The studies serve a twofold purpose. First, they provide a research example whereby a chapter's statistical methods are appropriately applied. Second, they provide the context for exercises making use of the simulated data. Problem Set C involves applying statistical concepts and techniques to the "Sesame Street" data base. These exercises exemplify an emphasis on relevance to research and evaluation in the real world. Answers to problems in Sets A and C are given in Appendix D.

1.5 THE "SESAME STREET" DATA BASE[1]

At its inception, "Sesame Street" was developed as a television series aimed primarily at teaching preschool-related skills to children in the 3–5-year-old age range with special emphasis on reaching 4-year-old disadvantaged children. The program was funded cooperatively in 1968 by several federal and private agencies. The Children's Television Workshop (CTW) was created as a nonprofit organization to produce this program. The success of its programming is evident: "Sesame Street" continues to be produced and shown 16 years later.

The initial format for the program was designed to hold young children's attention through action-oriented, short-duration presentations teaching specific preschool cognitive skills and some social skills. The interaction of humans, puppets, and animated characters was the vehicle for delivery of the desired concepts. Each show was of one hour's duration and involved much repetition of concepts both within and across shows. Some of these initial characteristics may still be found in current programs, but the programming sophistication, purpose, and curricular content have been expanded tremendously since the initial conceptualization. If you have not seen this television program recently, take the time to view a couple of hours of its programming. We think you will find it informative and enjoyable.

Evaluation of the First Year of "Sesame Street"

The evaluation of the impact of the first year of "Sesame Street" was conducted by Educational Testing Service (ETS) under the direction of Samuel Ball and Gerry Ann Bogatz. The goals of this evaluation were focused appropriately on

---

[1]Permission to use the data from the first-year evaluation of "Sesame Street" within the confines of this text was granted by the Children's Television Workshop (CTW). The permission granted by CTW constitutes a limited release of the data and even though a subset of these data is reproduced in Appendix A, it does not constitute the transfer or release of the data into the public domain. We are indebted to Drs. Edward L. Palmer at CTW and Gerry Ann Bogatz at ETS for their special efforts in initially providing the data for our use and to Dr. Keith W. Mielke at CTW for suggesting modifications. We appreciate the cooperation, time, and effort expended by these three individuals on our behalf.

the initial purposes of "Sesame Street" programming. However, it should be noted that these goals constituted a focus only in the early years of "Sesame Street" and do not reflect the present-day curriculum, which has been expanded considerably from the early years. The evaluation during this first year also concentrated on the impact of "Sesame Street" on disadvantaged preschool children. While children from other backgrounds were sampled and data collected, special care was taken to present results for subgroups separately with no intent of making comparisons across subgroups based on demographic characteristics. With this in mind, six major questions guided the first-year evaluation activities (Ball and Bogatz, 1970, p. 349):

1. What was the overall impact of "Sesame Street"?
2. What were the moderating effects of age, former achievement, sex and socioeconomic status on the impact of "Sesame Street"?
3. Did children at home who watched "Sesame Street" benefit in comparison to children at home who did not watch?
4. Did children in preschool classrooms benefit from watching "Sesame Street" as part of their school curriculum?
5. Did children from Spanish-speaking homes benefit from "Sesame Street"?
6. What were the effects of home background conditions on the impact of "Sesame Street"?

A major concern in the sampling for the evaluation was that it would permit generalization to the populations of children of most interest to CTW. The population of major interest consisted of 3–5-year-old disadvantaged children from inner-city areas in various parts of the country. Other populations of interest included 4-year-old advantaged suburban children, advantaged and disadvantaged rural children, and disadvantaged Spanish-speaking children. Children representative of these populations were sampled from five different locales in the United States for data collection purposes. The sites chosen for data collection were in areas served by VHF television stations with sizable groupings of one or more of the populations described. Children sampled within each group were either at home or were attending some form of preschool.

In addition to the population centers and advantaged/disadvantaged pupils sampled, several other groupings were identified within the context of the evaluation design. These included

1. sex designation: female, male
2. viewing location: home, school
3. viewing status: encouraged, not encouraged
4. observational status: observed, not observed
5. predominant language spoken in home: English, Spanish

Both before viewing (pretest) and after viewing (posttest), data were collected over a variety of variables for all groups. In all, 1124 children were selected to participate in the evaluation. Complete, usable data were reported for 943 children.

The Text Data Base

Because of the large number of children and variables included in the ETS evaluation of "Sesame Street," it was not feasible to provide the complete data set in the text. Rather, we have sampled data for subgroups of individuals to meet our own requirements. To create a more workable set, 240 individuals were selected to make up the data base used for the text. A description of these 240 individuals will be given in illustrations in subsequent chapters. Table 1.1 shows the variables selected and included in the text's data base. Appendix A contains a complete description of each variable, the numerical scales used to code the data, and the raw data on variables selected for each of the 240 children.

TABLE 1.1
Variables included in the text data base from the "Sesame Street" evaluation.*

Background Information	Achievement Variables
1. Sampling site	Pre- and postviewing knowledge of
2. Sex	1. Body parts
3. Age in months	2. Letters
4. Viewing category	3. Forms
5. Where "Sesame Street" was viewed	4. Numbers
6. Viewing encouragement	5. Relations
7. Frequency with which the child was read to	6. Classifications
8. Whether the child attended school and the type of school attended	Postviewing scores only for
9. Peabody mental age	7. Which comes first
10. Home Education Index	8. "Sesame Street" Test
11. Home Stimulation-Affluence Index	
12. Mother-Child Socialization Index	

*A complete description is given in Appendix A.

It should be stressed that the data in Appendix A, while real data, are for illustrative purposes only and do not exhaust the vast amount and type of data collected for the evaluation of the first year of "Sesame Street." These data should *not* be used with an intent to replicate findings from the ETS evaluation or to perform secondary analyses on the data. Our reduction of the complete ETS data set through arbitrary elimination of children has introduced bias in the Appendix A data set. Because of the sampling bias, the results obtained from the data in Appendix A could conceivably deviate from those reported in the *First Year of Sesame Street: An Evaluation* (Ball and Bogatz, 1970).

In concluding the description of the "Sesame Street" data base used, it is *not* our purpose to criticize any phase of the ETS evaluation—neither the objectives, methods, instrumentation, sampling, data collection procedures, data analysis, results, nor the conclusions. Rather, we intend to use the data base as a vehicle for teaching concepts within the domain of an introductory statistics course. If you are interested in detailed discussions of the results reported by ETS, you should read the final evaluation report, the critique and secondary analysis of the entire ETS data set by Cook et al. (1975) and the review of the Cook et al. work by Schramm (1976).

1.6 SUMMARY

In plural form, statistics represents a domain of content knowledge and skills. Singularly, a statistic is a numerical index calculated from a sample set of data that summarizes and communicates information about the sample. Your grasp of both concepts should increase tremendously as you progress through this course.

This introductory chapter has introduced the main features of the text. It also outlined a desirable learning set for approaching the study of statistics. The basic points covered were

1. Remember that the role of statistics is as a tool in the overall research and evaluation process. The validity of data analysis results depends on the application of appropriate statistical techniques. However, the *quality* of the data analyzed is the prime determinant of valid or invalid results and conclusions.
2. Get involved as an active learner. The format and style of the text will allow you many opportunities.
3. Use the instructional approach in the text to your advantage. Take advantage of the cues that identify important concepts, the opportunities for problem solving and applications, and the provision for self-evaluation and feedback.
4. Become familiar with the "Sesame Street" data base. This data set provides relevant applications and interpretations.

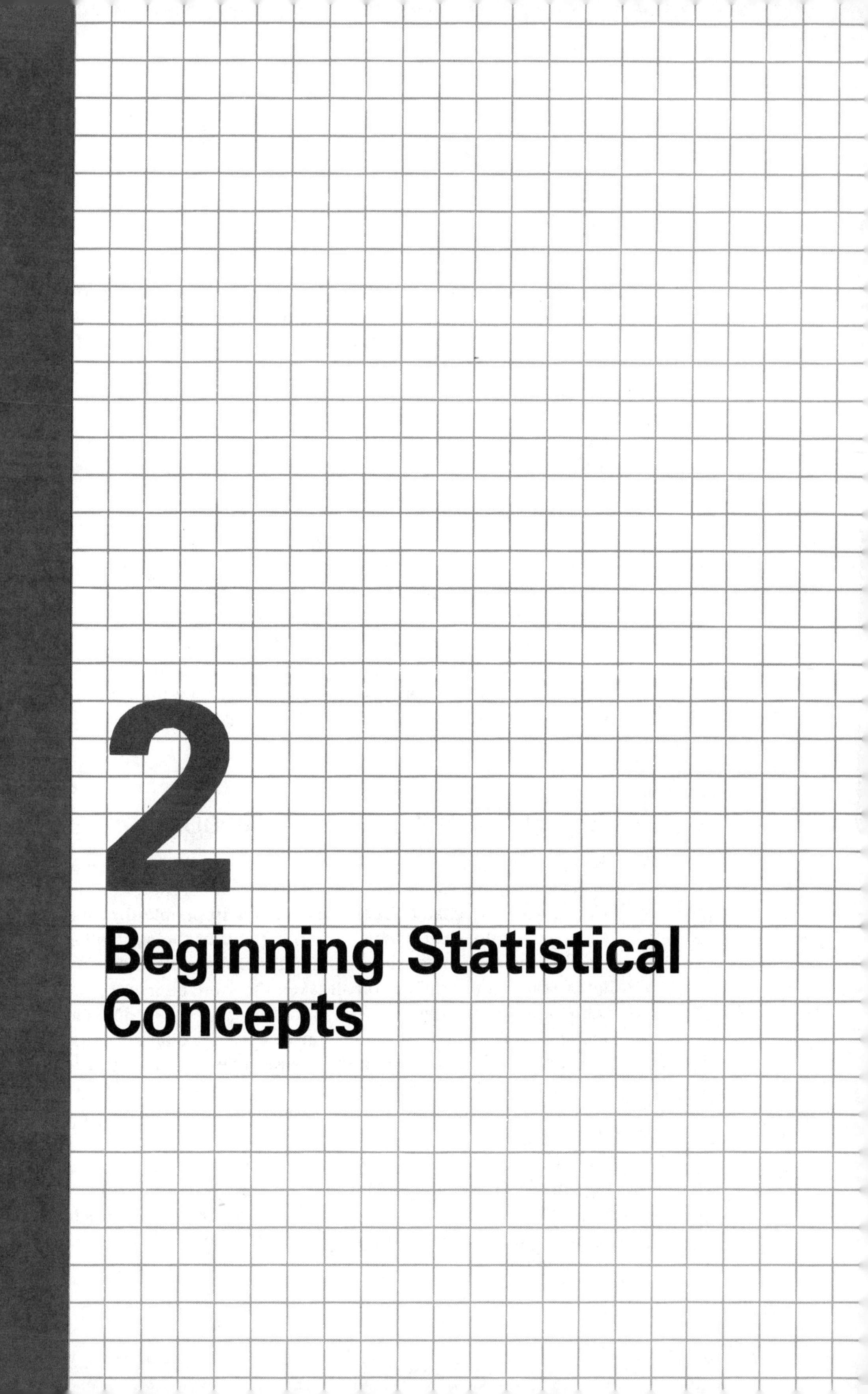

2
Beginning Statistical Concepts

We come into contact with statistical information daily. Newspaper and magazine articles, advertisements, and radio and television reports convey information that is filled with statistics. Consider the newspaper article that reports "The President's popularity has increased by as much as 10% in rural areas," or election-night coverage in which the commentator announces "With 2% of the vote counted we predict the incumbent will be reelected by a landslide," or the headline stating "Average Test Scores Show Decline Over Elementary Grades," or the warning that "the surgeon general has determined that cigarette smoking is dangerous to your health." Each illustrates how statistics provide information and conclusions.

Applying statistical procedures in situations such as these is termed **statistical analysis. Data** are the information on which statistical analyses are performed. The word *data* is a plural reference to group information. A specific piece of information for one member of the group is a **data point** or **datum.** The "Sesame Street" evaluation information in Appendix A is an example of a large set of data. In most instances data will take the form of numbers.

Statistical manipulation of data results in a numerical summarization of the information first gathered. This numerical summarization is called a statistic; it is then interpreted and a judgment is made.

2.1 THE BASIC INGREDIENT: INDIVIDUAL DIFFERENCES

Why does the discipline of statistics exist and attract so much attention in the behavioral sciences? Because of individual differences. Individual differences account for any characteristic or behavior in people that is not the same for all persons under study. For example, think of the students in your statistics class. Do you have equal verbal ability? Do you all share the same degree of interest in politics? Have you all taken the same courses or performed comparably on examinations? Are you all of the same sex? Do you all have the same income? These few examples point to some specific characteristics on which individuals differ. We cannot expect behavior or characteristics to be exactly the same for all persons. To study behaviors or characteristics requires statistical procedures that allow us to describe, predict, or compare while accommodating differences in individuals.

Characteristics or behaviors on which a set of persons differs are **variables.** Those on which a set of persons is the same are **constants.** Think again of your classmates. What are some other variables for this group of students? (Now is the time to get involved. Answer the questions!) Other variables might be how much people weigh, their ethnic background, how fast they could run the 100-meter dash, or how extroverted they are. Variables are things on which persons differ.

Can you think of any characteristics that are constants? The constants in the group are probably more difficult to identify. Level of educational attainment for the group is a constant if everyone in the class is at the same level. The person responsible for judging students' achievement in the class is an example of another constant, if you are taught by the same instructor. Another example is that this may be the first course in statistics for all students in the class. If so, data on the number of statistics courses previously taken would result in a constant. If even one person in your class had previously taken a course in statistics, however, then this characteristic would be a variable for the group.

The number of examples of variables in the behavioral sciences is virtually infinite. Practically any behavior, characteristic, trait, or construct studied in the various disciplines might be classified as a variable. A few examples from the literature include the following:

attentiveness
achievement in an area
questioning behavior
amount of eye contact
empathy
intelligence
impulsivity
time to task completion
self-concept
amount of classroom structure
assertiveness
need for affiliation
leadership capabilities
job satisfaction
occupation
perceptual discrimination
teacher effectiveness
motivation
organizational climate
systolic blood pressure

In the research process constants are very often elements that the investigator controls. For example, rather than permitting a person's sex to be a variable, the experimenter selects only females for a study, thus sex is a constant. When only persons who are classified as high in socioeconomic status

(SES) are surveyed, SES becomes a constant for that study. Given this description, recognize that constants can help precisely define a group while variables are characteristics on which members of the group differ.

2.2 POPULATION AND SAMPLES

The terms **population** and **sample** are used frequently in research and statistics. A population is understood to include the entire collection of objects or observations that share qualities about which one may make conclusions. This definition does not indicate size. Instead, a population is defined by its constants. Any member of a defined population is an **element** of the population, which could include people, events, or things. Examples of populations include third-grade students, first-year teachers, persons over 65 years of age, persons with a specific handicap, small businesses, organizational units with a specific administrative structure, and test items measuring survival skills for adults; also seventh-grade students taught by tenured teachers, persons aged 35–45 earning more than $35,000 annually, and learning disabled youth living in rural areas served by itinerant teachers. It is important to realize that the definition of a population restricts the group about whom conclusions can be drawn. In designing a research study the definition of the population is not a statistical decision. It is a pragmatic decision of the investigator based on the particular group about whom conclusions are to be drawn.

Populations themselves may be thought of as infinite or finite. Some populations, while finite, are so large that they may as well be considered infinite. For example, the researcher who begins by defining as a population all elementary school–age children in the United States who have lived in this country since birth is defining an extremely large group, although finite at any moment. Pause for a moment and think of characteristics that would define the members of your statistics class as a unique, finite population. How would this description change if it considered all students studying statistics in this country?

Measuring some characteristic or behavior in a population in some instances can be rather easy. At other times it may be practically impossible. The larger the population, the more difficult it becomes to gather information on all elements. Consider a study to determine if "Sesame Street" is capable of achieving its objectives for all 3-, 4-, and 5-year olds. To measure all such individuals would undoubtedly be time-consuming, costly, and difficult. The solution when it is not possible to assess all members of the population is to select a *sample* from the population. The process is one of drawing a sample, making observations and measurements in the sample, tabulating the data, and then *inferring* that what is observed in the sample is what would be found had all elements of the population been measured. Think about this for a moment. It is an important concept in statistical reasoning. A sample is defined as a subset of a population. The selected sample is observed and used for data collection, not the population. What is observed in the sample is assumed to be characteristic of the population. If the assumption is sound, then the conclusions drawn from the sample also apply to the population. The accuracy of this process depends upon the sampling process. The next chapter considers sampling methods.

For now, understand that a sample represents a subset of population elements. Sample size can range from one element up to one less than the total number of elements in the population. Most research is carried out on samples with findings then generalized to the population. The sample's **representativeness** thus becomes a crucial aspect of the research process.

2.3 PARAMETERS AND ESTIMATES

When all elements of a defined population are measured, any resulting numerical index based on the population data is referred to as a **parameter.** A numerical index computed on data from a sample of observations is termed a **statistic.** The statistic computed from a sample set of data provides an **estimate** of the population parameter. This jargon requires some getting used to. We infer something about a population based on what we find in the sample. We compute a sample statistic and use it to infer the population parameter. What we observe for the sample can only *estimate* what may actually be the state of affairs for the population.

An example from the "Sesame Street" evaluation might be useful at this point. Rather than measure all 3-, 4-, and 5-year olds in the United States for the first-year evaluation, samples were selected from locations in the country containing populations of primary interest to CTW. Five locations were selected. In all, 1124 children were selected to participate in the first-year evaluation. Table 2.1 shows the number of children selected from each location.

TABLE 2.1
Number and percentage of children participating in the first-year "Sesame Street" evaluation from five selected locations.

Sampling Site	Number Selected for ETS Evaluation	%	Number Selected for Text	%
I	402	36	60	25
II	177	16	55	23
III	218	19	64	27
IV	262	23	43	18
V	65	6	18	7
TOTAL	1124	100	240	100

In preparing the data base for this text it was impractical to provide information for all the children and their families in the evaluation. Thus, we decided to include data on a sample of all children who had participated. Table 2.1 presents the number of children included in the text data base from each location. Also given in the table is the percentage of the total number selected by location for the evaluation itself.

Let us begin by defining the children who participated in the ETS first-year "Sesame Street" evaluation as a unique population. We have then drawn a sample from this population. In this case, the percentage of children at each location participating in the ETS evaluation is a parameter of the population, while the percentage in the text data base is an estimate of the very same parameter. Each of the values in the last column of Table 2.1 is a sample index called a statistic.

In a somewhat different vein, we might define children who participated in the ETS evaluation as a sample, that is, as members of a population defined as all 3-, 4-, and 5-year olds living at the locations chosen. In this situation, the ETS evaluation percentages are all estimators of unknown population parameters. Often the investigator does not know the numerical value of the population parameter being estimated by the sample statistic. You are undoubtedly asking yourself how much confidence you should place in a process that selects a sample from the population and then forms conclusions about the population based on that sample. For now let us say that the dependability of this process is a function of the sample's representativeness. This point will be considered again in Chapter 3.

2.4 DESCRIPTIVE AND INFERENTIAL STATISTICS

When an investigator infers the properties, characteristics, or behaviors of the population, based on what was observed in the sample, this is referred to as **inferential statistics** or **sampling statistics.** Inferential statistics is based on the principles and theories of probability. The later portion of this text introduces the methods of inferential statistics.

When information is gathered and used to describe, tabulate, or depict specific objects, the process employs **descriptive statistics.** The primary purpose of descriptive statistics is to summarize collected data into a convenient, manageable form. Descriptive statistical analyses may be applied to data from either a population or a sample. The number of boys in the "Sesame Street" data base, a graph of scores on the Body Parts test, or the average score on the "Sesame Street" Test all represent descriptive statistical analyses. Each requires manipulation of data into forms that are more convenient and interpretable than the large array of data presented in the data base itself.

The use of descriptive statistics is commonplace. Your local newspaper probably contains stories about, for example, the number of residents favoring the use of nuclear energy as a power source, the unemployment rate, a graph or profile of standardized test scores for students in the community, or the rate of buying and selling of stocks over the past week or year. In each case, descriptive statistics summarized a large amount of data into an interpretable format.

2.5 THE MEASUREMENT PROCESS

Chapter 1 made the point that statistical formulas provide nothing more than a convenient shorthand for processes when working with data. Of and by themselves, formulas have little if any value. The information gathered in an investigation is the crucial element. Before statistical formulas can be applied, this information is quantified; that is, numbers are used to represent the amount of feelings, attitudes, achievements, ability, behavior, thoughts, and characteristics we choose to study. Pause for a moment and think about this transformation process. It is akin to translating words from one language to another. If we gather data on a survey that indicates a person's political affiliation, we can label persons who identify themselves as Democrats with the number 1, persons who identify themselves as Republicans number 2, those who indicate they are Independents number 3 and all remaining types number 4. What are you? A 1, 2, 3, or 4? A Democrat, Republican, Independent, or other? The meaning is the same whether you identify your political grouping by name or number. Recognize, however, that what the numbers represent has meaning only for persons who know the translation key.

The process described, called *measurement,* is the assignment of numbers to objects, events, or characteristics according to rules. Assigning numbers operationally defines a variable in terms of a measurement scale. The numbers take on meaning relative to the variable being measured. The specific characteristics of the numbers and the mathematical operations that can be applied to them depend on the type of measurement used.

Measurement defines a variable by a numerical scale. The process quantifies a variable, defining it in terms of numbers. Statistical analyses then are carried out on the numbers to summarize them. The resulting statistical indices are in the form of numbers that must be referenced back to the measurement scale defining the variable under study. The numbers are not merely symbols; they represent either category membership or a quantity of some variable. They can and should be interpreted relative to the variable being measured given the measurement rules applied.

2.6 MEASUREMENT LEVELS

If two persons had achievement test scores of 15 and 30, could you conclude that one individual knew twice as much as the other? Or if it takes you 30 minutes to get to work and a friend 15 minutes, can you be more precise than if you concluded that it takes you longer? Or, from the earlier illustration, if Democrats are labeled 1s and Republicans 2s, is it fair to say Democrats are half as good as Republicans? For each of these questions, the possible interpretations will depend on the measurements. Stevens (1951) defined four levels of measurement: nominal, ordinal, interval, and ratio. Other measurement scales have been proposed, but Stevens' levels are commonly accepted.

Nominal Measurement

When numbers are assigned to observations so that we can only make statements of sameness or difference, measurement is at the **nominal** level. Nominal measurement is the simplest form. With nominal measurement, the number assigned to an observation serves as its name. Numbers are assigned to each category only to identify similar objects within a category and to differentiate these similar objects within a category from elements in another category that are different. For example, in the previous coding of Democrats, Republicans, and Independents as 1s, 2s, and 3s, respectively, the numbers only substitute for the labels, and all persons in a specific category are thought of as equivalent, yet different from those in the other categories on the political affiliation variable. Consider the "Sesame Street" data. The location variable is coded 1, 2, 3, 4, or 5 for each child. While these numbers refer to specific locations, the only arithmetic interpretation possible for these data is that persons were living in the same or a different locale. The numbers do not imply that one site is bigger or that Site I is closer to Site II than Site II is to Site IV. When measurement is nominal, the numbers assigned only have the property of being distinct from each other. The numbers do not allow comparison in terms of magnitude of the variable measured. Consider the description of the viewing category included in the Appendix A data base. Would you judge it to be a nominal-level variable?[1]

Ordinal Measurement

Nominal-level measurement is the simplest form of measurement. The next-complex level is **ordinal** measurement. In addition to categorizing elements, ordinal measurement ranks observations relative to each other. Runners completing a race are categorized as finishing first, second, third, and so on. The order in which they finish is ordinal measurement. Students' ranks in their graduating class are ordinal measurements. A classification system that categorizes persons as above average, average, or below average represents ordinal measurement. It is possible to compare ordinal measurements using state-

[1]Viewing category is *not* a nominal-level variable. Categories do exist and numbers are assigned to the categories, but the numbers *have comparative meaning* in terms of "frequency with which a child watched the show." A score of 3 does indicate a greater frequency of viewing than a score of 2 or 1. Nominal-level variables do not allow us to rank order the categories.

ments such as "greater than," "less than," and "equal to." The runner who finishes first ran faster than the person who finished second. The student ranked 12th in the class achieved at a level below the student ranked 11th. Notice, however, that ordinal measurement lacks specificity. The amount separating observations is not indicated. While the runner who finished first ran faster, we don't know *how much* faster than the person who finished second, third, and so on. Ordinal-level numbers do not represent equal intervals on the measurement scale. The viewing category variable (4) from the "Sesame Street" data illustrates this concept. The characteristic measured is the "extent to which a child watched the show on the average per week." Rather than scale this variable by assigning a score for the exact number of times on the average a child watched, responses were coded in four categories. Children were given a 1 if they watched the show rarely or never; 2 if they watched the show an average of two to three times a week; 3 if they watched the show an average of four to five times a week; and 4 if they watched the show an average of more than five times a week. Three characteristics identify the scale used to measure viewing amount as ordinal. First, a number represents specific categories. Second, the numbers represent more or less of the variable. Thus, the scale allows us to rank the categories meaningfully. Finally, the categories (and therefore the numbers) do not represent equal viewing amounts. Which other variable in the "Sesame Street" data has measurement characteristics similar to viewing category?[2]

Interval Measurement

The third category is **interval** measurement. Measurements at the interval level possess all the properties of measurement at the nominal and ordinal levels. In addition, the intervals between score categories are assumed to be equal. To consider the score values 50, 70, and 90 as measurements at the interval level, for example, the 20-unit difference between 50 and 70 must represent the same amount of the variable as the difference between 70 and 90. In effect, in interval measurement, differences in score units are precisely defined and represent the same amount of the variable measured at all points on the scale.

An example of a variable that uses interval measurement is a Celsius scale in which the unit of measurement is one degree. An increase in temperature from 30° to 35° is equivalent to increases from 0° to 5° or from 8° to 13°. One unit of measurement (in this instance, one degree of heat) represents the same amount of heat or absence of heat no matter where the measurement is taken

[2]Variable 13, Frequency of Mother Reading to Child, has similar characteristics. Categories of unequal frequencies have been formed and coded. The first five categories and the numbers assigned do represent greater or less frequency of reading and therefore, can be rank ordered. They do not, however, represent points on the scale equal distances apart relative to reading frequency. The sixth category, Don't Know, represents missing data and is not a point on the scale continuum.

on the Celsius scale. The scale units are equal and the numbers assigned to represent each scale point reflect that equality.

Another characteristic of interval scales is the absence of a true zero point. With interval measurement, the numbers assigned to the scale are arbitrary. Even though some point on the scale has been assigned a value of zero, zero does not represent the complete absence of the characteristic being measured. The zero points on the Celsius and Fahrenheit temperature scales illustrate this. The temperatures 0°F and 0°C do not represent the complete absence of heat. Rather, they are arbitrary points that do not represent the same amount of heat. The intervals between numbers in the scales differ (1°F does not equal 1°C). However, within each scale, the intervals are equal.

The measurement scale used with many variables in the behavioral sciences is unclear. Methods of scaling ability and achievement variables, attitudes, values, and personality dimensions result in scores that are more precise and detailed than ordinal measurement, but still do not fit the exact specifications of equal intervals for interval measurement. Consider the measurement of achievement using a multiple-choice test. The number of items correct is generally the score assigned, representing an individual's knowledge level. Does each additional score unit on the scale mean the same amount of knowledge? If one individual received a score of 33 and one received a score of 30, the difference in their performance levels would be three units of knowledge. Would these three units imply the same amount of knowledge as the three units separating scores of 17 and 20? Only if the items were of equal difficulty and from the same specific content domain could we legitimately answer yes to either question. Measuring variables in other domains causes the same kind of dilemma. Assume an attitude scale has scores ranging from 20 to 100. Does each score unit represent the same amount of attitude?

While complete agreement does not exist among researchers in the behavioral sciences on how to treat measurement of these variables, some practices are generally accepted. Most researchers will treat the variables in behavioral domains as though measurement is at the interval level. Even though the measurement units do not strictly represent equal intervals, the research literature indicates that assuming so does not distort results much. All of the achievement variables in the "Sesame Street" data base are measured using a quasi-interval measurement scale. The Peabody mental age scores (23) and the home background/socioeconomic variables (24, 25, and 26) provide other examples of quasi-interval level variables.

For any of these examples, a true zero point does not exist even though zero is a point on the scale. With interval measurement a score of zero on an ability test does not mean the total absence of ability, and an attitude score of zero does not imply a total lack of attitude. Zero exists on the scale only as an arbitrary point. Without a true or absolute zero point, we cannot make certain comparative statements about the scores. For example, a test score of 30 does represent 15 more units than a score of 15. However, a score of 30 does *not* represent twice as much of the variable as a score of 15. Comparative statements such as "twice as much" or "four times more" should not be made when comparing specific score values measured at the interval level. However, the equal intervals permit meaningful comparison between intervals formed by score value differences. For example, it is appropriate to state that the difference between 70 and 50 units of a variable (20 equal units) is twice as much as the difference between the scale values of 25 and 15 (10 of the same equal units).

Ratio Measurement

The final level of measurement is the **ratio** level. Ratio scales have all the properties of interval scales plus a true zero point. Examples of ratio-level measurements are height, weight, speed, and elapsed time. We can make any comparative statement about the scores for variables measured on a ratio scale. Thus, if one person weighs 100 pounds and another 200 pounds, we can conclude that the second is twice the weight of the first. That is, ratios of numbers on a ratio scale are meaningful. The numbers themselves are anchored to an absolute zero point. The zero point is not necessarily an attainable score; it need only exist on the scale. Actual time in the 50-meter dash is measurement at the ratio level even though no one could actually run it in 0.00 seconds.

Few behavioral science variables are measured using ratio-level scales. The characteristics of the variables themselves and limitations of the psychometric procedures prohibit the development of instrumentation to collect data at the ratio level. An absolute zero point suggests the total absence of the trait or characteristic being measured. Think how difficult it would be to define *no* ability in an area, *no* knowledge in a given content domain, or the *complete absence* of a personality trait. These definitions would all be needed for ratio-level measurement.

The majority of variables in the behavioral sciences are measured at nominal, ordinal, and interval levels. Often, measurements are somewhere between ordinal and interval levels. Research has shown that treating such measures as though they were interval is permissible and acceptable. Finally, the measurement scales are hierarchical. Nominal measurement is the least manipulative, ordinal measurement is an improvement over nominal and has all the properties of nominal, interval measurement has all the properties of ordinal scales, and ratio scales have the characteristics of interval measurement.

One final distinction must be made. When the measurement scale used is nominal or ordinal, the variable measured is *discrete.* Interval and ratio scales measure *continuous* variables. For discrete variables only specified numbers can occur. For example, if the variable sex (nominal) has been coded 1 for female and 2 for male, no numbers other than 1 and 2 can be assigned the observations. The same holds for ordinal scales. Ranks are 1, 2, 3, 4, and

TABLE 2.2
Summary of measurement levels.

Level of Measurement	Type of Scale	Primary Characteristic
Nominal	discrete	categorization only for group identification—no comparisons can be made among categories
Ordinal	discrete	categories can be rank ordered in term of "moreness" of the variable measured—categories do not represent equal amounts of the variable
Interval	continuous	units of measurement are assumed to represent equal intervals or amounts of the variable—no absolute zero point exists
Ratio	continuous	absolute zero points exists on the scale

so on as assigned and only specific numbers can be assigned. Interval and ratio levels of measurement assume that an infinite number of possible measurements can occur between specified scale extremes. Thus, we can speak of the average number of children (ratio) per family as a number like 2.046 and we are able to conceptualize this position on a continuum. The continuum provides a reference for the underlying continuous nature of the variable. Most quasi-interval measurement can be conceptualized on a continuous scale. Table 2.2 summarizes the characteristics of each level of measurement.

2.7 SUMMARY

Familiarity with the statistical terminology presented will facilitate future learning. Certainly, the concept of data is fundamental to research and statistical analysis. The quantification of attributes, traits, characteristics, or behaviors through measurement results in data in a numerical form appropriate for analysis. Statistical procedures provide the means for analyzing the data.

Statistical methods summarize and describe data. The characteristics we study that can vary from observation to observation are called variables. Characteristics in a study that do not vary from observation to observation are termed constants. Data are gathered from elements of a population. A population is the group about which we wish to draw conclusions. When all elements of the population are measured, a resulting index is termed a parameter. When only a subset of the population is measured the group measured is referred to as a sample. An index computed on data from a sample is termed a statistic or estimate. Descriptive statistics summarize either population or sample data. We use inferential statistics to infer something about the population based on sample measurements.

As Chapter 1 noted, measurement and statistical methods are necessary, interrelated components of the research process. The type (level) of data resulting from our measurements is an important consideration in choosing a statistical procedure. Different statistical techniques are available for treating data at nominal, ordinal, interval, or ratio levels. The characteristics of each measurement level summarized in Table 2.2 should be kept in mind as you attempt to measure variables or as you examine variables in a study.

PROBLEMS

Note: Answers to Problem Sets A and C appear in Appendix D.

Problem Set A

1. The Board of Education for a local school system wants to survey the community about attitudes toward high school graduation requirements. The board has available for its use in forming a sample (a) voter registration records, (b) the PTA membership list, (c) a listing of parents whose children are currently enrolled, (d) homeowner property tax roll, and (e) the yellow pages telephone directory. What characteristics of each source could make it inappropriate for surveying the attitudes identified? What source might be better than any of the five given?

2. A researcher surveys a sample to determine average salaries of teachers statewide. While gathering data, the researcher is denied access to teachers in one large urban district, and 22% of the people contacted refuse to reveal their salaries. Describe the population that the sample represents.

3. Identify three different types of data that could be collected to reflect the popularity of six similar magazines. What is the level of measurement of the data collected?

4. What is the level of measurement typically used for each of the following?
   a. miles per gallon
   b. room numbers (e.g., 102, 108, 212, etc.)
   c. room size in square feet
   d. Mister Universe ratings
   e. price of gasoline
   f. poll of best dressed women
   g. television program ratings
   h. pollen count
   i. typing speed (words per minute)
   j. IQ scores
   k. travel guide motel ratings
   l. license plate numbers
   m. football jersey numbers
   n. coat size

Problem Set B

I. Section 402 of the 1964 Civil Rights Act mandated that a survey be conducted concerning the lack of equal educational opportunities for individuals by reason of race, color, religion, or national origin in public educational institutions at all levels in the United States, its territories, and possessions. The findings from a study, directed by James S. Coleman, were published and widely disseminated in a report entitled *Equality of Educational Opportunity* (1966). The investigation was a massive undertaking, gathering information from thousands of individuals including students, teachers, principals, and school district superintendents. Findings from this investigation have affected many of the policies and practices of the nation's schools, most notably their racial balance.

The primary vehicle used in gathering data for some facets of the study was questionnaires. A select subset of the questions for 12th-grade students follows. Items were chosen and edited only for illustrative purposes. The published report can be obtained if more information is desired.

Selected Grade 12 Questionnaire Items

1. How old were you on your last birthday?
- a. 14
- b. 15
- c. 16
- d. 17
- e. 18
- f. 19
- g. 20

2. Where have you spent most of your life?
- a. In this city, town or county
- b. In this state but outside this city, town or county
- c. In another state in the U.S.
- d. In Puerto Rico or another U.S. possession
- e. In Mexico
- f. In Canada
- g. In a country other than the U.S., Canada or Mexico

3. How many brothers and sisters do you have who are older than you are? Include stepbrothers and stepsisters and half brothers and half sisters, if any.
- a. None
- b. 1
- c. 2
- d. 3
- e. 4
- f. 5
- g. 6

4. Do you speak a foreign language other than English outside of school?
- a. Yes, frequently
- b. Yes, occasionally

c. Yes, rarely
d. No

5. How far in school did your father go?
a. None, or some grade school
b. Completed grade school
c. Some high school, but did not graduate
d. Graduated from high school
e. Technical or business school after high school
f. Some college but less than 4 years
g. Graduated from a 4 year college
h. Attended graduate or professional school

6. Does your mother have a job outside your home?
a. Yes, full-time
b. Yes, part-time
c. No

7. How good a student does your father want you to be in school?
a. One of the best students in my class
b. Above the middle of the class
c. In the middle of the class
d. Just good enough to get by
e. Don't know

8. On an average school day, how much time do you spend watching TV outside of school?
a. None or almost none
b. About ½ hour a day
c. About 1 hour a day
d. About 1½ hours a day
e. About 2 hours a day
f. About 3 hours a day
g. 4 or more hours a day

9. About how many days were you absent from school last year?
a. None
b. 1 or 2 days
c. 3 to 6 days
d. 7 to 15 days
e. 16 or more days

10. What was the first grade you attended with students from another race in your classes?
a. 1st, 2nd or 3rd
b. 4th, 5th, or 6th
c. 7th, 8th, or 9th
d. 10th, 11th, or 12th

11. How bright do you think you are in comparison with the other students in your grade?
a. Among the brightest
b. Above average
c. Average
d. Below average
e. Among the lowest

12. What is your grade average for all your high school work?
a. A (either A−, A or A+)
b. B (either B−, B or B+)
c. C (either C−, C or C+)
d. D (either D−, D or D+)

1. Review the list of questionnaire items and identify the level of measurement of each item as worded.

2. How could items 3, 5, and 9 be reworded to change the level of measurement of the information being obtained?

II. The statistical techniques used by members of a discipline can provide useful insights. Based on the patterns and frequency of use of each statistical procedure, information can be obtained regarding curricular offerings, the scope of textbooks, and to what extent a field is interdisciplinary. Willson (1980) investigated the frequency of use of specific statistical methods in data-based research published in the *American Educational Research Journal (AERJ)* over a 10-year period (1969 through 1978). *AERJ,* the only publication surveyed by Willson, is a respected outlet for research in education. For the period studied, the author identified 280 empirically based articles published in this source. Excluded from consideration were expositions, statistical derivations, book reviews, and simulation studies. Willson found 481 separate applications of statistical procedures in the articles. Table 2.3 reports the methods used and the proportion for each.

1. What are the variables in this study?

2. What is/are the constant(s) in this investigation?

3. How can this study provide information about certain population parameters?

4. Under what conditions could data resulting from this investigation be treated as sample statistics?

Problem Set C

Use the "Sesame Street" data base to answer the following.

1. Look at the data on sampling site location, sex, and the Forms pretest for individuals 2, 3, 13, 15, 29, 38, 46, and 54. Would you say the measurements define a variable or a constant for each characteristic?

2. At what level of measurement would you classify the identification (ID) numbers?

TABLE 2.3
Statistical methods used in data-based research for 280 *American Educational Research Journal* articles.

Statistical Method	%	Statistical Method	%
Analysis of variance	31	Chi square	3
Analysis of covariance	3	Factor analysis	9
Correlation	16	Guttman scaling	.5
Multiple regression	10	Multidimensional scaling	1
Discriminant analysis	2	Cluster analysis	.6
Canonical analysis	1.5	Time series analysis	.5
Path analysis	.6	Latent partition analysis	.2
Survival rate	.2	Structural analysis	1
Multivariate analysis of variance	6	Multivariate analysis of covariance	.6
Multiple comparisons	9	Power	.2
Nonparametric analysis	2	Jackknife	.2
Descriptive statistics	1	Secondary analysis	.2

Source: From "Research techniques in AERJ articles: 1969–1978" by V.L. Willson, 1980, *Educational Researcher, 9,* pp. 5–10. Copyright 1980 by *Educational Researcher.*

3. Assuming that the individuals in the data base represent a population, what would be the primary descriptor of the constant(s) defining the population?

4. Under the conditions in Problem 3, would the proportion of females constitute a parameter or a statistic?

3
Sampling

The importance of sampling to research and to statistics as a research tool cannot be overstated. In practice, very little research collects data from every unit in a population. Rather, data are collected from sample elements intended for generalization to a population. This generalization process makes sampling so important. Statistical methods provide the means for summarizing the sample information and the models for making the inferences to the population, but the validity of the information and the inferences depend upon the adequacy of the sampling procedure.

3.1 SAMPLING, SAMPLING UNITS, AND SAMPLING PLANS

Sampling is a process of selection. A sampling plan describes the procedure used to select a subset (sample) of elements from a population of elements. Remember that an element of the population is not always a person; it can be an object, an event, a place, or an occasion. The elements of a population define the sampling units. If a population consists of individuals, then a person is the sampling unit. With a population of test items measuring a content domain, the sampling unit would be an individual test item. Other examples of sampling units include fourth-grade classes, countries, states, years, boxes of detergent, time intervals, cars, and sentences. Sampling units need to be well defined prior to actual sampling for two reasons. First, a complete definition is necessary to identify potential units. Second, the definition of sampling units limits the population to which generalizations will be made.

Why sample at all? Generally, to economize on time, effort, and money. Also, statistical theory allows us to obtain desired information accurately about a population without measuring every element in the population. The sample information may then be generalized to the population.

We have said that sample data can be used to infer characteristics of a population, if the sample elements as a group are *representative* of the population elements. In a way, we would like the sample to be a miniature replica of the population, mirroring the relevant characteristics of the population, but being considerably smaller in size. How will we know if any one sample is representative of the desired population? We never will unless we also collect measurements on every element in the population. To do so would defeat the purpose of sampling. In reality we never will be 100% confident that the sample we have selected is representative of the population to which we want to generalize. What are we to do? We institute a sampling strategy that, if followed, gives us a prespecified level of statistical confidence that our sample is representative of the population. Sampling theory provides the model on which statistical inference is based.

What is an adequate sampling plan? Obviously, it is one that will produce a representative sample with a prespecified level of confidence following sampling theory. Any one sample may or may not be representative but sampling theory allows us to expect a proportion of samples selected to be representative if we follow a specific sampling plan. The sampling plan in general—not a specific application—produces our degree of confidence.

3.2 ERRORS IN SAMPLING

Any sampling plan results in samples with varying representativeness. Sampling error, which always exists, determines the degree of representativeness. Two kinds of sampling errors can occur as a result of a specific sampling plan. One type, **unsystematic** sampling error, is acceptable within a sampling plan and forms the basis for sampling theory expectations. Unsystematic sampling errors are haphazard or unpredictable. Unsystematic errors occur by chance. They are termed **random** errors and tend to balance out over the long run. Random errors are "okay" errors, if any error can be classified as okay. Sampling plans are judged by the extent to which they result in representative and nonrepresentative samples following expectations based on random sampling error theory. The concept of randomness and random sampling are explored in more detail in the next two sections.

The other type of sampling error is **systematic.** Systematic sampling error produces a bias. As an example, suppose you need a sample of third-grade students representative of the third-grade population in a school district. Your sampling plan instructs each third-grade teacher in the district to select four students. Research indicates that without further sampling directions, the teachers will more than likely select their "best" students, who are high achievers, responsible, well behaved, and self-confident. Would this group of students be representative of the population of typical third graders? No, because a systematic bias is operating in the selection process. Your sample would not be representative or typical for many variables, such as achievement, aptitude, attitudes, personality characteristics, or interests. Other familiar examples of systematic error are mail surveys in which the return rate is not 100% (respondents may differ from nonrespondents) and in the collection of home-to-home interview data during the day (a large portion of the potential population is not at home during daytime hours). To the extent that systematic errors may exist as a result of a sampling plan, the sample data are of limited use when generalizing to the intended population. All sampling plans need to be evaluated for the absence of systematic errors. There must be no logical relationship between the sampling method and the characteristic under investigation in the sample.

"No logical relationship" implies that no systematic error occurs in the sample *for the characteristic under study* as a result of the sampling plan used. Discussions of sampling often overlook this point. The focus tends to be on the sampling unit because it is the physical entity selected. However, a sample is selected for one purpose, so that data can be collected on variables of interest. The representativeness of a sample is not determined by the physical sampling units themselves. *The representativeness of the measurements taken from the sampling units is important.* If we were studying achievement motivation, our sampling plan that allowed teachers to select only the "best" third-grade students would result in systematic error, a biased sample of the district's third-grade achievement motivation scores. The level of the characteristic under study, achievement motivation, was connected to the sampling method. What if we were studying third graders' artistic ability? Distance from home to school? Shoe size? Maturation level? Sociability? Would a sample of "best" students be as representative as any other sample we could select? As long as no relationship exists between the sampling plan and the characteristic under study, then the sample selected has the potential to be representative. Remember, the generalizations we want to make from the sample to the population focus on the measured characteristic being investigated in the study.

Sampling error will always be an expected ingredient in research analyzing sample data. An adequate sampling procedure, however, is one that produces only random error across sampling units for the characteristic on which data are collected.

Sampling is an important determinant of the representativeness of sample data points, but it is not the only one. The procedures used to collect data from each sampling unit (the measurement process) also will introduce errors that affect the representativeness of the **sample set of measurements.** As with sampling error, measurement error may be either unsystematic (random) or systematic. Unsystematic measurement errors are due to the lack of reliability in the instrument used. These errors may be minimized using measurement procedures with high reliability. The systematic measurement errors have the same effect as systematic sampling errors. They produce systematic bias in the set of sample data so that inaccurate inferences are made about the population characteristic.

Contrary to what some think, statistics is not the remedy for inadequate sampling or measurement in the research process. Systematic error introduced by the sampling plan or measurement process causes "sick" data that will not be detected by statistical applications and therefore not cured. The generalizations made through statistical inference when these systematic errors exist in sample data are not healthy and will not be valid. Unsystematic errors, on the other hand, pose no real problem to the validity of generalizations made from sample data. Inferential statistical procedures have been developed assuming that random errors exist. Statistical theory uses these assumptions to establish the degree of confidence a researcher may have regarding inferences.

3.3 THE NOTION OF RANDOMNESS

Randomness might best be considered in the context of a specific pattern in which events occur. Specific events will occur a certain *number* of times depending on their frequency in the population of possible events, but *when* an event will occur in the pattern or *in what order* events will occur is completely unpredictable if randomness is operating. Consider, for example, the malfunctioning of a particular brand of toaster. Assuming equal use and maintenance, toaster malfunctioning for a given population of toasters might be considered a random event. We could collect data on the average life expectancy of this brand of toaster to give us some idea of how long they generally last, but when a specific toaster would break down would be a completely random event. The pattern of toaster malfunctioning for the group might be known in that a certain number would break down early, a certain number would break down late, and so on, but which specific toasters would break down when and in what order is altogether unpredictable.

Randomness is linked directly to probability. On the most elementary level, probability can be defined as the frequency with which an event will occur in relation to the total number of events possible. Random events are equally probable on any one occasion. Let's reconsider the toaster malfunctioning example. Assume that the quality of workmanship is lower on Mondays and Fridays. Lower-quality workmanship implies a higher risk (probability) of the product malfunctioning. If these assumptions are true, then the pattern of toaster malfunctioning would no longer be random. For this population of toasters, then, each toaster would not have the same probability of malfunctioning nor would the order in which toasters will malfunction be unpredictable. Toasters constructed on Mondays and Fridays would have a higher probability of malfunctioning sooner, thus toaster malfunctioning would not be random.

Although unsystematic errors are considered random, systematic errors are not. This concept may be most easily illustrated by considering a missile's accuracy in hitting its intended target. An error occurs if a direct hit is not made. The magnitude of the error is the distance between the target and the actual landing site of the missile. For any given set of events (missile firings), the pattern of errors will be random. Some will land beyond the target, some below the target, some to the left, and some to the right. What a given missile will do is unpredictable. The probabilities that it will land beyond, below, to the right, or to the left of the target on any one launching are equal. The pattern of errors also is unpredictable over a series of launchings. In contrast to the random error pattern displayed by an accurately fired missile at an intended target, an error could be made in programming the missile so that it always veers to the right and overshoots the intended target. The programming error would introduce systematic error into the process. Missile landings would still exhibit a random error pattern, but the landings would fall around a point different from the intended target. The distance of a missile landing from the actual target in the latter case would be determined by both systematic and random error.

Sampling error resembles the errors in missile landings. The person responsible for programming the missile wants to avoid systematic errors that will push all missiles off target by a certain amount. The person responsible for devising a sampling plan wants to avoid systematic errors for the same reason. Both want their sampling unit data points to follow patterns of randomness, but around the population target point. With only random errors existing for a sample of events, error theory ensures that both persons will end up "on target" with their samples.

3.4 RANDOM SAMPLING

Simple random sampling is a basic design used to ensure representativeness. In a **random sample,** each element in the population has an equal chance of being selected for the sample and each sample of elements that forms a set has the same chance of being selected. Each event (whether a population element is selected for inclusion in the sample) will have the same probability of occurring on any one occasion and the pattern in which events occur (selection of a fixed number of elements) will have the same probability of occurring for patterns of fixed size over repeated samplings. As an illustration of these concepts, assume we have a population of four elements—W, X, Y, and Z—from which we want to select a sample of two elements. If our sample is to be considered random, then we must devise a method of selection such that on each of the two occasions when we select an element, the four letters have an equal chance of being drawn. Because there are only four elements in the population, the probability that any one letter will be drawn at random is ¼ or .25. Each letter has a 25% chance of being drawn at random on any one of the two occasions. The last condition of randomness involves the makeup of the sample itself. Each possible sample of two elements must have the same opportunity to be selected. There are only 16 possible patterns for selecting two elements from a population of four elements: (W,W), (W,X), (W,Y), (W,Z), (X,W), (X,X), (X,Y), (X,Z), (Y,W), (Y,X), (Y,Y), (Y,Z), (Z,W), (Z,X), (Z,Y), and (Z,Z). The second condition means that over repeated samplings, each pattern will be selected the same number of times if the process is random.

One point to note in the last example is that sampling was done **with replacement.** The element selected on the first occasion was placed back into the population and had an opportunity to be selected again. In theory, random sampling assumes sampling with replacement. In practice, however, many statistical methods require that sampled units be independent of each other, that is, an element cannot be included in the sample more than once. This restriction on the sampling plan usually does not distort true random sampling much because the number of elements in the population is assumed to be large. Using a random sampling plan **without replacement** in large populations is still considered to meet all criteria for randomness.

Random selection most commonly begins by assigning an identification number to each element in the defined population. This assumes that the population is finite and can actually be counted. Then, ID numbers usually are selected randomly using a table of random numbers or a computer program that will generate a fixed set of random numbers within the range given. You choose as many ID numbers as you want in your sample. If 20 elements are desired as the sample size, then the first 20 ID numbers selected at random identify the elements from your population that constitute your sample. Because the 20 ID numbers were chosen from a set of random numbers, the sample is random.

For our random sampling we will use the table of random numbers reproduced in Appendix B. In using this table, the largest ID number in the population determines the number of digits needed for any one random number selected from the table. For example, if 7538 elements are in the population, ID numbers might be assigned starting with the four-digit ID 0001 and concluding with the last element in the population receiving a four-digit ID of 7538. A row and column are identified as the starting place in the table. Our table contains 50 rows and 40 columns of 5 random digits each. One should not always start at the same place in the table. This would introduce a systematic bias in the selection process. The beginning row and column should be chosen at random (unsystematically). Because the table is finite, limited to the space available on a page, decision rules for moving through the table should be identified in advance. These include

1. the direction to be followed in moving through the table, which continues until enough numbers are chosen for the sample size needed
2. the positions in the grouping of random digits to be read; for example, if numbers are in groups of five and you need to read three-digit ID numbers, then decide which positions you will be using (i.e., the leftmost three, the rightmost three, the first, third, and fifth, etc.)

Once these decisions are made, they need to be followed consistently for the specific sampling activity undertaken on that occasion. Numbers in the table that are not in the range of population ID numbers are ignored.

The random digits in Appendix B have been grouped in columns of five digits for visual clarity and convenience. Each digit or group of digits represents a random number. To illustrate, part of the sequence of five digits starting in row 36, column 6 is reproduced here. Assume we want to sample four elements from a population containing 66 elements. The ID numbers that could be sampled range from 01 to 66, so we would sample two-digit numbers from the table. We decide to use the second and third digits from each five-digit group in column 6. Based on these decisions, the four elements from the population that are selected to comprise our random sample are the elements with ID numbers 16, 42, 21, and 63. Boxes have been drawn around these ID numbers in the set of random digits.

37172
08727
5[16]78
46897
6[42]33
2[21]43
8[63]70
⟶ Sampling stops at this point.
83453
03120
80608

Note that ID numbers 71, 87, and 68 occurred in the sequence as we moved down the column, but were ignored because they are out of the range of population IDs. The use of the second and third digits in the five-digit group was arbitrary. What population elements would have been sampled if we had decided to use the first and fifth digits to select our sample?[1]

While ID numbers become the values selected in random sampling, it should always be remembered that we are choosing elements from the population. Furthermore, the physical entity in the form of a person or other sampled element is not of primary interest. The whole basis for research is to sample the elements so that the measured characteristic represents the characteristic parameters in the population. The "Sesame Street" data can be used to illustrate this point. Assume that the Letters Pretest scores are of interest as the characteristic to be studied in this population. To select a random sample of 15 scores, we would actually sample 15 individual children from Appendix A. The children serve as the sampling unit, but we are interested in their Letters Pretest scores. Use the following sampling plan to select 15 Letters Pretest scores at random from Appendix A:

1. Start at the 39th row and 18th column of the five-digit sets.
2. Use the first, third, and fifth random digits from a five-digit group to identify the three-digit ID to be selected. Read these positions from left to right.

---

[1] [3]717[2], or 32
[0]872[7], or 07
[5]167[8], or 58
[4]689[7], or 47

3. Move through the table on a page going from row to row in the following pattern:

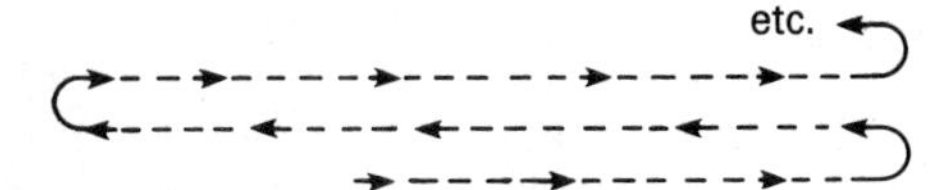

What are the 15 ID numbers you selected and the corresponding Letters Pretest scores for persons assigned these ID numbers?[2] These 15 scores constitute a random sample of Letters Pretest scores.

3.5 STRATIFIED AND PROPORTIONAL STRATIFIED SAMPLING

Simple random sampling is obviously the most basic sampling plan and one of the simplest to implement. An extension of simple random sampling is **stratified** random sampling. Sampling based on a stratified random model offers increased accuracy and representativeness beyond that of randomness alone. When the sample to be drawn is large relative to the population size, simple random sampling gives us confidence that our sample contains elements that represent the characteristics in the population. By chance, however, some samples might contain a disproportionate number of elements from a particular group or with a particular level of some characteristic. For example, one sample may include a greater proportion of females than exists in the population. If females and males differ on the characteristic being studied, then the sample will not give an accurate estimate of the population parameter. The sample is still random, and in that sense representative, but the accuracy could be increased by separating the data by sex using this stratification as an additional control in our sampling plan.

A stratified sampling model separates the population into subpopulations or strata based on known levels of some characteristic. The goal is to be able to control this characteristic as it occurs in the sample. Examples of stratification variables often considered include sex, age, socioeconomic status, ability level, educational background, occupation, ethnic group, or school size. Any variable has the potential to be so crucial that even the random errors resulting from simple random sampling cannot be tolerated in the study. In a stratified

[2]

ID	Letters Pretest Scores	ID	Letters Pretest Scores	ID	Letters Pretest Scores
049	15	001	23	077	13
208	15	089	29	087	10
084	12	078	27	064	16
099	13	006	26	181	13
043	12	214	19	057	14

sampling model, the sampling plan is employed within each strata or subpopulation identified. If a stratified *random* sampling model is used, the procedures in section 3.4 are employed to select random samples from each strata. These random strata subsamples are then combined to make up a single stratified random sample.

Two decisions must be made with a stratified random sampling plan: (a) Which characteristic or characteristics does one use as the stratification variable(s)? and (b) What is the relative size to be selected from each strata? Identifying variables to use as a stratification factor usually depends on group differences. The accuracy of sample statistics in making inferences about the total population improves through stratified random sampling if the subpopulations defined by the various strata differ on the characteristic to be studied. Suppose you were surveying teachers about their attitudes toward increasing the curricular emphasis on basic skills instruction. You know that including in your sample substantially more or fewer teachers in basic skills areas would result in an inaccurate sample estimate of attitude levels in the general population of teachers. This *potential* inaccuracy could be controlled through stratified random sampling. You might stratify your sampling of teachers according to the content area they teach, assuming that attitudes would differ for subpopulations of teachers teaching different content. A simple random sample might contain a disproportionate number of mathematics or language arts teachers (either more than or less than exist in the population).

While stratification variables are usually selected for control purposes or to analyze subgroup differences, on occasion stratification is used for political purposes to ensure that the sample appears representative. Even though there are no differences in the responses for difference groups, you may still want to use stratification in sampling so that you can state that data were collected from a specific group. Particularly in survey research, it allows each constituency to think its group has had input. Stratified sampling enhances the image of the study. In these instances, it may make no difference statistically whether sampling is stratified, but the data appear more representative and accurate. Stratification in this case is merely cosmetic, but perhaps necessary.

Each stratum can be proportionally sampled or over/undersampled. **Proportional stratified** random sampling ensures that the sample has the same percentage of elements from each stratum as the percentage existing in the population. If sex is the stratification variable and the population consists of 60% females and 40% males, then a stratified random sample would be employed to select 60% of the sample desired from the female stratum and 40% from the male stratum. If total sample size is 50 individuals, then 30 females and 20 males would be randomly selected from the separate female and male subpopulations. Proportional stratified random sampling exactly duplicates the relative proportions of each subpopulation stratum. Random sampling within strata controls for the representativeness of a characteristic in the sample.

While proportional stratified random sampling would appear to be the perfect way to ensure representativeness, at times we want to oversample or undersample from certain strata. Oversampling or undersampling from a stratum gives more or less weight, respectively, to the responses in that stratum in estimating the population parameter. It may be that one group's responses are more accurate than others for the parameter we are trying to estimate. Polling to predict election outcomes is a good example of selective oversampling. Rather than randomly sampling precincts from across a state or the nation, pollers stratify precincts based on their past voting history. Precincts that have always voted for the winning candidate are oversampled because they are seen as more representative of the entire voting public. If history indicates that blue-collar workers are prime determinants of an election outcome, that stratum might be oversampled to ensure a good estimate of its voting preference. Certain groups may influence elections disproportionately to their numbers in the population. Over- and undersampling would help equalize the effects of these groups. For example, on some issues, rural dwellers' opinions might be two, three, or four times more important than those in urban or suburban communities. Neither simple random sampling nor proportional stratified random sampling would provide the information desired. The rural communities should be oversampled.

While stratification is most evident as a control mechanism in survey research, it is also very important in comparative research studies as a control for apparent initial group differences due to sampling. Stratification variables often are used also as factors in designs that compare groups defined by the strata (e.g., do older persons differ from younger persons?). In these cases sampling through stratification is not intended as a control mechanism, but becomes a research variable to be investigated. When certain groups are the focus of a study, it often is necessary to oversample them to ensure a sufficient sample size because they are scarce. For example, suppose we wanted to study the achievement expectations for low socioeconomic students of female elementary teachers versus male elementary teachers. Using simple random sampling or proportional stratified random sampling would produce a large discrepancy in group sizes for male and female elementary teachers. To ensure sufficient and economical sample sizes for the two groups, we would proportionally undersample female elementary teachers and oversample male elementary teachers relative to their frequency in the population.

Stratification is an important, desirable sampling strategy in many situations. It can increase accuracy over simple random sampling, particularly when smaller samples are to be used. Sampling through stratification alone, however, is not sufficient, even when it is proportional. Randomness is still key to any sampling plan. Stratification only enhances it. Random sampling from different strata is the critical ingredient to promote valid statistical inferences.

Practical experience with proportional stratified random sampling might be beneficial. Of the variables in the "Sesame Street" data base, sampling site

TABLE 3.1
Random subsamples of 40 children from "Sesame Street" data base stratified by location.

Location	Number in Population	Proportion (÷ by 240)	Number to Be Sampled
I	60	.25	10
II	55	.23	9
III	64	.27	11
IV	43	.18	7
V	18	.07	3
TOTAL	240	1.00	40

would be an important variable on which to stratify if we wanted to obtain a representative sample of 40 individuals. Because of the diversity of characteristics differentiating the subpopulations at each site, it is important to represent them proportionally if we want to estimate some general characteristics such as specific pretest scores in the population of all 240 cases.

How many different random subsamples would we need to select if we stratified by sampling site?[3] Assume we want to have 40 children in the total sample. To determine the number of individuals we would select from each stratum with a proportional stratified random sampling plan, we use the proportions in Table 3.1. The numbers of children in the "Sesame Street" data base from each location appear in Table 3.1 along with the proportion and the number from each site to make up the sample of 40 individuals. The population proportion is obtained by dividing each location frequency by 240. The population proportion is then multiplied by 40, the desired sample size, to obtain the number from each location that would need to be randomly sampled. The actual sampling at random within strata would follow the procedures in the last section. ID numbers in subpopulations could either be arranged starting with 01, or the existing ID numbers could be used. If we reassigned ID numbers with 01 and ending with 18, which three Site V children would be selected randomly under the following plan?[4]

1. Enter the table of random numbers at row 4, column 22.
2. Use the fourth and fifth digits as the random digits for each ID selected.

---

[3]Five random subsamples would need to be selected; one from each of the subpopulations of children in locations I, II, III, IV, and V.

[4]Many of the IDs selected following these procedures in the table are out of the range of IDs between 01 and 18. The random IDs in the range and the three that would be selected are 07 (row 9), 02 (row 18), and 15 (row 25). The 2nd, 7th, and 15th children from the Site V subpopulation of 18 children would be included in our sample of 40. These three children's data base ID numbers are 224, 229, and 237.

3. Proceed down the column to select random digits. (If necessary, when the numbers in column 22 are exhausted, move to column 23 and proceed up the column, then to column 24 and down, etc.)

If we had used the ID numbers assigned to the 18 Site V children in Appendix A, which three would be selected entering the table at the same place, following the same pattern, but using the third, fourth, and fifth digits as the potential IDs to be selected?[5]

3.6 OTHER SAMPLING PROCEDURES

Modern sampling theory distinguishes designs based on probability from those based on nonprobability sampling. In probability designs, randomness is the fundamental control. Probability sampling lets us specify the probability that any sampling unit will be selected from the population. Simple random sampling, stratified, and proportional stratified random sampling are all examples of probability designs. Even in oversampling and undersampling of strata, we can estimate the population characteristics through weighting if sampling has been random within strata.

Nonprobability designs do not systematically introduce randomness into the selection process. Selection of elements from a population is based more on the researcher's judgment, convenience, economical criteria, or sample availability. Nonprobability designs follow several forms.

Systematic Sampling

Systematic sampling involves selecting every *i*th member of a population until the desired number of elements in the sample has been selected (e.g., list the population and sample every tenth person). This is generally an accepted sampling strategy as long as it meets certain conditions. First, one must choose a starting element at random to ensure that every element has an equal opportunity to be included in the pattern of selection. Second, it is inappropriate if the population has a cyclic pattern for the variable being investigated. Finally, the population cannot be ordered on the variable of interest or on any variable related to the one of interest. Basically, if the population is arranged unsystematically, preferably in random order, and we select the starting point at random, selecting every *i*th person will result in a random sample.

Accidental Sampling

In accidental sampling, individuals who happen to be available are selected. This procedure results in what is termed a convenient sample. Using classes

[5]The population ID numbers from which we are sampling range from 223 to 240. The three children selected would be those with ID numbers of 232 (row 50, column 24), 230 (row 46, column 26), and 235 (row 7, column 28).

of students, interviewing the first 100 people met on the street, getting friends to respond to a questionnaire, and so on are all examples of accidental sampling. Convenience and availability are the primary criteria for sample selection. Biases introduced in such samples cannot be evaluated, however. This is probably the poorest sampling procedure available, but it is often used.

Purposive Sampling

In purposive sampling, representativeness is based totally on the researcher's judgment. Cases to be included in the sample are hand-picked based on their judged representativeness for the population to which generalizations will be made. There is no way to evaluate any bias in the sampler's judgment about which elements are typical of the population.

Quota Sampling

Quota sampling employs characteristics of both purposive sampling and stratified sampling (note that *randomness* has been omitted). In quota sampling, different strata are purposively sampled to guarantee that the sample characteristics mirror those thought to be in the population. Representativeness of elements within strata is often ignored. The task is to collect data from a *quota* of cases within a stratum so that the sample will look representative. The adequacy of a sample resulting from quota sampling depends on the type of sampling used within strata. Proportional stratified random sampling would obviously be best and would result in the intended outcome of quota sampling—perceived representativeness through inclusion of different strata in the sample. Technically, the term quota sampling applies to stratified sampling designs in which the investigator's judgment rather than random sampling is used to select elements.

Cluster Sampling

Cluster sampling is different from other sampling plans in that the initial unit sampled contains the elements that will be analyzed. Groupings of elements such as school districts, school buildings, intact classes, families, communities, or city blocks are sampling units. In cluster sampling, the set of elements in the sample is determined by sampling larger groupings or clusters. For example, to assess the competency level of fourth-grade students in a state, we might sample school districts across the state. The elements in the sample would be all fourth-grade students in districts selected for participation, but the districts would be the sampling units.

The adequacy of cluster sampling depends on how the clusters were sampled. Any plan can be used, but a simple random or stratified random plan would be best. Introducing randomness into the cluster sampling process results in an approximation of a probability sampling plan. Lack of randomness in selecting clusters results in a nonprobability sampling design. Under either plan, cluster sampling is particularly attractive in large-scale data collection efforts in which easy accumulation of large samples is desirable.

3.7 IMPORTANCE OF SAMPLING IN STATISTICS

Statistics can be applied to any set of data points as long as it has been quantified. Statistical manipulation of numbers is independent of how or where the numbers were obtained. Within the context of the entire research process outlined in Chapter 1, however, there is more to statistical applications than mere number crunching. Statistical methods allow us to summarize the results of analyses on sample data. The interpretations, inferences, generalizations, and conclusions drawn from these results become important in the research process. These outcomes are *not* independent of either the sampling or the measurement processes employed in the research design. They are totally dependent on the adequacy of each. Sampling and measurement directly affect the validity of the results from statistical applications. Statistical procedures cannot improve bad data that result from inappropriate sampling or inadequate measurement.

Valid interpretations of statistical results rely on assumptions that statistical procedures make. Most statistical methods assume randomness of data points, since statistical inference is based on random error theory. The interdependence of the sampling plan and the validity of results from statistical analysis is obvious. Statistical inference assumes randomness whether it exists or not. The validity of such inferences and conclusions depends on the extent to which the sampling procedure produced random or approximately random data. As mentioned earlier, many of the sampling plans implemented in research are not probability designs based on randomness. The nonprobability designs focus more on establishing apparent representativeness in the sample elements through the sampling process. A design based on random selection is the best approach to obtaining a sample that is representative of the population. Unfortunately, the reverse is not true. An apparently representative sample does *not* imply randomness. All sampling designs—even random ones—have the potential to result in a biased sample. Using statistical inference, the probability of a biased sample can be determined for random sampling designs but not for biased samples from nonprobability designs.

3.8 CONCLUSION

Keep sampling in mind while reading about statistical techniques and their applications and interpretations in the remainder of this text. Become enthralled with statistics, but not so enthralled that you worship it as a savior in the research process. Statistics is one component in a sequence of interdependent components. If any one part breaks down, the product will be flawed.

3.9 SUMMARY

In sampling, we select a subset of elements from a population of interest. The goal of sampling is to allow us to observe specified characteristics such as performance, behavior, or outcomes in the sample and then to infer or generalize about the same characteristic for the population. The elements selected for the sample may be persons, objects, events, places, or occasions. These elements are referred to as sampling units. The actual units sampled define the population about whom generalizations can be made.

The validity of generalizations for the population depends on the sample's representativeness, which, in turn, depends on the sampling plan. Sampling plans may be characterized as yielding probability or nonprobability samples. Probability sampling approaches are most likely to produce a sample representative of the population. Central to probability-based sampling is the notion of random sampling. In this context, random means that inclusion in the sample is left to chance. A basic probability sampling method is simple random sampling. Each element in the target population has an equal chance of being chosen for membership in a simple random sample. Stratified and proportional stratified random sampling are sampling methods that include random sampling procedures and additionally control the sampling process to increase representativeness. Random sampling, regardless of the particular method used, eliminates systematic errors that could otherwise bias the conclusions drawn from the sample data. Approaches like proportional stratified sampling attempt to control a portion of the unsystematic (or random) error that affects all sampling methods.

Nonprobability sampling procedures are deficient from a statistical viewpoint. Results obtained from nonprobability samples only describe the elements of the sample. Inferences offered or implied from such samples have little validity. Data analysis methods neither improve nor alter the representativeness of the sample and thus they are unable to salvage an inadequate sampling plan.

PROBLEMS

Problem Set A

1. Suppose we are devising a sampling plan to estimate the total number of words in this text. Although you have probably already done so, glance through the text and recognize that there are many figures, tables, and illustrations.

a. Is there any problem in defining the population?
b. What are the advantages and disadvantages of treating each of the following as the sampling unit: the page? the line? the paragraph? the chapter?

2. A researcher wants to survey the membership of a state association of school psychologists. The survey will ask if they endorse an increase in annual dues. Names and addresses of each member are maintained on a computer file arranged by zip code and by alphabetical order within zip code. Offer a critique of each of the following methods for drawing a sample of the membership:
a. Every 10th name is selected beginning with the first name on the list.
b. For the sample size desired, a random start is identified and from that point consecutive names are selected up to the needed sample size.
c. Within each zip code, the person in the middle of that subsample is chosen.
d. Only persons who had attended that year's annual meeting are surveyed.
e. Persons who have been members of the association for fewer than two years are sampled.

3. Stratified sampling can be an efficient, justifiable approach to sampling. Choosing appropriate strata to define subsamples can present a problem, however. In each of the following situations, identify what might be a useful stratification variable and any weakness that might result from using it:
a. a study to determine the annual income of physicians
b. an investigation of voters' willingness to vote in federal elections
c. a survey of incoming first-year students to provide enrollment projections for college majors
d. an analysis of letters of recommendation for admission to graduate school

4. A section of a table of random numbers is reproduced here. Use it to respond to questions a through d.

3923	1686	7273	5246	1362
1815	7064	1733	7650	0159
1281	5821	4867	6590	9473
8108	9816	7430	2964	0797
1918	8512	0530	4310	0143
1756	1371	9244	2437	2856
9933	6445	1570	2827	0019
0291	3271	8029	9405	2771
6451	8867	2106	5668	5896
5264	3464	9906	2781	0293

a. Assume that you enter this table at the second column, second row (7064). Reading down the second column from this position using the two rightmost digits *only,* how many cases would you select from this column? What are the ID numbers for the cases chosen?

b. Suppose a target population contained 20,000 cases, each identified by a five-digit ID number, starting with ID 00001 and continuing through ID 20000. Describe two different ways one could use the random-numbers table to draw a random sample so that each member of the population would have a chance to be included in the sample.
c. A researcher selects 30 cases at random from a target population. The investigator wants to assign 10 cases randomly to each of three treatment groups. Describe two approaches using the random-numbers table to assign 10 cases to each treatment randomly.
d. In part c, identify and then use one of your methods to assign cases to the three groups. Assume each case is numbered 01 to 30. List the cases you assigned to each group.

Problem Set B

I. Shepard, Smith, and Vojir (1983) conducted a recent study aimed at describing the characteristics of school-aged children whom educators had identified as learning disabled (LD). A description of the sampling plan follows.

> A stratified, two-stage cluster sampling design was used to sample the population of identified LDs. Of the state's 48 special education administrative units, 22 were sampled at random from strata representing size and type of unit (district or cooperative board serving several districts). All 22 units agreed to participate. The number of units to be selected from each stratum was chosen to keep the number of students in the sample roughly proportional to corresponding population sizes. Exact proportionality was achieved by post hoc weighting. (The sampling frame and weighting procedures are described in Technical Appendix A, Shepard & Smith, 1981). From the administrative units sampled in the first stage, probability samples of LD case files were selected in the second stage from population lists submitted to the Colorado Department of Education to obtain reimbursement of expenses for services provided to handicapped children. The resulting sample of 1,000 LD cases (3.8 percent of the Colorado LD population) was randomly subdivided into 200 cases to be used in a qualitative study and 800 cases for quantitative study (eventually 790 because of clerical errors or logistical problems) are the subjects of this research. (p. 312)

Refer to this description to respond to the following problem exercises.

1. For which of the following is there evidence that the sampling procedure identified was employed? Cite the evidence on which you base your conclusion.
a. random sampling
b. stratified sampling
c. proportional sampling
d. systematic sampling
e. accidental sampling
f. purposive sampling
g. quota sampling
h. cluster sampling

2. Following the sampling plan described, 22 units were sampled at random from the state's 48 special education administrative units, but sampling was stratified based on size and type of unit. Assume there were three sizes (large, medium, small) and two types (single district board versus cooperative board serving several districts) of units. Assume also that the number of the state's 48 units classified into each strata are as follows:

Size	*Type*	*Number of Units*
Large	Single District	8
	Cooperative Board	4
Medium	Single District	8
	Cooperative Board	8
Small	Single District	4
	Cooperative Board	16

a. If the 22 units are to be sampled proportionally relative to number of units in each strata, approximately how many units would you randomly sample from each of the six strata?

b. Describe two different random sampling plans for assigning ID numbers to the 48 units and then, using the table of random numbers in Appendix B.1, select the number of units needed from each strata.

c. Describe an appropriate procedure for randomly subdividing the 1000 LD cases into the subsample of 200 cases to be used in the qualitative study and 800 cases for the quantitative analyses.

Problem Set C

Use the "Sesame Street" data base in Appendix A to answer the following:

1. The "Sesame Street" data base in Appendix A is arranged by sampling location, but cases are listed at random within location. Suppose you wanted to determine if the proportion of males and females in the data base was similar across locations. Describe the sampling plan you would follow to address this question.

2. Chapter 1, section 1.5, described how children were sampled for the "Sesame Street" evaluation conducted by ETS. Review this description, then answer the following questions.

a. What was the sampling unit for the ETS evaluation?

b. In light of the sampling unit, define the target populations about whom inferences would be possible.

c. Characterize the sampling plan used. Which label would you attach to the method of sampling used?

3. In section 3.4 of this chapter, you were asked to select 15 children at random and to record their Letters Pretest scores. Using a simple random sampling approach and the table of random numbers in Appendix B, select 15 additional children from the "Sesame Street" data base. Compare this second sample to that selected based on the exercise in section 3.4. Were any of the same children sampled? Is the range of scores similar in the two samples? Do the scores in the two samples appear comparable?

4. Suppose you wanted to draw a proportional stratified sample from the data base on the basis of viewing category. The total sample size is to be 80 children. How many children are to be included from each viewing category in the sample?

4

Data Representation: Frequency Distributions and Graphs

When data are collected, whether from a sample or population, one of the first tasks is examining the data for the group as a whole. Each person or sampling unit will have a score that represents the measurement for that individual on the variable being studied. If we list the scores obtained by each individual in an array of data points for the group, it is almost impossible to get any meaningful information from the data. For example, the scores on the "Sesame Street" Test for the 240 individuals in Appendix A are reproduced in Table 4.1. Obviously, it is difficult to make any sense of the data unless we systematically organize it.

To begin to understand these data, we might ask: "With what frequency was a particular score obtained for the group under study?" Systematically organizing and summarizing data into a form that identifies a score frequency requires constructing a **frequency distribution.** A frequency distribution conveys information about the distribution of individuals' scores across the score categories. The frequency distribution for the "Sesame Street" test scores is given in Table 4.2. This summarization concisely presents how the scores of 240 individuals are distributed across the scale continuum; for example, 76 individuals have scores of 10, 26 scored 9, and 12 individuals marked only one item correctly.

The measurement level of the variable dictates the statistical procedures appropriate for representing data in frequency distribution form. For data at the nominal level of measurement, scores in a data set do not have numerical meaning other than as labels for a category. Take the sampling site variable from the "Sesame Street" data as an example. The coded values are 1 through 5. The number 3 has no numerical meaning in terms of quantity; it merely identifies an individual as a member of the sample drawn from Site III. Data of this type are defined as **qualitative** data. In contrast, **quantitative** data are numerical indices that represent some magnitude of a characteristic being measured. The scale values represent an underlying continuum for the characteristic. The pre- or posttest achievement scores from the "Sesame Street" evaluation are examples of quantitative data. Each type of data will be considered in this chapter.

4.1 FREQUENCY DISTRIBUTIONS FOR QUALITATIVE DATA

The structure, development, and definition of a frequency distribution remains constant for any type of data. The information to be communicated is the

TABLE 4.1
Scores from 240 participants on the "Sesame Street" Test.

4	4	10	10	10	7	7	5	4	9
10	6	5	10	7	8	8	5	8	9
6	9	2	10	9	10	10	9	5	8
2	9	7	10	2	1	10	8	3	2
10	10	3	10	10	2	4	7	1	5
6	4	6	10	10	8	8	5	Missing	10
7	6	9	9	10	5	10	2	Missing	4
9	1	4	10	10	Missing	3	8	2	5
10	9	8	9	10	2	10	1	3	5
10	9	Missing	10	10	10	10	1	10	1
6	10	5	10	10	10	3	5	7	1
8	5	7	9	9	3	9	Missing	2	Missing
3	10	8	10	10	4	8	2	10	2
8	3	1	10	9	3	8	Missing	10	1
10	4	10	10	10	10	7	1	8	2
10	5	5	6	10	5	10	2	Missing	10
6	8	7	9	9	10	3	Missing	2	10
9	10	8	10	10	10	10	Missing	7	10
10	8	10	9	4	6	10	10	3	9
5	1	10	10	3	5	10	4	6	6
10	7	7	9	Missing	10	9	8	1	9
10	10	9	10	7	8	7	10	8	5
8	10	10	10	4	5	Missing	6	8	10
9	7	8	10	5	3	3	4	4	10

TABLE 4.2
Frequency distribution of scores from the "Sesame Street" Test.

Score Value	Frequency (f)
10	76
9	26
8	24
7	16
6	12
5	20
4	14
3	14
2	14
1	12
Missing Data	12
TOTAL	240

number (frequency) of individuals within a specific category. From variable to variable the meaning that the scores represent changes. For qualitative data, a frequency distribution is constructed by simply counting the individuals in a category. Variables in the "Sesame Street" data defined by qualitative measurements include sex, sampling site, and viewing location and encouragement.

Are any other variables in Appendix A measured qualitatively? One variable is the type of preschool attended (variable 14). Two other variables, viewing category and frequency of reading to child (variables 4 and 13), are less clear. While all the variables identified previously as qualitative were measured at nominal level, codes for these two variables represent at least rank orders of the categories. The numbers themselves, however, still do not represent quantities of some characteristic, but are merely labels for categories. As such, these data are best represented qualitatively. A rule of thumb to differentiate qualitative from quantitative data is that data need to be at the interval or quasi-interval (treated as interval) level to be classified as quantitative. An example of a quasi-interval level variable is the Mother-Child Socialization Index. The scores assigned to item categories are treated as interval measures of an underlying continuum (degree of mother-child interaction) and as such are combined to create a scale. The measurements of all other variables in the "Sesame Street" data base may be treated quantitatively.

Table 4.3 represents the frequency distribution for sample site. The steps in constructing the frequency distribution include

1. Label the categories by name or number. If using numbers, define them in the table when reporting the frequency distribution.
2. Tally the frequencies in each category while going through the data once. Do *not* search the data counting all 1s, 2s, and so on. This requires passing through the data five times. It is more efficient to tally all categories as you pass through the data once. This step is usually not reported in a frequency distribution table.

TABLE 4.3

Frequency distribution for children sampled from each site in the "Sesame Street" data base.

Sampling Site	Assigned Code	Tally*	f	Prop. f	$\%f$
I	1	卌 卌 卌 卌 卌 卌 卌 卌 卌 卌 卌 卌	60	.25	25
II	2	卌 卌 卌 卌 卌 卌 卌 卌 卌 卌 卌	55	.23	23
III	3	卌 卌 卌 卌 卌 卌 卌 卌 卌 卌 卌 卌 IIII	64	.27	27
IV	4	卌 卌 卌 卌 卌 卌 卌 卌 III	43	.18	18
V	5	卌 卌 卌 III	18	.07	7
		TOTAL	240	1.00	100%

*Actual tallies are presented here only as an illustration. They are not generally included in this type of table.

TABLE 4.4
Frequency of students viewing "Sesame Street" at home or in school at various sampling sites.

Sampling Site	Home			School		
	f	Prop. *f*	%*f*	*f*	Prop. *f*	%*f*
I	28	.20	20	32	.33	33
II	31	.22	22	24	.25	25
III	47	.33	33	17	.17	17
IV	19	.13	13	24	.25	25
V	18	.12	12	0	.00	0
TOTAL	143	1.00	100	97	1.00	100

3. Count the tallies and record the frequency under a column labeled *f*. Note that the tallies are grouped in units of five to facilitate counting at this step.

In addition to the simple frequencies, the relative proportion and percentage frequency distributions are given in Table 4.3. The relative proportions for a category are obtained by dividing each category's raw frequency (f) by the total number of individuals across all categories. For example, the relative proportion of individuals from Site I (category code = 1) is 60 ÷ 240, or .25. The relative proportion for a category is then multiplied by 100 to find the relative percent. The frequency distribution in Table 4.3 shows that the sample of 240 individuals in Appendix A contains an approximately equal number of children from Sites I, II, and III, a somewhat lesser number from Site IV, and a relatively low percentage from Site V.

The relative proportions and percentages are particularly useful when comparing frequency distributions for two or more different sized groups of individuals. By converting raw frequencies to relative proportions or percentages, meaningful comparisons can be made across groups. Otherwise, the apparent differences in category frequencies may be due to the varying number of observations in the groups. Frequency, relative proportion, and relative percentage frequency distributions for the sampling site variable can illustrate this concept if done separately for students who viewed "Sesame Street" at home (coded 1 on variable 5) and those who viewed it at school (code 2). Table 4.4 provides an illustration.

4.2 GRAPHIC DISPLAY OF QUALITATIVE FREQUENCY DISTRIBUTIONS

The bar graph is the format for visually displaying the information in a qualitative frequency distribution. Figure 4.1 illustrates two forms of a bar graph for

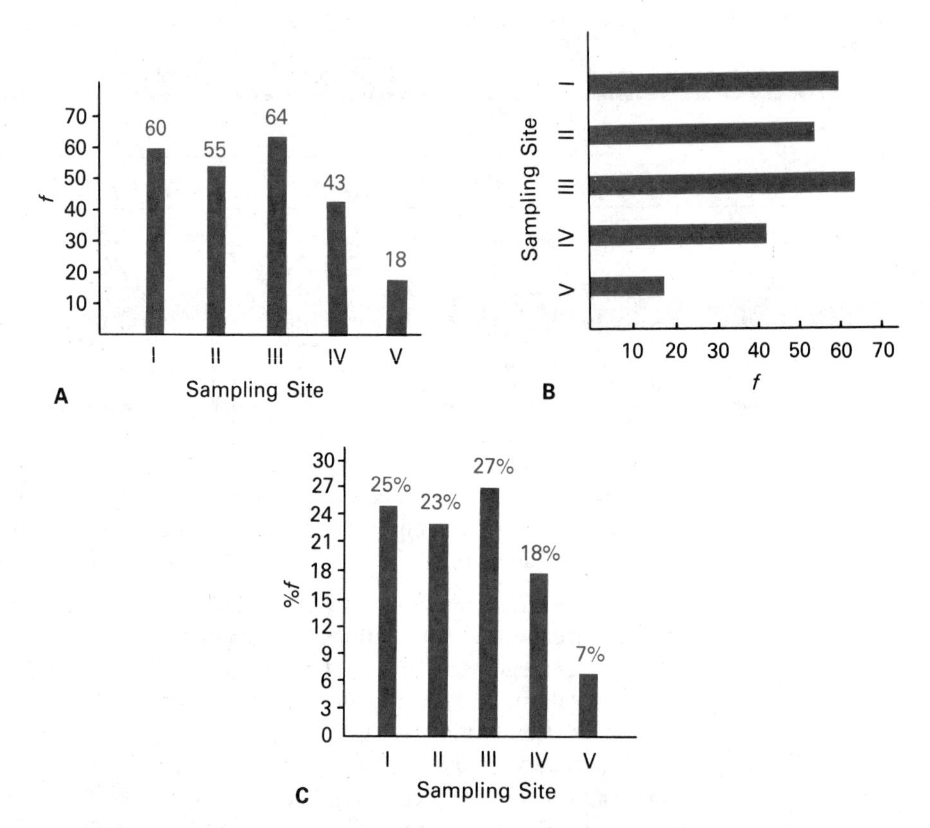

FIGURE 4.1
Bar graphs indicating number of children sampled from each site. **A,** raw frequencies displayed vertically; **B,** raw frequencies displayed horizontally; **C,** frequencies transformed to percentages.

the raw frequencies from Table 4.3 (*A* and *B*) and an example of these data represented by percentages (*C*). The raw-score distribution information is identified in cases *A* and *B*. The two raw-score frequency graphs differ only in that labels for the horizontal and vertical axes have been reversed. In the other illustration, *C*, the graphed data represent the percentage of cases included from each location. In a bar graph, one axis represents the specific categories for the variable and the other axis represents the raw frequencies or the relative proportion/percentage frequencies. What each axis represents is a matter of personal preference and convenience for the particular application.

One basic rule in constructing any graph is to avoid intentionally distorting differences or similarities. We can easily misrepresent large or small differences between groups by drawing axes of widely different lengths. Visual distortion has been a prime reason for statistics' bad reputation (see Huff, 1954, and Wainer, 1984).

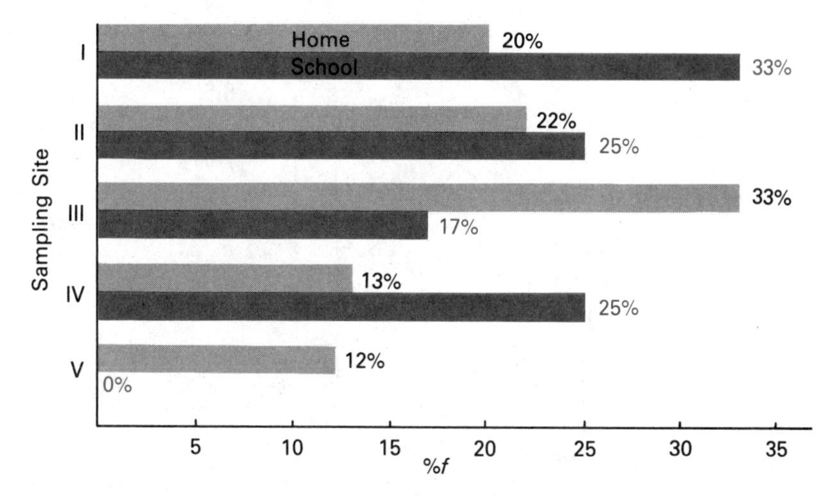

FIGURE 4.2
Relative percentage frequencies of students viewing "Sesame Street" at home or school at various sampling sites.

Table 4.4, the qualitative frequency distributions for the sampling site variable for students who viewed "Sesame Street" in the home or at school, is represented graphically in Figure 4.2. Because of the unequal sample sizes, relative percentages have been graphed. These values or the relative proportions based on each group's total size are legitimate values for comparative purposes. If you note the raw frequencies for Site II children viewing "Sesame Street" at home versus at school (31 vs. 24), you might conclude that a proportionately greater number of children watched at home. However, 22% of the Site II children viewed "Sesame Street" at home, while 25% of the Site II children viewed "Sesame Street" at school. The relative percentages adjust for disparities in sample size and allow more accurate qualitative comparisons.

4.3 ORGANIZING QUANTITATIVE DATA: ASSUMPTIONS

As has been discussed, numbers that represent quantitative data express the magnitude of some characteristic under study. As such, they can be used to group people into score categories (those receiving the same score or a score in the same interval) to rank the scores or score categories from low to high. As quantitative data points, the scores are assumed to represent equal intervals along an underlying continuous scale. The continuous nature of the variable and assumed equal intervals of the measurement scale are prerequisites for classifying data quantitatively. Figure 4.3 illustrates both assumptions for a limited range of scores on the "Sesame Street" test score continuum. On the scale, measurements were made using only whole numbers between 0 and 10,

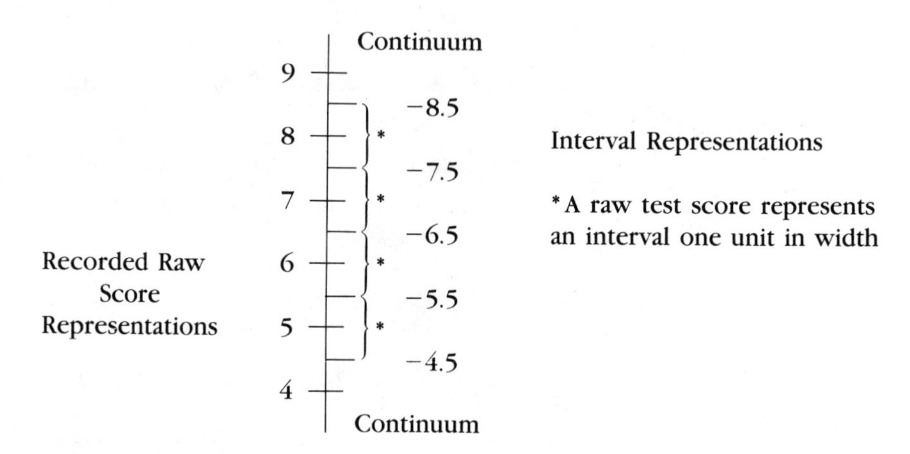

FIGURE 4.3
Continous scale: Degree of knowledge of "Sesame Street" concepts.

implying measurement accuracy to the nearest whole number. However, each whole number actually represents a one-unit interval on the continuous scale of "Sesame Street" knowledge. A score of 6 represents the interval of knowledge from 5.5 to 6.5, a score of 7 represents the interval from 6.5 to 7.5, and so on. Under these conditions the recorded units of measurement, which otherwise appear to be discrete numbers, reflect the continuous nature of the variable and equal intervals of the scale.

The accuracy of the measurement process dictates the length of an interval represented by a data point value. Instead of measurement to the nearest whole number as earlier, one could measure to the nearest one-tenth of a second in a reaction-time experiment or to the nearest $500 in a study of yearly earnings. In the reaction-time experiment, a data point of 5.3 would represent the interval from 5.25 to 5.35, while in the earnings study, a data point of $13,500 would represent the interval from $13,250 to $13,750. The recorded unit resulting from the measurement process is an important consideration in organizing quantitative data. The interval represented by the recorded data point will always be one-half a unit of measurement (i.e., whole units, tenths, thousands, etc.) below the data point to one-half a unit of measurement above the data point.

The simplest way to organize quantitative data is to form a frequency distribution using the consecutive single raw-score units to represent the categories. The frequency distribution in Table 4.2 illustrates this procedure. Each category for which frequencies are tallied is represented by a single whole number (the unit of measurement). This display of data easily communicates the distribution of persons across the "Sesame Street" knowledge continuum based on their performance on the "Sesame Street" test.

This form of a frequency distribution is very efficient and communicates maximum information *when the range of unit score categories is not large* (less than 20). However, when the score continuum encompasses more than 20 consecutive data points, the listing of categories by single units consumes too much space and is not an efficient way to organize and summarize a set of data. The range of Peabody mental age scores from Appendix A is a good example of this point. The raw scores range from a low of 27 to a high of 99. Obviously we would need many single unit score categories to represent the entire range of mental ages on a continuous score scale.

Frequency distributions in which each raw-score unit is used to label categories are called **ungrouped** frequency distributions. In these distributions, score values have *not* been combined to form category intervals encompassing more than one unit of measurement on the scale continuum. Distributions that do encompass more than one unit of measurement are called **grouped** frequency distributions. The remainder of this chapter treats organization of data into either grouped or ungrouped frequency distributions.

4.4 GROUPED FREQUENCY DISTRIBUTIONS

Class Intervals

In a grouped frequency distribution, score units are combined to form **class intervals.** Class intervals are categories into which different data points are classified. Three rules guide the identification of class intervals for a given set of data.

1. The categories of class intervals must be *exhaustive* of the scale continuum for the range of scores in the data set.
2. The categories must be *mutually exclusive,* or nonoverlapping.
3. The class intervals must be of equal width (an exception is made for the upper or lower category if extreme scores exist at the ends of the distribution).

Table 4.5 gives sets of class intervals for three different variables from the "Sesame Street" data base that satisfy the preceding rules.

The exhaustiveness requirement of the class intervals ensures that the categories span the entire scale continuum and reflect the scores' underlying continuity. The upper and lower score values of each class interval represent real, obtainable numbers. These limits are called the **apparent score limits** because they *appear* to be the upper and lower bounds to the class interval. Using the top class interval from Table 4.5 for the Letters pretest variable, 56 is the **upper apparent score limit** for the class interval and 53 is the **lower apparent score limit.** We can see from Figure 4.4 that using the apparent score limits as the boundaries for class intervals does not satisfy the condition

TABLE 4.5
Grouped frequency distributions for Age, Letters pretest scores, and Classification posttest scores.

Age (in months)		Letters Pretest		Classification Posttest	
Interval	*f*	Interval	*f*	Interval	*f*
67–69	3	53–56	1	24–25	8
64–66	4	49–52	0	22–23	29
61–63	3	45–48	4	20–21	28
58–60	31	41–44	3	18–19	33
55–57	41	37–40	4	16–17	35
52–54	43	33–36	2	14–15	25
49–51	46	29–32	4	12–13	21
46–48	32	25–28	9	10–11	28
43–45	18	21–24	11	8–9	18
40–42	7	17–20	37	6–7	9
37–39	7	13–16	79	4–5	5
34–36	5	9–12	62	TOTAL	239
TOTAL	240	5–8	18		
		1–4	6		
		TOTAL	240		

of exhaustiveness, however. Certain portions of the scale are not contained in any class interval, for example, 44–45, 48–49, 52–53, and 56–57.

The condition of exhaustiveness is met by assuming underlying continuity of the variable. The upper apparent score limit of 56 for the interval 53–56 does not just include the single point of 56 on the scale. Rather, it represents the interval from 55.5 to 56.5. All other scores in the continuous distribution also represent intervals. Therefore, while the apparent score limits encompass the interval 53–56, the *exact score limits* are 52.5–56.5. Using the exact score limits as true boundaries of class intervals ensures that all points on the vari-

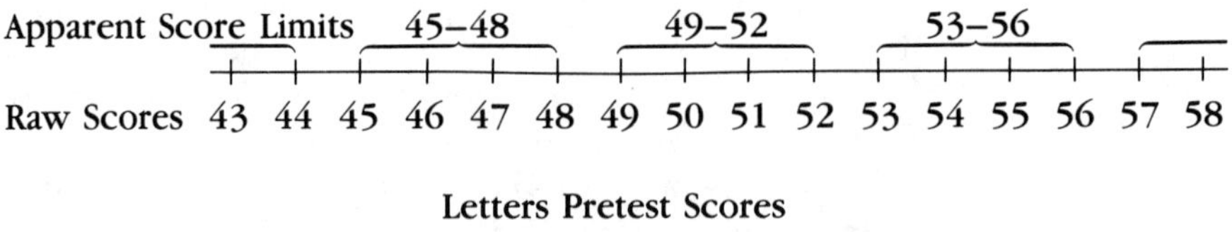

Apparent Score Limits 45–48 49–52 53–56

Raw Scores 43 44 45 46 47 48 49 50 51 52 53 54 55 56 57 58

Letters Pretest Scores

FIGURE 4.4
Apparent score limits for Letters pretest scores.

TABLE 4.6
Frequency distribution of Peabody mental age data.

Apparent Score Limits	Exact Score Limits	f
95–99	94.5–99.5	1
90–94	89.5–94.5	6
85–89	84.5–89.5	3
80–84	79.5–84.5	2
75–79	74.5–79.5	5
70–74	69.5–74.5	2
65–69	64.5–69.5	18
60–64	59.5–64.5	10
55–59	54.5–59.5	27
50–54	49.5–54.5	6
45–49	44.5–49.5	27
40–44	39.5–44.5	26
35–39	34.5–39.5	49
30–34	29.5–34.5	38
25–29	24.5–29.5	20
	TOTAL	240

able continuum are exhausted. Apparent and exact score limits for a frequency distribution of the Peabody mental age data are illustrated in Table 4.6. Note that throughout the scale continuum, the *upper* exact score limit for one class interval is also the *lower* exact score limit for the next highest, adjacent class interval. The value 29.5, for example, is both the upper exact score limit for the interval 25–29 and the lower exact score limit for the interval 30–34.

The condition of mutually exclusive class intervals requires that the intervals be nonoverlapping. Under this condition, an individual data point can be categorized into one and only one class interval. The only points on the scale continuum that appear to violate this condition are the exact score limits. The exact score limits, however, are merely separation points that preserve the exhaustiveness of the categories and reflect the variable's continuity. They are not score values actually recorded as data points for any individual in the group. For example, the unit of measurement for any of the variables in Tables 4.5 and 4.6 is the nearest whole number. The exact score limits are given in .5 of a score unit (e.g., 29.5), and as such are not obtainable through the measurement process.

With one exception, the class interval widths must be identical across all categories of the grouped frequency distribution. As mentioned, the exception is for categories at the upper or lower ends of the distribution. Either of two methods can determine interval width.

The definition of interval width (labeled i) is the difference between the upper and lower *exact* score limits of a class interval, or

$$i = \begin{pmatrix}\text{upper exact}\\ \text{score limit}\end{pmatrix} - \begin{pmatrix}\text{lower exact}\\ \text{score limit}\end{pmatrix}$$

Another method of calculating interval width uses the apparent score limits.

$$i = \begin{pmatrix}\text{upper apparent}\\ \text{score limit}\end{pmatrix} - \begin{pmatrix}\text{lower apparent}\\ \text{score limit}\end{pmatrix} + \text{one unit of measurement}$$

In this formula, the "one unit of measurement" would equal 1 if the data were measured to the nearest whole unit, .1 if to the nearest tenth, or 1000 if measurement were to the nearest thousand. Given any two consecutive class intervals, we should be able to determine the interval width. Look at each of the distributions in Tables 4.5 and 4.6 and determine the interval widths.[1] The class intervals within each distribution are the same width.

Constructing the Grouped Frequency Distribution

In addition to the previously specified rules, construction of grouped frequency distributions follows several other conventions. These conventions are guidelines, not hard, fast rules for all data sets. The number of class intervals in the frequency distribution should be somewhere between 10 and 20 for most data sets. The actual number of class intervals depends on the range of scores in the group, the space available for reporting the distribution, and the group size. The convention of 10–20 class intervals makes sense only if the scale range permits that many intervals and the sample size is sufficiently large (greater than 40). In some instances fewer than 10 categories will suffice, or it might be desirable to have more than 20.

The most difficult part (if there is one!) in constructing a grouped frequency distribution is determining the interval width. The range of scores (highest to lowest) in the data set and the interval width determine the number of class intervals. Since the range of scores is fixed, the interval size chosen should result in an appropriate number of class intervals (normally 10–20). The first step in the process is to determine the range of scores in the group under study. For the scores from the "Sesame Street" data on the Letters pretest, the highest score for the 240 individuals was 55 and the lowest was 1. The difference between these two scores is a range of 54. Dividing this range by different possible interval widths yields the resulting number of class intervals needed to cover the range of scores. An interval size of 10 units would result in 5.4 or 6 class intervals, one of 5 units would give 10.8 or 11 class intervals, one of 3 units would give 18 class intervals, and so forth. Given the range of scores, one could use a 3-, 4-, or 5-unit interval and still be within the 10–20 class

[1]Interval widths for age, $i = 3$; Letters pretest, $i = 4$; Classification posttest, $i = 2$; Peabody mental age, $i = 5$.

interval range. Now try your skill in a couple of situations. The score range on the Body Parts posttest is 21, with a high score of 32 and a low score of 11. What would be an appropriate interval width?[2] What about for the distribution of Peabody mental age scores with a range of 72 score units?[3]

Once the decision on interval width has been made, we can determine the specific class intervals. Again, convention dictates that the class intervals be ordered with the highest interval at the top and the lowest at the bottom. The bottom class interval will naturally contain the lowest score in the data set and the top class interval will contain the highest score. For the distribution of scores on the Letters pretest, the lowest score was 1. By choosing an interval width of 4 units, we have the option of specifying the apparent score limits for the lowest class interval as 0–3 or 1–4. The apparent score limits chosen for this bottom class interval will determine the apparent score limits for the remaining intervals. Table 4.5 illustrates the class intervals generated for an interval width of 4 and a bottom class interval of 1–4.

With the class intervals identified, raw scores from the data set are then tallied to determine the class interval frequencies. This is accomplished in the same manner as for qualitative frequency distributions (section 4.1). As a reminder, the most efficient procedure is to tally each score while making a single pass through the data. The tallies are then converted to raw frequencies and recorded under the column labeled *f*.

Note that when grouped frequency distributions are constructed, the exact number of individual score values is lost. This loss of information, called **grouping error,** results from the presentation of information that is one step removed from the exact raw score values. For example, consider the frequency distribution in Table 4.6. There are 10 scores in the interval from 60–64. The information lost in grouping the scores is the exact number of 60s, 61s, 62s, 63s, and 64s in the data base. Because this exact information is not available from a grouped frequency distribution, it is sometimes convenient to assume that the frequency of scores in an interval is distributed equally across the entire continuum of the interval. That is, without knowledge of the exact raw scores, we would assume there are two 60s, two 61s, two 62s, two 63s, and two 64s in the interval 60–64.

A single score that can represent all scores in an interval is the **midpoint.** The midpoint of an interval is the scale point exactly in the middle of the interval. Midpoints of class intervals are calculated as follows:

$$\text{midpoint} = \frac{\text{upper apparent limit} + \text{lower apparent limit}}{2}$$

---

[2]An interval width of 2 units would give 21 ÷ 2 = 10.5 or 11 intervals, one 3 units wide would result in 7 intervals, and so on. An interval width of two would be best.

[3]Dividing 72 by 4, 5, 6, 7, or 8 results in 18, 15, 12, 11, and 9 intervals, respectively. Given the number of intervals that result, the interval width could be anywhere from 4 to 7 and still be in the 10–20 range. An interval width of 5 units is the easiest to use and was chosen for the frequency distribution in Table 4.6.

TABLE 4.7
Comparative frequency distributions of Classification posttest scores for children viewing "Sesame Street" in the home or school.

	Viewing in Home		Viewing in School	
Interval	*f*	%*f*	*f*	%*f*
24–25	3	2	5	5
22–23	15	11	14	15
20–21	19	13	9	9
18–19	21	15	12	13
16–17	19	13	16	17
14–15	12	8	13	14
12–13	13	9	8	8
10–11	19	13	9	9
8–9	11	8	7	7
6–7	7	5	2	2
4–5	4	3	1	1
TOTAL	143	100	96*	100

*There were 97 in this group, but one participant had a missing data point for this variable.

For the distribution in Table 4.6, 27 is the midpoint in the interval 25–29, 32 is the midpoint in the interval 30–34, and so on. Identify the midpoints for the upper three intervals in each of the distributions given in Table 4.5.[4] The midpoints will become useful in section 4.5 when grouped frequency distributions are graphed. Note that the midpoint is an integer when the interval width is an odd number. For this reason, many persons prefer to construct frequency distributions with odd values of i.

As with qualitative data, comparative frequency distributions can be formed for groups of unequal size by converting the raw frequencies to relative proportional or percentage frequencies. To illustrate, in Table 4.7 the distribution of classification posttest scores first given in Table 4.5 has been broken down into two distributions, one each for students who viewed "Sesame Street" at home (total = 143) and those who viewed the program at school (total = 97).

The Cumulative Frequency Distribution

An extension of the grouped frequency distribution that is often useful is the **cumulative frequency distribution.** The cumulative frequency distribution is used to answer questions such as "How many students in the 'Sesame Street' data base were 52 months of age or younger?" or "What were the number of

[4](a) Age intervals: (67–69), midpoint = 68; (64–66), 65; (61–63), 62.
(b) Letters pretest intervals: (53–56), 54.5; (49–52), 50.5; (45–48), 46.5.
(c) Classification posttest intervals: (24–25), 24.5; (22–23), 22.5; (20–21), 20.5.

TABLE 4.8
Frequency distribution of Peabody mental age scores.

Apparent Score Limits	Exact Score Limits	*f*	*cf*	*cf*%
95–99	94.5–99.5	1	240	100
90–94	89.5–94.5	6	239	99
85–89	84.5–89.5	3	233	97
80–84	79.5–84.5	2	230	96
75–79	74.5–79.5	5	228	95
70–74	69.5–74.5	2	223	93
65–69	64.5–69.5	18	221	92
60–64	59.5–64.5	10	203	85
55–59	54.5–59.5	27	193	80
50–54	49.5–54.5	6	166	69
45–49	44.5–49.5	27	160	67
40–44	39.5–44.5	26	133	55
35–39	34.5–39.5	49	107	45
30–34	29.5–34.5	38	58	24
25–29	24.5–29.5	20	20	8
	TOTAL	240		

students with Numbers pretest scores between 22 and 40?" or "What was the score on the Classification posttest that split the group into the top and bottom 50%?" Answers to these questions all involve cumulative frequencies.

A cumulative frequency (cf) is one that identifies the number of scores *at or below the upper exact limit of a class interval.* In Table 4.2, how many scores are equal to or less than a raw score of 4? To answer the question, we would sum the frequencies associated with the score values of 1 ($f = 12$), 2 ($f = 14$), 3 ($f = 14$), and 4 ($f = 14$) to get a cumulative frequency of 54. Thus, 54 individuals received scores of 4 or less. Because of the scale's assumed continuity, the correct interpretation is that 54 scores are at or below the upper exact limit of the interval represented by the number 4. The upper exact limit is 4.5. Using the information in Table 4.2, how many individuals have "Sesame Street" Test scores at or below a scale score of 7.5?[5] How many have score values between 7.5 and 4.5?[6]

The frequency and cumulative frequency distributions of Peabody mental age scores are given in Table 4.8. To calculate the cumulative frequencies, we begin with the lowest class interval. The frequency and cumulative frequency are equal for this category. The frequency of 20 for the category 25–29 indicates that 20 mental age scores are assumed to be distributed equally

[5] 102 individuals; add the frequencies of all intervals for score values of 7 and below.

[6] $102 - 54 = 48$ individuals have scores between 7.5 and 4.5.

across the score range from 24.5 to 29.5. The cumulative frequency is 20 scores at or below a scale value of 29.5, the upper exact score limit of the category. To get the cumulative frequency of 58 for the interval 30–34, we add the cumulative frequency for the first interval ($cf = 20$) and the frequency for the second interval ($f = 38$). Each subsequent cumulative frequency is obtained by adding the cumulative frequency for the prior interval to the frequency for the present interval. For example, the cumulative frequency for the interval 70–74 is obtained by adding 221, (cf for the interval 65–69) to 2, (f for the interval 70–74). How many students have scores at or below a mental age of 49.5?[7] What percentage of the students have a mental age at or below 49.5? To answer the latter question, we need to change the cumulative frequency to a cumulative percentage frequency ($cf\%$). As in previous distributions, this is accomplished by dividing each cumulative frequency by the total number of scores ($n = 240$) and multiplying by 100. Based on the data in Table 4.8, 67% ($160/240 \times 100$) of the 240 children had mental age scores of 49.5 or less. To check on your calculations, the cumulative frequency for the highest interval in the frequency table will always equal the total number of scores in the distribution.

4.5 GRAPHING QUANTITATIVE FREQUENCY DISTRIBUTIONS

A frequency table summarizes information about a distribution and the shape of the data. A graphic display of a frequency distribution also is an effective form of communication. Two approaches for graphically presenting the data in frequency tables are the **histogram** and the **frequency polygon.**

Histograms

A histogram represents the frequency in a score category much like the bar graph for qualitative data, except that the bars are connected to each other to reflect and reinforce recognition of the continuous nature of the underlying scale. Figure 4.5 displays the frequency distribution in the form of a histogram for the Letters pretest scores from Table 4.5. The histogram illustrates vividly the large frequency of pretest scores at the lower end of the scale. As a pretest, these data were collected prior to children viewing the "Sesame Street" programs. The low scores demonstrate clearly that few of the children had acquired letter recognition skills.

Construction of histograms follows four rules.

1. The vertical axis represents the frequency scale and the horizontal axis, the score continuum of the variable being displayed.

---

[7] $cf = 160$

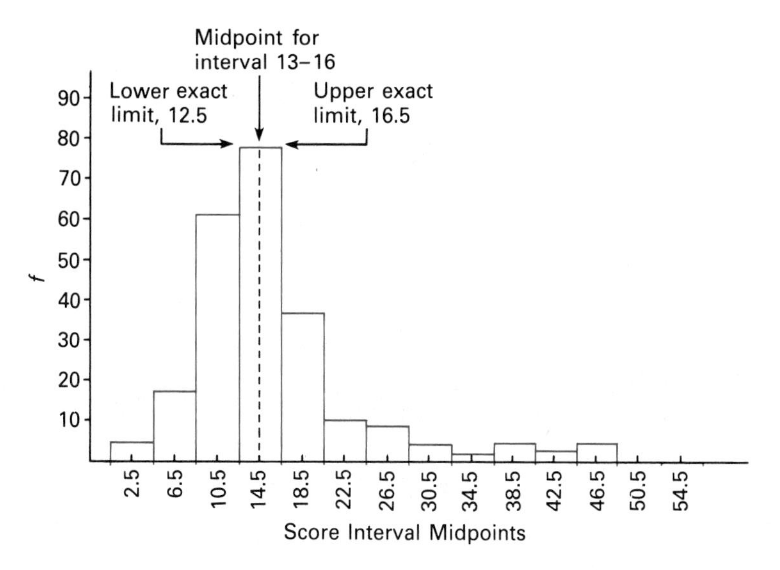

FIGURE 4.5
Histogram of Letters pretest scores.

2. The length of the vertical axis relative to the horizontal axis should be in approximately a 3-to-5 ratio.
3. The columns or bars of the histogram span the class intervals from the lower exact limits to the upper exact limits with the center of the bar over the midpoint of a class interval. This text will generally label the scale by its midpoints.
4. The height of each bar is drawn to represent the number of scores in the interval.

Figure 4.5 illustrates these rules. Another example is provided in Figure 4.6. The distribution of Peabody mental age scores is displayed in the form of a histogram. Note that the horizontal axis has been broken using the symbol —//—. This indicates that part of the scale is missing from the graph to conserve space and still maintain the zero point of the scale at the origin.

Frequency Polygons

The frequency polygon is constructed using the same procedures as those for the histogram with one exception. Instead of a bar or column, a point is placed above the midpoint of an interval to represent the frequency. These points are then connected to form an enclosed distribution shape. The construction of a frequency polygon is illustrated in Figure 4.7 for the same data as in Figure 4.6. It should be evident in graphing a frequency polygon why it is advisable to

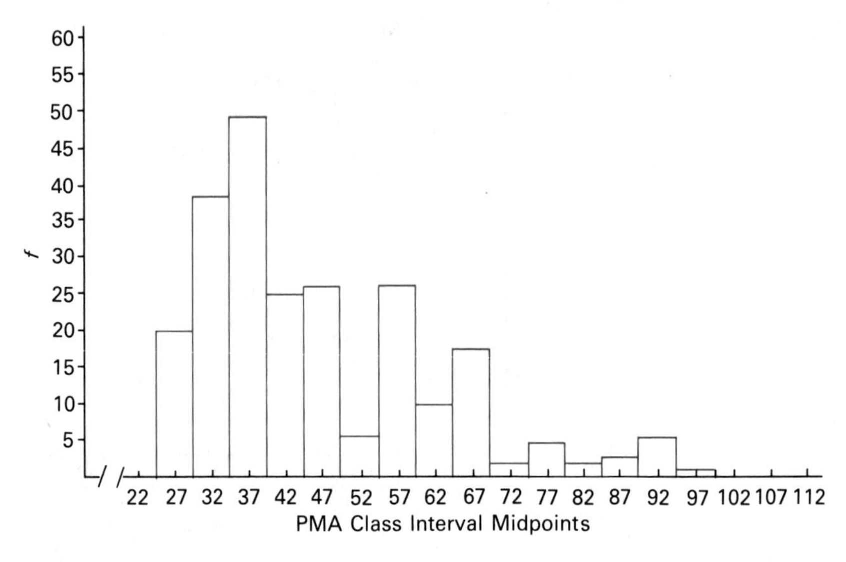

FIGURE 4.6
Histogram of Peabody mental age scores.

label the horizontal axis by the midpoints of the class intervals. The frequency points are placed directly above these values.

In Figure 4.7, the dotted lines at the ends of the distribution show the customary manner in which frequency polygons are closed at the extremes. Dotted lines illustrate this point; solid lines would normally be used. The poly-

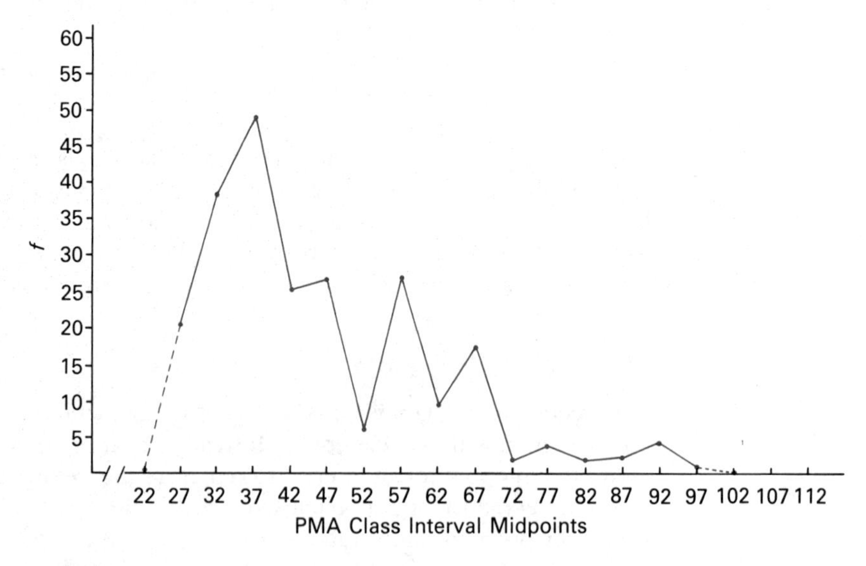

FIGURE 4.7
Frequency polygon of Peabody mental age scores.

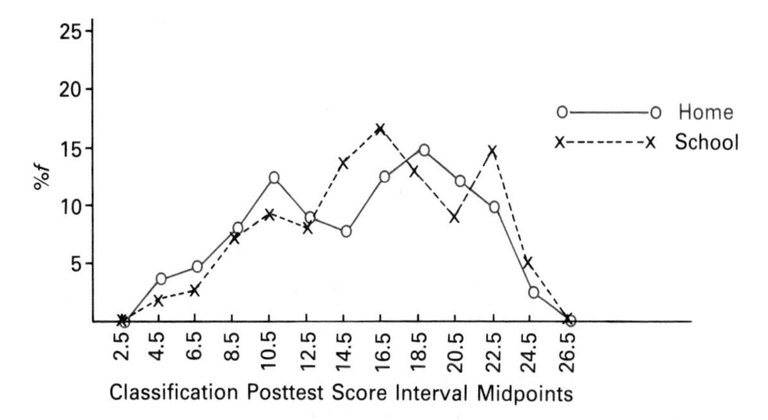

FIGURE 4.8
Comparative percentage frequency polygons of Classification posttest scores for children viewing "Sesame Street" at home or school.

gon is completed by extending solid lines to the midpoints of the intervals at the low and high extremes of the scale beyond the actual range of score values.

An advantage of the frequency polygon over the histogram is that distributions for two or more groups may be displayed on the same graph. If the groups are of unequal size, then the proportional or percentage frequencies are graphed. Figure 4.8 presents the percentage frequency polygon for the data in Table 4.7. The distribution shape tends to follow a similar pattern for each group with the distribution of Classification posttest scores moved slightly toward the higher end of the scale for children who viewed "Sesame Street" in the school setting.

4.6 PERCENTILES, DECILES, AND QUARTILES

While percentile points and percentile ranks are statistical indices, their primary usage is as interpretive indices in measurement, especially in standardized testing. Some statistical texts prefer the label "centile" in place of "percentile." However, we will use percentile to correspond with its use in the testing literature. Both percentile points and ranks are calculated from frequency distributions. A **percentile point** is a point on the score scale at or below which a specific proportion or percentage of the scores fall. The symbol for percentile point is P_p, where the subscript indicates the percent of scores at or below the percentile point. P_{42} refers specifically to the scale point at or below which 42% of the scores in the group fall. In contrast, a **percentile rank** is a percentage, not a scale point. Individual percentile ranks are commonly reported as normative scores from standardized tests. A percentile rank of 84 indicates that an individual's raw score on a test is equal to or higher than 84% of the scores for persons in the norm group. We need not know the exact raw score to interpret the percentile rank. If we wanted to know the raw score corresponding to a percentile rank of 84, we would need to calculate the percentile point, P_{84}. Each percentile point has a corresponding percentile rank and vice versa, but the former results in a point value on the variable scale while the latter represents a percentage.

These two indices take some getting used to before one can become comfortable with their interpretation. By way of continual examples, we may want to find the score point in a distribution at or below which 75% of the scores are located. This would require calculation of P_{75}, the 75th percentile point. Or, we may want to determine what percentage of those tested obtained a raw score of 120 or less. This requires calculating the percentile rank for the score 120. Also, if 25% of the scores in a distribution of scores were found to fall at or below a score point of 120.5, then $P_{25} = 120.5$, and the percentile rank of 120.5 is 25.

Other common partition points in addition to percentiles are **decile** and **quartile** points. Decile points divide a distribution into 10 equal parts with 10% of the scores falling between two decile points. Decile points are equivalent to the percentile points P_{10}, P_{20}, P_{30}, and so forth. Similarly, quartile points are defined as the percentile points P_{25}, P_{50}, and P_{75}. These three percentile points divide a distribution into four equal parts of 25% each. P_{25} defines the upper boundary for the first quartile range, P_{50}–P_{25} identifies the interval for the second quartile range, and so on. Percentiles, deciles, and quartiles define scale points that partition the distribution into 100, 10, or 4 equal areas, respectively, *in terms of the frequencies within the intervals.* The raw score interval widths will not necessarily be the same across percentile, decile, or quartile ranges, so that P_{30}–P_{20} need not equal the raw score interval range P_{60}–P_{50}, even though each interval range contains 10% of the scores.

Calculating Percentile Points

In calculating a percentile point it is important to remember that we start with a percentage to find a point on the scale. *Given a percentage, the task is to find a scale point.* As an example, let us find P_{85} from the data in Table 4.8. By definition, P_{85} is the Peabody mental age score that divides the distribution in Table 4.8 into the bottom 85% and the top 15% of the scores. Looking at the cumulative percentage frequency column, we note that 85% of the scores are at or below the upper exact limit of the score interval 60–64. P_{85} must then equal 64.5, or the upper exact limit of the interval. The 85th percentile point is 64.5. Equivalently, the score point 64.5 has a percentile rank of 85.

In the preceding example, we needed no formula to calculate the answer. Most situations will not be as easy, however, and it will help if we have a systematic method for finding P_p. The formula for calculating the pth percentile point from a frequency distribution is as follows:

$$P_p = L + \frac{pn - cf}{f}(i) \tag{4.1}$$

where

p = a percentage or proportion
n = group size
cf = cumulative frequency up to the lower exact limit of the interval containing the pnth score
f = frequency of scores for the interval containing the pnth score
i = interval width
L = lower exact limit of the interval containing the pnth score

Before you let this formula boggle your mind, remember what we said in Chapter 1. Formulas efficiently communicate statistical operations. Let's stop and analyze formula 4.1 through an application to the data in Table 4.8. Suppose we want to find the mental age score that represents the 50th percentile point, P_{50}. This is the scale point at or below which 50% of the scores fall. How many of the 240 scores in the distribution represent 50% of the scores? If p is 50%, or .5, and n is 240, then pn is 120, meaning 120 scores fall at or below P_{50}. Which interval contains the 120th score? The cumulative frequency distribution can help here. If 107 scores are at or below the scale point 39.5 and 133 are at or below 44.5, then the 120th score is in the interval 40–44. This percentile point, P_{50}, falls on the scale continuum between 39.5 and 44.5, the lower and upper exact limits of the interval.

Figure 4.9 illustrates the steps in the calculation process. Crucial to the calculation is the assumption that 26 scores in the 40–44 interval are distributed evenly across the 39.5–44.5 interval. How many of the 26 scores in the category are needed to give us the desired 120 scores? There are 107 scores up to 39.5. Therefore 13 (120 − 107) of these scores are needed to reach 120. Given that the 26 scores are evenly distributed over the interval, the proportion of the interval needed to reach P_{50} is 13/26 or .5 of the interval. The interval spans five score units, so the 120th score is .5(5) or 2.5 units into the interval. P_{50} is then equal to the lower exact limit of the interval (39.5) + 2.5, which equals 42.0. Fifty percent of the children had Peabody mental age scores of 42.0 or lower.

For the preceding example, the formula has the following values.

$$pn = 50\% \text{ of } 240 = 120$$
$$cf = 107$$
$$f = 26$$
$$i = 5$$
$$L = 39.5$$

In formula 4.1, the expression

$$\frac{pn - cf}{f} = \frac{120 - 107}{26} = \frac{13}{26} = .5$$

gives the proportion of the interval needed to reach the pnth (120th) score.

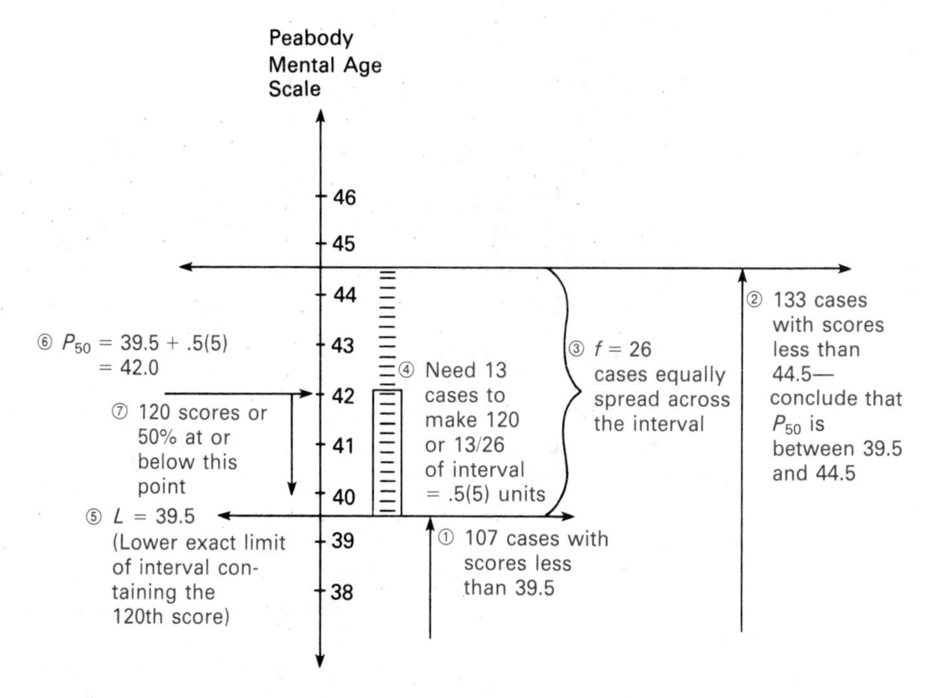

FIGURE 4.9
Steps in calculating percentile points.

To obtain the actual distance on the scale needed in the interval to arrive at the 120th score, we multiply this proportion by i, the interval width. We then add this portion of the scale to the lower exact limit of the interval, L.

Now for some practice. Follow this process to calculate P_{60}, P_{25}, and P_{58} from the data in Table 4.8.[8] In calculating the value of a percentile point, the

---

[8](a) P_{60} = ?

1. pn = 60% of 240 = 144
2. What interval contains the 144th case? The interval is 45–49
3. What proportion of the interval is needed?

$$\frac{pn - cf}{f} = \frac{144 - 133}{27} = \frac{11}{27} = .41$$

4. What is the scale distance in the interval needed?

$$.41(i) = .41(5) = 2.05$$

5. $P_{60} = 44.5 + 2.05 = 46.55$

(b) $P_{25} = 34.5 + \frac{60 - 58}{49}(5) = 34.70$

(c) $P_{58} = 44.5 + \frac{139.2 - 133}{27}(5) = 45.65$

Note: The percentile point P_{58} is the point on the scale at or below which 139.2 cases are to be found. This illustrates that percentiles are defined as specific points on the measurement scale and will not always correspond to whole units of the observations.

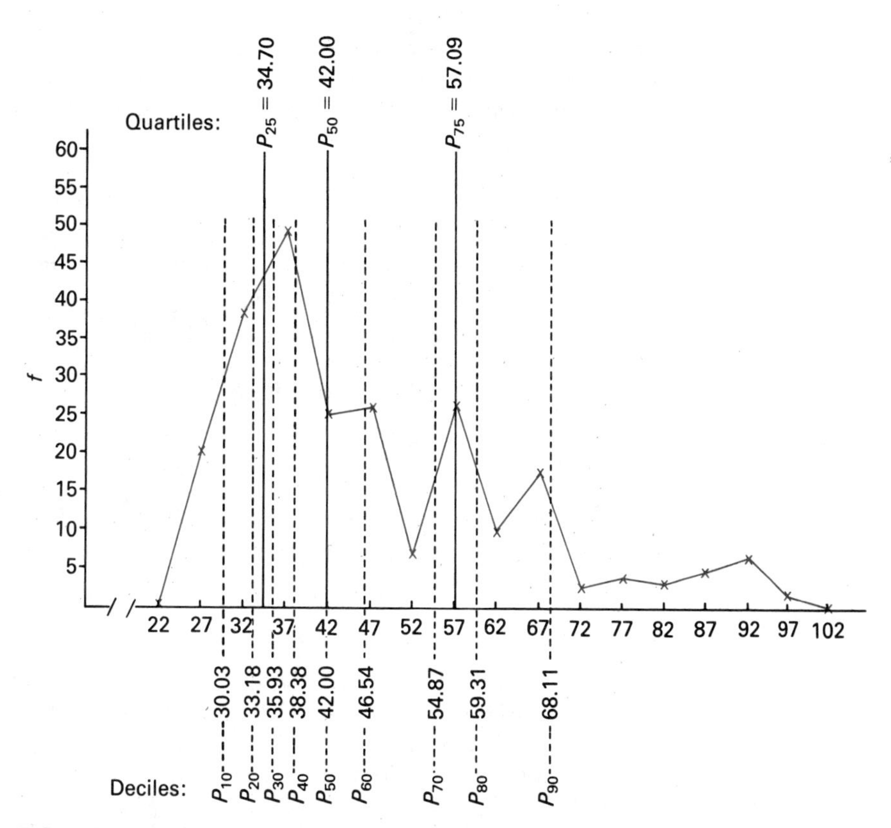

FIGURE 4.10
Deciles and quartiles for the frequency distribution of Peabody mental age scores.

answer is generally reported to the nearest tenth or hundredth. Even though it is impossible for an individual to obtain a raw score of 34.7, the answer for P_{25} reflects the mental age scale's assumed continuity.

All decile and quartile points are shown in the frequency polygon of Peabody mental age scores in Figure 4.10. Remember that the deciles and quartiles are percentile points and as such represent points on the scale. Notice that the range of raw score points needed for 10% of the frequencies (24 scores) changes across the distribution. The smallest mental age decile is the distance between P_{40} and P_{30}, which is $38.38 - 35.93$ or 2.45 mental age units. The largest range other than the last one is $P_{90} - P_{80}$, $68.11 - 59.31 = 8.80$ mental age units.

Calculating Percentile Ranks

Calculating percentile ranks is a reverse of the procedure for calculating percentile points. We start with a score on the scale and ask, "What percentage of the scores in the distribution are equal to or lower than this score value?" We will use the Peabody mental age data in Table 4.8 again for illustration. Suppose we are interested in finding the percentile rank (*PR*) for the score 42.0. The process is as follows:

- Step 1 Find the interval containing the score point. The score 42.0 is in the 40–44 interval.
- Step 2 Find the distance between the lower exact limit and the score point: $P_p - L = 42.0 - 39.5 = 2.5$.
- Step 3 Divide this distance by the interval width, *i*:

$$\frac{P_p - L}{i} = \frac{2.5}{5} = .5$$

to find the proportion of the interval represented by 2.5 units.

- Step 4 Multiply this proportion by the interval frequency to calculate the number of scores from P_p to L:

$$\frac{P_p - L}{i}(f) = .5(26) = 13$$

There are 13 scores from 39.5 to 42.0. How many scores are below 39.5? Look in the cumulative frequency column to find 107. Add $[(P_p - L)/i](f)$ to the cumulative frequency to find the number of scores at or below the score point 42.0.

$$cf + \frac{P_p - L}{i}(f) = 120$$

There are 120 scores in the distribution at or below the score of 42.0.

- Step 5 Convert the number of scores to a proportion by dividing by n: $120/240 = .50$.
- Step 6 Multiply the proportion by 100 to obtain the percentile rank [$.50(100) = 50$]. The percentile rank (*PR*) for the score 42.0 is 50. A percentile rank of 50 indicates that 50% of the individuals in the score distribution had scores equal to or less than 42.0.

The preceding steps can be summarized by the following formula:

$$PR = \frac{cf + \left[\frac{P_p - L}{i}(f)\right]}{n}(100) \qquad (4.2)$$

Follow through the step-by-step logic to calculate the percentile rank for a mental age score of 33.0.

1. What interval contains the score point? The interval is 30–34.

2. How many scores fall at or below that point in the interval?

$$\frac{33.0 - 29.5}{5}(38) = 26.6$$

3. How many mental age scores in the frequency distribution are at or below a score of 33.0? The cumulative frequency to 29.5 is 20, so 20 + 26.6 = 46.6.
4. What proportion of the scores does this represent? What percentage?

$$\text{Proportion} = \frac{46.6}{240} = .19$$

$$\text{Percentage} = .19(100) = 19$$

The percentile rank of the score point 33.0 is 19. The complete calculation is

$$PR = \frac{20 + \left[\frac{33.0 - 29.5}{5}(38)\right]}{240}(100)$$

$$= 19$$

As practice, calculate the percentile rank for a mental age score of 63.0.[9] The computation of percentile ranks for raw scores in a distribution is a common practice. Test result reports often convert raw scores to percentile ranks. Percentile ranks permit comparison of performance. For example, knowing that a person's raw score on the Peabody test was 33 is of little interpretative significance. Knowing that a person scored in the 19th percentile is more informative. This common transformation is further discussed in Chapter 7.

4.7 THE OGIVE

The frequency polygon graphically illustrates the frequencies in a frequency distribution plotted in relation to the midpoint of class intervals. The frequency polygon was discussed in section 4.5 and is useful for describing the shape of the distribution of a set of scores. Graphic presentation of the cumulative percentage distribution is also useful. A graph of the cumulative percentage distribution, called an **ogive,** is commonly used in educational and psychological measurement.

Recall that the cumulative percentage frequency is interpreted in relation to the exact upper limit of the class interval. For example, for the Peabody mental age data in Table 4.8, 96% of the children in the "Sesame Street" data

---

[9] $$PR = \frac{193 + \left[\frac{63.0 - 59.5}{5}(10)\right]}{240}(100) = 83$$

base received scores of 84.5 or lower. To construct an ogive, we plot the cumulative percentage frequency at the *exact upper limit* of the associated class interval. Figure 4.11 presents the ogive for the Peabody mental age scores given in Table 4.8.

In an ogive, the height or ordinate of the graph represents values for the cumulative percentage frequency, while the baseline or horizontal axis identifies the exact limits of the intervals. The height-to-width ratio for constructing a frequency polygon should be followed when preparing an ogive. When we construct an ogive we begin the graph by plotting a cumulative percentage frequency of zero at the exact lower limit of the lowest class interval. Then, we plot each cumulative percentage frequency at the exact upper limit of its interval. This procedure is different from the development of a frequency polygon, in which we plot frequencies at class interval midpoints. Once all points have been plotted on the graph, they are connected successively using straight line segments. Connecting adjacent points by a straight line reflects the assumption we made when computing percentile points and ranks—that the scores are distributed equally within the interval.

Once we prepare an ogive, we can use it to determine the score associated with any given percentile point or the percentile rank of a score. Dotted

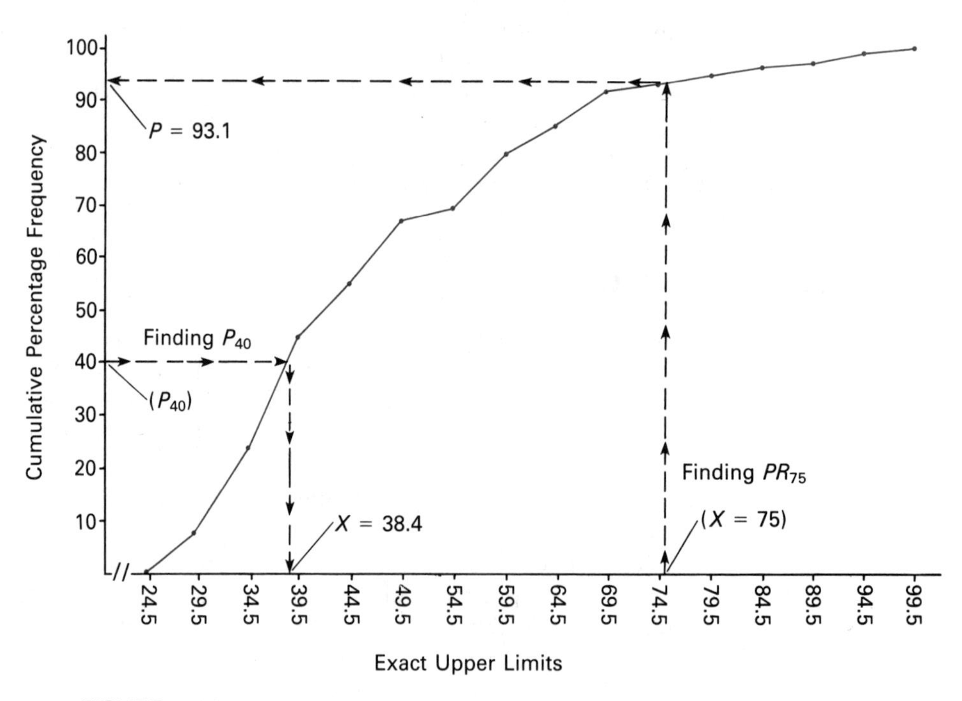

FIGURE 4.11
Ogive for PMA scores.

lines illustrate these operations in Figure 4.11. For example, to determine the percentile rank of a score of 75 we locate this score on the baseline, then extend a vertical line upward until the ogive is intersected. Then, we move straight across to the ordinate and read the percentile rank of the score 75. To find a specified percentile point we reverse this process. To find the score associated with P_{40}, we locate P_{40} on the cumulative percentage frequency axis, move across to where the ogive is intersected, then down to the baseline to find the score at the point P_{40}. The accuracy of using the ogive to determine percentile points or ranks is only limited by graphic accuracy. Ordinarily we expect the graphic method to yield results comparable to the computational procedures presented in section 4.6.

Ogives are S-shaped curves. Areas in the curve that are steep or rise sharply identify locations having proportionately many cases. Areas in which the curve tends to be relatively flat are locations with fewer observations.

4.8 DISTRIBUTION SHAPES

A frequency polygon or histogram reveals the shape of a distribution. From a descriptive statistics viewpoint, the shape of the distribution is one of the most important characteristics describing a set of data. Several common distributional shapes have been identified.

The shape with which practically everyone is familiar is the **normal** distribution. Normal distributions tend to be bell-shaped (see Figure 4.12A). The normal distribution is important to the foundation of inferential statistical concepts and Chapter 7 is largely devoted to its study.

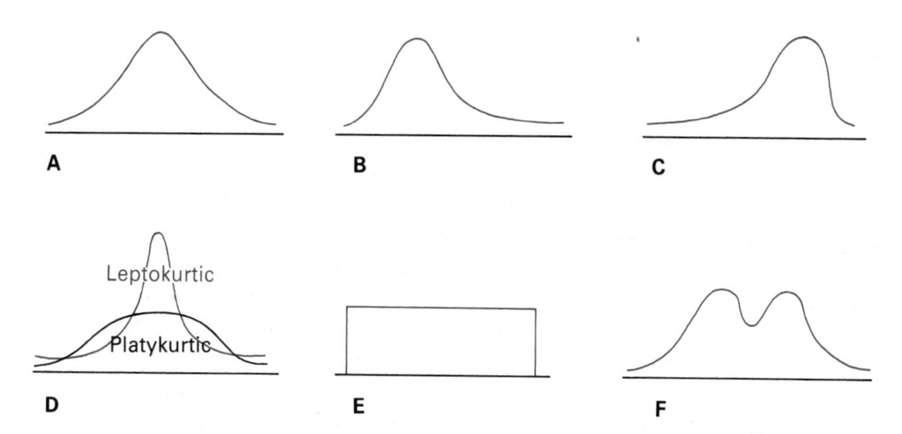

FIGURE 4.12
Various distribution shapes. **A,** Normal; **B,** positively skewed; **C,** negatively skewed; **D,** leptokurtic and platykurtic; **E,** rectangular; **F,** bimodal.

Skewed distributions represent another common family of shapes. A skewed distribution has scores that tend to pile up at one end of the distribution and taper off toward the other end. Figures 4.12B and C are examples of skewed distributions. If the tail, that is, the end that tapers off, tapers to the right or the high end of the scale, the distribution is **positively skewed.** Those skewed to the left or low end of the scale are **negatively skewed.**

In addition to skewness, another characteristic of a distribution is **kurtosis.** Kurtosis describes the peakedness of a distribution. The normal curve is the standard against which both degree of skewness and kurtosis are judged. A normal distribution has no skewness and is **mesokurtic.** If a distribution is more peaked (many score frequencies piling up in the center), it is said to be **leptokurtic.** If it tends to be flatter and spread out, it is said to be **platykurtic.** Figure 4.12D displays both.

Another quality of a distribution is its symmetry. A **symmetrical** distribution is one that would have identical halves if divided at its middle, P_{50}. The normal distribution, the rectangular distribution in Figure 4.12E, and the distribution in Figures 4.12D and F are all symmetrical. Skewed distributions are examples of nonsymmetrical distributions. The distribution in Figure 4.12F is called a bimodal distribution because of its two peaks. While the illustrated distribution is symmetrical, bimodal distributions generally are asymmetrical.

Looking at the frequency polygon in Figure 4.10, how would you characterize its shape?[10] Frequency distributions for small samples often will not conform to a common shape, and even large samples may only approximate or tend toward a particular shape. The two curves in Figure 4.8 typify real data. Simple descriptors for these asymmetrical distributions are difficult. The best may be that both tend to be bimodal and somewhat negatively skewed, the school distribution more so than the home distribution.

4.9 SUMMARY

Frequency distributions and graphs are methods for organizing and presenting data in forms that aid description and interpretation. With qualitative data, the resulting frequency distribution typically indicates the frequency of occurrence in specified categories as a count and as a proportion of the total. Bar graphs help graphically represent qualitative data. Such graphs are particularly useful when illustrating information for two or more groups. When group sizes differ, bar graphs report relative frequencies as percentages or proportions to allow more accurate comparison.

With quantitative variables, frequency distributions include intervals, interval limits, midpoints, frequency, relative frequency, cumulative frequency, and percentages. An interval represents a specified score range. In frequency

---

[10]Positively skewed.

distributions, each interval spans an identical range of scores, intervals do not overlap, and they span the entire observed range of scores. The exact limits of an interval are defined to reflect the continuity of quantitative data points. The midpoint of an interval is useful for identifying the most typical score. Histograms and frequency polygons help graphically represent data from a frequency distribution. Frequency polygons can be used when comparing distributions for two or more groups. In these cases, relative frequency percentages or proportions can adjust for differences in group sizes.

Once data from a quantitative scale have been assembled into a frequency distribution, we can calculate various percentile points and ranks. Percentiles, deciles, and quartiles define score points at or below which a designated percentage of the observations occur. A percentile rank specifies the percentage of cases a given score equals or exceeds. These score transformations are common for score reporting with norm-referenced tests. The ogive is a graphic representation of the cumulative percentage distribution. It is useful for estimating percentile points and ranks.

With graphically displayed quantitative data, it is common and useful to describe the shape of the distribution of raw scores. Distribution shapes may be characterized by degree of symmetry, peakedness or flatness, and modality. Asymmetric distributions are termed skewed. Positively skewed distributions are those in which scores appear to pile up on the left side of a frequency polygon; in negatively skewed distributions the mass of scores appears toward the right side of the distribution. Kurtosis refers to the width and height of a curve. When scores are located in one area of the distribution, the frequency polygon will appear tall or peaked in this range. In this situation the distribution is termed leptokurtic. When scores appear to spread out across the observed score range, the distribution is termed platykurtic. A normal distribution is symmetric (without evidence of skewness) and mesokurtic (neither particularly peaked nor flat). Finally, when data are assembled into a frequency distribution or are graphically displayed, different groupings for the same set of data are possible.

PROBLEMS

Problem Set A

1. The following are raw scores obtained by a group of 45 college students on a statistics examination:

38	37	33	24	34	37	25	23	40
36	31	32	39	30	35	34	36	37
33	35	27	27	28	31	32	33	35
35	28	36	38	37	30	35	30	34
22	34	35	33	38	36	29	35	32

a. Prepare a frequency distribution and a cumulative frequency distribution for their data. Use a class interval width of 2 and begin with the interval 21–22.
b. As part of this frequency distribution, report the exact limits and the midpoints for each interval.

2. Using the frequency distribution from problem 1, construct a histogram and a frequency polygon.

3. For each of the following intervals, identify the midpoints and the exact limits.

a. 85–87	b. 2.5–2.9	c. 19,000–24,000	d. 6–11
82–84	2.0–2.4	13,000–18,000	0–5

e. 131.50–131.74
131.25–131.49

4. Two researchers at different sites gathered data on job occupation ten years following high school graduation. The data coding scheme used by both investigations was as follows: 1 = professional, 2 = managerial, 3 = white collar, 4 = blue collar, 5 = farm work, 6 = other. The two studies produced the following data:

Site I				Site II							
5	3	3	2	6	4	2	5	6	3	1	5
4	2	4	3	3	5	4	1	2	2	5	5
4	2	3	4	5	5	1	2	6	2	2	1
2	4	4	2	2	6	1	4	1	1	1	2
1	4	5	3	5	6	5	5	5	2	2	6
1	2	1	2	1	1	3	2	5	3	6	3

Prepare a frequency distribution for the data from each site. Include the frequencies and the relative percentage frequencies. Construct a bar graph suitable for comparing occupations at the sites. What conclusions can you draw from inspecting these data?

5. An investigation was conducted to examine the effect of regular homework assignments versus no homework on final examination test scores. Table 4.9 reports the examination scores for students in each group. Prepare a frequency polygon for the group that did not have homework assigned. How would you describe the distribution shape of this group's examination scores?

6. Using the data from problem 5, convert the raw scores in each group to relative frequency percentages. Plot the values for each group on the same frequency polygon. How do the respective group polygons compare?

7. The frequency distribution in Table 4.10 was prepared from the English test scores obtained by eighth-grade students from one school district. Also given are the scores for a random sample of eighth graders from that region.

TABLE 4.9
Effect of homework on test scores.

Final Exam Score Interval	Homework Assigned (*f*)	No Homework Assigned Assigned (*f*)
90–99	22	8
80–89	36	18
70–79	31	35
60–69	11	38
50–59	5	32
40–49	2	15
30–39	1	9
20–29	0	3
10–19	0	2
0–9	1	2

TABLE 4.10
English test scores for one school district and a region.

Interval	District (*f*)	Region (*f*)
56–60	10	24
51–55	15	42
46–50	28	69
41–45	32	91
36–40	22	114
31–35	21	100
26–30	16	66
21–25	10	40
16–20	4	26
11–15	1	18
6–10	1	12
1–5	0	13

a. For the district, what is the score value at or below which 50% of the scores fall? For the region?
b. What is the percentile rank of a score of 23 in the district? In the region?
c. In the region, what is the score value associated with the first quartile, P_{25}? With the third quartile?
d. A student in the district received a raw score of 37. How does this student's performance compare with that of others in the district versus students in the region?

Problem Set B

I. Originally, a project of the Education Commission of the States called National Assessment of Educational Progress (NAEP) collected specific achievement data from large samples of individuals across the United States. The intent of NAEP was to systematically survey the knowledge, skills, understanding, and attitudes of persons in four age categories across several different subject areas. The survey sampled 9-, 13-, and 17-year olds plus young adults. Assessment and reassessment in the various content areas has followed a four-year cycle. The basic mechanism predominantly used by NAEP in reporting results is the percentage frequency distribution. Figure 4.13 illustrates the kinds of statistical information reported in the *NAEP Newsletter.* Use the information in the NAEP examples to answer the following questions and perform the identified tasks.

1. In example A, what is an appropriate label for the variable under study? What characteristics make it a qualitative variable?
2. If there were 8000 respondents in the age 17 group for example A, what raw frequency or number indicated that poverty was a problem for big cities?
3. Using every third category, such as pollution, economic problems, intergroup relations, and so forth, construct a bar graph comparing responses of 17-year olds and adults. Briefly describe the comparative results.
4. Identify the salient points communicated by the information in example B.
5. Construct a percentage frequency table to communicate the information given in example D.
6. Examples C and E illustrate other formats for presenting information in qualitative frequency distributions. For each, identify the form of the raw data necessary to allow compilation of the statistics (percentage frequencies) presented.

II. A study illustrating the utility of frequency distributions, percentile points, and percentile ranks was done by Salkind (1978). The intent of the Salkind study was to standardize the error and latency scores from the *Matching Familiar Figures Test* (MFF) across ages 5 to 12. In the MFF test, the child is shown a simple picture (standard) and asked to indicate which of several other pictures (variants) exactly matches that standard. The latency to the first response and the number of errors made before choosing the correct variant are recorded for each of 12 items. Each individual receives an error score representing the total number of errors made across all items and a latency score that is the average time before first response across all items. The error and latency scores are then used to classify individuals as impulsive or reflective. An impulsive individual is one who is *above* the 50th percentile point on the error scale and *below* the 50th percentile point on the latency scale (a faster, but inaccurate person). A reflective individual is defined as one who is *below* the 50th percentile point on the error scale and *above* the 50th percentile point on the latency scale (a slower, but accurate person).

Salkind and other researchers observed that in many studies of impulsive/reflective children, investigators were using their own samples of children to calculate the 50th percentile points for errors and latencies. Because frequency distributions of error and latency scores differed from study to study, P_{50} was different resulting in different definitions of impulsivity and reflectivity with the same age group. To correct this apparent flaw in the impulsivity/reflectivity research, Salkind acquired MFF data on a minimum number of children at each of the ages 5–12. Assuming these samples were representative of children at the specific age levels, he formed frequency distributions of both error and latency scores by age level. He then calculated the 50th percentile point for every distribution and proposed that these age level values for error and

Percentages of Respondents Listing Problems of Big Cities in the United States

Problem Category	*Age 13*	*Age 17*	*Adult*
Pollution	73%	69%	51%
Overpopulation	42	48	43
Crime	25	41	38
Economic problems	19	18	37
Drugs	18	18	16
Poverty	15	25	20
Intergroup relations	11	18	14
Transportation	8	9	17
Education	8	4	12
Social disorders	7	4	5
Health	4	4	4
Morals	4	5	2
Government	1	5	4
Natural disasters	1	0	0
Other	3	5	7

Respondents in the 1971–72 NAEP social studies assessment were asked to list three problems facing big cities in the United States. The majority of respondents listed pollution at least once.

A

	Percent Giving Correct Answer			
	Age 9	Age 13	Age 17	Adult
Computation				
Add: 38 +19	79	94	97	97
Subtract: 36 −19	55	89	92	92
Multiply: 38 × 9	25	83	88	81
Divide: 5 ⟌125	15	89	93	93

B

An average serving of which of the following foods would provide the most protein for building and repairing body tissues?

Boiled potatoes
Green beans
Lean meat
Oatmeal
White bread
I don't know.

This science exercise was administered in 1969 and again in 1972 to 13-year-olds. The correct answer is "Lean meat." In 1969, 53% of the 13-year-olds answered correctly; in 1972, 49% answered correctly.

C

Below is a layout of an office and shop area:

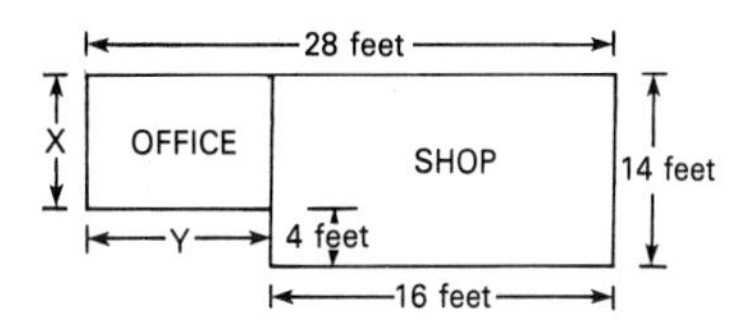

A. How many feet long is the office at Side Y?

ANSWER ______________

B. How many square feet of floor space are there in the office?

ANSWER ______________

This career and occupational development exercise was administered in 1974 to ages 13, 17 and adults. The correct answer for Part A is 12 feet; 58% at age 13, 84% at age 17 and 77% of the adults answered it correctly. The correct answer for Part B is 120 square feet; only 27% at age 13, 47% at age 17 and 35% of the adults gave correct answers.

D

	Percent at Age 13 Giving Each Response	
	1969	1972
Which of the following should you do when a person faints?		
Tightly bandage him.	0	0
Lay him down and keep him warm.*	32	34
Hold him up and apply hot packs.	1	1
Hold him up and apply cold packs.	12	14
Lay him down and apply cold packs.	48	45
I don't know.	7	6

**Correct answer.*

E

FIGURE 4.13
Examples of statistical information reported in the *NAEP Newsletter.*
Source: From *NAEP Newsletter*—A, 7(6), 1974; B and D, 9(3), 1976; C and E, 9(4), 1976. Reprinted by permission of the National Assessment of Educational Progress.

latency distributions be used as the standard values for error and latency cutoff scores to define impulsive and reflective behavior. Using a representative, constant set of scores for all studies on a specific age group would at least ensure constancy of definition within age groups for classifying children into impulsive and reflective categories.

Error and latency frequency distributions for 200 7-year olds and 300 10-year olds have been simulated in Table 4.11. Use these distributions when responding to the following problem exercises.

1. What are the interval widths for the two distributions?
2. The *score* range for error scores was from 0 to 34 and from 3 to 48 seconds for latency scores. What other interval widths would produce an acceptable number of score intervals for each distribution?
3. Construct two separate histograms of the error and latency distributions for the 7-year olds. Describe the shapes of these distributions.
4. Construct comparative frequency polygons for the 7- versus 10-year olds using separate graphs for error and latency distributions. Write a brief comparative description of the data as presented by the graphs.
5. Compute P_{50} for the 7- and 10-year old distributions for both errors and latencies. What descriptive statement(s) would be warranted when comparing these scale scores across the two groups?
6. Calculate the percentile rank for an error score of 16 in the 7-year old distribution and in the 10-year old distribution.

TABLE 4.11
Frequency distributions simulated for A, error scores; B, latency scores.

A			B		
Score Limits	Age = 7 *f*	Age = 10 *f*	Score Limits	Age = 7 *f*	Age = 10 *f*
33–35	2	0	47–50	0	3
30–32	4	0	43–46	0	4
27–29	6	0	39–42	0	8
24–26	6	2	35–38	2	8
21–23	18	7	31–34	2	15
18–20	26	15	27–30	6	12
15–17	40	18	23–26	14	21
12–14	38	45	19–22	18	45
9–11	30	60	15–18	20	48
6–8	14	75	11–14	46	60
3–5	10	42	7–10	66	58
0–2	6	36	3–6	26	18
	n 200	300		*n* 200	300

TABLE 4.12
Error and latency score percentile equivalents.

Age Group	Error Percentiles P_{25}	Error Percentiles P_{50}	Error Percentiles P_{75}	Latency Percentiles P_{25}	Latency Percentiles P_{50}	Latency Percentiles P_{75}
5	16.7	22.0	26.3	4.7	6.4	8.8
6	12.1	17.5	22.0	6.0	8.2	11.4
7	9.8	14.0	18.1	6.9	10.2	14.5
8	7.1	12.3	15.5	8.3	11.2	16.1
9	4.6	8.1	11.4	9.6	13.7	21.7
10	3.2	6.7	9.5	9.9	13.7	21.2
11	4.1	7.3	10.7	8.2	11.8	15.8
12	4.1	7.6	10.1	7.7	10.7	14.9

7. Calculate the percentile rank for a latency score of 20 in both distributions.

8. Table 4.12 gives the error and latency scale scores equivalent to P_{25}, P_{50}, and P_{75} for each of the age groups. What general descriptive trends are evident from the information provided relative to distribution shape, location on the scale, and range of the middle 50% of the scores?

Problem Set C

Use the "Sesame Street" data in Appendix A to answer the following questions.

1. The frequency distribution of pretest scores on the Letters test is given in Table 4.5. Using the same interval score limits, construct a frequency distribution of posttest scores on the Letters test for the first 100 individuals in Appendix A (IDs 1–100). Use the pretest distribution in Table 4.5 and the posttest distribution just constructed to respond to Exercises 2–5.

2. Construct a histogram of the raw frequencies for the posttest distribution.

3. Construct appropriate frequency polygons for visual comparisons of the pretest and posttest distributions of Letters test scores. How would you describe the shape of each distribution?

4. Calculate P_{50} and P_{75} for each distribution. Which distribution results in the largest scale interval for the third quartile range?

5. Calculate the percentile rank for a scale value of 20.0 in each distribution. Do these values correspond with differences you would expect in pre- and posttest distributions?

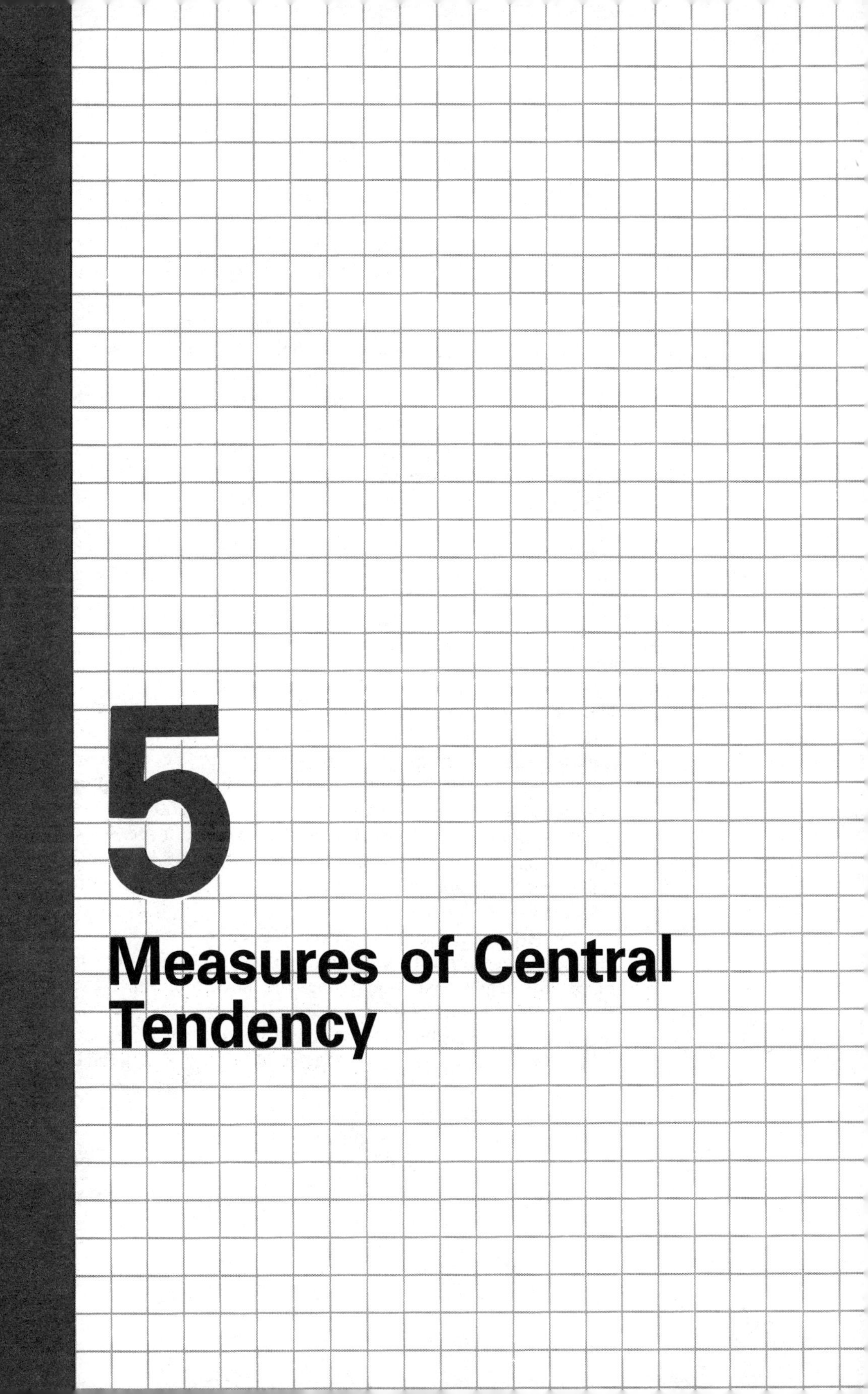

5

Measures of Central Tendency

Oh, I guess she's about *average* height." "Larry does less than *average* work." "Sally makes an *average* wage for people in her profession." "He has *average* ability." *Average, average, average!* What is it? What do people mean when they use it? Are you above or below *average?* The *average* person uses the term to mean "typical." If someone is average, some persons are below and some are above this individual on the scale of whatever is being measured. The concept of "typical" is important in describing a set of scores.

The term average itself can be confusing because three distinct statistical measures can define the average or typical score. The intent of each measure of average is to identify a score point that might represent the typical score of a data set. In general, these measures identify a point in the central part of a distribution and are therefore called measures of central tendency. The three measures of central tendency are the **mode, median,** and **mean.**

5.1 MODE

The **mode** is a simple, easily calculated index of central tendency. The mode is defined as the score in the data set obtained by the largest number of persons. It is the most frequently observed score in a set of raw score data. Identifying the mode merely involves counting. The ungrouped frequency distribution of Relational Terms pretest scores for the 240 students in Appendix A provides an illustration (see Table 5.1). As you can see, 11 was the most frequently obtained score. Thus, the mode for the Relational Terms pretest achievement variable is 11. Note that the mode is the score value, not the frequency.

When individual raw scores are either unavailable or unwieldy, a grouped frequency distribution with interval widths greater than 1 helps summarize the data. Convention defines the mode in grouped frequency distributions as the midpoint of the interval with the greatest frequency count. Look at the frequency distribution of Peabody mental age scores in Table 4.6. The mode is a mental age score of 37. Why? The interval with the largest frequency is 35–39 ($f = 49$). The midpoint of the 35–39 interval is 37. We do not know how many persons had scores of 35, 36, 37, 38, or 39, so we use the midpoint as the score value to represent the individual scores in the interval. Examine Table 4.5. What are the modes for the three distributions presented?[1]

---

[1]Age = 50 months
Letters Pretest = 14.5
Classification Posttest = 16.5.

TABLE 5.1
Frequency distribution of Relational Terms pretest scores.

Score Value	Frequency (f)
17	1
16	5
15	10
14	16
13	19
12	21
11	36
10	32
9	25
8	25
7	15
6	10
5	14
4	7
3	3
2	1
TOTAL	240

Characteristics of the Mode

The mode, as a statistical index of central tendency or average, is the point along the scale continuum obtained by the greatest number of persons. It is the scale value at which the frequency distribution of scores peaks in a histogram or frequency polygon. In many distributions, the frequency of scores to the left and right of the mode taper off, so it does represent a central value. The mode is not a very stable index of central tendency however. Different groupings of the same data can drastically change the mode. Table 5.2 illustrates this point using the classification posttest scores from the "Sesame Street" data base. Four different class interval groupings are used. The true mode (most frequently obtained raw score) is 19. More children received a raw score of 19 than any other score. When the data are grouped into score intervals with interval widths greater than 1, the exact frequency of individual scores is lost. Find the mode for distributions B, C, and D in Table 5.2. For distribution grouping B the mode is 16.5. Distribution grouping C has two modes! In distribution C, a frequency of 34 was obtained for score intervals 15–16 and 19–20. The two modes, by definition, are 15.5 and 19.5. For distribution D, the scores were grouped into three-unit intervals. Two intervals, 16–18 and 19–21, have the same frequency, 48. Because these two intervals are numerically consecutive, the mode for distribution D is midway between the midpoints of the two intervals. For two consecutive intervals, this point will always be the exact

TABLE 5.2
Frequency distributions for alternative groupings of the "Sesame Street" Classification posttest scores (n = 239).

A Score Interval (width = 1)	f	B Score Interval (width = 2)	f	C Score Interval (width = 2)	f	D Score Interval (width = 3)	f
24	8	24–25	8	23–24	24	22–24	37
23	16	22–23	29	21–22	27	19–21	48
22	13	20–21	28	19–20	34	16–18	48
21	14	18–19	33	17–18	30	13–15	38
20	14	16–17	35	15–16	34	10–12	36
19	20	14–15	25	13–14	22	7–9	25
18	13	12–13	21	11–12	21	4–6	7
17	17	10–11	28	9–10	28		
16	18	8–9	18	7–8	12		
15	16	6–7	9	5–6	6		
14	9	4–5	5	3–4	1		
13	13						
12	8						
11	13						
10	15						
9	13						
8	5						
7	7						
6	2						
5	4						
4	1						

limit separating them. For the two intervals with the greatest frequency in distribution D, 18.5 is the upper exact limit to the interval 16–18 and the lower exact limit to the interval 19–21. Thus, 18.5 is the mode for distribution D.

The findings from the three examples presented in Table 5.2 emphasize the major disadvantages of using the mode as the only measure of central tendency. First, it is unstable with grouped frequency distributions. Second, the mode may not be a distinct value. A distribution can have more than one mode. A distribution of scores is often characterized by its number of modes or peaks. A **unimodal** distribution refers to a score distribution with a single mode or peak. The distributions in Figures 5.1A, B, and C are all unimodal. A distribution with two peaks is **bimodal.** Distributions with three or more peaks are simply referred to as **multimodal.** Peaks in a distribution need not represent equal frequencies to be classified as bimodal or multimodal. The data points must merely tend to "bunch up" around two or more values. Fig-

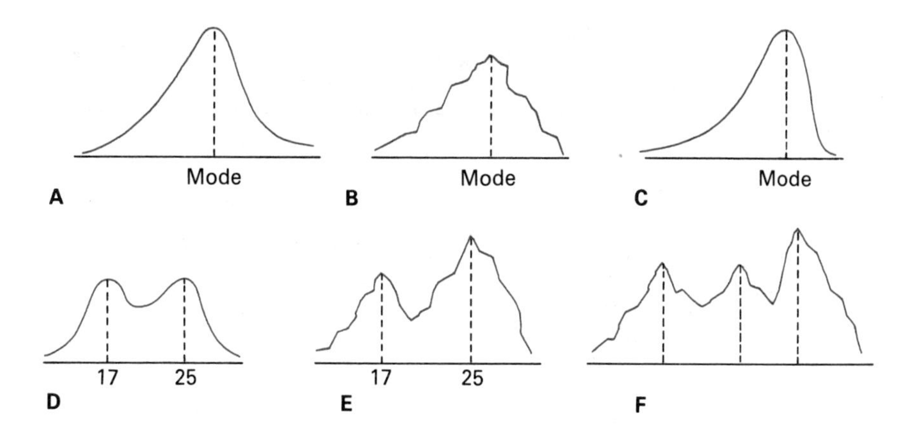

FIGURE 5.1
The mode and distribution shapes. A, unimodal; B, unimodal; C, unimodal; D, bimodal; E, bimodal; F, multimodal.

ures 5.1D, E, and F illustrate bimodal and multimodal distributions. D has two modes, 17 and 25. Each score has the same frequency. In E, however, 25 is the only mode, but because the distribution peaks in two places, we would classify the distribution as bimodal.

5.2 MEDIAN

The **median** (*Mdn*) is the exact middle point in a frequency distribution. It is defined as the scale point that divides the distribution of scores exactly in half. The median is equal to the 50th percentile point, P_{50}. The simplest approach to finding the median is to cast the scores into a frequency table, then compute P_{50} using the following formula:

$$Mdn = P_{50} = L + \frac{(.50)(n) - cf}{f}(i) \qquad (5.1)$$

where

L = lower exact limit of the interval containing the median or 50th percentile point

n = group size

cf = cumulative frequency up to the lower exact limit of the interval containing the median

f = frequency of scores for the interval containing the median

i = interval width

TABLE 5.3
Frequency distributions of Classification posttest scores for children viewing "Sesame Street" in the home versus at school.

Score Interval	Viewing in Home (n = 143)		Viewing in School (n = 96)	
	f	cf	f	cf
24–25	3	143	5	96
22–23	15	140	14	91
20–21	19	125	9	77
18–19	21	106	12	68
16–17	19	85	16	56
14–15	12	66	13	40
12–13	13	54	8	27
10–11	19	41	9	19
8–9	11	22	7	10
6–7	7	11	2	3
4–5	4	4	1	1

The preceding formula is a special case of formula 4.1 (in Chapter 4) used for calculating any percentile point value.

The Classification posttest scores for children viewing "Sesame Street" in the home versus at a school are given in the frequency distributions of Table 5.3. The median score for the distribution of children who viewed in the home may be computed as follows:

$$\begin{aligned} Mdn &= 15.5 + \frac{71.5 - 66}{19}(2) \\ &= 15.5 + .58 \\ &= 16.08 \end{aligned}$$

Using the data in Table 5.3, calculate the median classification posttest score for children viewing at school.[2] As measures of central tendency, do the modes for the two distributions in Table 5.3 differ from the medians?[3] The difference

---

[2] $Mdn = 15.5 + \frac{48.0 - 40}{16}(2) = 16.50$

[3]

Viewing at Home		Viewing at School
18.50	Mode	16.50
16.08	Median	16.50

between the median and mode is greater for the score distribution of children viewing at home. Looking at the frequencies, the latter distribution tends to be bimodal. The majority of scores "pile up" around the mode in the intervals 16–17, 18–19, 20–21. Fewer scores do "bunch up" around the score interval 10–11, however. (See Figure 4.8).

For small data sets, other conventional procedures can help us find the middle score (median). These procedures require a rank ordering of the scores and often ignore the continuity of the underlying measurement scale. For these reasons, it is advisable to cast data into a frequency distribution and calculate the median using formula 5.1—even if the class intervals are one raw score unit. For example, consider the data set consisting of 5, 6, 7, 8, and 9. The median is 7, the middle score. However, even for data such as these in which a median is apparent, the median conforms to its definition only if a continuous scale is assumed. Since the median was defined as the scale point to which half the scores are either equal or below in value, can we say half the scores in the set 5, 6, 7, 8, 9 are equal to or less than 7? Only if we assume that the scale value of 7 is not discrete, but rather represents the continuous interval 6.5–7.5. In that case, 2.5 scale units are equal to or below the scale point of 7 and 2.5 scale units are above it. What is the median for the set of scores if we add another 6 and 7 to the set? The only exact way to find the median for these scores is to treat them as continuous scale points and use a cumulative frequency distribution (see Table 5.4). Calculate the median for these seven scores.[4]

Characteristics of the Median

The median is the true middle score in the distribution based on frequencies below and above it. Because calculations of the median do not use the score values themselves, but rather the frequencies of scores, extreme scores do not affect the median. A few extreme scores in a data set will have a negligible effect on the value of the median. Consider the small set of data from earlier in the section: 5, 6, 7, 8, and 9. If the raw score of 5 had been a 0, the median would still be 7. Similarly, if the raw score of 9 had been a score of 25, the median value would not change. Calculation of the median is not affected by extreme scores because the magnitudes of raw scores are not figured into this calculation.

Unlike the mode, the median is relatively unaffected by different groupings of the data. It is a very stable index. Calculate the medians for the four

---

[4]
$$\begin{aligned} Mdn &= 6.5 + \frac{3.5 - 3}{2}(1) \\ &= 6.5 + .25 \\ &= 6.75 \end{aligned}$$

TABLE 5.4
Cumulative frequencies for a small data set.

Score	f	cf
9	1	7
8	1	6
7	2	5
6	2	3
5	1	1

different groupings of the data in Table 5.2. Note that test scores are available for only 239 students; one student's score for this variable was missing.[5] From these calculations, the medians differ little for different groupings of the data.

5.3 ARITHMETIC MEAN

The third measure of central tendency is the arithmetic mean, usually simply called the **mean.** The mean is the arithmetic average of the raw scores and may be computed for any continuous variable. When calculated for a *sample,* the mean is symbolized by $\overline{X}$ (read "X bar"). A *population* mean is symbolized by the Greek letter mu (μ). Both the sample and population means equal the *sum* of scores in the data set divided by the *number* of scores.

While the mean is a common statistical index, it is important to recognize the impact each score has on it. The mean treats each score in the distribution as an effective weight positioned along the scale continuum. The effective weight of a score is its distance in scale units from the mean; thus, the mean is the balance point for the effective weights of a given set of scores. The effective weights for the set of scores in a distribution below the mean and those above the mean balance each other out around the mean. Scores below the mean will have negative effective weights and summed will equal the sum of the positive weights above the mean.

---

[5]The cumulative frequencies for each distribution are needed before the medians can be calculated. *Distribution*

$$\text{A: } Mdn = 15.5 + \frac{(.50)(239) - 106}{18}(1) = 16.25$$

$$\text{B: } Mdn = 15.5 + \frac{(.50)(239) - 106}{35}(2) = 16.27$$

$$\text{C: } Mdn = 14.5 + \frac{(.50)(239) - 90}{34}(2) = 16.24$$

$$\text{D: } Mdn = 15.5 + \frac{(.50)(239) - 106}{48}(3) = 16.34$$

Consider the following scores: 9, 4, 8, 10, 4, 8, 3, and 10. The mean of this sample is

$$\bar{X} = \frac{9 + 4 + 8 + 10 + 4 + 8 + 3 + 10}{8}$$

$$= \frac{56}{8} = 7$$

Figure 5.2 demonstrates the concept of effective weights. Think of each score as having an absolute weight equal to each other score. A score of 4 represents a scale interval, 3.5–4.5, one unit wide. Its absolute weight can be thought of as a square block one unit per side. The scale score 9 also represents one unit of the scale, 8.5–9.5. Its absolute weight is the same as that for a score of 4, a one-unit square block. In this data set three scores (3, 4, 4) are below the mean of 7 and five scores (8, 8, 9, 10, 10) are above the mean of 7. The number of scores (blocks) above and below the mean do not balance the scale about the mean, however; rather, the effective weights or positions of the blocks relative to the mean (distance from the mean) balance the scale. Figure 5.2 illustrates the weighted balance. To rephrase the definition, the mean represents a point from which the score distances (effective weights) as a composite are equal on both sides. In that respect the mean is the balance point.

The concept of effective weights differentiates the mean from the median. An equal number of absolute score values fall above or below a median, regardless of their distance from it. In contrast, the mean is a result of the effects of these score distances. This difference is evident when calculating the median for the set of eight scores above. The mean is 7, but the median is 8. In the continuous distribution, four scores are at or below a score value of 8 and four scores are above the median value of 8.

Calculating the Mean and Summation Notation

While the formula for the mean can be simply written out or stated verbally, it is inefficient to do so for more complex statistical formulas. A mathematical

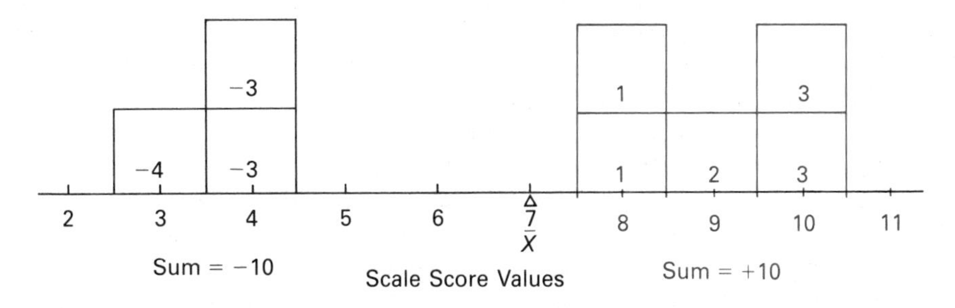

FIGURE 5.2
Effective score weights.

TABLE 5.5
The representation of raw scores in the context of summation notation.

Raw Scores Represented by Summation Notation		Actual Raw Scores	
X	Y	Body Parts Pretest Scores	Numbers Posttest Scores
X_1	Y_1	16	44
X_2	Y_2	30	39
X_3	Y_3	22	40
.	.	23	19
.	.	.	.
.	.	.	.
X_i	Y_i	X_i	Y_i
.	.	.	.
.	.	.	.
.	.	16	10
X_{239}	Y_{239}	23	50
X_n	Y_n	21	47

notation system called summation notation helps efficiently represent the computational procedures. Summation notation identifies a variable by a capital letter, usually X or Y. A subscript represents a specific raw score. For example, X_5 denotes the fifth number in the data set for variable X. Y_{17} is the 17th raw score in the data set for variable Y. The subscript i refers to any raw score (i.e., X_i represents the ith value in the data set). The number of scores in a data set for a single variable is represented by n, as mentioned in Chapter 4, so X_n refers to the last raw score in the data set for variable X. Similarly, X_{n-1} identifies the next to the last raw score in the data array on the variable X.

Table 5.5 illustrates actual raw scores and their symbolic match to variables using summation notation. The left side of the table lists the symbolic scores for variables X and Y. To the right the "Sesame Street" evaluation Body Parts pretest scores and Numbers posttest scores are listed. The first four and last three children's scores from the data base in Appendix A have been reproduced. In this format, X refers to Body Parts pretest scores and Y, Numbers posttest scores. The Numbers posttest scores of 50 and 47, for example, are designated Y_{239} and Y_{240}, respectively. As stated previously, X_i refers to any raw scores in the data set for the variable X (Body Parts pretest). Similarly, Y_i represents any raw score on the Numbers posttest. Using Table 5.5, what variable does the raw score of 30 on the Body Parts pretest represent?[6] How are the

[6]The raw score 30 is represented by X_2.

scores of 19 and 10 on the Numbers posttest identified?[7] Which scores on the Body Parts pretest are associated with the labels X_1, X_4, and X_{238}?[8]

An important summation symbol is the capital Greek letter sigma (Σ). In summation notation, Σ is read as "the sum of." ΣX_i is a common symbol in statistics indicating that "the sum of all" raw scores for variable X is to be calculated. For most data sets we will want to sum all scores for a particular variable. Therefore, we will omit the subscript i. By convention, ΣX will still mean to sum all X_i scores in a data set.

This brief introduction to some rules of summation notation will suffice for formulas to be introduced in this and the next chapter. Summation notation is useful for presenting and understanding derivations of statistical formulas. Derivations, however, are not themselves crucial to understanding the concepts of statistical analysis and reasoning. It is important to recognize and be able to perform the operations defined by the summation notation symbols within the statistical formulas. For those who might be interested in following future derivations of statistical formulas using summation notation, Appendix C provides the rules governing algebraic manipulations within the summation notation system.

Using summation notation, we can calculate the mean:

$$\bar{X} = \frac{\Sigma X}{n} \tag{5.2}$$

As an example, consider the data set from Figure 5.2 (3, 4, 4, 8, 8, 9, 10, 10). For this sample of scores, the mean would be calculated as follows:

$$\bar{X} = \frac{\Sigma X}{n} = \frac{3 + 4 + 4 + 8 + 8 + 9 + 10 + 10}{8} = \frac{56}{8} = 7$$

As a descriptive statistic, the mean is used to represent the distribution center for a single group of scores. Means can be used to compare the location of several groups' scores along a continuum. For example, we might want to compare children's Letters pretest scores grouped on the basis of the frequency with which the mother reads to them. Table 5.6 shows the sums (ΣX), sample sizes (n), and means ($\bar{X}$) for the Letters pretest scores grouped by frequency of reading scores.

The coded values under the frequency of reading variable were used to identify students in each group. For example, to calculate the mean of the Letters pretest scores for students from Group 1 (mother reports never reading to child), the raw scores for the 12 children in this group were summed.

Data for the frequency of reading variable are missing for participants 001, 122, 164, 199, 205, 206, and 222. Calculate the mean Letters pretest score for this group (Group 0). Descriptively, how do the means compare

[7]The scores 19 and 10 on the Numbers posttest are observations Y_4 and Y_{238}, respectively.
[8]$X_1 = 16$; $X_4 = 23$; $X_{238} = 16$.

TABLE 5.6
Letters pretest scores by frequency of home reading groups.

	Frequency of Reading				
	Never (1)	Less Than Once a Week (2)	Once a Week (3)	Several Times a Week (4)	Once a Day (5)
Sum of Letters pretest scores	130	229	619	1412	1114
Sample size	12	12	39	91	63
Mean	10.83	19.08	15.87	15.52	17.68

among the six groups?[9] Note that only one mean is a whole number. The others are computed to the nearest hundredth, but are typical of values found with real data.

Characteristics of the Mean

As discussed, the effective weights for scores below and above the mean are equally balanced. A score's effective weight is the distance it deviates from the

[9]
$$\overline{X}_{\text{missing}} = \frac{X_{001} + X_{122} + X_{164} + X_{199} + X_{205} + X_{206} + X_{222}}{7}$$
$$= \frac{23 + 13 + 10 + 13 + 14 + 15 + 17}{7}$$
$$= \frac{105}{7}$$
$$\overline{X} = 15.00$$

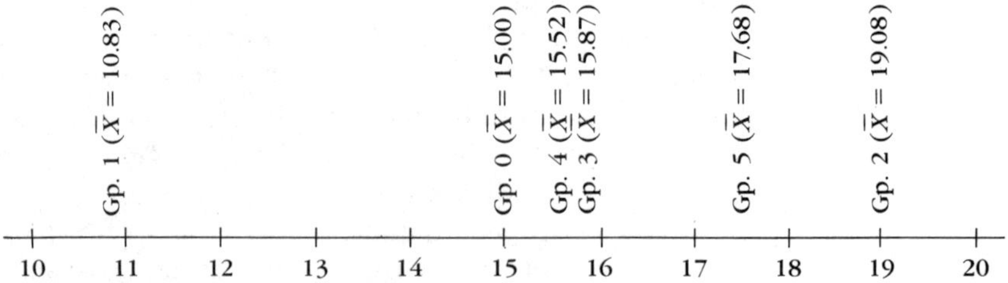

Gp. 1 ($\overline{X}$ = 10.83)
Gp. 0 ($\overline{X}$ = 15.00)
Gp. 4 ($\overline{X}$ = 15.52)
Gp. 3 ($\overline{X}$ = 15.87)
Gp. 5 ($\overline{X}$ = 17.68)
Gp. 2 ($\overline{X}$ = 19.08)

10 11 12 13 14 15 16 17 18 19 20

Group Mean Locations on Letters Pretest Scale

Knowledge of letters demonstrated by the mean performance on the Letters pretest for the groups differs considerably. Group 1 (mother reported never reading to child) had the lowest mean performance. The group with missing codes and Groups 3 and 4 scored higher than Group 1 and at approximately the same level on the scale. Group 5 had the next highest mean number correct, getting approximately two more items on the test correct *on the average* than Groups 0, 3, and 4. Surprisingly, Group 2 had the highest mean performance level. If the frequency with which the mother reported reading to the child was highly predictive of group knowledge level prior to viewing "Sesame Street," the Group 2 mean should have been located at a lower point between Groups 1 and Groups 3 and 4 on the scale continuum.

mean. A score's deviation from the mean is symbolized by $X_i - \overline{X}$. The deviation has a negative value if the raw score is below the mean and a positive value if the raw score is above the mean. Because of equally balanced composite effective weights above and below the mean, the sum of the deviations of raw scores from the mean equals zero, or $\Sigma(X_i - \overline{X}) = 0$. Using the small data set from the beginning of this section as an illustration,

$$\begin{aligned}\Sigma(X_i - \overline{X}) &= (3 - 7) + (4 - 7) + (4 - 7) + (8 - 7) + (8 - 7) \\ &\quad + (9 - 7) + (10 - 7) + (10 - 7) \\ &= (-4) + (-3) + (-3) + 1 + 1 + 2 + 3 + 3 \\ &= -10 + 10 \\ &= 0\end{aligned}$$

Another characteristic of the mean is that each score in the data set directly influences its magnitude. If there are extreme numbers in a data set, this property can be a disadvantage. Extreme scores will disproportionately influence the mean. Suppose one of the scores of 8 was instead 40 (i.e., 3, 4, 4, 8, 40, 9, 10, 10). This set of scores would not be unrealistic if one person already had received instruction and knew the information (score of 40). For this set of scores, then,

$$\overline{X} = \frac{88}{8} = 11$$

Replacing one score of 8 by a single score raises the mean four units! Seven of the eight scores are now below the mean. Most data sets, however, do not contain extreme numbers and using every number usually is considered an asset rather than a liability. Using all values in calculating the mean increases the mean's stability if it is calculated for different representative samples. In the vast majority of research situations, the mean is the most stable index of central tendency. This stability is important for consistent description of a variable across studies, but its stability is a particularly attractive attribute in inferential statistics in which the sample mean, $\overline{X}$, is used to estimate the mean of the population that the sample represents.

The mean has additional properties important to both descriptive and inferential statistics. Two measures of variability to be discussed in Chapter 6—variance and standard deviation—depend on a characteristic of the mean as an anchor point in the distribution. The mean also has additional properties besides its stability that make it an excellent estimator of the population mean in inferential statistics. These properties are covered in Chapter 11.

Calculating the Mean for Combined Groups

Often the means for separate groups are known and we want to know the mean for all the combined data. If the group sample sizes are different, we would be incorrect in computing the arithmetic mean of the individual group means to represent the mean for all groups combined. For example, we may know the *mean* for males and the mean for females on a given variable, but the *number* of males and females differ. We want to find the mean for males and females combined into a single group. We could compute this value by summing the scores across both groups, then dividing by the total number of males and females together. Symbolically, the sum of all male and female scores is

$$(\Sigma X)_{\text{total}} = (\Sigma X)_{\text{males}} + (\Sigma X)_{\text{females}}$$

The combined sample size is

$$n_{\text{males}} + n_{\text{females}} = n_{\text{total}}$$

The mean for the two combined groups can then be expressed as

$$\overline{X}_{\text{total}} = \frac{(\Sigma X)_{\text{total}}}{n_{\text{total}}} = \frac{(\Sigma X)_{\text{males}} + (\Sigma X)_{\text{females}}}{n_{\text{males}} + n_{\text{females}}} \tag{5.3}$$

If the group means and group sample sizes are the only values given, the sum of the scores may be found by algebraically rearranging the formula for the mean.

$$\overline{X} = \frac{\Sigma X}{n}, \quad \text{so} \quad n\overline{X} = \Sigma X$$

Multiplying the group sample size times the group mean ($\overline{X}$) will produce the sum of scores for that specific group. Substituting in formula 5.3,

$$\overline{X}_{\text{total}} = \frac{n_{\text{males}}\overline{X}_{\text{males}} + n_{\text{females}}\overline{X}_{\text{females}}}{n_{\text{males}} + n_{\text{females}}}$$

The formula may be extended easily for more than two groups. Let the letters A, B, C, D, and so on represent different groups. Using these group labels as subscripts, the mean for group C would be $\overline{X}_{\text{C}}$, the sample size, n_{C} and the sum of the scores, $(\Sigma X)_{\text{C}}$. The equivalent generalized formulas to calculate the mean for combined groups ($\overline{X}_{\text{total}}$) are

$$\overline{X}_{\text{total}} = \frac{(\Sigma X)_{\text{A}} + (\Sigma X)_{\text{B}} + (\Sigma X)_{\text{C}} + \cdots}{n_{\text{A}} + n_{\text{B}} + n_{\text{C}} + \cdots} \tag{5.4}$$

$$\overline{X}_{\text{total}} = \frac{n_{\text{A}}\overline{X}_{\text{A}} + n_{\text{B}}\overline{X}_{\text{B}} + n_{\text{C}}\overline{X}_{\text{C}} + \cdots}{n_{\text{A}} + n_{\text{B}} + n_{\text{C}} + \cdots} \tag{5.5}$$

Suppose the following means and sample sizes are available for four groups. We want to find the mean value for all individuals.

	Group A	Group B	Group C	Group D	Total
$\overline{X}$	30	25	32	30	?
n	10	20	15	16	?

Using formula 5.5,

$$\bar{X}_{total} = \frac{10(30) + 20(25) + 15(32) + 16(30)}{10 + 20 + 15 + 16}$$

$$= \frac{300 + 500 + 480 + 480}{61}$$

$$= \frac{1760}{61} = 28.85$$

In this example, $(\Sigma X)_{total} = 1760$ and $n_{total} = 61$. Use the mean Letters pretest scores for the six groups given in this section to calculate the mean score for the children in the "Sesame Street" data base.[10]

5.4 CHOOSING MEASURES OF CENTRAL TENDENCY

Each measure of central tendency results in a scale point that acts as a single, representative score for the group. Which measure *best* represents the typical score for a given set of data? If the data distribution is *unimodal and symmetrical,* the answer is simple. The mode, median, and mean apply equally well. For nonsymmetrical distributions, however, the three measures will result in different values. In fact, we can predict the relationship between the mean, median, and mode for typical skewed distributions. If the distribution is negatively skewed the mean will be numerically less than the median, which will be numerically less than the mode. For positively skewed distributions, the mean will be numerically greater than the median, which will be numerically greater than the mode. Figure 5.3 illustrates these relationships.[11]

The mode is the only appropriate measure for qualitative data. Otherwise, the mode should only be used as the single index of central tendency if we are looking for the most probable score or score category. In all other situations, the mode provides useful information and may be reported, but only as a supplement to the median or mean.

---

[10] $\bar{X}_{Gp.\,0} = 15.00\ (n = 7)$; $\bar{X}_{Gp.\,1} = 10.83\ (n = 12)$; $\bar{X}_{Gp.\,2} = 19.08\ (n = 12)$

$\bar{X}_{Gp.\,3} = 15.87\ (n = 39)$; $\bar{X}_{Gp.\,4} = 15.52\ (n = 91)$; $\bar{X}_{Gp.\,5} = 17.68\ (n = 63)$

$$\bar{X}_{combined} = \frac{7(15.00) + 12(10.83) + 12(19.08) + 39(15.87) + 91(15.52) + 63(17.68)}{7 + 12 + 12 + 39 + 91 + 63}$$

$$= \frac{3609.01}{224} = 16.11$$

Note: Total $n = 224$. The mothers for 16 children in the data base responded in coded category 6 (don't know) to the question about reading frequency. The scores for these 16 children were not involved when calculating any of the preceding group means.

[11] These relationships hold for typical skewed distribution of the type illustrated in Figure 5.3. In some atypical skewed distributions the mode is between the mean and median, but the relationships identified between the mean and median are always consistent.

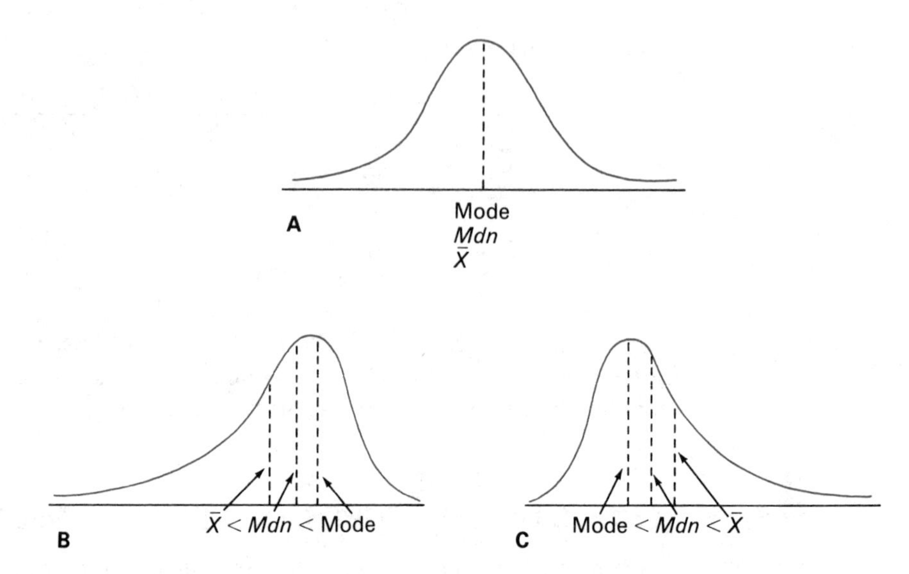

FIGURE 5.3
Relationships among the measures of central tendency for different distribution shapes. **A**, unimodal and symmetrical; **B**, negatively skewed; **C**, positively skewed.

The mean and median both assume the variable is measured on a continuous scale; both can be computed as long as the variable is quantifiable into equal units. Thus, the mean and median require interval or ratio levels of measurement. This assumption is often violated without undue effects when the scale appears to be interval, but the underlying distribution is ordinal. Indeed, the measurement devices primarily used in behavioral research—paper and pencil tests, rating scales, self-report inventories of interest, personality and attitudes, surveys, and questionnaires—frequently result in measurement at this quasi-interval level. The mean and median can adequately summarize and describe the quantified observations for these types of data.

The main characteristic that determines whether the mean or median is more appropriate to describe a data set is the absence or presence of extreme numbers. Because extreme scores influence the mean more than the median, the median may be more representative as an average value in these cases (e.g., in extremely skewed distributions).

When the distribution of observations is not extremely skewed, the preferred measure of central tendency is the mean. This preference is based on three interdependent considerations:

1. Both the median and mode values depend solely on the frequency of the observed scores. The mean depends on the magnitude of each score. The mean thus uses all the information available in the data.

2. The mean is the most stable measure of central tendency. Much research is carried out on samples. If we repeated our studies using a different sample drawn from the same population, the mean would vary less than the median or mode would from sample to sample.
3. The properties of the mean allow its use with many other statistics. (Later chapters will introduce these topics and show how the mean contributes to these other measures.)

When possible, all three measures of central tendency should be reported. The intent of descriptive statistics is to summarize a set of scores. Each index of central tendency provides unique information that facilitates a mental picture of the data. However, the mean is generally the most desirable measure for use in additional analysis. It is the most stable index and is most amenable to mathematical manipulations and applications.

5.5 SUMMARY

Measures of central tendency have two uses: (a) to describe the central location of a set of scores on a scale and (b) as sample statistics, to provide estimates of population parameters. Three measures, the mode, median and mean, were covered in this chapter. Each measure has individual properties, summarized here.

The mode

- is defined for a set of raw scores as the most frequently occurring value.
- is defined for a set of grouped values in a frequency distribution as the midpoint of the interval containing the highest frequency.
- may be affected drastically by different groupings of the same data.
- is the most unstable of the three measures for different samples from the same population of data.
- does not always result in a single point in the distribution (e.g., bimodal distributions).
- is of little use other than as a descriptive index.
- is the only measure of central tendency appropriate for data on qualitative variables.
- will have the largest value of the three indices in a typical negatively skewed distribution and the smallest value in a typical positively skewed distribution.

The median

- is defined as the point along the scale at or below which half the scores in the distribution fall. It is the 50th percentile point. The formula used in calculating the median is

$$Mdn = P_{50} = L + \frac{(.50)(n) - cf}{f}(i)$$

- is affected only by the number of data points and not by their actual magnitude.
- is less affected than the mean by extreme data points.
- is a better choice as a descriptive measure in distributions with extreme skew.
- is not as stable as the mean for different samples from the same population of data.
- is not as useful as the mean beyond the descriptive statistics level.
- can be compared with the mean to indicate skewness.
- should be used with open-ended grouped distributions in which a constant minimum or maximum score value is assigned all extreme values.

The mean

- is the focal scale point in a distribution around which the effective weights for the data balance perfectly. The formula for the mean is

$$\overline{X} = \frac{\Sigma X}{n}$$

- is the most stable measure of central tendency across different samples from the same population of data.
- is affected by the magnitude of each value and is thus affected more than the median or the mode by extreme values.
- is used frequently in inferential statistical procedures and may therefore be considered the most important of the three measures.
- indicates skewness when compared to the median.

PROBLEMS

Problem Set A

1. Find the mean, median, and mode for the following numbers: 8, 20, 11, 20, 12, 15.
2. Demonstrate how each measure of central tendency would be affected if *one* of the scores of 20 was changed to a score of 12 for the data in problem 1.
3. How would adding the number 55 to the set in problem 1 change each measure of central tendency?

4. Demonstrate using the data set in problem 1 that the sum of the effective weights for each score relative to the mean is equal to zero, that is, $\Sigma(X_i - \overline{X}) = 0$.

5. Given the following information, identify the shape of each distribution.
   a. $\overline{X} = 42$; *Mdn* = 42; Mode = 42
   b. $\overline{X} = 30$; *Mdn* = 27; Mode = 25
   c. $\overline{X} = 10.3$; *Mdn* = 10.5; Mode = 10.6
   d. $\overline{X} = 25$; *Mdn* = 25; Mode = 17 and 33

6. Assume that amount of time studying is related to achievement outcome and that most high achievers study approximately the same amount of time during a week. A few, however, can study considerably less while still achieving the same level as the others. Which measure of central tendency would best represent the average amount of time per week spent studying by a group of high achievers?

7. The mean entry levels of assertiveness and sample sizes for three groups of participants in an assertiveness training workshop were as follows:

	Group		
	1	2	3
$\overline{X}$	20	25	22
n	15	10	12

   What is the mean entry level of assertiveness for the 37 individuals as a group?

8. Given three of four numbers in a set (3, 7, 2, ?) and $\overline{X} = 4$, what must be the value of the fourth? Use effective weights to demonstrate how you arrived at your answer.

9. What would happen to the value of each measure of central tendency if we added 30 points to each number in a data set?

Problem Set B

1. Recently, considerable attention has been given to sex-related differences in mathematics achievement, but few studies have looked at sex-related differences in electing to study mathematics during high school. A study by Sherman and Fennema (1977) asked females and males enrolled in mathematics classes whether they intended to enroll in additional mathematics classes, comparing selected affective variables related to their intent. Near the time of registration for the following year's classes, students in tenth-grade mathematics classes were polled. Based on the responses, students were placed into two groups: (a) those definitely planning to study mathematics in the next year and (b) those who definitely were not or who were still undecided. Sherman and Fennema then reported results for males and females in the two intent groups on verbal ability, spatial relations, and eight affective variables measured by the Fennema-Sherman Mathematics Attitudes Scales. Mathematical ability was controlled by further dividing respondents into two groups—those who scored above the median of the distribution of mathematics achievement scores (sexes combined) and those who scored below the median.

   We have attempted to simulate the method of analysis used by Sherman and Fennema in Table 5.7. Simulated data are given for two variables, mathematics achieve-

TABLE 5.7
Simulated data for mathematical achievement and confidence by sex and intent to study mathematics.

Intend to Study Math				Do Not Intend to Study Math			
Females		Males		Females		Males	
Math Ach.	Confidence	Math Ach.	Confidence	Math Ach.	Confidence	Math Ach.	Confidence
60	52	60	45	44	32	46	38
42	44	46	44	64	52	52	48
58	48	49	50	31	35	37	35
68	50	47	41	52	39	56	42
52	41	59	48	38	40	45	40
54	48	65	53	69	46	51	43
46	47	54	42	35	43	36	38
72	55	51	48	67	44	53	45
45	41	43	45	55	38	39	37
59	47	71	56	49	50	47	38
39	40	61	48	66	43	37	39
52	43	51	46	52	41	55	47
69	56	50	44	53	43	41	35
50	46	46	40	57	43	49	46
55	49	58	49	51	39	32	37
54	46	50	51	63	45	53	38
56	47	49	47	29	32	30	32
49	42	70	52	45	34	44	36
57	51	58	43	33	38	48	40
66	49	65	47	54	42	52	40

ment and confidence in learning math, for students grouped by sex and intent to study mathematics.

a. Categorize the respondents into the two mathematics achievement groups (above and below the median) using the mathematics achievement scores for all 80 students to calculate the median. When calculating the median, group the scores into a frequency distribution with an interval size of 3 starting with a lower class interval of 29–31. You should end up with eight groups, high- (above median) and low- (below median) achieving students within each of the four existing sex-by-intent groups. Descriptively, is any trend in the intended enrollment pattern for males and females based on the group frequencies?
b. For each of the eight groups, calculate the mean "confidence in learning math" score. Do any of the groups appear more or less confident on the average in their abilities to learn math based on a descriptive comparison of group means?
c. Using the eight group means calculated in (b), find the mean "confidence in learning math" score for males and females separately, for those intending to

enroll in math and those not separately, and for high achieving and low achieving students separately. (Hint: Use the formula for finding the mean of combined groups.)

Problem Set C

Use the "Sesame Street" data in Appendix A to answer the following questions.

1. Table 5.8 lists pretest and posttest scores on the classification skills variable for two groups of children from the "Sesame Street" data base. Group A contains scores for a random sample of 30 children who had coded scores of 1 on the viewing category

TABLE 5.8
Pretest and posttest scores on classification skills by viewing categories 1 and 4.

Viewing Category 1			Viewing Category 4		
Child	Classification Skills		Child	Classification Skills	
ID	Pre	Post	ID	Pre	Post
44	4	7	237	15	11
167	7	4	233	23	23
100	7	13	172	6	9
204	13	16	70	16	20
182	12	16	206	11	20
56	13	16	24	9	15
231	10	17	21	23	24
181	9	11	163	11	17
38	3	6	52	21	23
115	10	7	160	11	18
128	14	7	75	12	17
186	11	13	60	7	11
180	8	11	240	12	19
53	11	6	61	20	22
201	11	9	103	20	24
178	7	5	5	22	21
188	6	10	27	11	14
197	10	9	72	15	14
193	11	11	164	12	18
228	12	19	76	19	20
194	17	13	63	12	22
229	4	11	123	10	19
161	6	16	28	11	15
155	10	5	94	17	16
211	6	9	77	24	23
51	10	10	87	15	21
62	11	17	214	16	10
196	16	16	35	17	22
209	11	9	151	15	17
13	12	13	73	10	17

variable (watched the show rarely or never). Group B contains scores for a random sample of 30 children who had coded scores of 4 on the viewing category variable (watched the show more than five times a week).

a. Calculate the modes, medians, and means for each group's set of classification pretest and posttest scores. Use the ungrouped data points for calculating the modes and means, but group each set of data into a frequency distribution with an interval width equal to two ($i = 2$) for calculation of the median. Start the bottom interval in the grouped frequency distribution with the lowest score in the data set.
b. Compare shapes and average performance of the pretest and posttest distributions for each group. Which group appears to have learned more based on the descriptive statistical average values?

2. Table 5.9 shows the mean Peabody mental age score and sample sizes for all children from the "Sesame Street" data base grouped by viewing category code. Compute the PMA mean score for children in viewing categories 1 and 2 combined, for children in viewing categories 3 and 4 combined, and for children from all four categories combined.

TABLE 5.9
Peabody mental age scores by viewing category.

	Viewing Categories			
	1	2	3	4
$\overline{X}$	40.70	44.92	47.59	52.04
n	54	60	64	62

6

Measures of Variability

Three primary pieces of information typically summarize a set of data. Two of these, distribution shape and central tendency, have been discussed in previous chapters. The third descriptive statistical concept is variability, or the dispersion of scores across a scale. Figure 6.1 illustrates the importance of variability measures. Two frequency polygons are shown. Both have the same shape (negative skew) and measures of central tendency. Would you conclude that both distributions are the same? Inspection of the distributions shows that they differ because the scores are more variable or spread out across the measurement scale for distribution B than for A.

Variability is a fundamental component in many statistical tests and is basic to the interpretation of research results. Consider the distribution of grade point averages for your fellow classmates. What about the distribution of scores measuring their attitudes toward statistics? Shoe size? Age? Leadership potential? Degree of assertiveness? For the variables mentioned, individuals in a group differ in degree or quantity. A measure of central tendency is one piece of information, a single representative score for the group. Obviously everyone does not have that score or perform at that "average" level. The extent to which scores spread out provides valuable information about the representativeness of the "average" score for the group and the degree to which individuals differ.

This chapter presents four measures of variability: range, semi-interquartile range, variance, and standard deviation. Each measure of variability is a unique statistical index to summarize the score dispersion for a given data set.

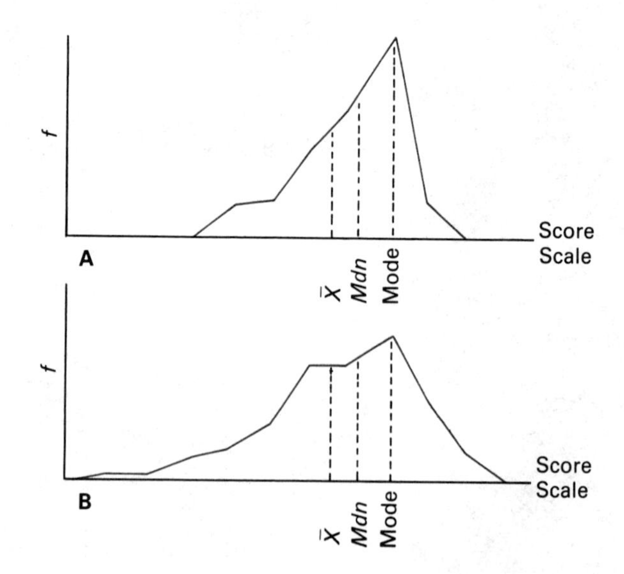

FIGURE 6.1
Distributions with the same shape and central tendency, but different amounts of variability (spread).

6.1 RANGE AND SEMI-INTERQUARTILE RANGE

As measures of variability, the range statistics indicate the distance between two points on the measurement continuum for a given distribution. These indices are rarely reported as the only measure of variability.

Range

The **range** is the simplest index of variability. It is the distance from the highest to the lowest scores in the data set, that is:

$$\text{range} = X_{\text{highest}} - X_{\text{lowest}} \tag{6.1}$$

The result is the distance in which the entire data set appears. Formula 6.1 produces the *apparent* range. The *exact* score range is the difference between the upper exact limit of the highest score and the lower exact limit of the lowest score. Because the range is primarily used as a general descriptive index, either the apparent or the exact score range conveys the desired information, but the apparent score range is easier to calculate.

To illustrate, look at the Age raw scores from the "Sesame Street" data in Appendix A. The oldest child was 69 months, the youngest was 34 months. The range is 69 − 34, or 35 months. If we presented the range of 35 months as a summary index, we would know that the 240 Age scores spread across a 35-month interval.

The greater the range for a set of data, the greater the dispersion of the scores. While the range is easy to calculate, its shortcomings outweigh this advantage. An index of variability is intended to convey information about an entire distribution. The range does so based on the values of only two scores in the distribution. It ignores information from all other scores. Thus, the range is very unstable from sample to sample, especially for small sample sizes. Additionally, studying a variable using samples of unequal size will affect the value of the range. Larger samples are likelier to contain lower and/or higher scores from the population of potential scores. In contrast to the range, indices that incorporate information from more than two score values will be more stable measures of variability.

Semi-Interquartile Range

The scale distance containing the middle 50% of a distribution is termed the **interquartile range.** The interquartile range is the distance between the first quartile point, P_{25}, and the third quartile point, P_{75}.

The **semi-interquartile range,** Q, is one-half of the distance represented by the interquartile range. Computationally,

$$Q = \frac{P_{75} - P_{25}}{2} \tag{6.2}$$

Displaying the frequency polygon of Peabody mental age scores from Appendix A, Figure 6.2 illustrates the values necessary to understand the semi-interquartile range. P_{75} and P_{25} are calculated using Formula 4.1 and the distribution in Table 4.8.

$$P_{75} = 54.5 + \frac{180 - 166}{27}(5) = 57.09 \quad \text{and} \quad P_{25} = 34.70$$

The interquartile range is 57.09 − 34.70, or 22.39. Then,

$$Q = \frac{22.39}{2} = 11.195$$

This variability information indicates that the middle 50% of the Peabody mental age scores can be found within a range of 22.39 mental age points. While Q is the index calculated and reported, the interpretive index actually is the interquartile range, or $2Q$. One way to think of Q is as the average (mean) positive distance of P_{75} and P_{25} from the median, P_{50}. That is,

$$Q = \frac{(P_{75} - P_{50}) + (P_{50} - P_{25})}{2}$$

Thus, since the median Peabody mental age score in Figure 6.2 is 42.00,

$$Q = \frac{(57.09 - 42.00) + (42.00 - 34.70)}{2} = \frac{15.09 + 7.30}{2} = 11.195$$

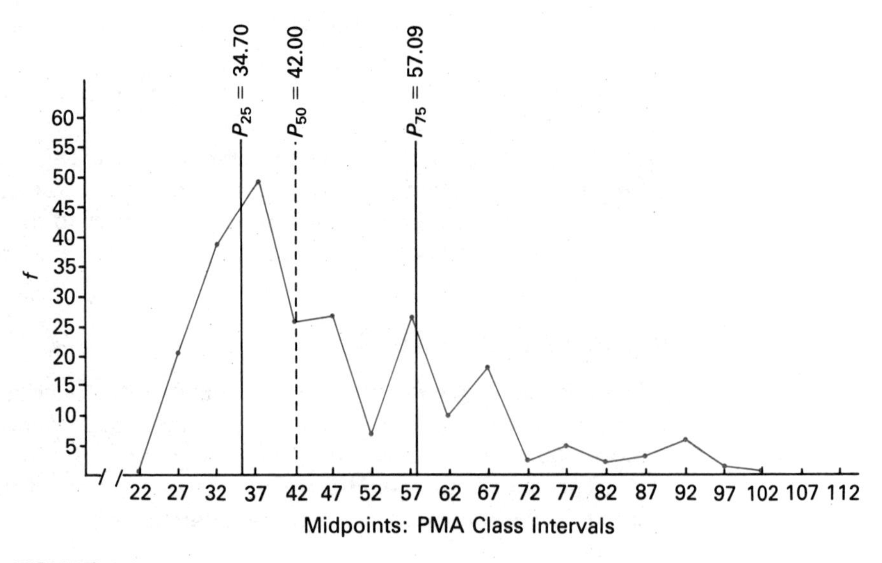

FIGURE 6.2
Frequency polygon of Peabody mental age scores.

TABLE 6.1
Frequency distribution of Peabody mental age scores for Site II children.

Score Interval	*f*	*cf*
95–99	1	55
90–94	2	54
85–89	3	52
80–84	1	49
75–79	4	48
70–74	1	44
65–69	11	43
60–64	3	32
55–59	12	29
50–54	4	17
45–49	5	13
40–44	6	8
35–39	2	2

While the distances of P_{75} and P_{25} from the median are different (15.09 vs. 7.30), the mean distance of the two quartile points is Q, or 11.195 units. In this manner, Q represents a measure of variability that uses the median as a reference point.

A frequency distribution of Peabody mental age scores for Site II individuals appears in Table 6.1. Using the semi-interquartile range as the index of variability, are the Peabody mental age scores for this subgroup more or less variable than for all 240 students?[1] How do the distances of P_{75} and P_{25} from the group median (the middle two quartile ranges) compare?[2]

As a measure of variability, the semi-interquartile range focuses on the section in the middle of the distribution containing half of the scores. In this respect, it is influenced by the distribution of scores in the part of the scale in

---

[1] $$P_{75} = 64.5 + \frac{41.25 - 32}{11}(5) = 68.70$$

$$P_{25} = 49.5 + \frac{13.75 - 13}{4}(5) = 50.44$$

$$Q = \frac{68.70 - 50.44}{2} = 9.13$$

The Site II mental age scores are slightly less variable. The middle 50% of these students have mental age scores within an 18.26 range compared to the 22.39 unit range for the total group.

[2] $$Mdn = P_{50} = 54.5 + \frac{27.50 - 17}{12}(5) = 58.88$$

$$P_{75} - P_{50} = 68.70 - 58.88 = 9.82$$

$$P_{50} - P_{25} = 58.88 - 50.44 = 8.44$$

The third quartile range ($P_{75} - P_{50}$) is slightly larger than that for the second quartile range ($P_{50} - P_{25}$), but the discrepancy is not as large as for the entire data set (15.09 vs. 7.30).

which scores are most heavily concentrated. Like the median, extreme scores do not affect Q. Because the shape of the middle portion of a distribution will not fluctuate much from sample to sample, the semi-interquartile range is more stable than the range.

6.2 VARIANCE AND STANDARD DEVIATION

In measuring variability, ideally we want to know how much the scores in a distribution differ from each other. Comparing each score to every other one is an enormous undertaking, however, especially when the number of scores in the data set increases. We can more easily compare all scores in a distribution to a single anchor point. With the mean as the anchor point, the **variance** and **standard deviation** represent the difference between raw scores and the mean. The variance and standard deviation are derived using the same computational process. These indices are the dominant measures of variability used in practice and provide a basis for much of the statistical reasoning and interpretation in inferential statistics. As measures of variability they may be used whenever the mean is an appropriate measure of central tendency.

Evaluating the distance of each raw score from the mean of the distribution is a very logical approach to discussing variability. The distance of any individual raw score from the mean was introduced in Chapter 5 as an effective weight. This distance for an individual score can be represented as

$$x_i = X_i - \overline{X}$$

The score x_i is termed a **deviation score** in statistics. A deviation score represents both the distance and location (above or below) of a raw score relative to the mean. Deviation scores may be positive or negative. Negative deviation scores identify raw scores below the mean. Raw scores above the mean will have positive deviation scores. Deviation scores cannot be averaged for the group to produce a measure of variability, however. When deviation scores are summed, the result will always be zero, since the deviation scores are equally balanced below and above the mean.

The last characteristic of the mean was discussed and demonstrated in section 5.5 of Chapter 5 using the concept of effective weights instead of deviation scores. A mathematical proof of this characteristic uses the rules of summation notation (see Appendix C):

$$\begin{aligned}
\Sigma x_i &= \Sigma(X_i - \overline{X}) \\
&= \Sigma X_i - \Sigma\overline{X} \\
&= \Sigma X_i - n\overline{X} \\
&= \Sigma X_i - \cancel{n}\left(\frac{\Sigma X_i}{\cancel{n}}\right) \\
&= \Sigma X_i - \Sigma X_i \\
&= 0
\end{aligned}$$

TABLE 6.2
Deviations and squared deviations from median.

Raw Score	Deviation from *Mdn*	Squared Deviation
11	0	0
19	8	64
15	4	16
10	−1	1
10	−1	1

Compute the deviation scores for each raw score in the following data set: 11, 19, 15, 10, 10. Add the deviation scores. Do they sum to zero?[3]

Obviously, a usable measure of variability cannot be produced from the sum of the deviation scores. It would always be zero. To avoid this difficulty, the deviation scores are squared. Squared deviation scores, when summed, are unique in computing a measure of variability. When deviation scores are squared and then summed the resulting value will be a **minimum.** If any point on the scale continuum other than the mean is used as an anchor, the sum of the squared deviation scores will be larger than when the mean is used. A squared deviation score is represented by x_i^2, where

$$x_i^2 = (X_i - \overline{X})^2$$

Summing all squared deviation scores for a group results in a minimum value. This relation is expressed as

$$\Sigma x_i^2 = \Sigma(X_i - \overline{X})^2 = \text{a minimum}$$

To illustrate this property, refer to the set of five scores above. The deviation scores around the mean were −2, 6, 2, −3, and −3.[3] The squared deviation scores are 4, 36, 4, 9, and 9, respectively. For these data,

$$\begin{aligned}\Sigma x_i^2 &= 4 + 36 + 4 + 9 + 9 \\ &= 62\end{aligned}$$

If we used a scale point other than the mean as anchor, the sum of the squared distances of raw scores from that point would be greater than 62. For example, if the scale median (*Mdn* = 11) were chosen as the anchor point, the deviation and squared deviation values would be as in Table 6.2. The sum of the

---

[3]

(X_i)	x_i
11	−2
19	6
15	2
10	−3
10	−3
$\Sigma X_i = 65$	$\Sigma x_i = 0$
$\overline{X} = 13$	

squared deviation scores around the median is 82. This value is obviously larger than 62, the sum of squared deviations about the mean. Choose any other score as an anchor point and calculate the sum of the squared deviation scores: the value will always be greater than 62. The quantity Σx_i^2 is very important to the concept of variability. The sum of squared deviations is the basis for understanding both the variance and standard deviation.

Variance

Variance for *population* and *sample* data is calculated differently. The variance for a population of size n is defined as σ^2 where

$$\sigma^2 = \frac{\Sigma x_i^2}{n} = \frac{\Sigma(X_i - \mu)^2}{n} \tag{6.3}$$

The Greek letter sigma (σ), when squared, represents the variance for population data. Formula 6.3 defines the population variance as the *average squared distance* (deviation) of raw scores from the population mean, μ. It is analogous to the average area of a series of squares distributed around the mean. Using the scores 3, 4, 5, 8, 10, and 12, Figure 6.3 displays the area concept of squared deviation scores and variance as an average area for scores along a scale continuum. The mean for the data set is 7. The deviation scores for 3, 4, 5, 8, 10, and 12, respectively, are −4, −3, −2, 1, 3, and 5. The sum of these numbers is zero, but the squared scores summed are 16 + 9 + 4 + 1 + 9 + 25 = 64. The

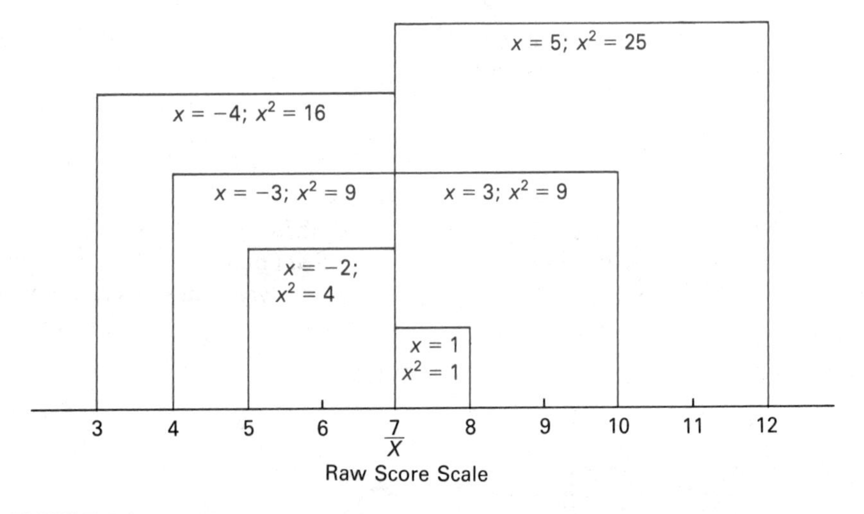

FIGURE 6.3
Concept of variance as an average area for averaged squared deviation scores.

average area of the squares in Figure 6.3 or the variance for the population of six scores is

$$\sigma^2 = \frac{\Sigma x_i^2}{n} = \frac{64}{6} = 10.67$$

As a statistic, the sample variance will most often be used as a substitute for or estimate of the population variance. Because data from a sample do not provide exact information, but result in sample statistics that differ from their counterpart population parameters, the computational formulas for sample data are derived to allow the most accurate estimate possible. For sample data, Σx_i^2 is divided by $n - 1$ rather than n. The sample variance is represented by s^2:

$$s^2 = \frac{\Sigma x_i^2}{n - 1} = \frac{\Sigma(X_i - \overline{X})^2}{n - 1} \tag{6.4}$$

Formula 6.4 will be used throughout the text unless the population variance is specified. Calculators and computer statistical programs usually assume sample data and use $n - 1$ when computing the variance. Chapters 11 and 13 contain more detailed discussions of the rationale for using $n - 1$ in computing sample variance.

While formula 6.4 bases variance on squared deviation scores, this formula is usually not the most efficient when actually computing Σx_i^2. An easier, more direct computational formula for Σx_i^2 using raw scores can be derived algebraically using the rules of summation notation (the derivation appears in Appendix C).

$$\Sigma x_i^2 = \Sigma X_i^2 - \frac{(\Sigma X_i)^2}{n} \tag{6.5}$$

Formula 6.5 does not require subtracting the mean from each raw score. Rather, the raw scores are used directly in the computation. Note the important distinction between ΣX_i^2 and $(\Sigma X_i)^2$. The first symbol, ΣX_i^2, means square each raw score, then sum. The term $(\Sigma X_i)^2$ means sum the raw scores, then square the sum. Table 6.3 outlines the process of calculating Σx_i^2 using both the deviation scores, $(X_i - \overline{X})$, and raw scores. For convenience, we will drop the subscripts in the formulas as we did in the formula for the mean. In this case,

$$\Sigma x^2 = \Sigma(X - \overline{X})^2 \qquad \text{(conceptual definition formula)}$$

$$\Sigma x^2 = \Sigma X^2 - \frac{(\Sigma X)^2}{n} \qquad \text{(raw score computational formula)}$$

Remember that both procedures result in the same value within rounding error. The quantities $\Sigma(X - \overline{X})^2$ and $\Sigma X^2 - (\Sigma X)^2/n$ are algebraically equiva-

TABLE 6.3
Computation of sum of squares using both the deviation(1) and raw-score(2) methods.

Raw Score (X)	X^2	$x = (X - \overline{X})$	x^2
6	36	$(6 - 9) = -3$	9
8	64	$(8 - 9) = -1$	1
14	196	etc. 5	25
10	100	1	1
8	64	−1	1
9	81	0	0
7	49	−2	4
11	121	2	4
9	81	0	0
8	64	−1	1
9	81	0	0
$\Sigma X = 99$	$\Sigma X^2 = 937$	$\Sigma(X - \overline{X}) = 0$	$\Sigma x^2 = 46$

$$\overline{X} = \frac{99}{11} = 9$$

$$(1)\ \Sigma x^2 = \Sigma(X - \overline{X})^2 = 46$$

$$(2)\ \Sigma x^2 = \Sigma X^2 - \frac{(\Sigma X)^2}{n} = 937 - \frac{(99)^2}{11} = 937 - 891 = 46$$

lent and either could be used to compute Σx^2. When the mean is not a whole number, the use of $\Sigma(X - \overline{X})^2$ becomes considerably more time consuming. Also, we can use most calculators to find ΣX^2 and ΣX by entering each raw score into the machine once.

Once Σx^2 has been calculated, the variance can be computed using formula 6.4. For the 11 scores in Table 6.3, $\Sigma x^2 = 46$ and

$$s^2 = \frac{\Sigma x^2}{n - 1} = \frac{46}{10} = 4.6$$

To indicate the time involved between the squared deviation and raw score computational formulas when real data form the data set, Peabody mental age scores were randomly sampled for 13 students from the "Sesame Street" data

TABLE 6.4
Peabody mental age scores for 13 students.

Child ID	Peabody Mental Age Raw Score	Child ID	Peabody Mental Age Raw Score
10	32	57	34
55	34	78	75
188	47	48	29
162	34	239	69
71	58	137	29
119	35	222	61
210	45		

base (see Table 6.4). Compute the variance, s^2, independently using both methods. Keep track of the time involved and compare.[4]

Standard Deviation

The standard deviation is easily computed once we know the variance. It is represented by σ (population) or s (sample) and is defined as the positive square root of the variance:

$$\sigma = \sqrt{\sigma^2} \quad \text{and} \quad s = \sqrt{s^2}$$

To compute the sample standard deviation, the sum of squared deviations (Σx^2) and the sample variance should be computed first. The process is as follows:

- Step 1 Compute the raw score sums, ΣX and ΣX^2.
- Step 2 Compute Σx^2.

$$\Sigma x^2 = \Sigma X^2 - \frac{(\Sigma X)^2}{n}$$

[4]Raw score method:

$$\Sigma X = 582;\ \Sigma X^2 = 29{,}104;\ \overline{X} = \frac{582}{13} = 44.77;\ \Sigma x^2 = 29{,}104 - \frac{(582)^2}{13} = 3048.31;$$

$$s^2 = \frac{3048.31}{12} = 254.03$$

Deviation score method:

$$\overline{X} = 44.77;\ \Sigma(X - \overline{X})^2 = (-12.77)^2 + (-10.77)^2 + \cdots + (16.23)^2 = 3048.31$$

$$s^2 = \frac{3048.31}{12} = 254.03$$

Using a calculator with accumulating capabilities, the raw score method took 3-minutes while the deviation method took 10 minutes.

TABLE 6.5
Numbers scores for a random sample.

Child ID	Numbers Pretest (X)	Numbers Posttest (Y)
5	51	54
22	22	45
52	41	50
74	23	37
91	18	32
138	25	38
160	23	42
171	16	29
195	33	44
203	19	39

- Step 3 Compute the sample variance.

$$s^2 = \frac{\Sigma x^2}{n - 1}$$

- Step 4 Compute the sample standard deviation.

$$s = \sqrt{s^2} = \sqrt{\frac{\Sigma x^2}{n - 1}} \qquad (6.6)$$

Using the data in Table 6.3 with $\Sigma x^2 = 46$ and a variance of 4.6, the standard deviation using Formula 6.6 is

$$\begin{aligned} s &= \sqrt{4.6} \\ &= 2.14 \end{aligned}$$

For the 13 Peabody mental age scores in Table 6.4, the standard deviation is $15.94(254.03)^{1/2}$. Pre- and posttest scores on the Numbers test for a random sample of 10 children from Appendix A appear in Table 6.5. The pretest mean was 27.1 and the variance was 126.10. What is the standard deviation of Numbers pretest scores?[5] Calculate the mean, variance, and standard deviation of the 10 Numbers posttest scores.[6]

---

[5] $s_x = \sqrt{126.10} = 11.23$

[6] $\Sigma Y = 410$; $\Sigma Y^2 = 17340$; $\overline{Y} = \frac{410}{10} = 41.0$

$$\Sigma y^2 = 17340 - \frac{410^2}{10} = 530.0$$

$$s_y^2 = \frac{530}{9} = 58.89$$

$$s_y = 7.67$$

The standard deviation and mean are the basic descriptive indices reported for most data. While numerically different, the sum of squared deviations, variance, and standard deviation can be thought of as the same basic index of variability derived from the same source, distance of raw scores from the mean. As indicated previously, the standard deviation cannot be calculated without first computing the sum of squared deviations and variance. The latter two indices, Σx^2 and s^2, are extremely important to understanding advanced statistical manipulations and interpretations, but as a basic descriptive index of variability, the standard deviation is the important index to report.

Like the mean, the variance and standard deviation are computed using all the scores in the data set. Each number thus affects the magnitude of these indices. This is a disadvantage only for data containing extreme scores, because extreme values will inflate the variance. The variance and standard deviation are both generally more stable across samples and situations than either the range or semi-interquartile range. When the mean is the primary measure of central tendency reported, the variance and standard deviation are the measures of variability to compute.

6.3 INTERPRETATIONS

The minimum computed value for variability is zero. A value of zero indicates no variability; all numbers are identical. As the scores spread out across the scale continuum (differ from each other), variability increases. A negative value is impossible to obtain.

The major question in interpretation becomes: "Given an index of variability, what information does it convey?" For a set of data, the range may be 32, Q may equal 4.5, or s might equal 6.37. Are these values good? bad? high? low? No standard interpretation applies to an index across all research settings. We might expect several factors to affect the degree of spread. Interpretation must be made relative to the research conditions under which data were collected.

Sampling is one component of the research process that can affect the variability for a data set. If problem-solving skills are being measured, for example, a study of academically gifted students would produce less variability than a study of all students. Research involving subgroups of the general population often will exhibit less variability on variables related to the subgroups. When an index of variability is reported, we can only interpret its importance based on our expectations for the population under study.

Another important factor that affects the variability for a study is the instrumentation used, particularly the range of the scale. Measuring job satisfaction using a single item with a range of 1 (Strongly Disagree) to 5 (Strongly Agree) will result in a very small index of variability value. If 10 items are used, however, and a single job satisfaction score is obtained by summing response scores for the 10 items, then possible scores would range from 10 to 50. We would expect a larger variability in the latter case only because the measurement scale allows scores to spread out more. A standard deviation of 5.2 might be typical in the latter instance, while for a single-item response, a standard deviation of .72 might be typical.

As a descriptive statistic, variability provides only one piece of information. Only in conjunction with a measure of central tendency and the shape of the score distribution can we begin to get an accurate picture for a given set of data. A measure of central tendency will anchor the distribution along the scale. The range gives the width of the interval containing an entire distribution. While the range is a useful index, an extreme score in a data set will affect it drastically. For this reason the range is infrequently the only index reported.

The semi-interquartile range is interpreted using $2Q$, or the scale distance containing the middle 50% of the distribution. Given an expected unimodal distribution and the median as the anchor point, we can visualize where the bulk of the scores are using $2Q$ as an index. While the median (P_{50}) is a useful anchor point for interpreting the semi-interquartile range, the middle 50% of the scores will not distribute across equal intervals on each side of the median unless the distribution is symmetrical. For positively skewed distributions, the distance containing the 25% of the scores above the median will be larger than the interval containing the 25% below the median $(P_{75} - P_{50}) > (P_{50} - P_{25})$. For negatively skewed distributions, the distance between P_{75} and P_{50} will be less than the distance between P_{50} and P_{25} (i.e., $P_{75} - P_{50} < P_{50} - P_{25}$). Only when distributions are symmetrical will the two intervals be equal. As an index of variability, Q represents the mean of these two interval distances.

The variance and standard deviation tend to be abstract concepts that are easily defined mathematically, but difficult to understand conceptually. We acquire a "feel" for the variance as a measure of individual differences on a trait or construct through analyses of data from a variety of research studies. Repetition in calculating and applying the variance increases our understanding.

The means and standard deviations for the continuous variables in the "Sesame Street" data base are given in Appendix A following the raw data. Look at the raw score means to see where the distribution is centered on the measurement scale. Then notice how the standard deviations vary across the different variables. The sample is the same, but the potential range of measurement values differs for each variable. For example, the Letters and Numbers tests have more items than the Body Parts or Forms tests. Students' raw scores can vary more on the former two tests. This is reflected in the posttest standard deviations of 13.19, 12.60, 5.41, and 3.92 for the four measures, respectively.

Knowing the shape of a distribution of scores often can facilitate interpretation of the standard deviation, particularly if the scores are normally distributed. Because the curve properties of the normal distribution are known and the curve is constant, a standard interpretation can be made assuming the data are approximately normally distributed. Figure 6.4 illustrates the percent-

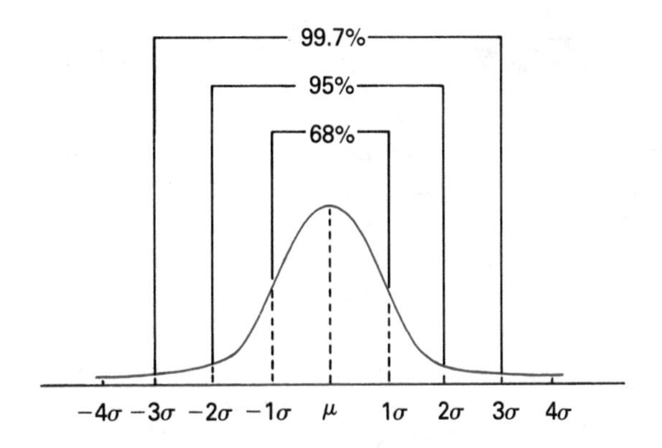

FIGURE 6.4
Proportion of scores in a normal distribution falling within specific standard deviation intervals.

age of scores in a *population* that can be expected to fall within specific intervals defined in standard deviation units. For the scale interval bounded by values equal to the mean plus or minus one standard deviation unit, we would expect 68% of the scores to fall within this interval. Approximately 95% should fall within an interval with $\mu + 2\sigma$ as the upper limit and $\mu - 2\sigma$ as the lower limit. The interval bounded by $\mu + 3\sigma$ and $\mu - 3\sigma$ includes 99.7% of all scores. Data on intelligence provide a familiar example. The mean IQ score for many intelligence instruments is 100 and the standard deviation is approximately 15. Table 6.6 illustrates the percentage of persons we expect to fall within specific IQ ranges if we assume that IQ scores are normally distributed. These interpretative guidelines are only appropriate for approximately normal distributions. However, they can be used as a general indication of the spread of scores along the scale. The age of children in the "Sesame Street" data base tends to be normally distributed. From Appendix A, the mean age in months is 51.53 and the standard deviation is 6.28. Calculate the age interval limits expected to include 68%, 95%, and 99.7% of the children.[7]

TABLE 6.6
Expected proportions for IQs in normal distribution ranges.

Lower Bound	Upper Bound	Percentage of Persons with IQs in Range
$(\mu - 1\sigma) = 85$	$(\mu + 1\sigma) = 115$	68%
$(\mu - 2\sigma) = 70$	$(\mu + 2\sigma) = 130$	95%
$(\mu - 3\sigma) = 55$	$(\mu + 3\sigma) = 145$	99.7%

[7] $\mu \pm 1\sigma = 45.25$ to 57.81
$\mu \pm 2\sigma = 38.97$ to 64.09
$\mu \pm 3\sigma = 32.69$ to 70.37

6.4 SCORE TRANSFORMATION EFFECTS ON $\overline{X}$, s^2 AND s

It is sometimes efficient to transform a set of data to simplify computations, to facilitate interpretations, or to make a set of data conform to specific assumptions. Often the transformations involve simply adding a constant to each raw score or multiplying each raw score by a constant. Each type of transformation on the mean and variance of the distribution has a predictable effect. When a constant, c, is added to each score in a distribution, then the mean of the new set of scores, $\overline{X}_{new}$, is equal to the mean of the original set of scores plus the value of c. The variance of the new set of scores, s^2_{new}, formed by adding c to each original raw score, equals the variance, s^2, of the original scores. Adding a constant to each score does not change the amount of variability in the set of scores, it only changes the central location of the score distribution. To illustrate these two principles, consider the data set consisting of 3, 4, 5, 8, 10, and 12. For this set $\overline{X}$ is 7 and s^2 is 12.8. If we added 3 to each score, the data set would consist of scores 6, 7, 8, 11, 13, and 15. For this set $\Sigma X = 60$, $\Sigma X^2 = 664$, $\overline{X}_{new} = 7 + 3 = 10$, and

$$s^2_{new} = \frac{664 - \dfrac{60^2}{6}}{5} = 12.8$$

Adding three units to every score moved the entire distribution (including the mean) three units to the right. Because the mean was moved three units to the right on the scale along with each score, the deviations of the scores from the mean did not change, thus not affecting the variability.

In contrast to the effect on s^2 and s when *adding* a constant, ***multiplying*** each raw score by a constant will affect the variability as well as the mean. When each score is multiplied by c, the mean of the new set of scores, $\overline{X}_{new}$ equals $c(\overline{X}_{old})$. The variance of the new set of scores is

$$s^2_{new} = c^2(s^2_{old})$$

As the positive square root of the variance, the standard deviation of the new set of scores, s_{new}, equals the absolute value of c times s_{old}, or $|c|(s_{old})$.

Using the previous original data set of six scores with $\overline{X} = 7$, $s^2 = 12.8$ and $s = 3.578$, a new set of data may be formed by multiplying each score by 5. The new data set would then be 15, 20, 25, 40, 50, and 60. For these data, $\Sigma X = 210$, $\Sigma X^2 = 8950$, $\overline{X} = 35$, $s^2 = 320$, and $s = 17.89$. The latter values would have been predictable.

$$\overline{X}_{new} = c(\overline{X}_{old}) \qquad 35 = 5(7)$$

$$s^2_{new} = c^2(s^2_{old}) \qquad 320 = 5^2(12.8)$$

$$s_{new} = |c|(s_{old}) \qquad 17.89 = 5(3.578)$$

6.5 SKEWNESS AND KURTOSIS

In addition to variability, distributions are characterized by skewness and kurtosis. This chapter and the last two introduced descriptive approaches to identifying the shape of a distribution. The degree of skewness and kurtosis, however, can be computed exactly using mathematical formulas. While the basic component for variance is squared deviations from the mean, $(X_i - \overline{X})^2$, the basis for the index of skewness is cubed deviations, $(X_i - \overline{X})^3$, and for kurtosis, deviation scores raised to the 4th power, $(X_i - \overline{X})^4$.

These indices are mentioned merely to show that it is possible to use such computational procedures. The statistical formulas for skewness and kurtosis produce exact comparative values. The index of skewness generally will range from -3 to $+3$ with 0 indicating a symmetrical distribution. A negatively skewed distribution will result in an index with a negative value; the index of skewness will exceed zero for a positively skewed distribution. For kurtosis, a value of 3 is the middle value, indicating a mesokurtic distribution. Distributions with values greater than 3 are leptokurtic and less than 3, platykurtic. The index of kurtosis has a lower boundary of zero.

6.6 SUMMARY

Statistical indices of variability indicate the amount of dispersion among the score values. They summarize in a single index the magnitude of individual differences. Of the four measures of variability presented, range indices are used infrequently while the related variance/standard deviation indices are fundamental to many statistical procedures. A summary of the characteristics of each follows.

The range

- is defined as the difference between the highest and lowest scores in a data set. As such, it identifies the apparent interval containing all the score values in the data set.
- depends on only two score values in the data set and therefore is unstable across samples from the same population.
- is affected drastically by extreme score values.

The semi-interquartile range

- is symbolically represented by Q, defined as

$$Q = \frac{P_{75} - P_{25}}{2}$$

- is represented interpretatively by $2Q$, the interval that includes the middle 50% of the scores. The median may be used as a reference point, but the P_{75} and P_{25} points are not equidistant from the median (P_{50}) in nonsymmetrical distributions.

- is determined by the frequency of scores in $2Q$ and therefore is not affected by extreme scores.

The variance and standard deviation

- are defined by the square distance of score values from the mean. The sum of squared deviation scores, $\Sigma x^2 = \Sigma(X - \overline{X})^2$, forms the basis for both the variance and the standard deviation.

$$\text{variance} = s^2 = \frac{\Sigma x^2}{n - 1} \quad \text{where } \Sigma x^2 = \Sigma X^2 - \frac{(\Sigma X)^2}{n}$$

$$\text{standard deviation} = s = \sqrt{\text{variance}}$$

- are more stable than the range indices for samples from the same population.
- are determined using each score in the data set.
- are frequently used with many other statistical procedures.
- are indicators of the relative magnitude of variability among a set of elements. The larger the variance or standard deviation, the more the observed scores vary. While $s^2 = s = 0$ is the lower bound indicating no variability, values for s^2 and s other than zero cannot be interpreted absolutely. They need to be interpreted (a) within the context of the measurement scale being used that restricts or expands the potential magnitude of the indices, (b) relative to expectations of variability that should exist based on results from other studies, or (c) comparatively to the variability obtained in other groups.

PROBLEMS

Problem Set A

Use Data Sets 1–4 to answer the problems that follow. Each set of data has the same number of scores (ns are equal) and the same range, 8 (all scores fall within the scale range between 10 and 2).

Data Set 1	Data Set 2	Data Set 3	Data Set 4
6	4	10	10
6	2	2	6
2	6	2	10
6	6	10	10
6	8	6	2
6	10	10	6
10	4	2	10
6	6	6	10
6	8	2	6
6	6	10	10

1. Cast the scores in each data set into an ungrouped frequency distribution (i.e., use each raw score unit as a class interval) and graph each distribution using a histogram. How would you describe the distributional shapes of each set of scores?

2. Confirm your description of shape for Data Sets 2 and 4 by comparing the values of P_{75}–P_{50} and P_{50}–P_{25}. From the visual shapes of the histograms for Data Sets 1 and 3, what would you anticipate as the comparative size of the values of P_{75}–P_{50} and P_{50}–P_{25} for these two distributions?

3. Compute the mean, median, and mode for each distribution. Do the comparative values of the three measures of central tendency conform to your expectations based on your descriptions of the distributional shapes for each data set?

4. When the range is used as the measure of variability, we know that each set of scores is equally variable (range = 8 for each data set). By inspecting the distributions and using only what you know about the properties of the other measures of variability (do not perform any calculations yet), would they also result in equal variability values across data sets? If not, how would you predict the four data sets would be rank ordered from least to most variable based on values of the semi-interquartile range? Based on values of the variance or standard deviation?

5. Calculate the semi-interquartile range (Q) for Data Sets 2 and 4. (Refer back to the calculated values in problem 2 to avoid duplication of computational effort.) Do the values of Q for these two data sets match your predictions in problem 4?

6. Compute the values for the sum of square deviations (Σx^2) using both the deviation formula and the raw score formula for each set.

7. Choose a raw score value different from the mean, but within the score range for the data in Data Set 4, and compute the sum of squared deviations around the score value selected. Is this latter sum of squared deviations larger than the value computed when using the mean in problem 6? (If not, you better check your computations; you have made an error!)

8. Using the values calculated in problem 6 for Σx^2, compute the variance and standard deviation for each data set of scores. Assume that each set of scores represents a *sample* data set. Do the values for s^2 and s confirm your predictions from problem 4?

9. If Data Set 1 were a population, how would the value for the standard deviation differ from that of the sample value?

10. Add a score of 3 to each score value in Data Set 1. Recompute the mean, variance, and standard deviation for the new set of scores. Do these values conform to your expectations based on the properties discussed in section 6.4? What would you predict as the values for $\overline{X}$, s^2, and s if you were to multiply each score value in Data Set 1 by 3? Confirm your predictions by performing the necessary calculations.

Problem Set B

Based on Carroll's (1963) model for school learning, Bloom (1968) developed a learning-for-mastery strategy. Using Carroll's definition of aptitude as the amount of time a learner requires to master a learning task, Bloom posited an instructional model that manipulated variables to reduce the amount of time needed for mastery. One of the basic assumptions underlying the mastery learning strategy is "Given enough time and appropriate instructional support, 95% of all students are capable of mastering the skills and knowledges taught at that age level." As support for the effectiveness of the mastery learning strategy, several studies have considered the variability of achievement scores at the end of instruction. For the mastery learning strategy to be effective, students should be more homogeneous in achievement over the skills and knowledges taught. Most should be able to function at or above mastery level.

Simulated achievement score data are given in Table 6.7 for two different classes, one taught using a learning-for-mastery approach, the other a traditional, less individualized approach. The same objectives were used to guide instruction and the same achievement test over the objectives was given at the end of a fixed period to both classes. The maximum score on the achievement test was 50. Use these data to answer the following questions.

1. If 80% correct is the mastery level set, did the proportion of students *not* attaining the mastery level or higher differ descriptively across the two classes? Rephrased, what is the percentile rank for a raw score of 39 for each class?

2. Does the range of scores differ descriptively for the two classes?

3. a. Calculate P_{25}, P_{50}, and P_{75}.
   b. Using the semi-interquartile range as the comparative descriptive statistic, do the data support the theory that mastery learning students will be more homogeneous in achievement?
   c. Descriptively, are the average (median) achievement scores for the two classes different or about the same?
   d. Do the shapes of the achievement distributions differ for the two classes? (Use the relationship between $P_{75}–P_{50}$ and $P_{50}–P_{25}$).

4. a. Calculate the mean, sum of squared deviations, variance, and standard deviation for each set of scores.
   b. Descriptively, are the average (mean) achievement scores for the two classes different or about the same?

TABLE 6.7
Simulated Achievement test data for two groups.

Achievement Test Scores					
Mastery Group			Nonmastery Group		
47	45	48	44	41	45
50	49	47	35	32	47
30	47	49	45	46	30
48	48	46	44	45	44
41	46	50	50	48	34
49	43	45	42	41	46
46	48	47	45	43	40
47	45	44	46	46	48
41	46	47	43	36	45
48	47	45	38	49	44
			47	45	

c. Does the relationship between the mean and median for each distribution support your conclusion in problem 3(d) about distribution shapes?
d. Are mastery learning students more homogeneous in achievement than nonmastery students using the standard deviations as descriptive measures of variability?

5. What general conclusion about the comparative effectiveness of the learning-for-mastery approach can be made based on your descriptive interpretation of the statistical indices calculated?

Problem Set C

The Classification Skills posttest scores for the two samples of children from Problem Set C in Chapter 5 are reproduced in Table 6.8. Use these data when responding to the following problems.

TABLE 6.8
Classification Skills posttest scores for two samples.

Viewing Category 1		Viewing Category 4	
Child ID	Posttest Score	Child ID	Posttest Score
44	7	237	11
167	4	233	23
100	13	172	9
204	16	70	20
182	16	206	20
56	16	24	15
231	17	21	24
181	11	163	17
38	6	52	23
115	7	160	18
128	7	75	17
186	13	60	11
180	11	240	19
53	6	61	22
201	9	103	24
178	5	5	21
188	10	27	14
197	9	72	14
193	11	164	18
228	19	76	20
194	13	63	22
229	11	123	19
161	16	28	15
155	5	94	16
211	9	77	23
51	10	87	21
62	17	214	10
196	16	35	22
209	9	151	17
13	13	73	17

1. Calculate the following Classification Skills posttest indices for children in each of the two viewing categories.
   a. range
   b. semi-interquartile range (Use the grouped frequency distribution with $i = 2$ calculated in Chapter 5, Problem Set C, 1(a).)
   c. sum of squared deviations
   d. variance
   e. standard deviation

2. Compare the distributions of Classification Skills posttest scores for the two viewing categories in terms of shape and degree of variability.

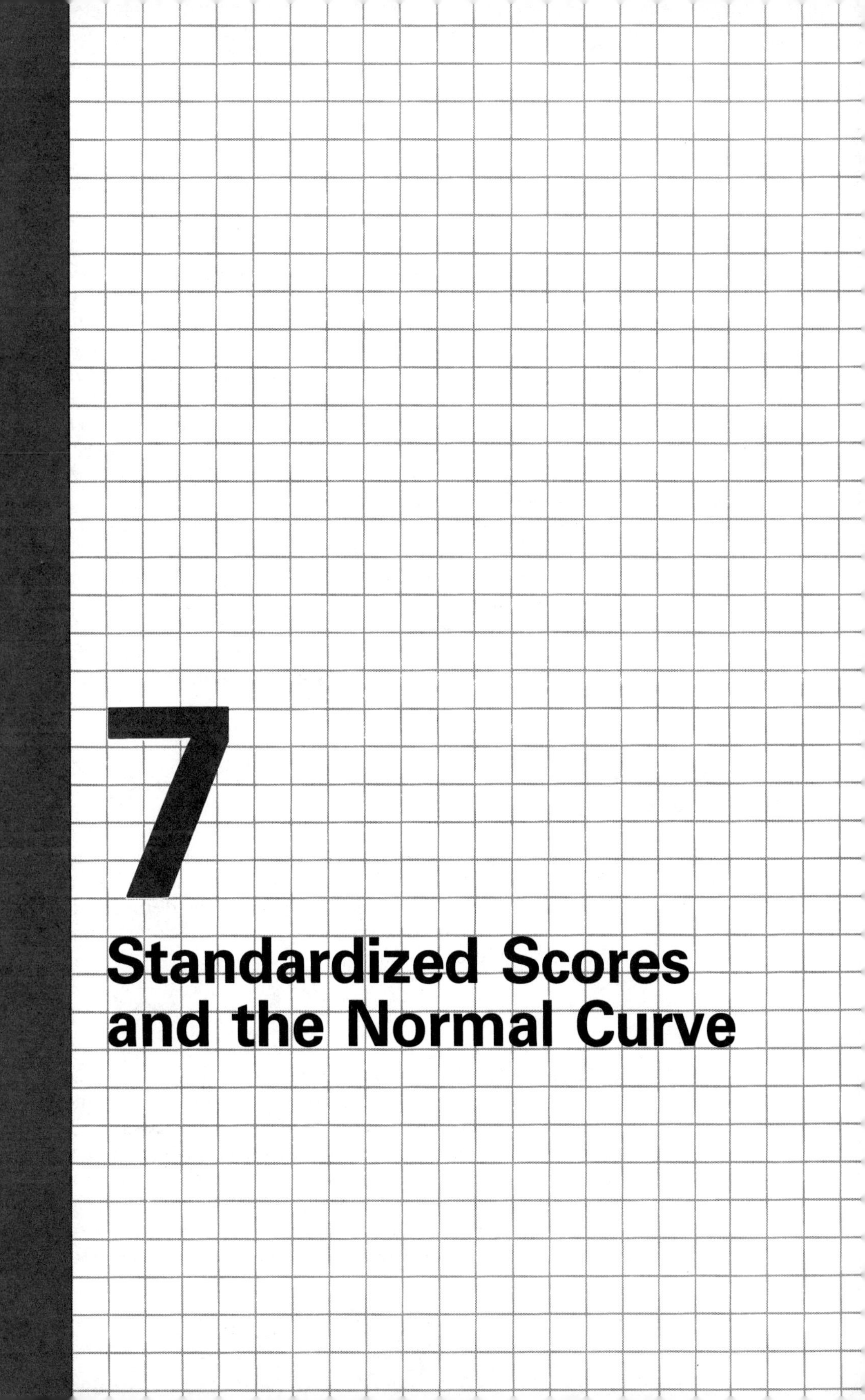

7

Standardized Scores and the Normal Curve

Interpreting a raw score requires additional information about the entire distribution. In most situations, we need some idea about the average score and an indication of how much the scores vary. Even with this information, it is difficult to compare scores from different distributions or from the same distribution on different variables. For example, assume that an individual took tests in reading and mathematics. The reading score was 32 and mathematics, 48. Is it correct to say that performance in mathematics was better than reading? Not without additional information. Did the individual do well on the mathematics test? Again, we can't say. Raw scores do not lend themselves to any interpretation other than as a point on a scale.

One method to increase the interpretability of raw scores is to transform them to standardized scores. Standardization means constancy across different times or situations. When interpretation of performance is the goal, standardized scores have a constant definition and interpretation attached to them. Whether the raw score represents creativity, assertiveness, intelligence, conformity, visual discrimination, blood pressure, or art appreciation, once it has been transformed, the standard score is readily interpretable.

7.1 *z* SCORES

Several standard score forms exist, but all are related to the most fundamental standard score, the ***z* score.** The *z* score is a statistically simple transformation that uses the mean as the reference point in the distribution of scores. Every raw score may be transformed (changed) to a *z* score. To find the *z* score for the *i*th raw score in a sample,

$$z_i = \frac{X_i - \overline{X}}{s_x} \tag{7.1}$$

In the formula, $X_i - \overline{X}$ is the distance in scale units of a specific raw score from the mean. This segment of the *z* score transformation was discussed in Chapters 5 and 6. Subtracting the mean ($\overline{X}$) from a raw score (X_i) produces a deviation score (x_i). When the mean ($\overline{X}$) is subtracted from a raw score (X_i), the level of performance is represented in relation to the mean of the raw scores, rather than some other arbitrary point.

Dividing the deviation score by the standard deviation of the distribution (s_x) transforms the deviation score from raw score units to standard deviation units. The transformation changes the measurement scale values. A z score unit no longer represents a raw score unit, but rather one standard deviation unit. If $s_x = 2.5$, one z score unit is the same as 2.5 raw score units. This process represents a change in the measurement unit, from measuring performance in original raw score units to measuring performance in standard deviation units. It is comparable to dividing inches by 12 and representing the measurement in terms of feet rather than inches.

The advantage of the z score transformation is that it takes into account both the average (mean) value and the amount of variability in a set of raw scores. The entire group becomes the internal norm group for interpreting an individual's raw score. No matter what the raw score measurement scale units, the value of the mean, or the degree of score variability, a z score always indicates the same thing: *the relative distance of an individual's raw score from the mean in standard deviation units.* Raw scores below the mean result in negative z scores. Raw scores above the mean will have positive z scores.

To illustrate, consider the following situation. Jim had a mathematics pretest score of 18 and a posttest score of 42 after instruction. His performance obviously changed. We can also ask, however, whether Jim learned at a rate equivalent to his classmates. Answering this question would require the following indices:

Pretest	*Posttest*
$\overline{X} = 17$	$\overline{X} = 49$
$s = 3$	$s = 5$
$z \text{ score} = \dfrac{18 - 17}{3}$	$z \text{ score} = \dfrac{42 - 49}{5}$
$= .33$	$= -1.40$

These data indicate that while Jim's raw score did increase from pre- to posttest, his relative position in the class decreased. His pretest z score of .33 indicates that his raw score was one-third of a standard deviation above the class pretest mean. His posttest z score of -1.40 indicates that compared with classmates, his performance was 1.4 standard deviation units *below* the average.

Because means and standard deviations are not identical for all variables or samples, raw scores on different measures do not have the same meaning. Transforming raw scores to z scores results in a scale in which identical z scores have the same meaning. In the preceding example, Jim's pretest score of 18 and posttest score of 42 were standardized, allowing us to conclude that his performance was relatively higher on the pretest. Using the same pre- and posttest means and standard deviations, how does a pretest score of 20 compare to a posttest score of 54?[1]

7.2 CHARACTERISTICS OF STANDARD SCORES

As noted in previous chapters, three characteristics are used primarily in describing a distribution: shape, central tendency, and variability. Transforming raw scores to z scores changes the distribution mean and standard deviation, but the shape of the z score distribution is identical to the shape of the raw score distribution. If the raw score distribution is positively skewed, then a frequency polygon of the z scores also will be positively skewed.

While the shape is unaltered, the transformation does change the mean and standard deviation. The effects on $\bar{X}$, s^2, and s for a set of scores of subtracting a constant from each score or dividing each score by a constant may be applied here (refer to section 6.4 in Chapter 6). When changing raw scores to z scores, $\bar{X}$ is the constant subtracted from each raw score and the constant divisor for each deviation score (x_i) is s_x. The manipulations that transform any set of raw scores to z scores always will result in a z score distribution with a mean ($\bar{z}$) of zero and a standard deviation (s_z) of one. Table 7.1 illustrates these properties using a sample of Numbers posttest scores and Classification posttest scores for 15 children from the "Sesame Street" data base. The raw scores and equivalent z scores are presented for each individual. The raw score means and standard deviations are given at the bottom of the table. The two sets of raw scores represent different measurement continuums with vastly different means and standard deviations. The values of $\bar{X}$ and s for the Numbers scores are 35.73 and 13.14, respectively. The values of $\bar{X}$ and s for the Classification Skills scores are 17.40 and 4.60, respectively. Check the two z scores for individual 218 to make sure you know how they were calculated.[2]

---

[1] On the pretest for a raw score of 20:

$$z = \frac{20 - 17}{3} = \frac{+3}{3} = +1.00$$

On the posttest for a raw score of 54:

$$z = \frac{54 - 49}{5} = \frac{+5}{5} = +1.00$$

The two raw scores reflect identical positions in their respective distributions. While the actual raw scores are different, both are located one standard deviation above their respective class means.

[2] Individual 218 had raw scores of 35 and 21 on the respective variables. The z score equivalencies were calculated as follows:

$$\text{For } X_{218},\ z = \frac{35 - 35.73}{13.14} = -.056$$

$$\text{For } Y_{218},\ z = \frac{21 - 17.40}{4.60} = .783$$

TABLE 7.1
Numbers and Classification Skills posttest raw scores and z scores for 15 "Sesame Street" viewers.

	Numbers Posttest		Classification Skills Posttest	
Individual Number	Raw Score (X)	z Score	Raw Score (Y)	z Score
005	54	1.390	21	.783
007	44	.629	15	−.522
022	45	.705	19	.348
023	19	−1.273	14	−.739
052	50	1.086	23	1.217
063	45	.705	22	1.000
090	50	1.086	23	1.217
110	45	.705	18	.130
112	21	−1.121	16	−.304
151	30	−.436	17	−.087
177	27	−.664	10	−1.609
180	16	−1.502	11	−1.391
216	17	−1.425	10	−1.609
218	35	−.056	21	.783
224	38	.173	21	.783
$n = 15$		$n = 15$	$n = 15$	$n = 15$
$\Sigma X = 536$		$\Sigma z = .00$	$\Sigma Y = 261$	$\Sigma z = .00$
$\Sigma X^2 = 21{,}572$		$\Sigma z^2 = 14.00$	$\Sigma Y^2 = 4837$	$\Sigma z^2 = 14.00$
$\overline{X} = 35.73$		$\overline{z} = .00$	$\overline{Y} = 17.40$	$\overline{z} = .00$
$s_x = 13.14$		$s_z = 1.00$	$s_y = 4.60$	$s_z = 1.00$

The z score means and standard deviations for the two variables also appear at the bottom of Table 7.1. Note what happens when you sum the z scores (Σz_i) and the squared z scores (Σz_i^2). The sum of the z scores will always equal zero; thus, the mean z score is zero. The sum of the squared z scores in sample data will always equal $n - 1$ ($\Sigma z^2 = n - 1$) within rounding error. In Table 7.1 there are 15 scores for each variable. The $\Sigma z^2 = 14.00$ or 14 for the Numbers scores and $\Sigma z^2 = 14.00$ or 14 for the Classification Skills scores. Substituting $\Sigma z = 0$ and $\Sigma z^2 = 14$ into the variance and standard deviation formulas,

$$s^2 = \frac{\Sigma X^2 - \dfrac{(\Sigma X)^2}{n}}{n - 1} = \frac{\Sigma z^2 - \dfrac{(\Sigma z)^2}{n}}{n - 1}$$

$$s_z^2 = \frac{14 - \dfrac{0^2}{15}}{14} = \frac{14}{14} = 1.00$$

$$s_z = \sqrt{1.00} = 1.00$$

The transformation of both variables to z scores resulted in standardized distributions with equal means ($\bar{z} = 0$) and standard deviations ($s_z = 1.00$).

Using the z scores for interpretation, it appears that individual 007 had the most inconsistent performance across the two variables. The Numbers posttest performance was considerably above the group mean ($z = .629$), while the Classification Skills score was below the group mean performance level ($z = -.522$). Which individuals appear to be performing at approximately the same level on the two tests?[3]

Raw scores are commonly converted to z scores in both statistics and measurement. Converting to z scores allows direct comparison of performance. Once raw scores have been converted, the interpretations are no longer exclusively tied to the properties of the original measurement scale. When converting raw scores to z scores, the transformed scores reflect both direction (the sign of the z score) and distance (magnitude of the z score) from the mean. For this reason it is important when calculating z scores to carry along the appropriate sign (+ or −) and to work all calculations to at least two decimal places.

7.3 OTHER SCALES BASED ON STANDARD SCORES

Other standardized scales exist, but the majority are formed by making a second transformation of z scores to a distribution with a particular mean and standard deviation. A simple computational procedure transforms raw scores to any specified standard score scale.

- Step 1 Obtain the raw score mean and standard deviation.
- Step 2 Transform a given raw score (X) to a z score using the formula:

$$z = \frac{X - \bar{X}}{s_x}$$

[3]The z scores for the following individuals indicate that their X and Y scores are approximately at the same distribution location.

ID	X z score	Y z score
052	1.086	1.217
090	1.086	1.217
180	−1.502	−1.391
216	−1.425	−1.609

- Step 3 Transform the z score from Step 2 using the following formula:

$$X_{new} = \overline{X}_{new} + s_{new}(z) \quad (7.2)$$

where

X_{new} is the equivalent new standard score

$\overline{X}_{new}$ is the mean of the new standard score distribution

s_{new} is the standard deviation of the new standard score distribution

To illustrate, consider a raw score of 29 in a distribution with a mean of 32 and a standard deviation of 6 (step 1). What would the standard score equivalent be in a standard score distribution with a mean of 50 and standard deviation of 10?[4] With a mean of 100 and standard deviation of 20?[5]

Many persons do not like to work with negative numbers or decimals—both are characteristics of z scores. To eliminate both properties, z scores are often transformed to a distribution with a mean of 50 and a standard deviation of 10. A score of 60 in such a distribution is one standard deviation unit *above* the mean. A score of 34 is 1.6 standard deviation units *below* the mean ($z = -1.6$) in the new distribution. Other examples of standard score scales are the Wechsler Intelligence IQ scale and the Graduate Record Examination subscales. The Wechsler IQ scale has a mean of 100 and a standard deviation of 15 not because the mean raw score is 100 on the test, but because the raw scores have been transformed to a standardized scale with a mean of 100 and a standard deviation of 15. For the Graduate Record Examination (GRE) subscales, raw scores are transformed to a standardized scale with a mean of 500 and a standard deviation of 100.

Remember: Any standard score indicates how many standard deviation units a score is from the mean of the distribution. Conceptually, the interpretation can always be compared with the basic z score distribution. What does a GRE verbal score of 560 or an IQ score of 109 mean? Each has a z score equivalent of .60 and indicates that the standard score is .6 of a standard deviation above the mean of the distribution for the characteristic being measured. In footnotes 4 and 5, when the raw score mean was specified as 32 with a standard deviation of 6 we found that the z score for a raw score of 29 was −.50; the standard score in a distribution with a mean of 50 and standard deviation of 10 was 45, and the standard score for a distribution with mean of 100 and standard deviation of 20 was 90. The scores 29, −.50, 45, and 90 are equivalent. Each score is one-half a standard deviation below the mean of its

---

[4] *Step 1* $\overline{X} = 32; s = 6$

Step 2 $z = \frac{29 - 32}{6} = -.50$

Step 3 $X_{new} = 50 + 10(-.50) = 45$

[5] *Step 3* $X_{new} = 100 + 20(-.50) = 90$

distribution. Raw scores with identical z scores are equivalent relative to their respective distribution means and standard deviations. As standard score distributions become familiar, the individual standard score values take on meaning, but the interpretation is always based on z scores. An IQ score of 70 may not be particularly desirable. Why? Because you now know that it is two standard deviation units *below* the mean and represents a low performance level when compared to others in the reference group.

Transforming scores in a distribution to z scores standardizes the interpretations of the relative position of scores by converting the original measurement scales to a standard scale with a mean of zero ($\bar{z} = 0$) and standard deviation of one ($s_z = 1$). While z score transformation standardizes the central location and variabilities of score distributions, their shapes are *not* changed. The distribution of transformed z scores will have exactly the same shape as the distribution of original scores. If two raw score distributions differ in shape, the distributions of their equivalent z scores also will differ in shape, but will have identical means of zero and standard deviations of one. Even though two scores in separate distributions can be at equivalent positions in their respective distributions based on their z scores, they will not have the same percentile rank equivalencies if the shapes of the distributions differ. The proportion of individuals in the distributions with z scores less than or equal to some fixed z score value will differ depending on how discrepant the distribution shapes are. For this reason it is advantageous when comparing z score equivalencies across distributions to have distributions with the same shape. When the distribution shapes are identical, equivalent z scores not only represent raw scores that are the same relative distance from their respective means: They also have the same percentile ranks. Thus, the same proportion of individuals will fall at or below the z score in each distribution.

This point is important to inferential procedures in statistics. Assume that we know that different variables have the same distribution shapes. We can use the shape plus the comparability of score values resulting from the z score transformation to define and compare expectations about the frequency with which scores will occur between equivalent points in the distributions. If, in addition, we can define a specific distribution shape mathematically, we can then predict the frequency of score values expected within z score intervals without examining an actual raw score distribution. Distribution shapes that can be defined by a mathematical function are known as theoretical distribution shapes. They identify mathematical models that provide a basis for expectations about frequency of occurrence of results from sample data. One such mathematical model that helps approximate the occurrence of events in practice is the theoretical normal distribution. The expectations formed from the theoretical normal distribution are basic to statistical reasoning. The standardization resulting from consistent interpretation of z scores plus the consistent expectations based on the theoretical normal distribution permit widespread application of this model. For this reason the remainder of the chapter introduces the normal curve as a distribution shape and describes the utility of z scores in the normal distribution. Chapter 11 elaborates on the use of the theoretical normal distribution as a mathematical model in the inferential statistical process.

7.4 THE NORMAL CURVE

A frequency polygon displaying normally distributed scores has particular characteristics. Visually, it resembles a bell-shaped curve (Figure 7.1A). The width, height, and location of the center of the distribution is a function of the mean and variance of the scores. The line that forms the curve is a function of the frequency at each point of the frequency distribution. A normal curve should be thought of as a smoothed frequency polygon. The "normal" shape of the polygon is a function of the distribution of scores. Keep this idea in mind throughout the discussion of the normal curve.

Figure 7.1B is the frequency polygon of children's ages recorded in months for the 240 observations included in the "Sesame Street" data base (Appendix A). Figure 7.1C is the frequency polygon for these 240 children on the Forms posttest. Compare the shapes of these two frequency polygons with Figure 7.1A. Notice that the age variable more closely approximates the shape of a normal distribution. How normally distributed is the variable? The normal curve functions principally as a statistical model to which things can be compared and consequently interpreted.

The normal curve depicts a mathematical model. Its specific shape has very definite characteristics. For this mathematical distribution to exist, it is not necessary for it to exist in nature. Research has shown, however, that many of the phenomena we examine are in fact distributed normally. The theoretical normal curve, therefore, has practical utility as a model representing real events. Knowing its theoretical properties, we can then determine if some characteristic takes on its form in a group, or if we know that a characteristic is distributed normally, we can use normal curve theory to understand or predict events.

Normal curve theory has evolved from the study of errors. An error may be thought of as the difference between any observed event and the correct outcome. In sampling error, if a case is sampled from a population and this case is used to estimate the mean of the population, the error is the extent to which the score differs from the population mean, $e = (X - \mu)$. Similarly, errors in prediction occur (see Chapter 9) when a predicted score does not equal a person's obtained score. A vast number of other kinds of errors exist, all representing the degree of inaccuracy of the statistic, measurement, or research process. Error theory has shown that errors, in general, are random and that

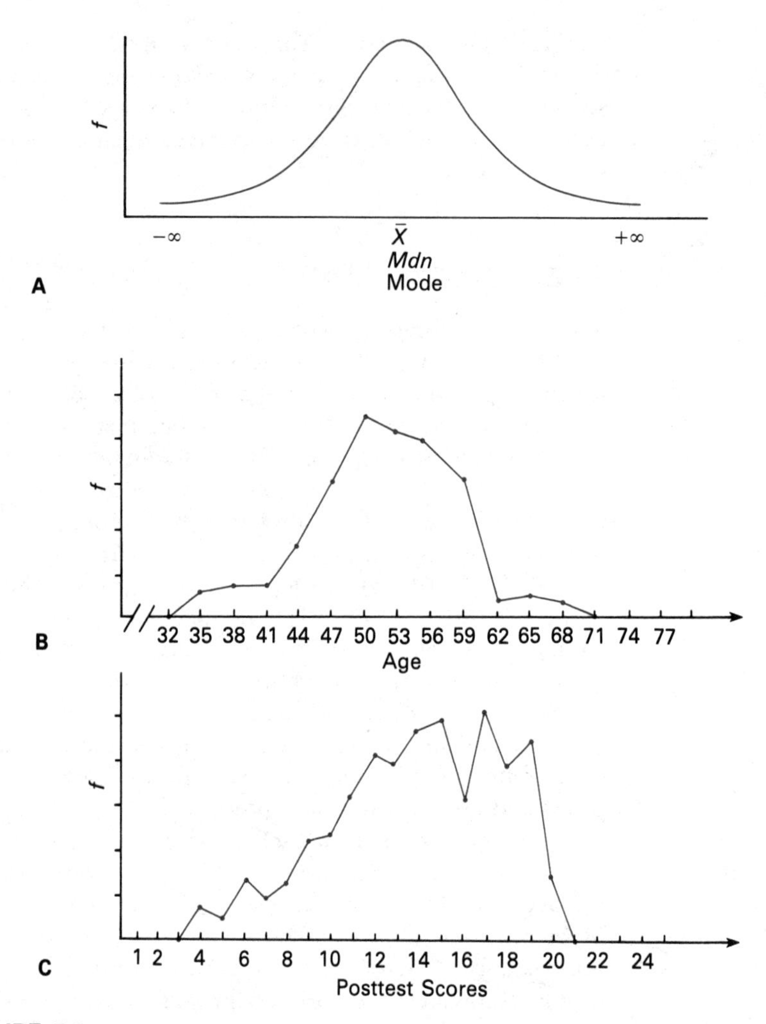

FIGURE 7.1
Normal and nonnormal curve illustrations. **A,** normal curve; **B,** frequency polygon of age; **C,** frequency polygon of Forms posttest.

they are distributed *normally.* This distribution property of errors allows us to examine the variability of errors and the expected frequency of occurrence of specific errors.

In reality, sample data cannot follow an exact normal curve. Limited sample size, sampling error, measurement error, and other factors often result in a distribution shape that only tends toward normality. For this reason, the sample data distribution shape usually is not of great importance to application of theoretical statistics. Rather, the theoretical underlying distribution shape is important. The normal curve serves as a theoretical model, a basis for expectations and probability statements about potential outcomes. For example, interpretation of the standard deviation is facilitated if scores are normally distributed (section 6.4). Based on the characteristics of a normal curve, we know, for example, that approximately 68% of the data points will fall from one standard deviation below the mean to one standard deviation above the mean. The normal curve allows us to form expectations about the frequency with which scores in a particular range will occur. These expectations can then be translated into probability statements.

7.5 THE NORMAL CURVE AS A MODEL

A smoothed frequency polygon depicting a normal distribution is defined by the following formula. It represents the normal curve as a mathematical model defined in terms of the population parameters μ and σ.

$$y = \frac{N}{\sigma\sqrt{2\pi}} e^{-(X-\mu)^2/2\sigma^2} \tag{7.3}$$

Here, y is the ordinate, representing a point on the line that forms the normal curve. It equals the height of the curve over a specific value of X on the horizontal axis. The value of y is a function of four terms: X, the raw score; μ, the mean of the distribution; σ, the standard deviation of the distribution; and N, the population size. Consider a population distribution with a mean of 50, a standard deviation of 10 and a size of 1000 observations. Figure 7.2 illustrates

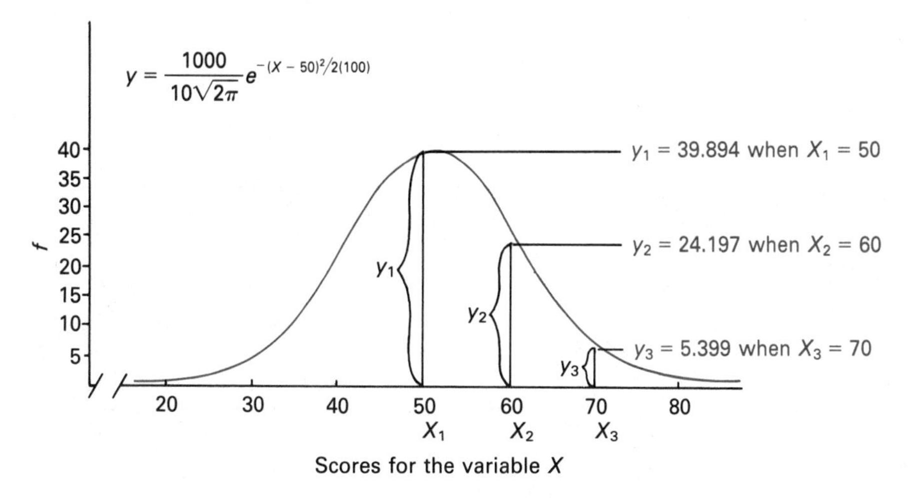

FIGURE 7.2
Values of y in the normal curve when μ = 50, σ = 10, and n = 1000.

that the more X deviates from μ, the smaller the value of y, or the height of the curve is less at X_3 than at X_2. The maximum height of the curve occurs when X is equal to the mean, μ. If you think of the values of y as representing frequencies in a frequency polygon, this makes sense. Scores that deviate further from the mean of a distribution have lower frequencies and the distribution tapers as scores move toward the extremes.

The degree to which the distribution spreads out is determined by the value of σ, the standard deviation. Because μ and σ can vary from one set of raw scores to another and each set of values determines a specific curve, there is no single, fixed normal curve. Rather, the formula represents a family of normal curves, each specific curve having the bell-shaped characteristics, but differing in variability and location depending on σ and μ, respectively. Figure 7.3 illustrates these differences in normal curves.

Although normal curves exist as a family, each curve can be standardized (transformed) to a single, standard curve by converting the raw score scales to the z score scale. This conversion results in a fixed normal curve with known properties called the *unit normal curve.* Because the score scale values for the unit normal curve are z scores, it has a mean of zero ($\mu = 0$) and a standard deviation of one ($\sigma = 1$). The general formula 7.3 can be easily changed to illustrate the transformation. Consider the normal curve formula exponent for e:

$$-\frac{(X - \mu)^2}{2\sigma^2}$$

The formula for calculating a z score is $(X - \overline{X})/s_x$. In population values,

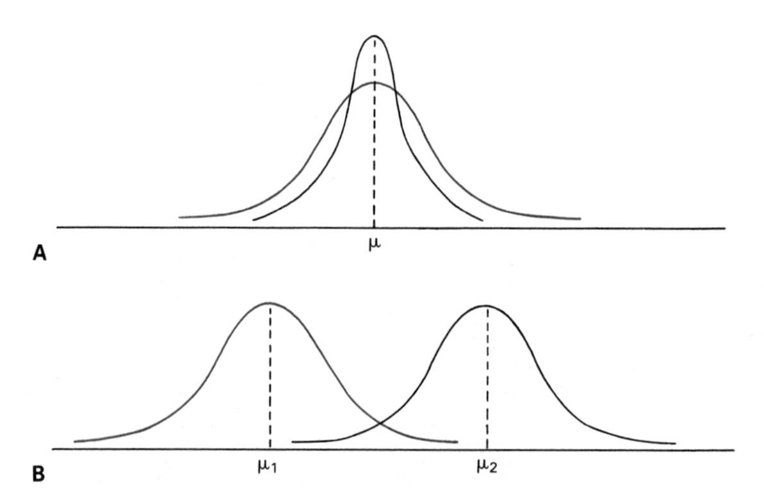

FIGURE 7.3
Normal curve with different values of μ and σ. **A,** same central location, different standard deviations; **B,** different central location, same standard deviations.

$z = (X - \mu)/\sigma$, implying that $z^2 = (X - \mu)^2/\sigma^2$. Once scores are transformed to z scores, $\sigma_z = 1$ and the general formula for the normal curve becomes the formula for the unit normal curve with fixed characteristics.

$$y = \frac{n}{\sigma\sqrt{2\pi}} e^{-(X-\mu)^2/2\sigma^2} = \frac{n}{\sqrt{2\pi}} e^{-z^2/2}$$

The importance of the standardized formula is that it represents a single curve with known, constant values. This unit normal curve is what the normal curve as a single distribution refers to.

In Chapter 4, the normal curve was characterized as *symmetrical, unimodal,* and *mesokurtic.* The theoretical normal curve is also asymptotic with a theoretical distribution scale ranging from $-\infty$ to $+\infty$. In the unit normal curve, however, almost all (99.7%) of the transformed z scores will fall in the range between $z = -3.00$ and $z = 3.00$. Frequency polygons from real data will never exactly fit these characteristics, but some variables will tend to approximate normality, particularly as sample size increases. Whether sample data are distributed as an exact normal curve is only important descriptively. More important is whether a population or theoretical distribution is normal.

7.6 AREAS UNDER THE NORMAL CURVE

The normal curve should be viewed as a smoothed frequency polygon representing the shape of the distribution for a set of scores. As a frequency polygon, the height of the curve over an interval represents the frequency or number of scores we would expect to lie within the interval. Identifying this expected frequency is the primary value of the normal curve as a model distribution. If scores are approximately normally distributed, the theoretical unit normal curve can indicate the expected number or proportion of scores that would fall above or below a particular scale point or in an interval range. A given area under the unit normal curve indicates the proportion of scores expected to lie in the scale interval bounding it. Using the curve we can answer such questions as "How many scores in a normal distribution are expected to have z scores

greater than 1.54?"
less than $-.40$?"
between the mean and a z score of .75?"
in the interval range between z scores of -1.3 and .62?"
beyond the z scores of -1.96 and $+1.96$?"

Before we answer these questions, we must review certain characteristics of the unit normal curve. First, the total area under the unit normal curve should be considered as just that, *one unit* of area. Any portion of this unit area will be expressed as a proportion or percentage. The limits are from .00 to 1.00 or 0% to 100%. The area under the curve represents the proportion of scores from a frequency distribution expected to fall within that interval, defined by z scores on the scale continuum. Because the curve is symmetrical, the left half is identical to the right, each accounting for 50% of the area under the curve. A line drawn vertically from the mean of the distribution ($\bar{z} = 0$) divides the area under the curve exactly in half.

Because the unit normal curve is a fixed, standard curve, the areas under it associated with specific z scores are constant. These areas as proportions of the unit normal curve are shown in Appendix B, Table B.2. Area values for only positive z scores or the right half of the curve are tabled. Because the curve is symmetrical, negative z scores represent the same areas as positive z scores. It is not possible to find the area under the curve for a discrete scale point or the proportion of persons who would be expected to get a discrete z score value. The scale is assumed to be continuous, with each z score actually representing an interval, and only areas for *intervals* on the scale can be obtained. A portion of Table B.2 is reproduced here for illustration.

z or $-z$	a	b
—	—	—
—	—	—
—	—	—
1.00	.3413	.1587
—	—	—
—	—	—

For each z score value, the area of the curve between the mean and that z score is given in column a. For the z score of 1.00, the proportion in column a of .3413 is the proportion of the area between the mean and a z score of 1.00. This area is illustrated by the shaded portion in Figure 7.4.

Because of the normal curve's symmetry, Table B.2 provides the area in the normal curve between the mean and a z score of -1.00 also (use the information for $z = 1.00$). This area is .3413 and is illustrated in Figure 7.5.

The information in column a of Table B.2 can also be used to find the area between a negative z score and a positive z score. Keeping the example simple, what is the proportional area between z scores of -1.00 and $+1.00$? The areas obtained in the two preceding examples can be added together to get the

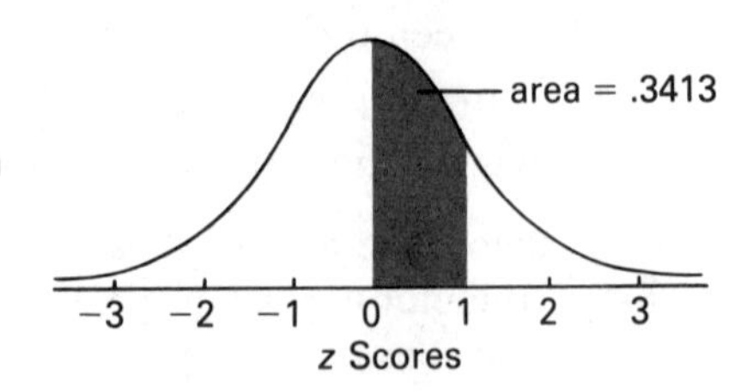

FIGURE 7.4 Proportion of area under the normal curve between the mean ($\bar{z} = 0$) and $z = 1$.

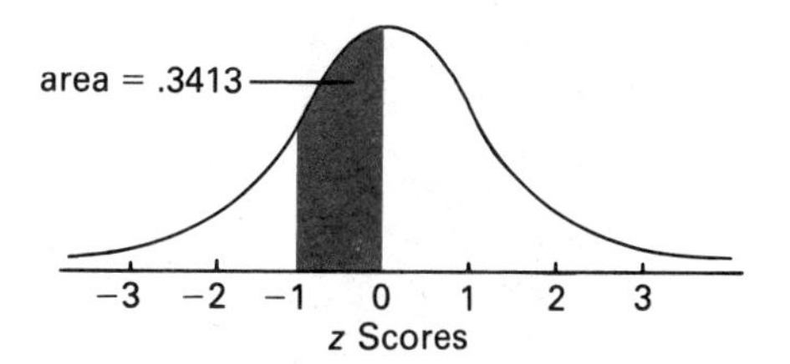

FIGURE 7.5
Proportion of area under the normal curve between the mean ($\bar{z} = 0$) and $z = -1$.

answer: .3413 + .3413 = .6826. Does this value seem familiar? It should! In Chapter 6 we saw that in a normally distributed data set, approximately 68% of the scores will fall within the range of −1.00 standard deviation unit and +1.00 standard deviation unit from the mean. The z scores of −1.00 and 1.00 represent $\pm 1\sigma$ from the mean. The proportion, .6826, is approximately 68%.

To change the problem slightly, what is the area between a z score of −2.14 and a z score of .85? The area between the mean and a z score of −2.14 is .4838. The area between the mean and z of .85 is .3023. Adding the two areas, .4838 + .3023 = .7861. Approximately 79% of the scores in a normal distribution would be expected to lie in the interval bounded by z scores of −2.14 and .85. Figure 7.6 illustrates the calculations.

Column b of Table B.2 indicates the proportion of the curve above a positive z score or below a negative z score. For the z score of 1.00 above, the tabled value of .1587 in column b indicates that about 16% of the scores in a normal distribution should have z score equivalents greater than 1.00. Similarly, .1587 is the proportion of the curve below a z score of −1.00. The two normal curves in Figure 7.7 illustrate these areas. What proportion of a normally distributed set of scores would be expected to have z scores greater than .35? less than −1.96?[6]

Many areas of the normal curve are not given directly, but may be found simply by adding or subtracting proportions. Two characteristics of the normal curve are useful in many calculations: (a) half the curve represents an area of .5000, and (b) the whole curve represents an area of 1.0000. To illustrate the use of these characteristics, suppose we want to find the area of the curve less

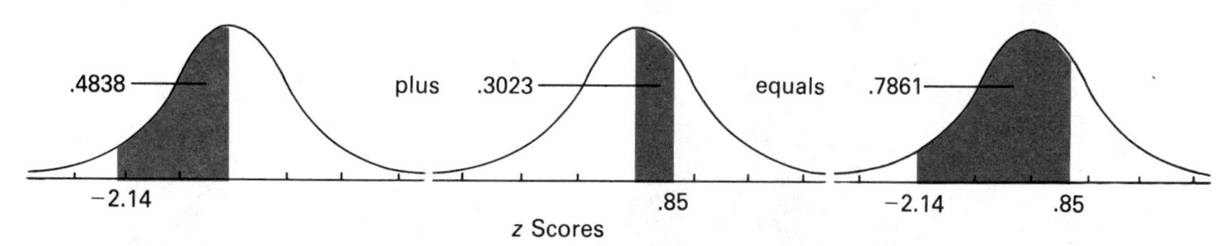

FIGURE 7.6
Adding areas under the normal curve between positive and negative z scores.

---

[6]Area greater than a z score of .35 is .3632; area less than a z score of −1.96 is .0250.

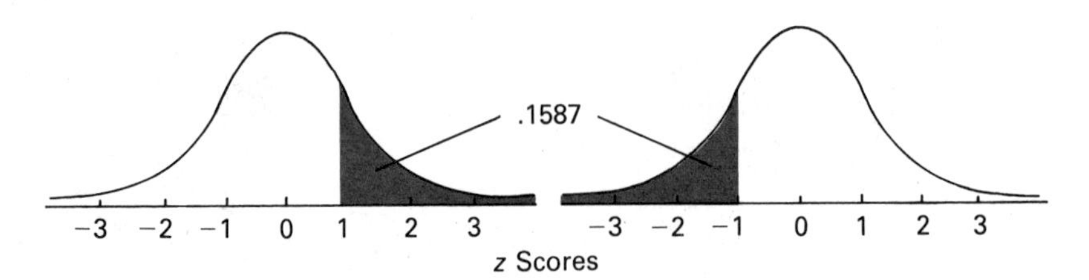

FIGURE 7.7
Areas above z = 1 and below z = −1.

than z = 1.00. The area between the mean and a z score of 1.00 was found to be .3413. The area below the mean is .5000. The area of the curve less than z = 1.00 is .5000 + .3413 = .8413. This same value could have been found by subtracting the proportion in Table B.2, column b from 1.0000. The area of the curve beyond z = 1.00 is .1587. The area below z = 1.00 is the total area, 1.000, minus .1587, which equals .8413. These concepts are illustrated in Figure 7.8.

The same approach applies in finding an area of the curve greater than a

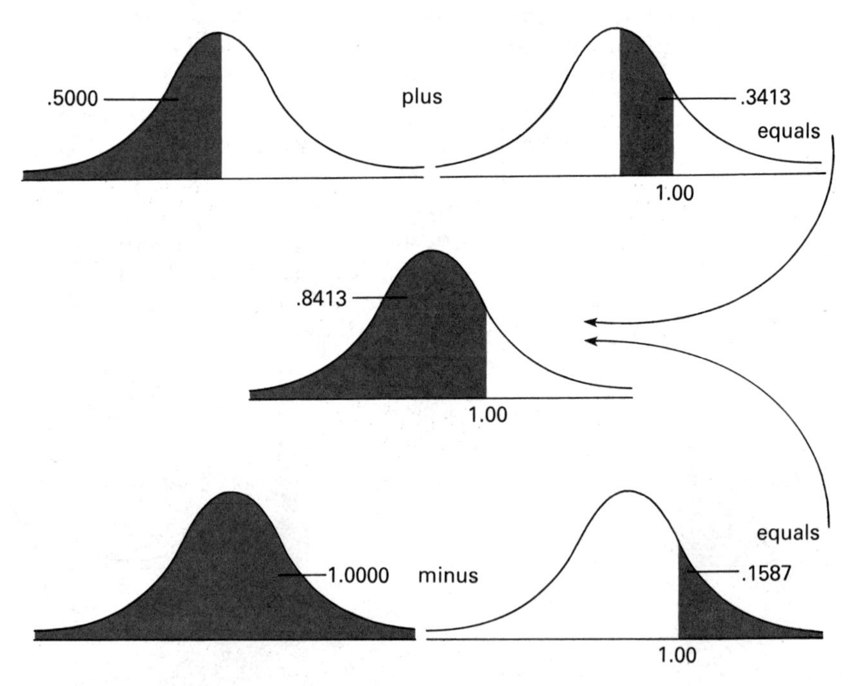

FIGURE 7.8
Adding and subtracting proportions to find the area under a normal curve.

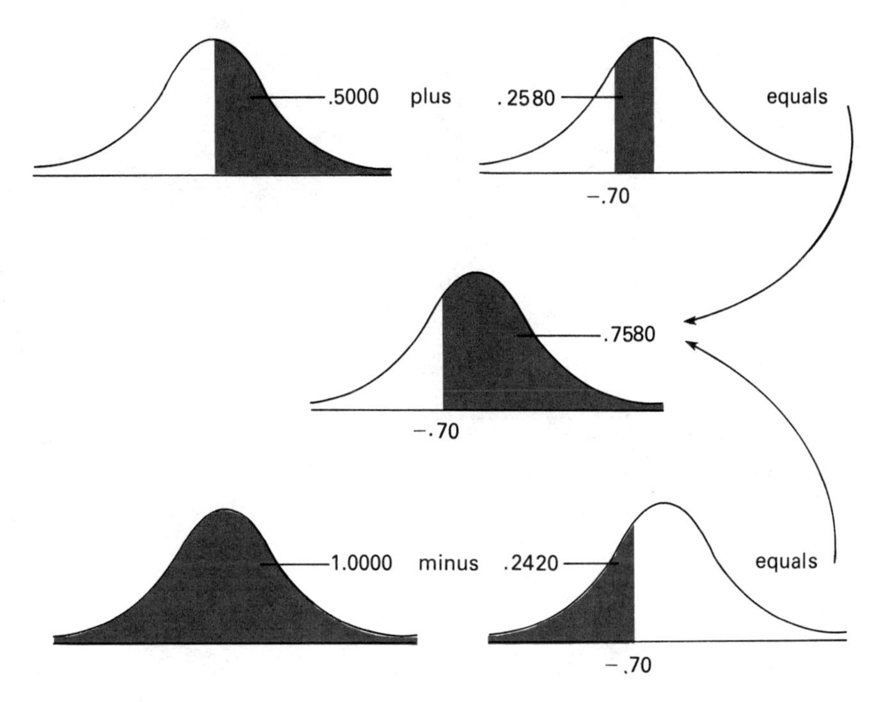

FIGURE 7.9
Finding the area of a curve greater than a negative z score.

negative z score. Let's find the area greater than $z = -.70$. Both approaches are illustrated in Figure 7.9. The proportion of scores in a normal distribution expected to have z scores greater than $-.70$ is .7580.

With the addition or subtraction of proportions, any z score interval area can be obtained from the information in Table B.2. The shaded areas in Figure 7.10A–H provide different possibilities for problems. You should be able to compute the areas in A–F in this figure using the information provided so far. Try it.[7] Figure 7.10G provides a different type of problem. We need to find the area between the mean and each of the z scores, then subtract.

---

[7]A: Area between $z = .4$ and the mean is .1554.
B: Area less than $z = -1.8$ is the same as the area greater than $z = 1.8$. This area is .0359.
C: Area less than $z = +1.5$ is $.5000 + .4332 = .9332$.
D: Area between $z = .7$ and the mean is .2580.
Area between $z = -.3$ and the mean is .1179.
Total area $= .2580 + .1179 = .3759$.
E: Area outside the interval $z = \pm 1.96$ is 2(.0250) or .0500. This indicates that 95% of the normal distribution lies between ± 1.96 standard deviation units. The $\pm 2.00s$ used previously was only approximate. The $\pm 1.96s$ is exact.
F: Area beyond $z = 1.1$ is .1357. Area less than $z = -2.4$ is .0082.
Total area $= .1357 + .0082 = .1439$.

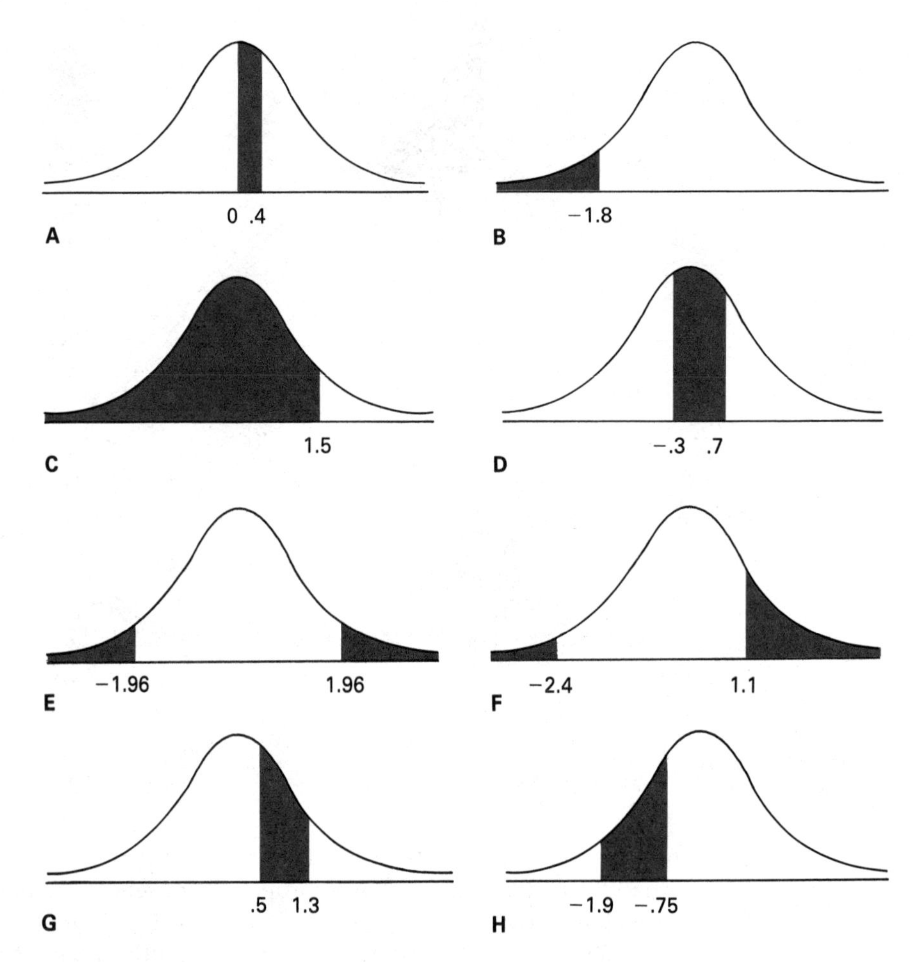

FIGURE 7.10
Other normal curve areas.

$$
\begin{array}{rl}
 & \text{area between the mean and } z = 1.30 \text{ is } .4032 \\
- & \underline{\text{area between the mean and } z = \ \ .50 \text{ is } .1915} \\
= & \text{area between } z = 1.30 \text{ and } .50 \qquad \text{is } .2117
\end{array}
$$

Often, it is convenient to find the area of the unshaded portion and then subtract from 1.000 to get the area of the shaded portion. Using this approach for Figure 7.10G provides the following:

area less than $z = .50$ is $.5000 + .1915 = .6915$
area beyond $z = 1.30$ is $.0968$
area of unshaded portion is $.6915 + .0968 = .7883$
area of shaded portion is $1.0000 - .7883 = .2117$

Find the shaded area in Figure 7.10H.[8]

[8]The area is .1979.

7.7 APPLYING NORMAL CURVE THEORY TO RAW DATA

Section 7.6 emphasized that the area under the unit normal curve represents the proportion of raw scores that we expect to fall within the specified scale range. The scale values we use with the unit normal curve are z scores. If we know the mean and standard deviation of the raw scores, then we can easily transform the raw score distribution to the standard normal curve distribution and back to the raw score distribution. Consider the following situation.

> Marty had a raw score of 44 on an achievement test in mathematics. The test was given to 75 students and the scores approximated a normal distribution in shape. If the mean was 40 and the standard deviation was 5, what proportion of the students performed better than Marty?

In analyzing this problem, recognize that the solution requires a proportion, the proportion who performed better than Marty. To obtain this proportion we refer to Table B.2 of the unit normal curve, which requires a z score. Therefore, we must transform the raw score information to a z score before proceeding. What is the z score equivalent for Marty's raw score of 44?

$$z = \frac{X - \overline{X}}{s}$$

$$= \frac{44 - 40}{5}$$

$$= \frac{4}{5} = .80$$

Marty's raw score is .8 of the raw score standard deviation above the mean. Now we can ask "Given a normally distributed set of scores (a very important point), what is the area (proportion) under the curve beyond a z score of .80?" Using Table B.2, we find that .2119 of the area under the curve lies above a z score of +.80. How many students does this represent? (.2119)(75) = 15.89. Thus, we would expect approximately 15 or 16 students out of the 75 to have scores higher than Marty's raw score. Assume that Beth had a raw score of 34. First, how many students had lower scores than Beth?[9] Then, how

[9]Beth's z score equivalent for a raw score of 34 is

$$z = \frac{X - \overline{X}}{s}$$

$$= \frac{34 - 40}{5}$$

$$= \frac{-6}{5} = -1.20$$

Then from the table of the normal curve, the proportion of the normal curve *below* a z of -1.20 is .1151. We find this proportion in column b, area beyond z, for a z of 1.20. Now, the proportion of the group having scores less than Beth is (75)(.1151) = 8.63 or approximately nine students.

many students would be expected to score in the range between Marty and Beth?[10]

Occasionally we must work from a proportion of the unit normal curve back to the raw score distribution. For example, assume a clerical aptitude test screens applicants for a job. Eighty applicants are expected. Time and other resources permit interviewing only 10 applicants following testing. What cut-off score should we use in deciding whether to invite an applicant back? From past data, we know that scores on the clerical aptitude test are approximately normally distributed and that for a typical group of applicants, the mean is 27 and the standard deviation is 4. We begin to solve this problem by analyzing the information given. In previous applications, the raw score was given. In this problem, the raw score is to be found. Since we can assume that the scores on the clerical aptitude test are normally distributed, we can use the unit normal curve. We want to categorize the 80 applicants' scores into the 70 below some cutoff point and the 10 above it. What proportions do 70 scores and 10 scores represent? 70/80 = .8750 and 10/80 = .1250, respectively. We want to identify a point in the raw score distribution where 12.5% of the cases fall above it and 87.5% fall below it. Once we know the proportion, we can identify the z score equivalent for the cut-off score point. Again, we can use Table B.2 to assist us. Now, however, we look at the table for a proportion, rather than a z score. We want to identify a particular standard score given the proportion of the curve it represents. For this problem we want to find the z score above which .125 of the curve appears. Using column b, we find that .1251 of the normal curve falls above a z of +1.15 in a normally distributed set of scores. Having worked from the proportion, we now know that 1.15 is the z score equivalent for the raw cutoff score. The mean and standard deviation are specified by the problem so we can use the z score formula to solve for the desired raw score cutoff. To facilitate the computation, an algebraic manipulation of $z = (X - \overline{X})/s_x$ results in the following raw score solution:

$$X = \overline{X} + zs_x \tag{7.4}$$

This formula produces the raw score for a specified z score given the raw

---

[10]Beth's z score equivalent is −1.2, while Marty's was .80. The number of scores in the group (n = 75) between Beth and Marty is represented by the shaded area:

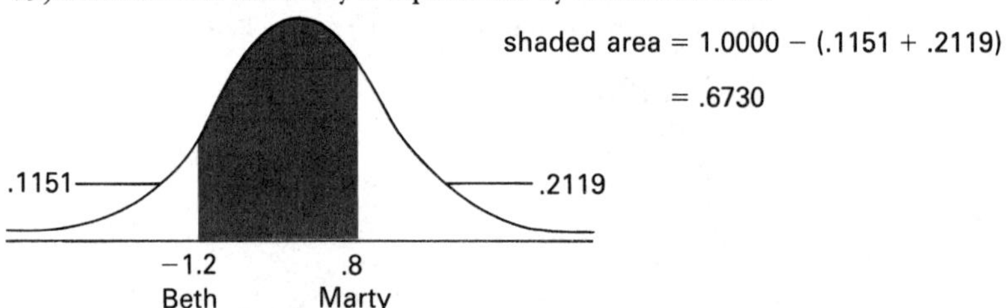

We would expect approximately 50 or 51 (i.e., 75 × .6730) students to score between Beth and Marty.

score distribution mean and standard deviation. Substituting the known values in 7.4 yields

$$X = 27 + 1.15(4)$$
$$= 31.60$$

Thus, a raw score of either 31 or 32 on the clerical aptitude test will identify approximately 10 persons for interviewing from a typical group of applicants.

Let's try this procedure once more. Assume that a mental health agency provides service to needy members of a particular community. Services can be offered to any resident whose gross income is below that of 40% of those in the community. Gross income in the community is normally distributed with a mean of $14,350 and a standard deviation of $1,850. Given these data, what gross income qualifies someone to obtain services from the agency? Here again, we can find the proportion because the scores, in this case gross income, are normally distributed. The task is to find the gross income. The problem is illustrated in Figure 7.11.

To find the gross income figure that will serve as a cutoff point, we need to identify the z score in the normal curve for which the proportional area is .40 below and .60 above. We need to enter the table for the normal curve with a proportion. Illustrated in Figure 7.12 are two ways to conceptualize this problem in terms of the information available in the normal curve table.

Part A in Figure 7.12 parallels the solution used in the preceding problem. In this illustration the z score equivalent is determined using column b of the normal curve table, looking for a specific proportion. We find the z score for the proportion and use it to solve the problem. We can also use column a, area of the normal curve, between the mean and z. This is illustrated in B of Figure 7.12. If we want to identify the lowest 40% of the distribution, then 10% of the distribution falls between that value and the mean. We then look at column a to locate .10; the z score associated with it is then used in the solution to the problem. Obviously, one may use either approach.

For the problem at hand, what is the maximum total income for the lowest 40% in the community? Using the normal curve table and searching column b for the proportion closest to .40 identifies the value of .4013, which we can use as an approximate value for .40. Notice as you read across this row to obtain the z score equivalent that the entry in column a is .0987, the compli-

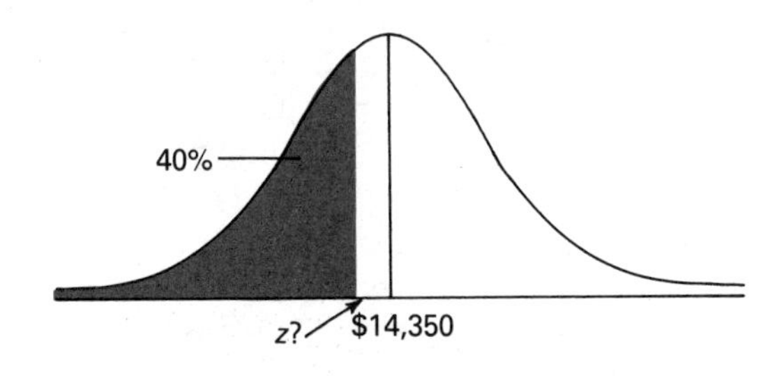

FIGURE 7.11
Area of the normal curve representing the lowest 40% incomes.

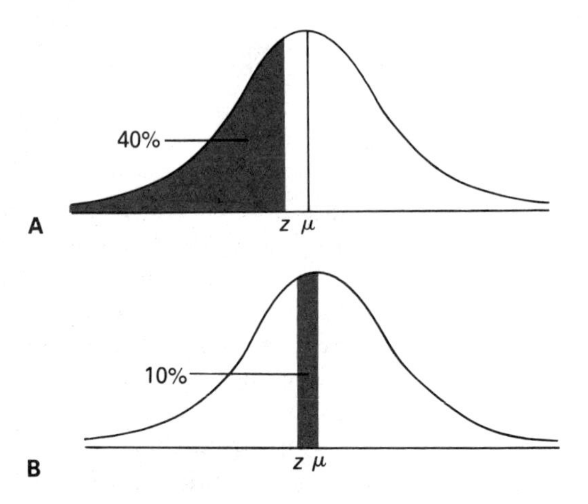

FIGURE 7.12
Finding z score values when only the area is known. **A,** a z score relative to the entire normal curve; **B,** a z score relative to the mean.

mentary value to that in column b. We would have located this value had we chosen to use column a for this problem. The z score that identifies the point below which 40% of the observations in the normal curve falls is $-.25$. Note that this value is *negative* because the area is *below* the mean of the distribution. The z score can now be used with formula 7.4 to solve the problem.

$$\begin{aligned} X &= \overline{X} + zs_x \\ &= \$14{,}350 + (-.25)(\$1{,}850) \\ &= \$14{,}350 + (-\$462.50) \\ &= \$13{,}887.50 \end{aligned}$$

Thus, any person whose total income is below \$13,887.50 qualifies for services provided by the agency.

Now try a problem on your own. From past administrations, a course final examination is known to have a mean of 85 and a standard deviation of 7. The instructor decides that the highest 10% of the scores on the test will receive grades of A, the next highest 20% will get grades of B, the next highest 40% will be given grades of C, the next highest 25% will be assigned grades of D, and the lowest 5% will be given grades of F. Assuming that test scores are normally distributed, what are the raw score ranges for each grade? Try this problem. It may take some time, but it should be a useful experience.

To begin to solve this problem recognize the basic assumption: Test scores are normally distributed. Now conceptualize what the instructor wants. This is important. Figure 7.13 shows the grade categories as proportions of the normal curve. Until you become very familiar with the normal curve, it would probably help you to draw out the conditions. The values needed are the raw scores that separate each grade. To begin, persons getting a grade of A will receive scores higher than 90% of their classmates. At this point the problem can be stated: What is the z score in the normal distribution above which 10%

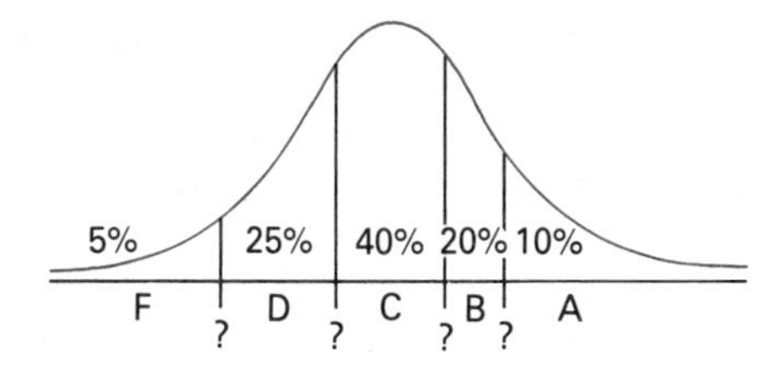

FIGURE 7.13
Final examination score ranges.

of the distribution lies? Using column b of the normal curve table, the answer is +1.28. Now we can find the raw score for this point, applying formula 7.4. Remember that the test mean is 85 and its standard deviation is 7.

$$\begin{aligned} X &= \overline{X} + zs_x \\ &= 85 + (1.28)(7) \\ &= 85 + (8.96) = 93.96 \end{aligned}$$

Thus, anyone whose test score is 94 or above receives a grade of A. This solution also provides the upper bound for a grade of B (i.e., 93). Now we need to figure the lower bound or lowest score eligible for a grade of B. Since we know the score above which the A grades are given, we use the *next* 20% of the distribution to find the interval receiving B grades. A grade of B is assigned when the test score is between the highest 30% and the highest 10% of those tested. To compute the lower bound for a grade of B, then, we need to identify the z score above which 30% of the observations are located. Again, using column b of the normal curve table we find a z score of approximately +.52. Substituting in the formula,

$$\begin{aligned} X &= 85 + (.52)(7) \\ &= 85 + (3.64) = 88.64 \end{aligned}$$

Thus, scores of 93–89 receive a grade of B.

A score of 88 then serves as the upper bound for a grade of C. The lowest score one could obtain and still receive a C grade is a score above 30% of those tested [5% (F) + 25% (D)]. Thus we need to identify the z score that exceeds 30% of the observations in a normal distribution. The table of the normal curve tells us that a z score of −.52 exceeds 30% of the scores in the distribution. The lower bound for a grade of C is

$$\begin{aligned} X &= 85 + (-.52)(7) \\ &= 85 - 3.64 = 81.36 \end{aligned}$$

Scores between 88 and 82 result in grades of C.

To compute the lower boundary for a grade of D, we need to identify the z score below which 5% of the observations fall. Using the normal curve table we find that 5% of the observations fall below a z of -1.645. Substituting and solving yields

$$X = 85 + (-1.645)(7)$$
$$= 85 - 11.52 = 73.48$$

Thus, raw scores in the range from 81 to 74 receive a D grade. By deduction scores of 73 and lower are assigned a grade of F. To summarize the solution,

Grade	Test Score
A	94 or higher
B	93–89
C	88–82
D	81–74
F	73 or lower

7.8 NORMALIZING SCORES

Converting standard scores as discussed in section 7.3 represents a linear transformation and does not alter the original distribution shape. Some methods of transforming a set of raw scores do alter the original distribution shape, however. For example, in Chapter 4 the conversion from raw scores to percentile ranks in a distribution results in a score index interpreted as the percentage of scores in the distribution that are less than or equal to a given raw score. Table 7.2 presents raw scores on the Forms posttest from the "Sesame Street" evaluation. In the table each raw score's percentile rank equivalent is shown. If the score frequencies were graphed using a percentile rank scale, the shape of the distribution would be rectangular. Converting to percentile ranks has the effect of representing each score as a percentage of the total number of observations. Once transformed, particular percentile positions define a fixed number of observations at a location. If each of 239 scores represents 1 of 239 or .0042 of the percentile scale range, then an equal number of scores will fall between P_{10} to P_{20} and P_{30} to P_{40} or P_{65} to P_{75}. The plot of percentile ranks is always rectangular, regardless of the shape of the raw score distribution.

A percentile rank indicates the percentage of scores equal to or less than the specified score. A unit normal curve's area to the left of a specific z score is its percentile rank equivalent. What is the percentile rank in a normally distributed set of scores for a raw score .75 standard deviation units *below* the mean ($z = -.75$)?[11] Table B.2 most efficiently provides the information for converting a raw score or z score to its percentile rank. The reverse operation is also possible. Here the percentile rank is given and the task is to find the z score.

[11]Percentile rank = Area × 100

$$= .2266 \times 100 = 23$$

TABLE 7.2
Scores and score transformations on the Forms posttest data (n = 239, $\overline{X}$ = 13.75, and s_x = 3.92).

			Transformation From Raw Scores		Transformation Based on Normal Curve	
X	f	cf	Nonnormalized z Score	Percentile Rank	Normalized z Score	Normalized T Score
20	7	239	1.59	99	2.33	73
19	22	232	1.34	92	1.41	64
18	18	210	1.08	84	.99	60
17	25	192	.83	75	.67	57
16	14	167	.57	67	.44	54
15	24	153	.32	59	.23	52
14	23	129	.06	49	−.02	50
13	19	106	−.19	40	−.25	48
12	21	87	−.45	32	−.47	45
11	17	66	−.70	24	−.71	43
10	13	49	−.96	18	−.92	41
9	12	36	−1.21	13	−1.13	39
8	7	24	−1.47	9	−1.34	37
7	5	17	−1.72	6	−1.55	35
6	7	12	−1.98	4	−1.75	33
5	0	5	−2.23	2	−2.05	30
4	2	5	−2.49	2	−2.05	30
3	3	3	−2.74	1	−2.33	27

For example, find the raw score at the 58th percentile in a normal distribution with $\overline{X}$ = 45 and s = 8.[12]

A raw score can be transformed so that the transformation depends upon characteristics of a second distribution rather than its own distribution. The second distribution may be real or hypothetical for the variable being measured. A common practice is to transform raw scores to an approximately normal distribution shape. Converting scores so that they appear to be normally distributed facilitates interpretation or fulfills requirements of a statistical procedure. Figure 7.14 is an often-used illustration of several derived scores and their relationship to a normal distribution and each other. The remainder of this section discusses normalizing of raw scores.

A conventional scale for normalized scores is a standard score scale, or T score scale, with a fixed mean of 50 and standard deviation of 10. Interpretations of T scores differ from the transformed scores in section 7.3 because of their fixed properties. If the raw score distribution is not normal, then the

[12]The z score equivalent is approximately .20. Using formula 7.4, X = 45 + .20(8). The raw score (percentile point) in the normal distribution with a percentile rank of 58 is 46.6.

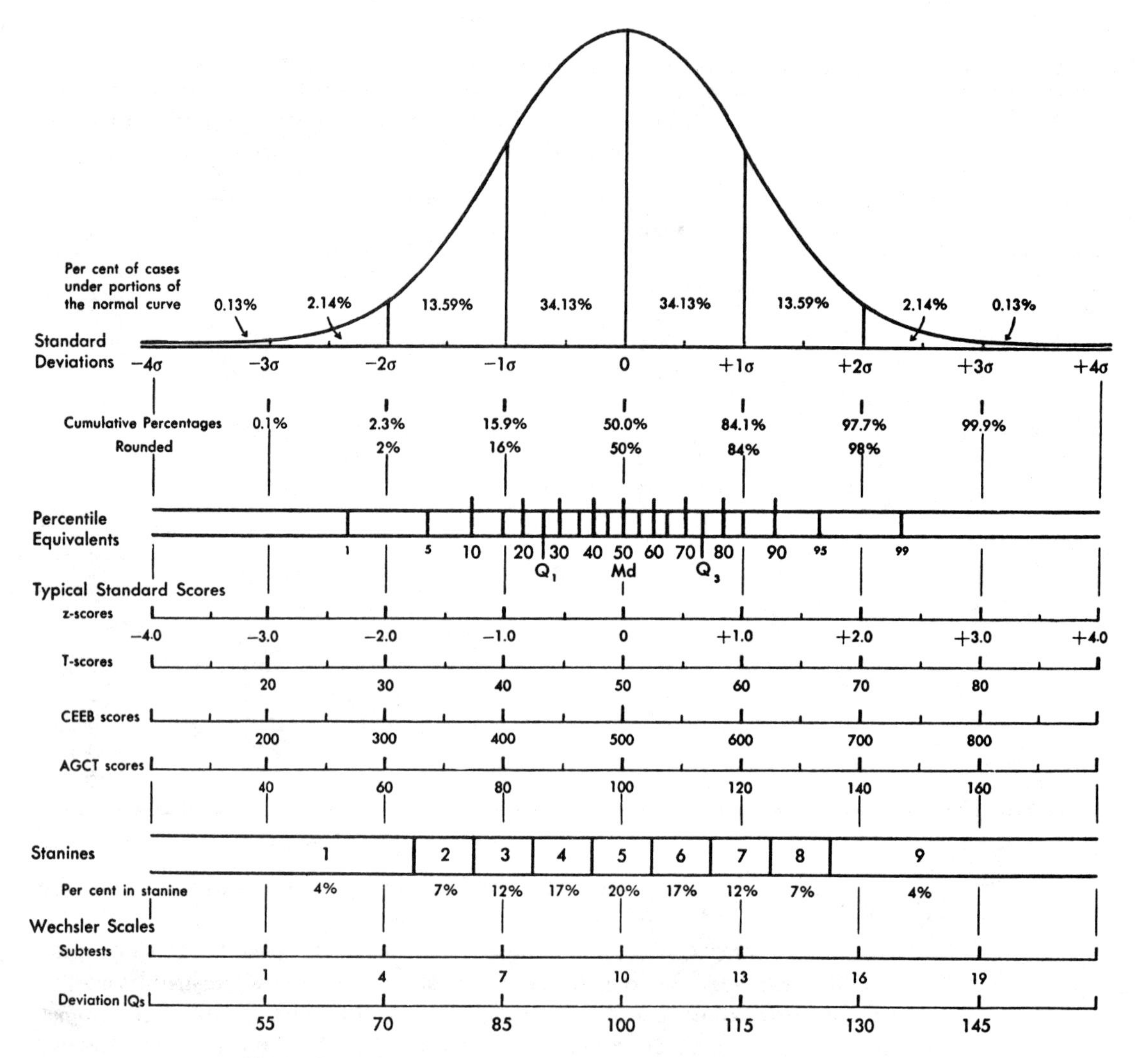

FIGURE 7.14
Normal curve, percentiles, and standard scores.
Note: From *Test Service Bulletin* No. 48, The Psychological Corp. Copyright by The Psychological Corp. Reprinted by permission.

distribution must first be *normalized* before using the procedures of section 7.4 to convert to T scores. To normalize a distribution of raw scores, the following steps are taken:

1. Find the percentile ranks for the score points in the raw score frequency distribution.
2. Convert each percentile rank to its equivalent z score in a normal

distribution using Table B.2. This step gives the z score equivalency *in a normal distribution* for the raw score.

Following these two steps will *normalize* any distribution and give unit normal curve z score equivalencies for raw scores. Transforming the unit normal curve z scores to T scores uses the formula 7.4.

$$T = 50 + 10(z)$$

Table 7.2 illustrates the result from the normalizing procedure. The raw scores on the Forms posttest are given in the frequency distribution. Also reported are the cumulative frequency for the raw score distribution, the percentile rank of each raw score, and the nonnormalized z score for each raw score based on the raw score mean and standard deviation. The normalization process for the raw score 19 follows.

- Step 1 Find the percentile rank for the raw score. Use information from the cumulative frequency distribution and apply formula 4.2 to compute the percentile rank of the raw score 19.

$$PR_{19} = \frac{210 + \frac{19 - 18.5}{1}(22)}{239} \times 100$$

$$= \frac{210 + (.5)(22)}{239} \times 100$$

$$= \frac{221}{239} \times 100 = 92$$

- Step 2 Convert the percentile rank to its normal curve z score equivalent.

$PR_{19} = 92$ means that 92% of the raw scores lie at or below a score of 19. The z score in the unit normal curve below which 92% of the observations fall is $+1.41$. The z score of 1.41 is the normalized z score equivalent for a raw score of 19.

- Step 3 Convert the normalized z score to a normalized T score.

$$T = 50 + (10)(+1.41)$$

$$= 50 + 14.1 = 64$$

All other normalized z and T scores in Table 7.2 were computed in this manner. Check one or two of the raw score entries to see if you are following the process correctly. The transformation that normalizes raw scores uses the normal curve and its characteristics as a reference point. A raw score is converted to a normalized z score based on its percentile position in the actual raw score distribution. The standard T score is then determined by the value of z at that percentile point in the normal curve.

Let's consider the consequence of having normalized the Forms posttest data. Figure 7.15 presents the frequency polygons for the smoothed raw score and the normalized, smoothed *T* score distributions. Viewed from these distributions, the conversion "tightens up" or presents as closer together scores at the low end of the scale continuum while providing greater separation be-

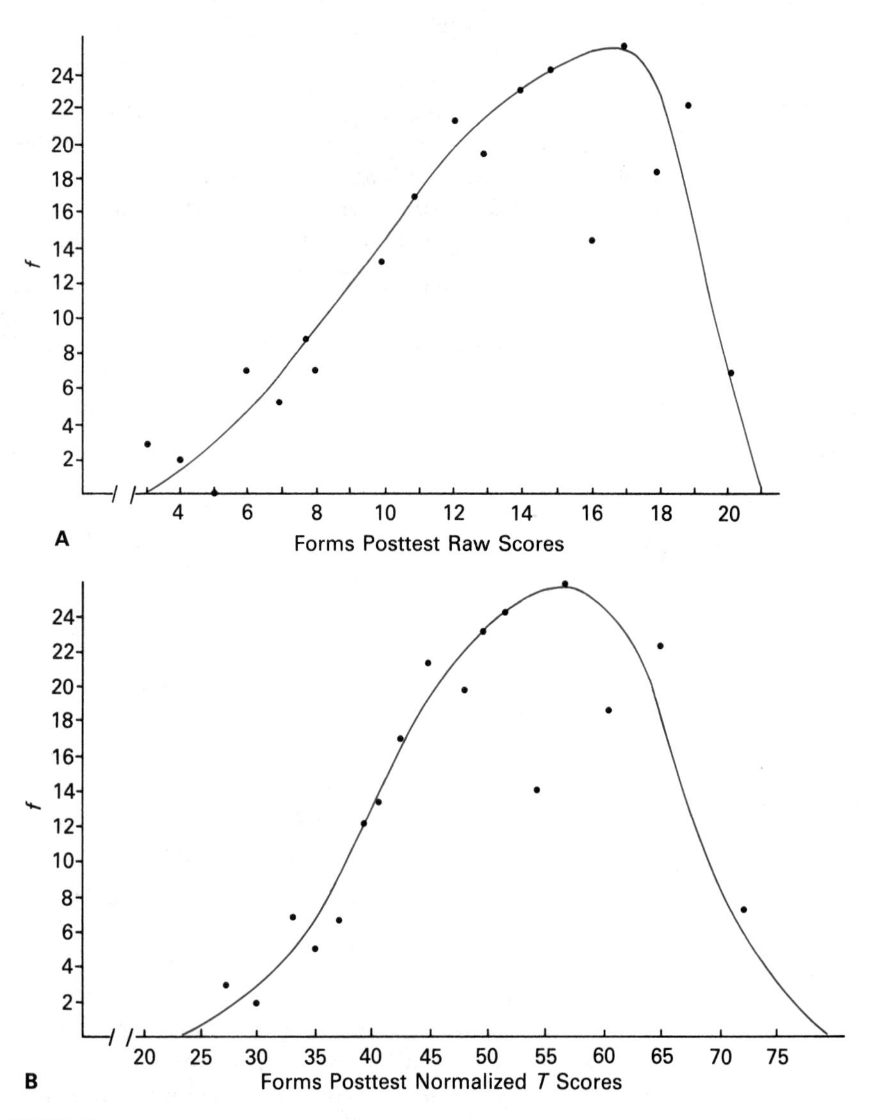

FIGURE 7.15
A, smoothed raw score; **B,** normalized *T* score distributions for the Forms posttest.

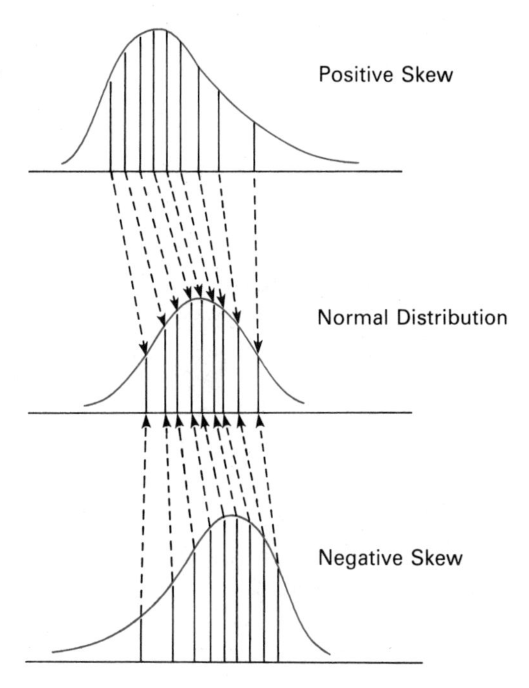

FIGURE 7.16
The effect of normalizing.

tween scores at the high end of the scale. Doing this eliminates the negative skew in the original score distribution; the transformed data become more symmetrically distributed about the scale mean. Figure 7.16 depicts the effect of normalizing on both positively and negatively skewed distributions.

Try your hand at normalizing. Using the information available in Table 4.8, we can convert mental age raw scores of 27, 38, 42, 47, and 89 to their normalized z and T equivalents. Results of the transformation appear in Table 7.3. The procedure is documented for a raw score of 47.

TABLE 7.3
Transformation of raw scores to z and T scores.

Mental Age Raw Score	Percentile Rank	Normal Curve z Score	Normalized T Score
89	97	1.88	69
47	61	.28	53
42	50	.00	50
38	38	−.31	47
27	4	−1.75	33

- Step 1 From the cumulative frequency information in Table 4.8,

$$PR_{47} = \frac{133 + \frac{47 - 44.5}{5}(27)}{240} \times 100$$

$$= .61 \times 100 = 61$$

- Step 2 $PR_{47} = 61$ means that 61% of the distribution lies at or below a score of 47. The z score equivalent in the unit normal curve for a proportion of .61 is .28.
- Step 3 $T = 50 + 10(.28)$

$$= 52.8 \text{ or } 53$$

7.9 SUMMARY

The z score transformation produces one of the most fundamental standard scores in statistics. The formula for a z score is

$$z = \frac{X - \overline{X}}{s_x}$$

Regardless of the raw score measurement scale units, the value of the mean, or the degree of raw score variability, a z score always indicates the same thing. It represents the relative distance of a raw score from the mean of the distribution expressed in standard deviation units. Transforming raw scores to z scores changes the distribution mean and standard deviation, but the shape of the z score distribution will match the shape of the raw score distribution. Once raw scores have been changed to z scores, the mean of the z scores always will equal zero ($\overline{z} = 0$) and the standard deviation of the z scores always will equal one ($s_z = 1$).

Other derived standardized score scales can be created by making a second transformation of z scores to a distribution with a particular mean and standard deviation. Raw scores can be transformed to equivalent standardized scores for any defined new values of $\overline{X}_{new}$ and s_{new} using the following procedure:

- Step 1 Calculate the raw score mean and standard deviation.
- Step 2 Transform the raw score to a z score
- Step 3 Transform the obtained z score to the new standardized score (X_{new}) using

$$X_{new} = \overline{X}_{new} + s_{new}(z)$$

While z score equivalencies for raw scores in any distribution are important descriptive and comparative statistical indices, z scores are helpful primarily as scale values in the theoretical unit normal distribution. The unit normal curve proportional areas identified as a function of z score scale values form the basis for expectations about the frequency with which scores or other events will occur. The curve provides a mathematical model with standardized properties approximating what often occurs in reality. The z score scale provides the necessary standardized scale to which any set of scores may be transformed and the unit normal curve properties applied. Because of its frequent application in statistical procedures it is important to have a thorough working knowledge of z scores and related unit normal curve properties.

PROBLEMS

Problem Set A

1. Identify the proportional area in a normal curve that would fall (a) below $z = -1.5$, (b) above $z = 2.3$, (c) between the mean and $z = -.32$, (d) below $z = .70$, (e) above $z = -1.35$, (f) between $z = -.65$ and $z = .65$, (g) between $z = 1.28$ and $z = 2.00$, (h) between $z = -.32$ and $z = -.10$, (i) beyond $z = 1.96$ and below $z = -1.96$, and (j) to the left of $z = -1.45$ and to the right of $z = -.20$.

2. If scores are normally distributed with a mean of 100 and a standard deviation of 10, what proportion of scores would we expect to be (a) greater than 115? (b) less than 75? (c) between 92 and 103? (d) between 87 and 92? (e) greater than 120 or less than 110?

3. In a normal distribution of 500 scores with $\overline{X} = 36$ and $s_x = 4$, how many scores would you expect to be (a) less than 40? (b) greater than 30? (c) between 35 and 38? (d) less than 32 or greater than 35?

4. For the distribution described in problem 3, what are the raw scores (a) at or below which 300 scores would fall? (b) above which 250 scores would fall? (c) that identify the limits for the middle 200 scores? (d) that identify the limits for the bottom 25 scores and the top 50 scores?

5. What z score is equivalent to the 20th percentile point in a normal distribution of scores? The 63rd percentile? The 50th percentile?

6. a. Jim's raw score is 104 in a normally distributed set of scores with $\overline{X} = 110$ and $s_x = 8$. What proportion of the distribution would you expect to have scores greater than Jim?
   b. Joe's raw score in the distribution referred to in (a) was 113. Given a class size of 60, approximately how many individuals would you expect to have raw scores between Jim's and Joe's?

7. Given that a raw score of 35 is at the 74th percentile in a normally distributed set of scores with a variance of 169, find the distribution's mean.

8. An individually administered intelligence test is being used to identify academically gifted second-grade students. The resources available can only support a program

for about 30 of the 600 second graders in the school district. Assuming that the IQ scores for the 600 students are normally distributed with a mean of 100 and standard deviation of 15, what is the IQ cutoff score you might use to place students in the program? If you wanted to build a waiting list of about 20 students who did not attain the cutoff score, but were in the next IQ category range, what would you use as the IQ cutoff scores?

Problem Set B

I. In a continuation of their study of impulsivity, Salkind and Wright (1977) proposed an integrated model of development involving two statistically independent constructs, impulsivity and cognitive efficiency. Prior to this time, cognitive efficiency had not been considered in the scaling of reflectivity/impulsivity. Rather, Error and Latency scores from the *Matching Familiar Figures Test* (MFF) were used to categorize students into reflective or impulsive groups.

The proposed Salkind-Wright model used the same error and latency scores from the MFF, but defined the speed/accuracy domain in terms of two constructs, impulsivity and efficiency. Impulsivity was defined as a continuous dimension of individual differences ranging from fast-inaccurate to slow-accurate performance. Efficiency was defined as a continuous dimension conceptually orthogonal to impulsivity along which individual differences range from slow-inaccurate to fast-accurate performance. The formulas proposed to scale impulsivity and efficiency are

$$\text{Impulsivity Score} = I = z_e - z_l$$

$$\text{Efficiency Score} = E = z_e + z_l$$

where

z_e = the standard score for an individual's total errors on the MFF

z_l = the standard score for an individual's mean latency on the MFF

Large positive I scores indicate impulsivity; negative scores, reflectivity. Positive E scores indicate inefficiency and negative E scores indicate efficiency. Figure 7.17 pictorially displays the scaling of I and E scores.

The data in Table 7.4 are simulated error and latency raw scores for 20 individuals. These data provide the basis for answering the questions. Use the group means and standard deviations when computing z_e and z_l.

a. Compute the I score and E score for individuals 4, 12, 15, and 17. Where would you classify them on the impulsivity and efficiency dimensions?

b. I scores and E scores are assumed to have normal population distributions with means of 0 ($\mu_I = \mu_E = 0$), and standard deviation of 1.4 ($\sigma_I = \sigma_E = 1.4$). What proportion of the I scores would you expect to be greater than 1.0? less than −1.4? between .32 and −1.8? between .75 and 1.75?

c. Assuming the population distribution characteristics in part b, find the E score cut-off for defining an extremely efficient person if only 12% of the population is thought to be extremely efficient. Find the I score cut-off if 30% of the population is thought to be impulsive.

FIGURE 7.17
Scaling for Impulsivity and Reflectivity scores.
Note: From "The Development of Reflection-Impulsivity and Cognitive Efficiency: An Integrated Model" by N.J. Salkind and J.C. Wright, 1977, *Human Development, 20*(6), p. 382. Copyright 1977 by S. Karger. Reprinted by permission.

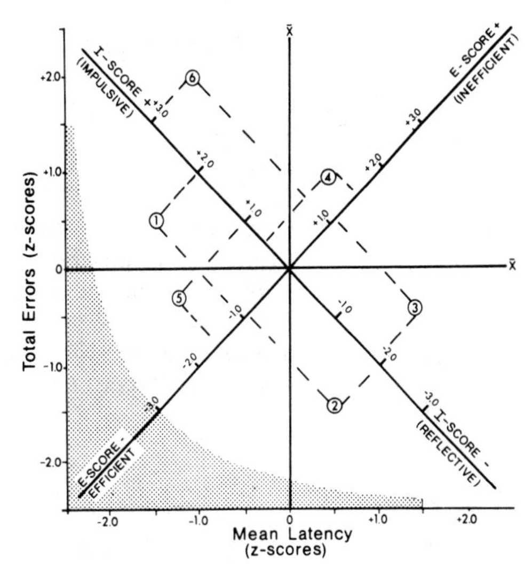

TABLE 7.4
MFF Error and Latency scores.

Individual	Error Score	Latency Score
01	8	14
02	11	10
03	7	14
04	18	34
05	6	17
06	14	11
07	9	8
08	7	12
09	4	21
10	8	19
11	7	8
12	15	7
13	6	14
14	10	5
15	5	32
16	8	13
17	3	9
18	1	15
19	4	11
20	2	22

Problem Set C

I. The means and standard deviations for all the variables in the "Sesame Street" data base are given at the end of the raw data in Appendix A. The values given are based on the data for all 240 students. Use these values to transform raw scores to z scores for students 35, 148, and 179 on both the pretests and posttests for the following variables: Body Parts, Letters, Numbers, and Classification Skills. Assess the comparative profiles for pre- versus posttest scores for each of the three students. Who made the biggest increase from pre- to posttest relative to the group's performance? For which variable did that occur?

II. Using the following frequency distribution for Classification Skills posttest scores, what is the T score for a raw score of 13? for a raw score of 18?

Score Interval	f
24–25	8
22–23	29
20–21	28
18–19	33
16–17	35
14–15	25
12–13	21
10–11	28
8–9	18
6–7	9
4–5	5

8
Measures of Relationship

Is student achievement related to parental involvement in education? Do people who exhibit aggressive behavior also exhibit increased amounts of peer alienation? Is early career selection predictive of adult job satisfaction? Is per-pupil expenditure related to student achievement? These are examples of research questions that ask about the relationship between two variables. The statistical procedures covered in previous chapters allow us to summarize the scores for any one variable using descriptive indices, but no technique has been presented yet that will permit including two variables in the same summary index.

Statistical procedures for assessing the extent of a relationship fall into the category of "correlational analysis." A correlation index identifies the extent to which two variables correlate or covary, that is, how much they tend to go together. Do individuals with high scores on variable X tend also to be high on variable Y? Or is there no systematic score pattern between the two variables?

At this point we must distinguish between a correlation model and a prediction (regression) model approach to examining relationships between two variables. The correlation model does not distinguish between the function of the two variables. Its sole intent is to identify the degree of the relationship between the two variables. In contrast, if the intent of a study focuses on predicting one of the variables based on information about the other variable, then a regression model approach should be used. The regression model examines changes in one variable as a function of changes or differences in fixed values of the other variable. While the two models both examine relationships between variables, their purposes are quite different. Chapter 9 discusses the regression model. The remainder of this chapter introduces topics necessary to understanding correlation model relationships.

8.1 BIVARIATE FREQUENCY DISTRIBUTIONS

Correlation methods can be applied when data have been collected on two different variables. Scores on each of the two variables (X and Y scores) must exist for each sampling unit. Correlational data can be summarized in a **bivariate frequency distribution.** A graph called a **scatter plot** is one method of displaying a relationship evident in a bivariate frequency distribution. It shows the joint distribution of two variables and pictorially displays the data point for each individual on a two-dimensional graph. Consider the following data set for five individuals. Variable X represents a measure of test performance, while variable Y is ratings of instructor effectiveness.

Individual	X	Y
A	12	7
B	10	5
C	9	5
D	7	3
E	5	2

Each individual has two scores, one for variable X, achievement, and one for variable Y, effectiveness rating. These two scores are used to locate each individual on the scatter plot in Figure 8.1. The horizontal axis represents the scale continuum for variable X; the vertical axis represents the scale continuum for variable Y. To locate an individual's joint data point for the two variables in the scatter plot, lines can be drawn perpendicular to the X axis from an individual's score on the X scale and to the Y axis from an individual's score on the Y scale. An individual's bivariate data point is located where the two perpendicular lines intersect. The scatter plot data point for individual C was found by drawing perpendicular lines from 9 on the X axis and 5 on the Y axis. The perpendicular lines in Figure 8.1 illustrate the process. In practice, the lines

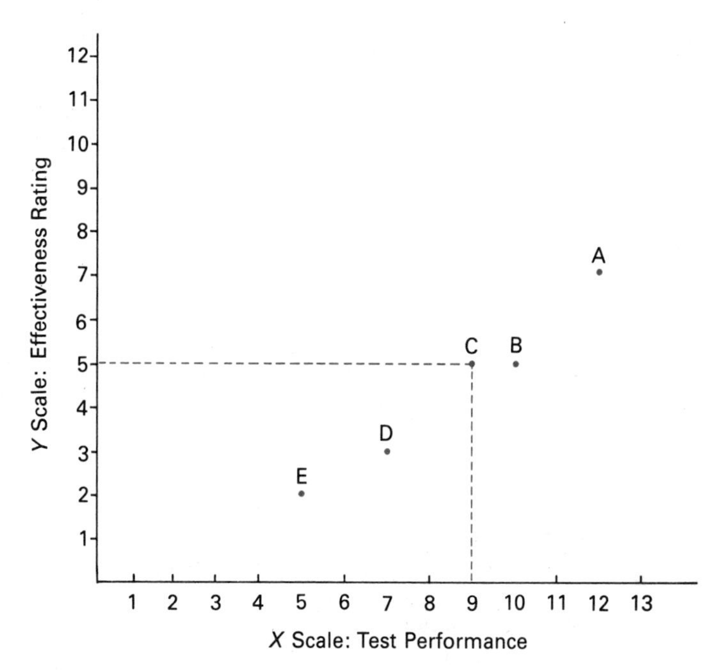

FIGURE 8.1
Scatter diagram of five bivariate data points.

TABLE 8.1
Letters (X) and Numbers (Y) posttest scores for 20 individuals.

ID	Letters (X)	Numbers (Y)
44	6	6
67	43	52
82	53	52
84	21	32
95	46	51
96	43	42
100	19	18
110	48	45
114	32	23
128	13	17
137	22	38
159	16	22
179	19	13
184	11	13
199	18	13
208	45	38
210	14	35
211	15	23
237	26	20
238	19	10

are not drawn. Take the time to graph the bivariate frequency distribution for the sample data on Letters posttest (X) and Numbers posttest (Y) raw scores in Table 8.1. The scatter plot for these data appears in Figure 8.2. The data points in the scatter plot appear to spread out considerably. This occurs because the variability of scores for the 20 individuals is quite large on both variables ($s_x = 14.50$; $s_y = 15.02$).

The relationship in a scatter plot demonstrates either a *linear* or *curvilinear* trend in the paired data points. A linear trend is represented by a scatter plot of points that tend to cluster in a straight line. Figures 8.3A and B represent linear relationships. A curvilinear trend is represented by data points that tend to form a curved rather than a straight line. Figures 8.3C–F provide examples.

8.2 PEARSON PRODUCT MOMENT CORRELATION COEFFICIENT

While curvilinear relationships between variables do exist, many relationships are linear. The single, most widely used index of linear relationship is the **Pearson product moment correlation coefficient.** This coefficient, developed by Karl Pearson, is a single summary index of the extent to which two

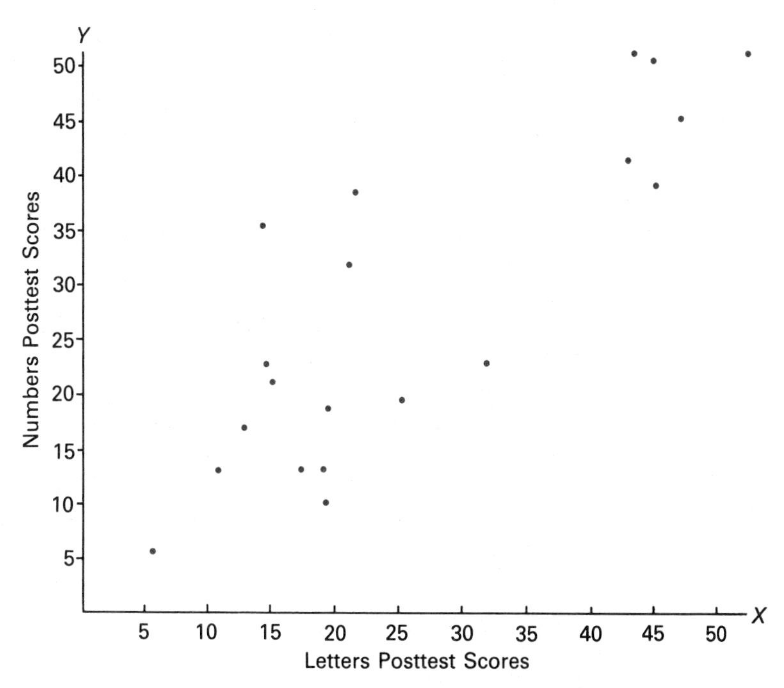

FIGURE 8.2
Scatter plot of Letters and Numbers posttest scores for the sample in Table 8.1.

variables are linearly related. The standard symbol for the sample Pearson product moment correlation coefficient is r with subscripts identifying the two variables whose correlation is represented. If X and Y are the two variables, then r_{xy} would represent the Pearson product moment correlation coefficient indicating the relationship between X and Y scores. If variables 4 and 7 are to be correlated, then r_{47} would be the appropriate symbol. Because the Pearson product moment correlation coefficient is so widely used, it is usually shortened to "Pearson r" or just the "correlation coefficient." As with the mean and variance, the Pearson r coefficient is represented by a Greek letter ρ_{xy} (rho) for a population value.

In describing the magnitude of the linear relationship between two variables, the Pearson r ranges in value from -1.00 to $+1.00$. A correlation coefficient of .00 is an anchor point, indicating no linear relationship. Any deviation from $r_{xy} = .00$ in either the positive or negative direction indicates that a relationship exists. The larger the absolute value of a coefficient (ignoring the sign), the greater the relationship between the variables. A correlation coefficient of either -1.00 or $+1.00$ represents a perfect linear relationship between two variables. This does not mean that raw scores on the variables for *individuals* are identical. Rather, it means that the data points in a scatter plot

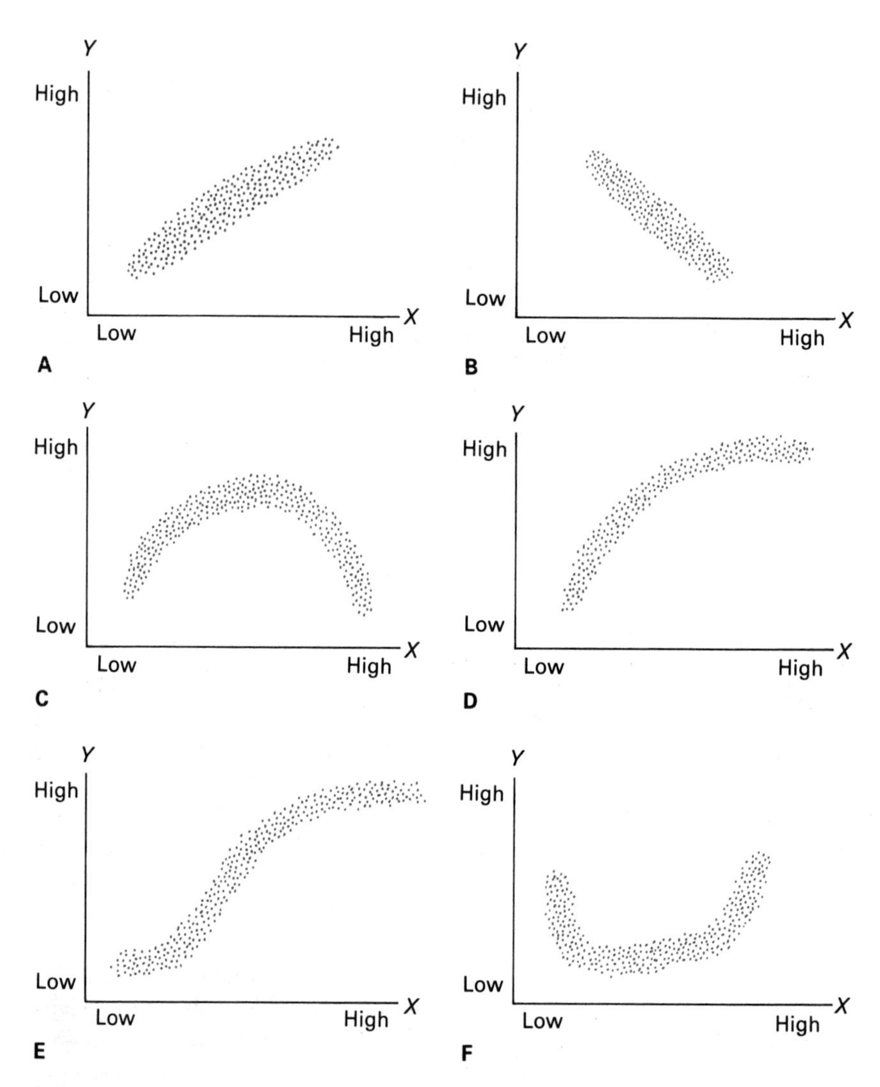

FIGURE 8.3
Linear and curvilinear relationships. **A,** linear; **B,** linear; **C,** curvilinear; **D,** curvilinear; **E,** curvilinear; **F,** curvilinear.

form a straight line. If a relationship is less than perfect, the scatter plot of data points will deviate from a perfect straight line. For a correlation of $r_{xy} = .00$, the boundary of the scatter plot will be roughly circular. As the relationship increases, the data points tend to cluster more in the form of an ellipse (football shape). As a relationship approaches -1.00 or 1.00, the data points cluster closely around a straight line.

Because the Pearson r is a measure of the linear relationship between two variables, a low value means that little, if any, of the relationship between the variables can be described by a linear or straight line. It does not allow any conclusion about the possibility of a *curvilinear* relationship between the variables; another index is used for that (see section 8.7).

The sign of a correlation coefficient identifies the direction of the relationship between the two variables. Positive correlation coefficients identify a direct relationship between the two variables. Those whose scores are high, in the midrange, or low on one variable scale tend to have high, middle, or low scores on the other variable scale, respectively. The larger the positive correlation, the greater the location similarity between pairs of scores for individuals. An opposite or inverse relationship will produce a correlation coefficient with a negative sign. For instance, persons scoring high on one variable tend to score low on the other variable and vice versa. Figure 8.4 illustrates scatter diagrams for positive, zero, and negative relationships. Table 8.2 outlines the score pattern for direct (positive) and inverse (negative) relationships. The magnitude of a relationship does not depend on the sign of the correlation coefficient. A correlation coefficient of $-.56$ indicates the same *degree* of relationship as $r_{xy} = .56$. The only difference is that $r_{xy} = -.56$ identifies a negative (inverse) relationship between two sets of scores while $r_{xy} = .56$ indicates that the score pairs vary in the same scale direction (a positive or direct relationship).

Conceptually, the Pearson r can be explained by examining z scores. Recall from Chapter 7 that transforming raw scores to z scores yields the relative distance of a score from the group mean. If raw scores on two variables are changed to z scores, the relationship between the two variables is high when z scores for individuals resemble each other. If the z scores for variables X and Y are identical for each individual, then the relationship between X and Y will be perfect: $r_{xy} = 1.00$. Identical z scores indicate that an individual's X and Y scores are exactly the same distance from the respective group means. If the relationship were inverse (negative) and perfect, each individual would have a positive z score for one variable and the same negative z score for the other variable.

Any deviation from identically matched z scores will lower the correlation coefficient. For high, positive correlation coefficients, persons with positive z scores on one variable will tend to have positive z scores on the other variable. Persons with negative z scores on variable X will tend to have negative z scores on Y. As the correlation coefficient approaches zero, the pairs of z scores for individuals will be randomly matched. Some z score pairs will both be positive and some both negative, but an equal number will be positive-negative and negative-positive with no systematic pattern. This latter mix of z scores indicates that individuals' raw scores are not systematically located in the same place relative to the mean. Rather, scores on the two variables (X and Y) tend to be independent of each other.

Carefully examine the z scores for the Letters (X) and Numbers (Y)

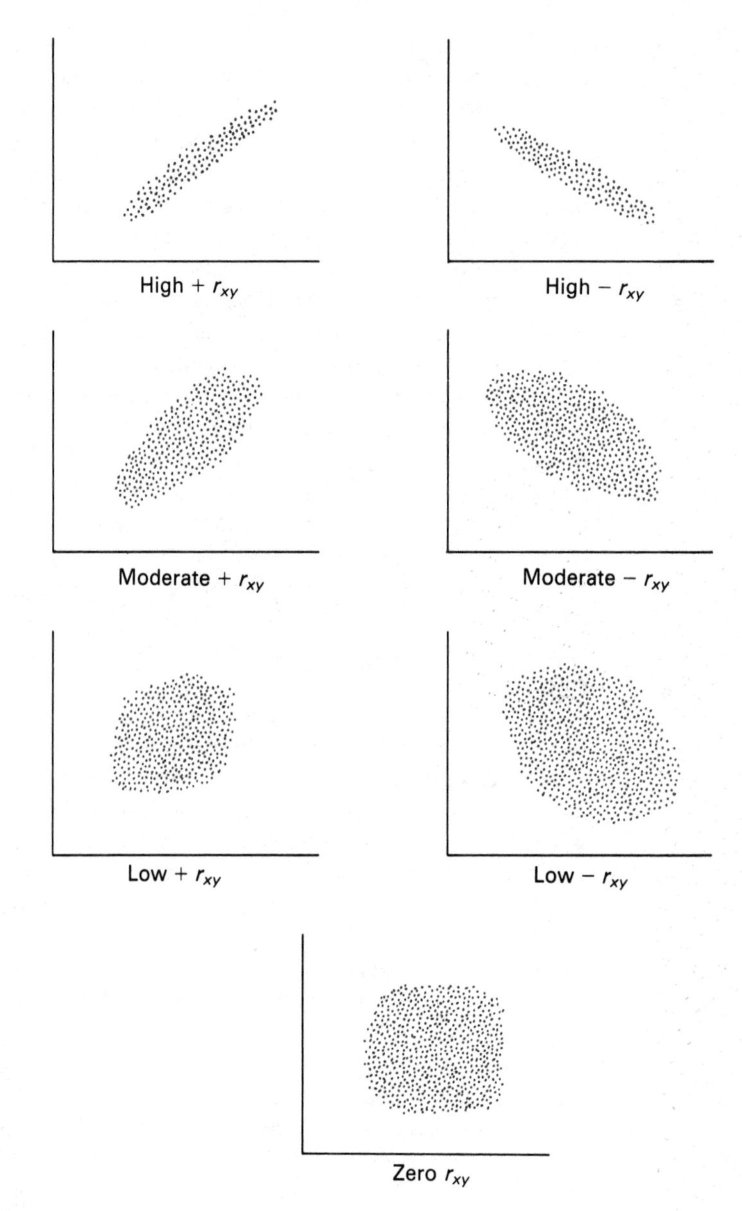

FIGURE 8.4
Positive, zero, and negative relationships.

posttest scores in Table 8.3. If no systematic response pattern exists for a group of individuals, r_{xy} will be approximately zero. Notice that the majority of the pairs of z scores (z_x and z_y) for individuals tend to have the same signs, indicating the presence of a positive relationship. The magnitude of the z scores for individual pairs also seems to be reasonably consistent with the z scores for

TABLE 8.2
Scores associated with direct and inverse relationships between variables.

Score on Variable *Y*		Score on Variable *X*		Score on Variable *Y*
	Direct Relationship		Inverse Relationship	
High	←——→	High	←—(crosses)—→	High
Medium	←——→	Medium	←——→	Medium
Low	←——→	Low	←—(crosses)—→	Low

individuals 44 and 184 having the closest match. The extent to which the other pairs of z scores differ from each other lowers the correlation between the two variables. The individual whose score pattern was most different from the rest was 210. This individual's X score was .859 standard deviation units *below* the X mean and Y score was .456 standard deviation units *above* the Y mean. This type of performance is inconsistent from the majority of patterns in the data set and lowers the group correlation coefficient.

TABLE 8.3
Correlation example: Letters (X) and Numbers (Y) posttest scores for 20 individuals.

Individual	Raw Posttest Scores		z scores		
ID	Letters (X)	Numbers (Y)	z_x	z_y	$z_x z_y$
44	6	6	−1.410	−1.475	2.080
67	43	52	1.141	1.588	1.812
82	53	52	1.831	1.588	2.908
84	21	32	−.376	.256	−.096
95	46	51	1.348	1.521	2.050
96	43	42	1.141	.922	1.052
100	19	18	−.514	−.676	.347
110	48	45	1.486	1.122	1.667
114	32	23	.383	−.343	−.131
128	13	17	−.928	−.742	.689
137	22	38	−.307	.656	−.201
159	16	22	−.721	−.409	.295
179	19	13	−.514	−1.009	.519
184	11	13	−1.066	−1.009	1.076
199	18	13	−.583	−1.009	.588
208	45	38	1.279	.656	.839
210	14	35	−.859	.456	−.392
211	15	23	−.790	−.343	.271
237	26	20	−.031	−.543	.017
238	19	10	−.514	−1.208	.621
TOTAL	529	563			16.011

8.3 CALCULATING THE PEARSON CORRELATION COEFFICIENT

The easiest-looking formula for calculating the Pearson product moment correlation coefficient uses z scores as the score elements. The z score formula for computing r_{xy} is

$$r_{xy} = \frac{\Sigma z_{x_i} z_{y_i}}{n - 1} \tag{8.1}$$

This formula requires transforming X and Y raw scores to z scores. Then the product of the X z scores (z_x) and the Y z scores (z_y) are taken for each individual in the group (z_{x_i} times z_{y_i}). These products are termed **cross-products** and indicate the multiplication of scores together *across* two variables for each individual. The z score cross-products are then summed over all individuals in the group ($\Sigma z_{x_i} z_{y_i}$). The subscript i indicates that different individuals' z scores are included in the summation. As in previous formula notation, we will drop the subscript i, but it is to be inferred in the summation $\Sigma z_x z_y$. The sum of the z score cross-products divided by $n - 1$ results in the value of r_{xy} for a given sample.

While the mathematical proof is not presented, the maximum value for $\Sigma z_x z_y$ is $n - 1$. If the X z scores and Y z scores are identically matched for each individual, then $\Sigma z_x z_y$ will equal $n - 1$. Recall from Chapter 7 that the sum of the squared z scores, Σz^2, equals $n - 1$. If X and Y z scores are identical, then $\Sigma z_x z_y = \Sigma z^2 = n - 1$ and $r = \pm 1$, indicating a perfect relationship. Therefore, taking the sum of the z score cross-products makes sense. The following set of X and Y scores for five individuals will introduce the procedures for calculating r_{xy}. These data are labeled Example A. First, calculate the means and standard deviations for each set of scores.[1]

---

[1] $\Sigma X = 175$; $\Sigma X^2 = 7725$

$\overline{X} = \frac{175}{5} = 35$

$\Sigma x^2 = 7725 - \frac{175^2}{5} = 1600$

$s_x^2 = \frac{1600}{4} = 400$

$s_x = 20$

$\Sigma Y = 55$; $\Sigma Y^2 = 669$

$\overline{Y} = \frac{55}{5} = 11$

$\Sigma y^2 = 669 - \frac{55^2}{5} = 64$

$s_y^2 = \frac{64}{4} = 16$

$s_y = 4$

Example A:

Individual	Raw Scores X	Y
A	60	16
B	45	13
C	40	12
D	20	8
E	10	6

To compute the Pearson correlation coefficient for variables X and Y, convert the raw scores to z scores, then compute the cross-product ($z_x z_y$) value for each individual. Make sure you know how to transform each raw score to a z score.[2]

Individual	z scores z_x	z_y	Cross-products $z_x z_y$
A	1.25	1.25	1.5625
B	.50	.50	.2500
C	.25	.25	.0625
D	−.75	−.75	.5625
E	−1.25	−1.25	1.5625

Note that the X and Y z scores are identical for each individual. When the products of z_x and z_y are taken, all values are positive, thereby increasing $\Sigma z_x z_y$. The larger the sum of the cross-products, the larger the correlation coefficient. As noted earlier, when z scores on two variables are identical for individuals, the correlation coefficient is perfect; $r_{xy} = 1.00$, as it is in this example:

$$r_{xy} = \frac{\Sigma z_x z_y}{n - 1}$$

$$= \frac{1.5625 + .2500 + .0625 + .5625 + 1.5625}{5 - 1}$$

$$= \frac{4.00}{4} = 1.00$$

[2] z scores for Individual B:

$$z_x = \frac{45 - 35}{20} = .50; \; z_y = \frac{13 - 11}{4} = .50$$

z scores for Individual E:

$$z_x = \frac{10 - 35}{20} = -1.25; \; z_y = \frac{6 - 11}{4} = -1.25$$

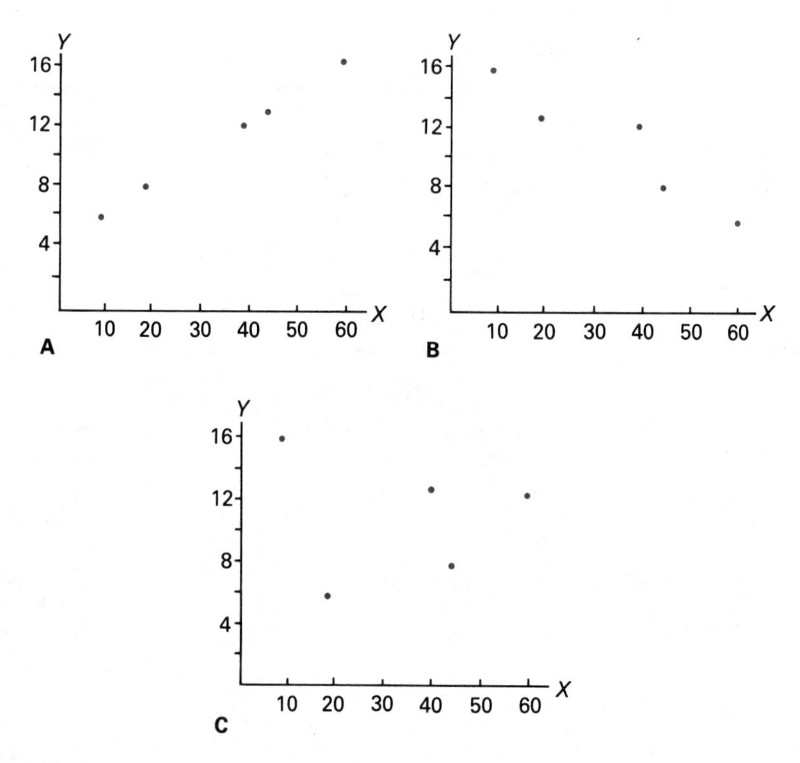

FIGURE 8.5
Scatter diagrams for different matchings of the same data. **A**, perfect positive relationship (r_{xy} = 1.00); **B**, high negative relationship (r_{xy} = −.953); **C**, low negative relationship (r_{xy} = −.141).

Figure 8.5A illustrates the five plotted data points. Notice that they all fall on a straight line providing visual evidence of a perfect linear relationship.

What would happen to the relationship between X and Y scores in the preceding illustration if X and Y scores had been different for individuals? Two different arrangements of the data appear in Examples B and C. Calculate r_{xy} in each instance using formula 8.1.[3]

---

[3]For Example B,

$$\Sigma z_x z_y = -3.8125$$

$$r_{xy} = \frac{-3.8125}{4} = -.953$$

For Example C,

$$\Sigma z_x z_y = -.5625$$

$$r_{xy} = \frac{-.5625}{4} = -.141$$

Example B:

Individual	X	Y	z_x	z_y	$z_x z_y$
A	60	6	1.25	−1.25	−1.5625
B	45	8	.50	−.75	−.3750
C	40	12	.25	.25	.0625
D	20	13	−.75	.50	−.3750
E	10	16	−1.25	1.25	−1.5625

Example C:

Individual	X	Y	z_x	z_y	$z_x z_y$
A	60	12	1.25	.25	.3125
B	45	8	.50	−.75	−.3750
C	40	13	.25	.50	.1250
D	20	6	−.75	−1.25	.9375
E	10	16	−1.25	1.25	−1.5625

In Example B the relationship is a highly negative. In Example C the relationship is very low and negative. Scatter plots for these data sets appear in Figures 8.5B and C, respectively.

As the z score examples illustrate, the magnitude of r_{xy} depends on the extent to which pairs of X and Y scores tend to "go together" or "covary." In z score terminology, do the pairs of scores tend to be the same distance from the means of the score distributions in standard deviation units? If score pairs tend to be the same distance and in the same direction, the correlation is positive and high. If pairs of scores tend to be the same distance, but in opposite directions, the correlation is negative and high. When score pairs are different distances and the directions from the mean are mixed, then r_{xy} will tend to be closer to zero, indicating the lack of a linear relationship.

Look at the data in Table 8.3. It was noted previously that the X and Y z scores tend to have the same sign and their distances from the means are in moderate agreement. Summing the z score cross-products and substituting in formula 8.1:

$$r_{xy} = \frac{16.011}{19} = .843$$

Performance on the Letters posttest and Numbers posttest is highly related based on the sample of 20 students. Students who scored high on the Letters posttest also tended to score high on the Numbers posttest. Students who performed otherwise on the Letters posttest performed in comparable ranges on the Numbers test. While a correlation of .843 is very high, the variables still do not agree perfectly. The performance scores for individuals 137 and 210 illustrate the most glaring inconsistencies between X and Y scores. Were these two inconsistent scores not in the data set, the correlation based on the remaining 18 pairs of scores would be higher.

Using z scores might appear to be an efficient method of calculating r_{xy}. It is, in fact, quite time consuming. Once the z scores for each individual are obtained, the remaining calculations do not take very long. However, transforming each raw score to its z score equivalent does take time. The z score approach is best used to conceptualize and understand how the Pearson correlation coefficient measures the extent to which two sets of scores covary in direction and distance on their respective measurement scales. For direct computation of r_{xy}, however, transformation to z scores is not necessary. Raw scores can be directly substituted in a formula. Formula 8.2 is the raw score formula typically used to compute r_{xy}. It looks complex, but with the exception of one sum, ΣXY, all other quantities should be familiar.

$$r_{xy} = \frac{n\Sigma XY - (\Sigma X)(\Sigma Y)}{\sqrt{[n\Sigma X^2 - (\Sigma X)^2][n\Sigma Y^2 - (\Sigma Y)^2]}} \qquad (8.2)$$

Let's look at the terms in this formula more closely. Six quantities are necessary in the computation: n, ΣX, ΣY, ΣX^2, ΣY^2, and ΣXY. The first five should be familiar. You would need to calculate these indices if you were computing the variance or standard deviation of X and Y using the procedures in Chapter 6. To calculate ΣXY, the X score and Y score are multiplied together for each individual. Then the XY products are summed for all individuals:

$$\Sigma XY = X_1Y_1 + X_2Y_2 + X_3Y_3 + \cdots + X_nY_n$$

The Example C data illustrate the calculation of all necessary sums for use in Formula 8.2:

Individual	X	Y	X^2	Y^2	XY
A	60	12	3600	144	720
B	45	8	2025	64	360
C	40	13	1600	169	520
D	20	6	400	36	120
E	10	16	100	256	160
Sums:	175	55	7725	669	1880

The quantity ΣXY was obtained as follows:

$$\begin{aligned}\Sigma XY &= 60(12) + 45(8) + 40(13) + 20(6) + 10(16)\\ &= 720 + 360 + 520 + 120 + 160 \text{ (values in } XY \text{ column)}\\ &= 1880\end{aligned}$$

Using z scores to calculate r_{xy} from Formula 8.1 for the Example C data results in a correlation coefficient of $-.141$. Substituting the raw score sums in Formula 8.2 will yield the same value.

$$\begin{aligned}r_{xy} &= \frac{5(1880) - (175)(55)}{\sqrt{[5(7725) - (175)^2][5(669) - (55)^2]}}\\ &= \frac{-225}{\sqrt{(8000)(320)}}\\ &= \frac{-225}{1600}\\ &= -.141\end{aligned}$$

Use formula 8.2 to compute r_{xy} for the five pairs of scores in Example A. The X and Y scores are the same but have been paired differently for individuals A–E. ΣXY is the only sum that will change from the Example C data. The ΣX, ΣY, ΣX^2, and ΣY^2 are identical for both data sets:

$$\begin{aligned}\Sigma XY &= 60(16) + 45(13) + 40(12) + 20(8) + 10(6)\\ &= 2245\end{aligned}$$

$$\begin{aligned}r_{xy} &= \frac{5(2245) - (175)(55)}{\sqrt{[5(7725) - (175)^2][5(669) - (55)^2]}}\\ &= \frac{11225 - 9625}{\sqrt{(38625 - 30625)(3345 - 3025)}} = \frac{1600}{\sqrt{(8000)(320)}}\\ &= \frac{1600}{\sqrt{2,560,000}} = \frac{1600}{1600}\\ &= 1.00\end{aligned}$$

Formula 8.2 was developed through the algebraic manipulation of formula 8.1. Several other useful formulas result as intermediary steps in finding formula 8.2. These formulas are developed here to demonstrate the importance of covariability and variability in the determination of r_{xy}. Starting with the z score formula,

$$r_{xy} = \frac{\Sigma z_x z_y}{n - 1}$$

Substituting the computational formula from Chapter 7 for z_x and z_y,

$$r_{xy} = \frac{\dfrac{\Sigma(X - \overline{X})(Y - \overline{Y})}{(s_x)(s_y)}}{n - 1}$$

$$= \frac{\Sigma(X - \overline{X})(Y - \overline{Y})}{(n - 1)s_x s_y}$$

The quantities $(X - \overline{X})$ and $(Y - \overline{Y})$ are deviation scores represented by x and y. Then,

$$r_{xy} = \frac{\Sigma xy}{(n - 1)s_x s_y} \tag{8.3}$$

The numerator Σxy is called the sum of the deviation score cross-products. It may be computed using the following raw score formula:

$$\Sigma xy = \Sigma XY - \frac{(\Sigma X)(\Sigma Y)}{n} \tag{8.4}$$

The sum of the deviation score cross-products, Σxy, provides an index whose magnitude is influenced by the extent to which two sets of scores covary or relate. In fact, if we divide Σxy by $n - 1$, we get a statistical index called the **covariance.** Symbolically, the sample covariance is represented by s_{xy}. Computationally,

$$\text{covariance} = s_{xy} = \frac{\Sigma xy}{n - 1} \tag{8.5}$$

Conceptually, the covariance is much like the variance. While the variance deals only with the variability in *one* set of scores, the covariance identifies the amount of shared variability between *two* variables. It might help to think of the variance as the covariance of a variable with itself. The deviation score formula for the variance is

$$s_x^2 = \frac{\Sigma x^2}{n - 1}$$

This can also be rewritten in the form of formula 8.5:

$$s_x^2 = s_{xx} = \frac{\Sigma x^2}{n - 1} = \frac{\Sigma xx}{n - 1}$$

The numerators of the raw score computational formulas are compatible. For the variance formula,

$$\Sigma x^2 = \Sigma X^2 - \frac{(\Sigma X)^2}{n}$$

$$\Sigma xx = \Sigma XX - \frac{(\Sigma X)(\Sigma X)}{n}$$

Compare this expression to the quantity in formula 8.4,

$$\Sigma xy = \Sigma XY - \frac{(\Sigma X)(\Sigma Y)}{n}$$

Going back to formula 8.3 and substituting s_{xy} in place of $\Sigma xy/(n - 1)$ gives the covariance formula for r_{xy}.

$$r_{xy} = \frac{s_{xy}}{s_x s_y} \tag{8.6}$$

The correlation coefficient, r_{xy}, may be defined as the covariance (s_{xy}) divided by the product of the standard deviations for X and Y. Alternatively, formula 8.3 can be written using deviation scores:

$$r_{xy} = \frac{\Sigma xy}{(n - 1)s_x s_y}$$

$$= \frac{\Sigma xy}{(n - 1)\sqrt{\frac{(\Sigma x^2)}{(n - 1)}\frac{(\Sigma y^2)}{(n - 1)}}}$$

$$= \frac{\Sigma xy}{\sqrt{\Sigma x^2 \, \Sigma y^2}} \tag{8.7}$$

where the raw score formulas for Σx^2, Σy^2, and Σxy are

$$\Sigma x^2 = \Sigma X^2 - \frac{(\Sigma X)^2}{n}$$

$$\Sigma y^2 = \Sigma Y^2 - \frac{(\Sigma Y)^2}{n}$$

$$\Sigma xy = \Sigma XY - \frac{(\Sigma X)(\Sigma Y)}{n}$$

Substituting the raw score formulas for the deviation scores and simplifying gives formula 8.2, the raw score formula for computing r_{xy}.

$$r_{xy} = \frac{\Sigma xy}{\sqrt{\Sigma x^2 \, \Sigma y^2}}$$

$$= \frac{\Sigma XY - \frac{(\Sigma X)(\Sigma Y)}{n}}{\sqrt{\left[\Sigma X^2 - \frac{(\Sigma X)^2}{n}\right]\left[\Sigma Y^2 - \frac{(\Sigma Y)^2}{n}\right]}}$$

$$= \frac{n\Sigma XY - (\Sigma X)(\Sigma Y)}{\sqrt{[n\Sigma X^2 - (\Sigma X)^2][n\Sigma Y^2 - (\Sigma Y)^2]}} \tag{8.2}$$

You have just been bombarded with a lot of information about various formulas for computing r_{xy}. You might possibly be a little bewildered and frustrated about which formula you use when. Think through what you have just read as we try to clarify and summarize the important points. In any practical situation, we would start with the raw score data. If this is the case, the six raw score quantities (n, ΣX, ΣY, ΣX^2, ΣY^2, and ΣXY) are necessary to compute r_{xy} from any of the formulas. It is very important that you realize this. Formulas 8.1, 8.2, 8.6, and 8.7 are algebraically equivalent to each other and all use the same information to compute r_{xy}. The important indices and formulas for computing r_{xy} are as follows:

Variance for X | *Variance for Y* | *Covariance for XY*

$$\Sigma x^2 = \Sigma X^2 - \frac{(\Sigma X)^2}{n} \qquad \Sigma y^2 = \Sigma Y^2 - \frac{(\Sigma Y)^2}{n} \qquad \Sigma xy = \Sigma XY - \frac{(\Sigma X)(\Sigma Y)}{n}$$

$$s_x^2 = \frac{\Sigma x^2}{n - 1} \qquad s_y^2 = \frac{\Sigma y^2}{n - 1} \qquad s_{xy} = \frac{\Sigma xy}{n - 1}$$

Correlation formulas

(z score Formula 8.1) $$r_{xy} = \frac{\Sigma z_x z_y}{n - 1}$$

(Covariance Formula 8.6) $$r_{xy} = \frac{s_{xy}}{s_x s_y}$$

(Deviation Score Formula 8.7) $$r_{xy} = \frac{\Sigma xy}{\sqrt{\Sigma x^2 \, \Sigma y^2}}$$

(Raw Score Formula 8.2) $$r_{xy} = \frac{n\Sigma XY - (\Sigma X)(\Sigma Y)}{\sqrt{[n\Sigma X^2 - (\Sigma X)^2][n\Sigma Y^2 - (\Sigma Y)^2]}}$$

TABLE 8.4
Values used to compute a correlation coefficient.

Individual	Letters Posttest (X)	Which Comes First (Y)	X^2	Y^2	XY	z_x	z_y	$z_x z_y$
9	50	7	2500	49	350	1.51	0	0
45	8	5	64	25	40	−1.13	−1.08	1.22
56	14	5	196	25	70	−.75	−1.08	.81
73	23	9	529	81	207	−.19	1.08	−.21
92	36	6	1296	36	216	.63	−.54	−.34
129	17	6	289	36	102	−.56	−.54	.30
168	14	8	196	64	112	−.75	.54	−.41
222	46	10	2116	100	460	1.25	1.62	2.03
TOTAL	208	56	7186	416	1557	0	0	3.423

Scores on the Letters posttest (X) and the Which Comes First test (Y) for eight individuals from the data base provide a practice set. We can compute the correlation coefficient using each of the four methods, formulas 8.1, 8.2, 8.6, and 8.7. Table 8.4 lists the raw-score values, z scores, and sums used.

$$\Sigma x^2 = 7186 - \frac{208^2}{8} = 1778 \qquad \Sigma y^2 = 416 - \frac{56^2}{8} = 24 \qquad \Sigma xy = 1557 - \frac{(208)(56)}{8} = 101$$

$$s_x^2 = \frac{1778}{7} = 254 \qquad s_y^2 = \frac{24}{7} = 3.43 \qquad s_{xy} = \frac{101}{7} = 14.43$$

$$s_x = 15.94 \qquad s_y = 1.85$$

$$\Sigma z_x z_y = 3.423; \; r_{xy} = \frac{3.423}{7} = .489 \tag{8.1}$$

$$r_{xy} = \frac{s_{xy}}{s_x s_y} = \frac{14.43}{15.94(1.85)} = .489 \tag{8.6}$$

$$r_{xy} = \frac{\Sigma xy}{\sqrt{\Sigma x^2 \, \Sigma y^2}} = \frac{101}{\sqrt{(1778)(24)}} = \frac{101}{206.57} = .489 \tag{8.7}$$

$$r_{xy} = \frac{8(1557) - (208)(56)}{\sqrt{[8(7186) - 208^2][8(416) - 56^2]}} = \frac{808}{\sqrt{(14224)(192)}} \tag{8.2}$$

$$= \frac{808}{1652.58} = .489$$

8.4 COVARIANCE: A NONSTANDARDIZED MEASURE OF ASSOCIATION

As defined in section 8.3, the covariance is an index whose magnitude depends on information from both sets of scores. The covariance formula requires that the cross-products of the deviation scores, x and y, be found for each individual and then summed over individuals, Σxy.

$$s_{xy} = \frac{\Sigma xy}{n - 1}$$

Consider the deviation scores. They provide important information. Each deviation score identifies the distance and direction of a raw score from the mean. If both x and y are positive or if both are negative, then the cross-product, Σxy, will be positive and all xy products will contribute to increasing Σxy. In this instance there is a systematic match between X and Y scores and the correlation coefficient will be positive. When x and y are of opposite signs for all individuals, then xy will be negative and Σxy will be large, but negative. The correlation will be negative, indicating an inverse relationship. The greater the number of mixes of x and y paired combinations (some both positive, some negative, and some of opposite signs), the lower the value of r_{xy}.

The covariance between two sets of scores is the index that includes the information on how related two variables are. The covariance term includes the cross-product or extent of matching information. However, the covariance does not have constant upper or lower bounds across different data sets. The size of the covariance depends entirely on the amount of variability in the X and Y scores. If X or Y have large variances, then s_{xy} can be relatively large. If there is little variability in X and Y, then s_{xy} will be smaller. It can be shown that the upper bound to the covariance value for any set of data is the product of the X and Y score standard deviations, $s_x s_y$. The lower bound is the negative value of this product, $-s_x s_y$. If the correlation is less than perfect, the covariance will be somewhere between $s_x s_y$ and $-s_x s_y$, as limits.

Covariance is a nonstandardized measure of relationship. Nonstandardized means that its limits change depending on the scale used to measure X and Y and the amount of variability on the X and Y scales. Due to this lack of constancy across situations, covariance cannot be universally interpreted. Rather, interpreting a covariance value depends on the variables and the situation.

To get around this limitation of the covariance as an index of relationship, s_{xy} is standardized by considering the amount of variability in the X and Y scores. The covariance is standardized by dividing it by the product of the standard deviations for X and Y: $s_{xy}/s_x s_y$. This standardized covariance index is the Pearson correlation coefficient, r_{xy}. Standardization produces an index, r_{xy}, that has constant interpretation. No matter what the scale is or how much the scores vary, a value for r_{xy} is interpreted the same from one research setting to another. This is the value of a standardized index, universal interpretation. The Pearson r is just such a standardized index.

8.5 INTERPRETING THE CORRELATION COEFFICIENT

Basic information about r_{xy} necessary to interpretation has already been covered. An r of ± 1, representing a perfect relationship, does *not* imply that X and Y scores for an individual are identical. Rather, it implies that they are the same distance in standard score units from the distribution means. If the X and Y scale means and standard deviations are different, the X and Y raw scores will be different for an individual. If $r_{xy} = 1.00$ or -1.00, however, a transformation of raw scores to z scores will result in z scores whose absolute values are identical. The z scores will differ only in sign if the relationship is inverse, that is, $r_{xy} = -1.00$. The midpoint of the correlation coefficient scale, $r_{xy} = .00$, is the value indicating no relationship in the score patterns between two variables: they are independent. As the values of r_{xy} deviate from .00, the magnitude of the relationship increases. The sign of r_{xy} only indicates the direction of the relationship and has nothing to do with its magnitude. While r_{xy} as a descriptive index has constant meaning across settings, judging the merit of a specific coefficient depends on the research situation. A correlation coefficient of .72 between two tests purporting to measure achievement in the same domain might not be considered a *good* coefficient even though it is in a moderate to high range. Alternatively, an r_{xy} value of .35 might be viewed as an impressive coefficient if previous research had consistently found no relationship. Table 8.5 lists some general guidelines for interpreting r_{xy}. Statistical results should always be interpreted within the context of the variables under study, the purpose of the study, and findings from previous research.

An important point when trying to conceptualize the Pearson correlation coefficient scale is that the coefficient scale values are not of equal intervals. The difference in a relationship of .40–.50 does *not* represent the same magnitude of relationship difference as .70–.80. The correlation scale is ordinal, not interval or ratio; a value of $r_{xy} = .80$ does not represent twice the relationship as $r_{xy} = .40$. However, if we square r_{xy}, descriptive comparative statements can be made. The correlation coefficient squared, r^2_{xy}, represents the proportion of variance in common between two variables. Think of each variable, X and Y, as having one unit of variance. Figure 8.6 provides some illustrations of r^2_{xy} as the proportion of variance overlap between X and Y. As the proportion of variability in common, r^2_{xy} indicates the amount of X score and Y score variability among individuals that appears to be related or due to the same sources. The correlation coefficient squared provides an added dimension for interpreting and understanding the magnitude of r_{xy}. A correlation of .5, for example, is in the moderate range. When squared, however (r^2_{xy}), the result is .25, indicating that the two variables share 25% of their respective variances. Thus, 75% of the differences among individuals' scores (as defined by the variances for each variable) is not accounted for between the variables. The development of r^2_{xy} as an interpretive index is presented in Chapter 9. A complete understanding is easier with a basic knowledge of regression.

TABLE 8.5
Guidelines for interpreting r_{xy}.

Range of Positive Coefficients	Type of Relationship	Range of Negative Coefficients
1.00–.80	High	–.80–(–1.00)
.80–.60	Moderate to High	–.60–(– .80)
.60–.40	Moderate	–.40–(– .60)
.40–.20	Low to Moderate	–.20–(– .40)
.20–.00	Low	.00–(– .20)

FIGURE 8.6
The squared correlation coefficient as the variance shared by *X* and *Y*.

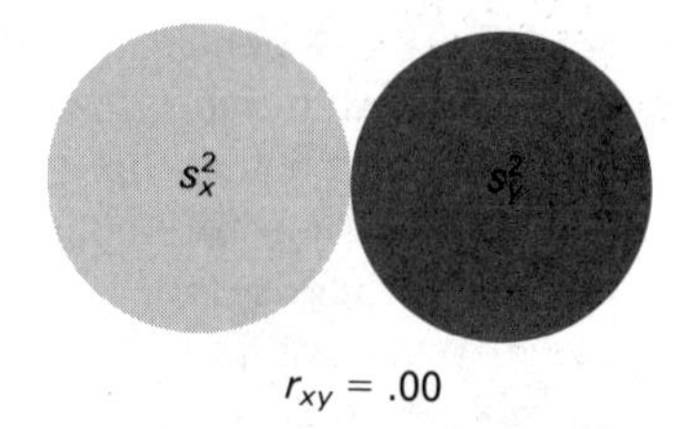

$r_{xy} = .00$

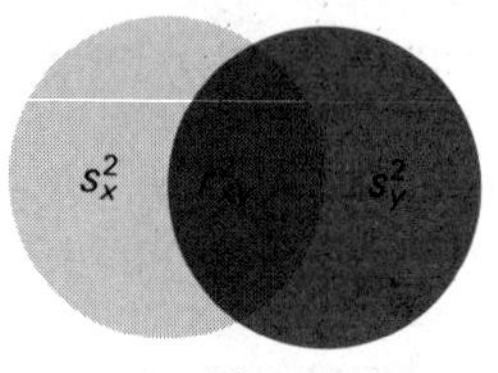

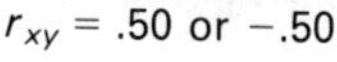

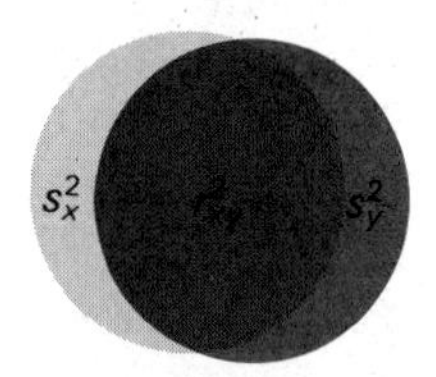

$r_{xy} = .90$ or $-.90$

Causation and Correlation

A Pearson r value other than zero indicates only that a linear relationship exists between two variables. It is incorrect to infer cause-and-effect relationships based on a nonzero value of r_{xy}. That two variables are related is no basis for concluding that one variable causes another. This interpretation should not be made for most cases. As an example, attitude toward school and mathematics achievement are related. The relationship reported in the literature is an r_{xy} value of around .40. While this relationship is not large, it does indicate that the two variables are positively related. Students with positive attitudes tend also to have higher mathematics achievement scores. The relationship exists, but does one variable cause the other? Do positive attitudes cause higher mathematics achievement? Does high mathematics achievement result in a more positive attitude toward school? The correlation coefficient alone cannot answer these questions. What it does show is that scores on the two variables moderately tend to go together. What causes persons to have high or low scores on either variable may be due to a third or combination of other variables. Indeed, the correlation coefficient may provide a clue to a possible causal connection. At this point, however, experimental manipulation of one of the variables would be required to test the hypothesis of a causal link.

A news release reported that a researcher had found a relationship between the weight of individuals and salaries in a sample of employees from several major firms. The implication in the news release was that persons who tend to be overweight have a better chance of getting jobs with a higher salary. This is an example of incorrect cause-and-effect inference based on correlational research. A single correlation coefficient does not provide sufficient evidence to establish a causal connection between two variables.

The causal connection between variables is very difficult to establish without strict experimental manipulation of one of the variables. However, a growing body of literature is examining methodology for inferring causation from correlational analyses. This methodology is called causal analysis or path analysis. The basic methodology is to examine the relationship among variables using correlational techniques within the context of a theoretical network of variables where the logic of the network defines the hypothesized causal links among variables. Analyzing relationships using correlation coefficients within this network supports conclusions on causation.

8.6 FACTORS AFFECTING THE CORRELATION COEFFICIENT

Curvilinear Effects

The Pearson correlation coefficient is a measure of only the *linear* component of a relationship. The higher the relationship, the more the data points in a scatter diagram tend to cluster in a straight line pattern. It is not sensitive to any higher order curvilinear trends in the data. An r_{xy} value around .00 *does* indicate that there is *no linear* relationship, but it *does not* imply that there is *no* relationship of any kind. The relationship between two variables could be quite strong, but entirely curvilinear. Figures 8.7A and B demonstrate two scatter diagrams in which r_{xy} equals zero. In A, there is no relationship of any kind. The boundary of the data points forms a circle and the points do not cluster in a straight or curved line. In B, there is no linear relationship, but a fairly strong curvilinear relationship. Similarly, one might compute an $r_{xy} = -.39$ for a set of data and conclude that the relationship is a low-to-moderate, inverse relationship. This statement would be true for the linear component. As Figure 8.7C illustrates, however, a much stronger curvilinear relationship could exist between the two variables.

While the majority of relationships tend to be linear and therefore appropriately described by r_{xy}, several curvilinear relationships are being reported in the literature. The Pearson r investigates only a linear relationship. A plot of the scatter diagram is suggested if any curvilinear relationship is suspected at

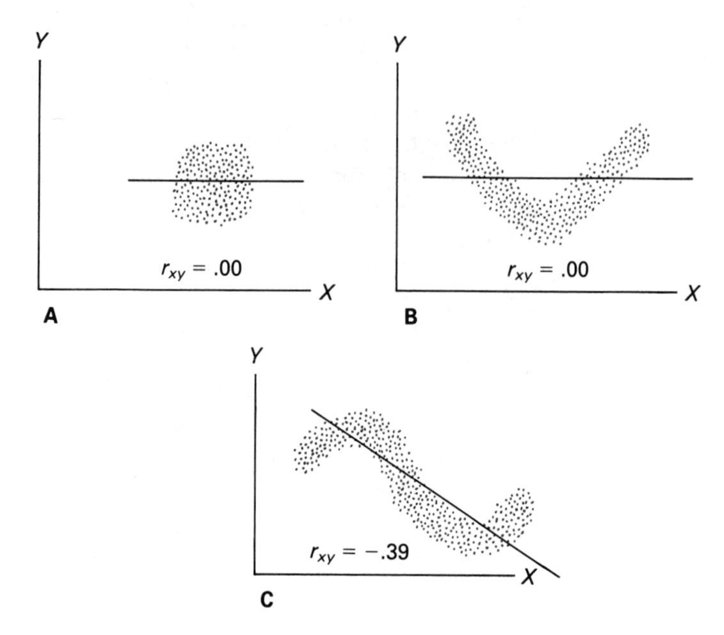

FIGURE 8.7
Linear and curvilinear scatter diagrams. **A,** no linear or curvilinear relationship; **B,** no linear, but strong curvilinear relationship; **C,** low-to-moderate linear and strong curvilinear relationship.

all. If the data exhibit a pattern other than that of a straight line, r_{xy} will underestimate the true magnitude of the relationship.

Distribution Shape Effects

The Pearson correlation coefficient was developed for continuous X and Y scores under the assumption that they were normally distributed. When these distributions have an approximately normal shape, r_{xy} can range across the full scale from -1.00 to 1.00. Actually, r_{xy} values can potentially range across the full scale whenever the X and Y score distributions are symmetrical. This is the necessary condition and not normality. Nonsymmetrical shapes will restrict the possible range of r_{xy}. In a scatter diagram, the X score and Y score distributions are called marginal. The distribution of data points in the scatter plot is called a bivariate distribution to represent the simultaneous distribution of two variables (bivariate). If the marginal distributions are normal, the bivariate distribution will take the shape of an ellipse (see Figure 8.8A). If the marginal distributions are both skewed in the same direction, the bivariate distribution will appear in a banjo shape (Figures 8.8B and C). Skewed distributions violate the assumption of homoscedasticity (discussed in Chapter 9) and restrict the range of r_{xy}.

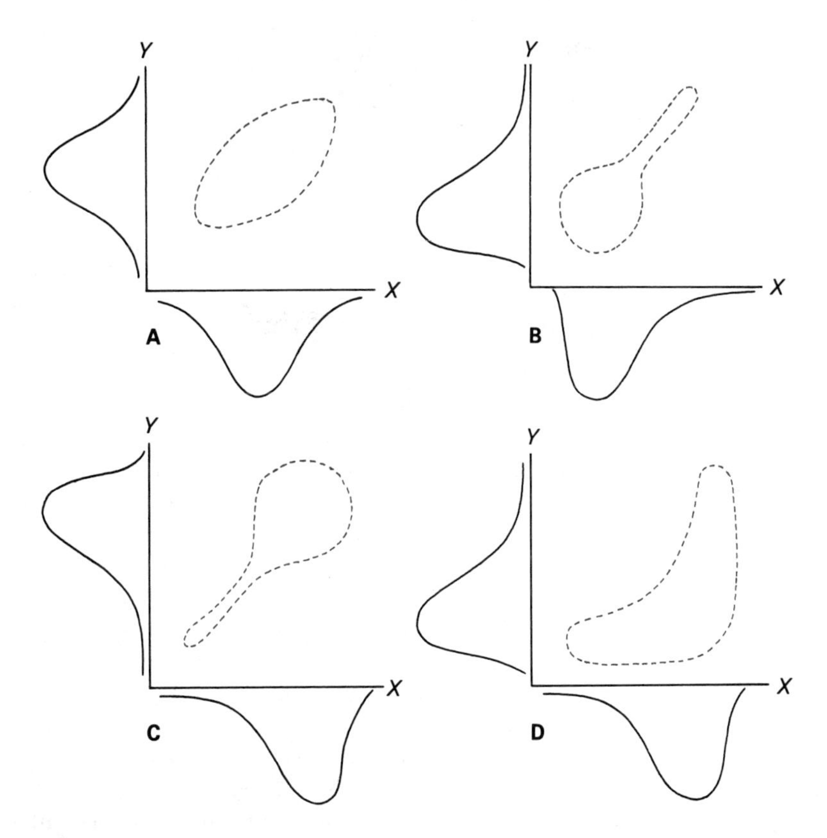

FIGURE 8.8
Marginal distribution shape effects on scatter diagrams.

Probably the most serious effect of the marginal distribution shapes on r_{xy} occurs when two distributions are skewed in opposite directions (Figure 8.8D). The result is an artificial, curvilinear effect in the relationship, thus severely restricting the potential range of r_{xy}. Consider the following set of five X and Y scores. The distribution of Y scores is the exact inverse of X scores. One might think r_{xy} would equal -1.00. Calculate r_{xy}, then plot the data points in a bivariate frequency distribution.[4]

---

[4] $r_{xy} = \dfrac{5(65) - (19)(19)}{\sqrt{[5(85) - (19)^2][5(85) - (19)^2]}} = -.56$

X	Y
5	1
5	3
5	5
3	5
1	5

$\Sigma X = 19 \quad \Sigma Y = 19$

$\Sigma X^2 = 85 \quad \Sigma Y^2 = 85$

$\Sigma XY = 65$

Restriction of Range Effects

Other things being equal, the less variability in the scores for a variable, the lower the correlation between that variable and other variables. Two common things which a researcher has control over may lessen the variability on a variable. The first is the homogeneity of the population selected for study and the second is the scale used to measure the variable under study. Looking at the relationship of aptitude scores to other variables across education levels illustrates the restriction of range concept. The correlation between aptitude and another variable will progressively decrease from an elementary school population to a high school population to an undergraduate population to a graduate school population. The reason for this decrease in the correlation coefficient is that each subsequent group is more homogenous in aptitude scores. A selection process at each level restricts (lessens) the range of aptitude scores and therefore lowers the correlation coefficient.

The relationship of high school grade point average (gpa) to college gpa's for a group of academic scholarship students will not be nearly as high as for the entire college student population. Figure 8.9A illustrates this effect. A high school gpa cutoff score is used for scholarship selection, thus restricting the gpa range. The elipse portion of the diagram (representing the scholarship population) is almost circular.

While you should be aware that group homogeneity will restrict the range of a variable or both variables and lower r_{xy}, the opposite effect can occur and artificially increase r_{xy}. If the sample under study contains two distinct groups in which the means on X and Y are greater for one group than the other, combining the groups will result in increased variability and a higher r_{xy} value than r_{xy} for each separate group. Male-female differences provide a natural example of this phenomenon. Many studies sample both females and males and treat them as a combined group. If the members of one group score higher on the variables under investigation, the correlation coefficient will be higher for the combined group than for the groups separately. Figure 8.9B illustrates a situation in which there is no relationship within groups (circular scatter plots), but the combined scatter plot would indicate a relationship.

The "Sesame Street" data illustrate the group difference effect and an increase in r_{xy} when two groups are combined. Two different groups were selected, those sampled from Viewing Category 1 and those from Viewing

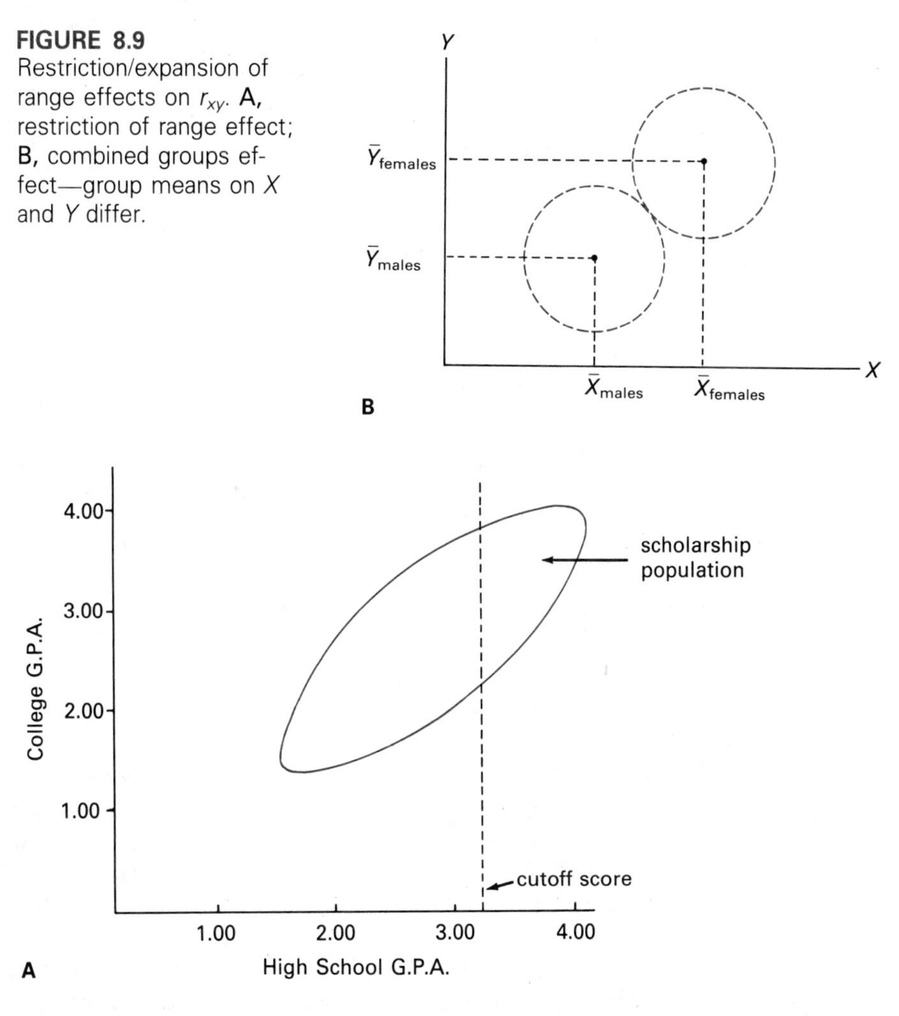

FIGURE 8.9
Restriction/expansion of range effects on r_{xy}. A, restriction of range effect; B, combined groups effect—group means on X and Y differ.

Category 4. Forms Posttest (X) and Peabody mental age (Y) scores were the variables selected for illustration. Table 8.6 shows the means, standard deviations, and correlation coefficients for each group and for the combined groups. As the data indicate, the means for the Viewing Category 4 group are considerably higher than for the other group. The correlations within each group are relatively low ($r_1 = .31$ and $r_4 = .28$). When the two groups are combined and r_{xy} calculated on all 116 students as a single group, the correlation coefficient increases to .40.

In addition to sampling effects, the researcher can affect the range of scores by the scale selected to measure the variable. Using one item to measure job satisfaction will reduce variability and restrict job satisfaction scores in comparison to using a 10-item scale with scores ranging from 0 to 10. The correlation between job satisfaction and other variables will be greater using

TABLE 8.6
Means, standard deviations, and correlation coefficients for two viewing categories.

Viewing Category 1 (n = 54)	Viewing Category 4 (n = 62)	Combined (n = 116)
$\overline{X}$ = 9.91	$\overline{X}$ = 13.15	$\overline{X}$ = 11.64
$\overline{Y}$ = 40.70	$\overline{Y}$ = 52.04	$\overline{Y}$ = 46.77
s_x = 3.04	s_x = 3.05	s_x = 3.02
s_y = 11.19	s_y = 17.74	s_y = 16.03
r_{xy} = .31	r_{xy} = .28	r_{xy} = .40

the scale with more values. Within limits, the more variability one can induce by increasing the scale values, the higher the correlation coefficient between that variable and other variables.

8.7 OTHER MEASURES OF RELATIONSHIPS

The Pearson correlation coefficient is by far the most widely used measure of relationship. Its use is so common that unless otherwise specified, any index of association reported can be assumed to be the Pearson r. While the appropriate use of r_{xy} as the correlation index is widespread, sometimes application of another index of relationship is more appropriate. Some of the correlational indices are special cases of the Pearson coefficient and actually are Pearson rs, but other names have been attached to them. The computational formulas for these indices are not given, but may be found in more advanced statistics texts.

The Pearson correlation coefficient originally was intended for use with continuously scaled X and Y variables assumed to be on an interval scale of measurement. If one or both of the variables is dichotomous (having only two scale points or categories), then other correlation coefficient indices are available. Sex is an example of a dichotomous variable. Smoker/nonsmoker is another example. A variable with only two categories or a continuous variable that has been reduced to two categories is dichotomous. The kind of dichotomy is very important in identifying which correlation index is defined. Two kinds of dichotomies exist, genuine and artificial. A genuine dichotomy exists naturally in the real world; it is a dichotomy not produced or defined by the experimenter. Examples of genuine dichotomies are sex (male/female), homeowner versus nonhomeowner, multiple-choice item score (right/wrong), and smoker versus nonsmoker. Each of these variables are sufficiently discrete and discontinuous to be genuine dichotomies. An artificial dichotomy results when a variable is continuously measured, but for some reason is reduced to two categories. Breaking a group at the median and creating two groupings (above and below the median) is an artificial dichotomy. On a criterion-referenced test, a total score standard is often set to identify students who are competent or not competent in the skills tested. The competent/noncompetent categorization creates an artificially dichotomized variable. Dichotomous variables are usually assigned score values of 1 and 0 to differentiate the two categories defining the variable.

Four coefficients have been named and special computational formulas developed for identifying the magnitude of relationships when at least one variable is dichotomous. The coefficients are the point biserial (r_{pb}); the biserial (r_{bis}); the phi coefficient (r_{phi} or φ), and the tetrachoric (r_t) correlation coefficient. Each coefficient is appropriately applied when the X and Y variables have the following characteristics.

Name of Correlation Coefficient	Variable Characteristics
Point Biserial (r_{pb})	one variable genuine dichotomy, other variable continuous
Biserial (r_{bis})	one variable artificial dichotomy, other variable continuous
Phi (φ or r_{phi})	both variables genuine dichotomies
Tetrachoric (r_t)	both variables artificial dichotomies

The point biserial and phi coefficients are Pearson correlation coefficients with special computational formulas. Applying any of the Pearson r computational formulas to data appropriate for r_{pb} or r_{phi} would produce the same numerical value as the index of relationship. The biserial and tetrachoric coefficients represent estimates of r_{xy} if continuous data were available, but they are not true Pearson correlation coefficients.

Other characteristics of the data will dictate the use of other coefficients of correlation. If one or both variables are measured at the ordinal level (ranked data) of measurement, a coefficient called Spearman's rank-order correlation (r_ρ) is appropriate. Like r_{pb} and r_{phi}, r_ρ is a special case of the Pearson r when applied to data whose score values are ranks. The Spearman formula is generally simpler and more efficient to apply when computing an index of the magnitude of the relationship between a pair of variables measured in ranks. The computational formula for Spearman's coefficient is

$$r_\rho = 1 - \frac{6\Sigma d^2}{n(n^2 - 1)}$$

where d is the difference in rank values between the X and Y measures for each individual and n is the number of individuals on which rank order data exist.

We have repeatedly indicated that the Pearson r is an index of the magnitude of the linear component in the relationship between X and Y. If data are curvilinearly related, other procedures and indices are available to describe the magnitude of this relationship. The correlation ratio represented by eta squared (η^2) is one such index. Several other coefficients exist for different data situations, but the ones already mentioned are the most commonly found. It is important to be aware that while the Pearson r is appropriate for use in the majority of situations, other coefficients apply in specific situations. Think about the characteristics of your data before you automatically apply the Pearson r.

8.8 SUMMARY

The Pearson r is a statistical index indicating the extent to which two variables are linearly related. In addition to r_{xy}, covariance (s_{xy}) was introduced as another statistical index of importance. Computationally, five raw score summation quantities are necessary to obtain the statistical indices of variability and covariability magnitude for two variables: ΣX, ΣX^2, ΣY, ΣY^2, ΣXY.

Using the five raw score quantities and knowledge of the sample size (n), we can summarize the most important formulas as follows:

- *Calculation of sample means*
  Quantities needed: ΣX, ΣY, n

$$\overline{X} = \frac{\Sigma X}{n} \qquad \overline{Y} = \frac{\Sigma Y}{n}$$

- *Calculation of sample (a) sum of squared deviations, (b) variance,* and *(c) standard deviation*
  Quantities needed: ΣX, ΣX^2, ΣY, ΣY^2, n

$$\Sigma x^2 = \Sigma X^2 - \frac{(\Sigma X)^2}{n} \qquad \Sigma y^2 = \Sigma Y^2 - \frac{(\Sigma Y)^2}{n}$$

$$s_x^2 = \frac{\Sigma x^2}{n-1} \qquad s_y^2 = \frac{\Sigma y^2}{n-1}$$

$$s_x = \sqrt{s_x^2} \qquad s_y = \sqrt{s_y^2}$$

- *Calculation of sample (a) sum of deviation cross-products* and *(b) covariance*
  Quantities needed: ΣXY, ΣX, ΣY, n

$$\Sigma xy = \Sigma XY - \frac{(\Sigma X)(\Sigma Y)}{n}$$

$$s_{xy} = \frac{\Sigma xy}{n-1}$$

- *Calculation of sample Pearson correlation coefficient*
  Quantities needed: ΣX, ΣX^2, ΣY, ΣY^2, ΣXY, n

$$\text{Raw score formula: } r_{xy} = \frac{n\,\Sigma XY - (\Sigma X)(\Sigma Y)}{\sqrt{[n\,\Sigma X^2 - (\Sigma X)^2][n\,\Sigma Y^2 - (\Sigma Y)^2]}}$$

$$\text{Deviation score formula: } r_{xy} = \frac{\Sigma xy}{\sqrt{\Sigma x^2\,\Sigma y^2}}$$

$$\text{covariance formula: } r_{xy} = \frac{s_{xy}}{s_x s_y}$$

$$z \text{ score formula: } r_{xy} = \frac{\Sigma z_x z_y}{n - 1}$$

The Pearson correlation coefficient ranges between -1.00 and $+1.00$, with the sign indicating the direction of the relationship. An r_{xy} value of .00 indicates the absence of a relationship between variables X and Y. Table 8.5 gives general guidelines for interpreting r_{xy} values of different magnitudes. A useful interpretive index is the squared correlation coefficient, r_{xy}^2. It represents the amount of X score and Y score variability that is related or due to the same sources.

PROBLEMS

Problem Set A

1. For these sets of paired scores, compute each of the following:
a. $\bar{X}$ b. Σx^2 c. s_x d. $\bar{Y}$ e. Σy^2 f. s_y g. Σxy h. s_{xy} i. r_{xy} (using the raw score formula) j. r_{xy} (using the deviation score formula) k. r_{xy} (using the covariance formula) l. r_{xy} (using the z score formula)

Individual	A	B	C	D	E	F
X Score	2	5	11	5	7	6
Y Score	3	4	5	6	9	9

2. For the paired scores in problem 1, what would you predict would happen to the value of r_{xy} if the Y scores for Individuals C and F were interchanged (i.e., C had a Y score of 9 and F had a Y score of 6)? Base your prediction on the change in the paired z_x and z_y scores, then calculate r_{xy} to see if your predicted increase or decrease was true.

3. The Y scores in the problem 1 data set are already arranged in order of increasing size. Rearrange the X scores in order of increasing size and compute r_{xy} for the paired data points. Why isn't the relationship a perfect one? Plot the relationship in a scatter diagram.

4. Rearrange the X score values in the problem 1 data set in descending order. Calculate r_{xy} from the set of paired scores resulting from this arrangement. Why does it differ in value (ignoring the sign) from the coefficient calculated in problem 3?

5. Add 10 points to each Y score in the problem 1 data set. Calculate $\overline{Y}$, s_y, s_{xy}, and r_{xy} for this new set of paired x and y scores. What effect does adding a constant seem to have on each of these statistical indices?

6. Multiply each Y score in the problem 1 data set by 4. Calculate $\overline{Y}$, s_y^2, s_y, s_{xy}, and r_{xy} for this new set of paired x and y scores. What effect does multiplying each score by a constant seem to have on the value of each statistical index?

Problem Set B

A study by Prawat (*Journal of Educational Psychology,* 1976) attempted to map the affective domain in young adolescents. Data were collected on locus of control, achievement motivation, and achievement. The Bialer Locus of Control Scale was used to measure the degree of internal versus external locus of control. Persons who attribute their successes and failures to fate, chance, or other people are designated *exter-*

TABLE 8.7
Simulated external/internal data.

Females			Males		
Locus of Control (X)	Achievement Motivation (Y)	Achievement (W)	Locus of Control (X)	Achievement Motivation (Y)	Achievement (W)
15	7	28	12	5	29
18	9	30	4	2	31
8	7	20	8	5	26
19	9	24	16	4	27
22	10	24	4	7	30
15	8	27	12	7	32
15	6	26	6	6	32
23	10	31	9	5	30
18	6	17	14	9	31
14	7	30	10	8	34
16	4	32	17	8	36
7	5	24	9	4	33
12	4	20	21	10	24
9	6	24	11	7	28
16	6	34	13	5	25
11	8	27			
21	7	31			
13	4	26			
10	3	21			
14	5	20			

nals. Those who accept personal responsibility for what happens to them are labeled *internals.* A high score on the Bialer scale indicates internality. The Herman's Prestatie Motivatie Test was used to measure achievement motivation. The achievement variable was measured by the language portion of the Stanford Achievement Test battery. One purpose of the study was to investigate the differential patterns of relationships among these three variables for males and females.

Simulated data on 20 females and 15 males are presented in Table 8.7 for all three variables. The three scores appear across rows for a single individual. Answer the problems that follow using the simulated scores.

1. Calculate each of the following indices separately for males and females:
   a. means for each variable
   b. variance and standard deviation for each variable
   c. covariance between Locus of Control (X) and Achievement (Y)
   d. Pearson correlation coefficient for each pair of variables. This problem requires six coefficients: r_{xy}, r_{xw}, and r_{yw} for females and for males.

2. Compare the score distributions between males and females based on the means and standard deviations.

3. Descriptively compare the relationships among the three variables between females and males. Can you see any characteristics of the data that might influence the magnitude of a relationship for one group and not the other? What conclusion(s) can you draw about the relationships among affective and achievement variables?

4. Plot the relationship between two variables for males and females on the same scatter diagram. Given the configuration of data points and the means for each group, would the correlations for the two groups combined be higher, lower, or about the same?

Problem Set C

We might hypothesize that a child's knowledge of letters at the time of the pretest in the "Sesame Street" data would be related to the Home Education Index. To investigate this relationship, use the data for the following students selected from Appendix A.

Student ID Number	Student ID Number	Student ID Number
7	91	166
12	96	182
19	109	188
21	119	189
30	126	194
40	146	202
56	150	226
67	153	230
71	160	234
82	163	238

Respond to the following exercises using the Letters pretest scores (X) and Home Education Index (HEI) scores (Y) for the 30 students listed.

1. Construct a frequency distribution for each variable using interval widths of three for Letters pretest and two for HEI. Then graph the frequencies in a frequency polygon. Also, plot the bivariate frequency distribution for the two variables. How would you classify the shapes of the distribution? Will the shapes adversely affect the relationship? What would you estimate the magnitude of r_{xy} to be based on the scatter plot?

2. Calculate the means and standard deviations for each variable.

3. Calculate the covariance between the two variables. Based on the value of the covariance, is the relationship between Letters pretest scores and HEI scores direct or inverse?

4. Calculate the Pearson correlation coefficient between the two variables. Interpret the relationship between pretest knowledge of letters and HEI based on the magnitude of r_{xy}.

9 Prediction and Regression

As tools in the research process, statistical methods help describe, explain, predict, and control certain phenomena—all of which are goals of scientific inquiry. So far, chapters have focused mainly on descriptive statistical indices that summarize information about phenomena for given groups of individuals or objects under study. These methods can answer descriptive questions such as:

1. How is the phenomenon distributed for this group?
2. What is the "average" value (amount of the phenomenon) for this group?
3. How much do the individuals in the group differ on the phenomenon?
4. How does an individual's measured value on the phenomenon compare to others' scores or scores on other phenomena in a standardized sense?
5. What is the extent of the association between two variables for a given group?

These questions are important and necessary to any research investigation. They are crucial to knowing "what your data look like." The answers to the questions all provide descriptive information, however. They do not attend to the "explanation," "prediction," or "control" goals of science.

To explain, predict, or control phenomena, we do not view variables in isolation. Rather, *how* variables relate or do not relate to other variables provides clues and information allowing us to explain, predict, and possibly control these variables. Examining these relationships leads to the formation of networks of variables that provide the basis for theories about a behavior or phenomenon. The study of the Pearson correlation coefficient gave us a statistical technique for efficiently summarizing the magnitude of the *linear* relationship between pairs of variables, but it did not provide the tools for trying to predict from the relationship. This purpose is accomplished through statistical procedures called **linear regression analysis.** Simple linear regression procedures are the focus of the current chapter. They are simple in the sense that the model to be introduced is the simplest linear regression model, examining the predictive or explanatory relationship for only two variables and only for linear relationships.

At the beginning of Chapter 8 we attempted briefly to differentiate between the correlation model and the regression model approaches to studying relationships between variables. The distinction is relevant and bears repeating. The correlation model makes no distinction between roles of the two variables in the study. Applying procedures based on a correlation model merely identifies the degree of the relationship between the variables. In contrast, a study that focuses on explaining or predicting one of the variables on the basis of information on the other variable should use a regression model approach. The regression model examines changes in one variable as a function of changes or differences in values of the other variable.

The regression model labels variables according to their roles. The variable whose variation we want to explain or predict is called the **criterion** (dependent) **variable.** Simple regression analysis details the functional relationship between the criterion variable and the other variable in the study, the **predictor** (independent) **variable.** The analysis aims to determine how, and to what extent, the criterion variable varies as a function of changes in the predictor variable. Simple linear regression describes how we can predict systematic changes in the criterion variable based on changes in the predictor variable. For example, suppose we want to study the variability in mathematics achievement as a function of the amount of on-task learning time. In this study, mathematics achievement is the criterion variable and time on-task is the predictor variable. A regression analysis of the relationship between the two variables would describe or estimate the expected change in achievement for a unit change in amount of on-task learning time. It also would permit prediction of a level of achievement for a fixed value of on-task learning time and provide some idea of the amount of expected error in that prediction.

Another study might focus primarily on predicting values for the criterion variable using the predictor variable, making no attempt to explain the existing relationship. For example, a university's first-year student enrollment is an important variable affecting various operations. Accurate predictions of a future year's enrollment level is critical to advance planning. The first step in predicting is identifying a variable with scale values that (a) are systematically related to the variable to be predicted and (b) are known so they are useful in prediction. In this example, the number of high-school graduates for the prior academic year might work. Suppose we examined the relationship between the two variables to determine the potential prediction capabilities. Which of the two variables—first-year student enrollment or number of recent high-school graduates—would you identify as the criterion variable and which as the predictor variable?[1] This study would try to derive a prediction equation based on the functional relationship established between first-year student enrollment and number of recent high-school graduates. Then, once we know the number of graduates in a given year, we could use that equation to predict the expected first-year student enrollment.

We can investigate a virtually limitless number of criterion variables by applying linear regression. Usually, the criterion variable in a study is easily identifiable. It is the variable of primary interest, the one that we want to explain or predict. The desire to gain more knowledge about the criterion

[1]The criterion variable is the first-year student enrollment. We want to predict specific yearly values on this variable based on the known fixed values of the predictor variable, number of recent high-school graduates.

variable is generally the impetus for conducting a study. Some recognizable examples of criterion variables include

1. achievement performance level in a content area (e.g., reading)
2. leadership ability
3. blood pressure level
4. teaching effectiveness
5. job satisfaction
6. employment longevity
7. foster home placement desirability
8. success in college
9. Letters posttest scores in the "Sesame Street" data
10. success in an introductory statistics class

While the criterion variable often is of primary interest, any study for which regression analysis is appropriate requires investigation of a predictor variable also. Identifying, defining, and measuring the predictor variable is a precondition to performing the analysis. The study produces more knowledge about the criterion variable by examining its relationship to the predictor variable. What variables might serve as predictor variables for the preceding list of criterion variables?[2]

A criterion variable does not always act as a criterion variable. The label is specific to the context of a specific research problem. In some situations all of the variables we just listed as criterion variables could function as effective predictor variables. For example, leadership ability could help predict organizational productivity (the criterion variable). Employee job satisfaction also could reverse its role and predict productivity level.

9.1 THE FUNCTIONAL RELATIONSHIP BETWEEN TWO VARIABLES

Several important points should be kept in mind in trying to conceptualize simple linear regression. First, data must be collected for every sampling unit

---

[2]A possible predictor variable is given for each of the criterion variables listed, respectively. These are not necessarily the most appropriate predictors.

1. intelligence
2. previous administrative experience
3. amount of daily exercise
4. degree of organization
5. level of congruency between perceived job role and actual practice
6. yearly income
7. age of foster parents
8. high-school GPA
9. amount of viewing time
10. abstract reasoning ability

on the two variables investigated. For consistency, the criterion variable will be designated Y, and X will represent the predictor variable. The data analyzed are the same as in a correlational analysis. We will still examine the covariability and variability of the two variables, but we will do so with different assumptions and intentions.

The relationship between Y and X that simple linear regression explores may be described by the mathematical **general linear model.** This model is "general": it applies to both experimental and nonexperimental settings and has both explanatory and predictive capabilities. The word *linear* indicates that the model produces a straight line. The mathematical equation for the general linear model defining the functional relationship between Y and X in a population is

$$Y_i = \alpha + \beta X_i + \epsilon_i \tag{9.1}$$

In this equation, Y_i and X_i represent the scores for individual i on the criterion and predictor variables, respectively. The parameters α (alpha) and β (beta) are constants; they describe the generalized functional relationship in the population. The value of β identifies the proportional change along the Y scale expected for every unit changed in fixed values of X. It represents the slope (degree of steepness) of the straight line. The value of α identifies an adjustment constant due to scale differences in measuring X and Y. The symbol ϵ_i (epsilon) represents an error component for each individual—the portion of Y score that cannot be accounted for by its systematic relationship with values of X.

It is important to note in formula 9.1 that each individual's raw score can be viewed as containing two separate parts: (a) the part that can be determined from knowing the individual's X score ($\alpha + \beta X_i$) and (b) the error component that cannot be determined by knowing an individual's specific X value (ϵ_i). The terms in formula 9.1 describing the relationship between Y and X ($\alpha + \beta X_i$) define the predictable part of any Y score for fixed values of X. Given values of α and β determined from the relationship of Y and X and a fixed value for X, $\alpha + \beta X$ will produce a predicted score, symbolically represented by Y' (Y-prime). Based on the form of the general linear model for a fixed value of X,

$$
\begin{aligned}
Y_i &= \underbrace{\alpha + \beta X_i} + \underbrace{\epsilon_i} \\
Y_i &= \begin{array}{l}\text{Part predictable from} \\ \text{the systematic relation-} \\ \text{ship between } Y \text{ and } X\end{array} + \begin{array}{l}\text{Part unrelated to} \\ \text{values of } Y\end{array} \\
Y_i &= Y_i' + \epsilon_i
\end{aligned}
$$

Remember that α and β are constants summarizing the general functional relationship between Y and X. Different values of X would produce different predicted Y scores.

The general linear model is presented in Equation 9.1 using population parameters. In practice, sample data will be used to calculate estimates of the values of α and β necessary to define the model. The sample general linear model may be represented as follows:

$$Y_i = a + bX_i + e_i \tag{9.2}$$

The statistical task ahead is to determine the values of a and b that *best* estimate the values of the parameters α and β, but that also *best* represent the functional relationship between X and Y.

Thorough familiarity with the two components of individual Y scores in the general linear model is important. These components in the context of the relationship between X and Y are best depicted in a scatter diagram. The component describing the functional relationship between Y and X, the predicted score component Y' is equal to $a + bX$. This general mathematical equation defines a straight line that may be fitted to the data points in a scatter diagram. The extent to which the data points do not lie on the straight line represented by $Y' = a + bX$ indicates individual errors. Figure 9.1 illustrates the two components. Keep these graphic qualities in mind as we discuss regression analysis.

9.2 THE ERROR COMPONENT AND THE PRINCIPLE OF LEAST SQUARES

The values of a and b in formula 9.2 that describe the *best* functional relationship between X and Y may be thought of as values that maximize either the explanatory power or predictive accuracy of X in relation to Y. In either case, maximizing explanatory power or prediction accuracy results in an opposite effect, that of *minimizing the error component* in the general linear model. When we let $Y' = a + bX$ in Equation 9.2,

$$Y = Y' + e$$

It stands to reason that the larger the portion of Y that is predictable (Y') in general across individuals, the more potent is our knowledge about Y and how it may change as a function of X. To maximize this knowledge, we will focus on minimizing errors as the criterion for determining the *best* value of a and b.

What does it mean to *minimize errors?* For simplicity, we will show how to do so for prediction, but the same concepts apply to explanation. For any data to which we apply regression procedures, measurements on both X and Y must exist. Once the functional relationship defined by $a + bX$ is developed, it can be used to calculate a predicted Y score (Y') for each individual, based on the individual's known X score. In these data, then, three scores potentially exist for each individual: an original X score, an original Y score, and a predicted Y' score based on the functional relationship between the group's original X and Y scores. The value of the predicted score depends on the level or

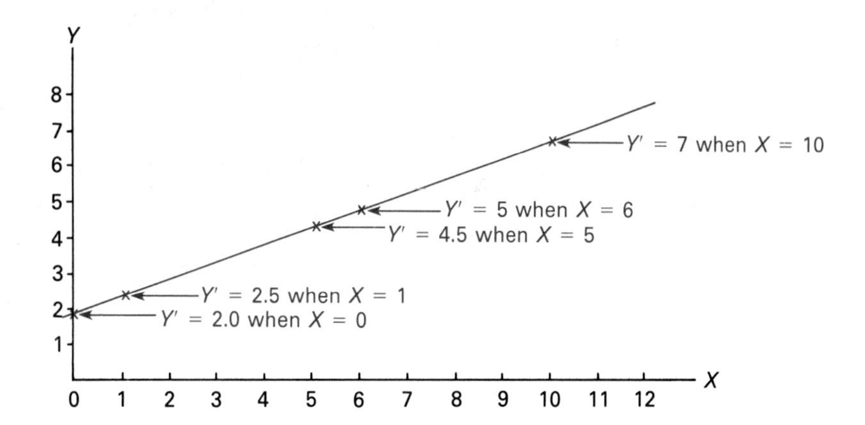

Predicted scores for X values of 0, 1, 5, 6, and 10 from the equation $Y' = 2 + .5X$. These predicted scores lie on a straight line.

A

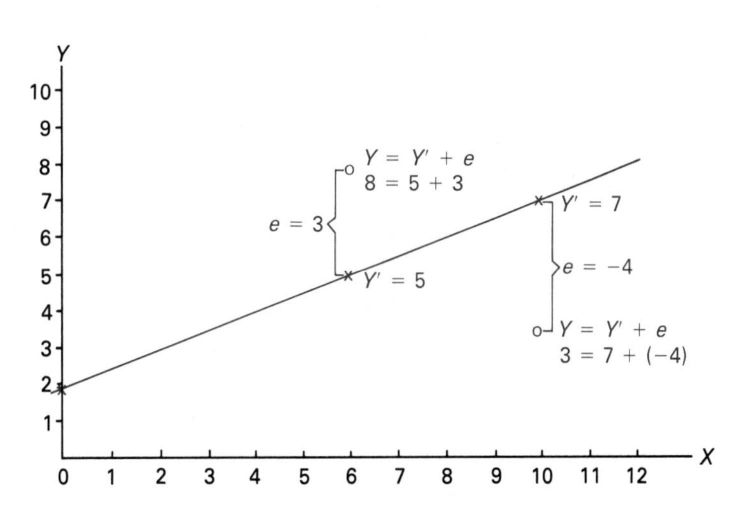

In the general linear equation $Y = 2 + .5X + e$, a Y score of 8 for an individual with an X score of 6 may be represented as the sum of their predicted scores ($Y' = 5$) and the distance of the Y score from Y' ($e = Y - Y' = 3$). The Y score of 3 for an X score of 10 is also represented by the general linear model, $Y = Y' + e = 7 + (-4) = 3$.

B

FIGURE 9.1
Components of the general linear model. **A,** $Y' = 2 + .5X$; **B,** $Y = 2 + .5X + e$.

value of an individual's X score, but the accuracy of Y' in predicting Y depends on the covariability and variability of the X and Y scores. We want to maximize the accuracy of Y' across individuals.

To maximize prediction accuracy, we try to minimize prediction error or inaccuracy. If Y represents an individual's score on the criterion variable and Y' is the predicted score, then $Y - Y'$, the error score (e), represents the discrepancy between the actual and predicted scores. The error score identifies that individual's error component in the general linear model.

Our goal is to minimize error scores across individuals. Some error scores will be large and some small, depending on the values of Y and Y'. Because error score values depend on the values of Y', the task statistically is to develop an equation to produce Y' scores that will minimize error scores for the group as a whole. To do so, we use the score variability concept (presented in Chapter 6). The "best" prediction equation will be the one that produces the smallest variability among error scores. This criterion is called the **principle of least squares.** Values of a and b for defining the functional relationship between X and Y will be determined using this principle.

To understand the principle of least squares, a brief review of variance is in order. Chapter 6 defined variance as a summary index of variability in terms of squared deviations from the mean of the score distribution:

$$s_x^2 = \frac{\Sigma(X - \overline{X})^2}{n - 1}$$

As a statistical index, the concept of variance (and its formula) never changes. It is an index for which the squared distances of raw scores from the mean $(X - \overline{X})^2$ is a minimum when summed over individuals in the group. Using error scores as the raw score distribution, the variance of error scores (s_e^2) can be represented as follows:

$$s_e^2 = \frac{\Sigma\left(\begin{matrix}\text{individual} \\ \text{error scores}\end{matrix} - \begin{matrix}\text{mean error score} \\ \text{for the group}\end{matrix}\right)^2}{n - 1}$$

$$= \frac{\Sigma(e - \overline{e})^2}{n - 1}$$

$$= \frac{\Sigma[(Y - Y') - \overline{e}]^2}{n - 1}$$

To minimize the variance of error scores, we need to minimize the quantity $\Sigma[(Y - Y') - \overline{e}]^2$. Mathematically, it can be shown that the mean error score is zero (the overestimations balance the underestimations). The quantity to be minimized then reduces to

$$\Sigma[(Y - Y') - \overline{e}]^2 = \Sigma[(Y - Y') - 0]^2$$
$$= \Sigma(Y - Y')^2$$

This quantity is the basis for the principle of least squares. The goal is to find a prediction equation that produces predicted Y scores to minimize the sum of the *squared* prediction errors, $\Sigma(Y - Y')^2$.

9.3 THE LINEAR REGRESSION EQUATION: LINE OF BEST FIT

The simplest prediction situation is one with information only on one variable for a group of individuals. What prediction equation would satisfy the principle of least squares in this situation? What score would we predict for each individual to minimize errors? If we have group information on a single variable, Y, and we want to use this information to predict an individual's score on this variable, the best prediction equation would be

$$Y' = \overline{Y}$$

In other words, we would predict that the individual's score would be the mean of the distribution for the Y variable. Doing so would satisfy the principle of least squares

$$\Sigma(Y - Y')^2 = \Sigma(Y - \overline{Y})^2 = \text{Minimum}$$

Using the mean as the predicted score for everyone would minimize the amount (variability) of error across individuals.

What if we had information on two variables for a group rather than on just a single variable? If the two variables are functionally related, information on the second variable (X) should enable us to predict scores on Y more accurately than using the Y variable mean as the predicted score for each individual. As you might expect, the prediction equation becomes more complex because it must incorporate information about each variable *and* any relationship between the variables. The question is no longer "What is the best single predicted score for *all individuals?*" Rather, the question becomes conditional: "What is the best single predicted score for *all individuals who have the same value on variable X?*" While only one predicted score, $\overline{Y}$, occurs in the first case, several predicted scores occur in the second case. All individuals who have the same score on the predictor variable (X) will have the same predicted score (Y'), but individuals with different X scores will have different Y' scores. This additional information about the score value on variable X will increase our ability to predict scores on variable Y.

Figure 9.2 illustrates the basis for describing the predictive relationship between X and Y. The scatter diagram plots pretest scores (X) and posttest scores (Y) on the Body Parts test for all 240 individuals from the "Sesame Street" data. Descriptively, $\overline{X} = 21.40$, $s_x = 6.39$, $\overline{Y} = 25.26$, $s_y = 5.41$, and $r_{xy} = .65$. Using knowledge of how pretest scores and posttest scores relate will improve our prediction accuracy over just using the mean of the posttest scores. Because the two variables are not perfectly correlated, however, every-

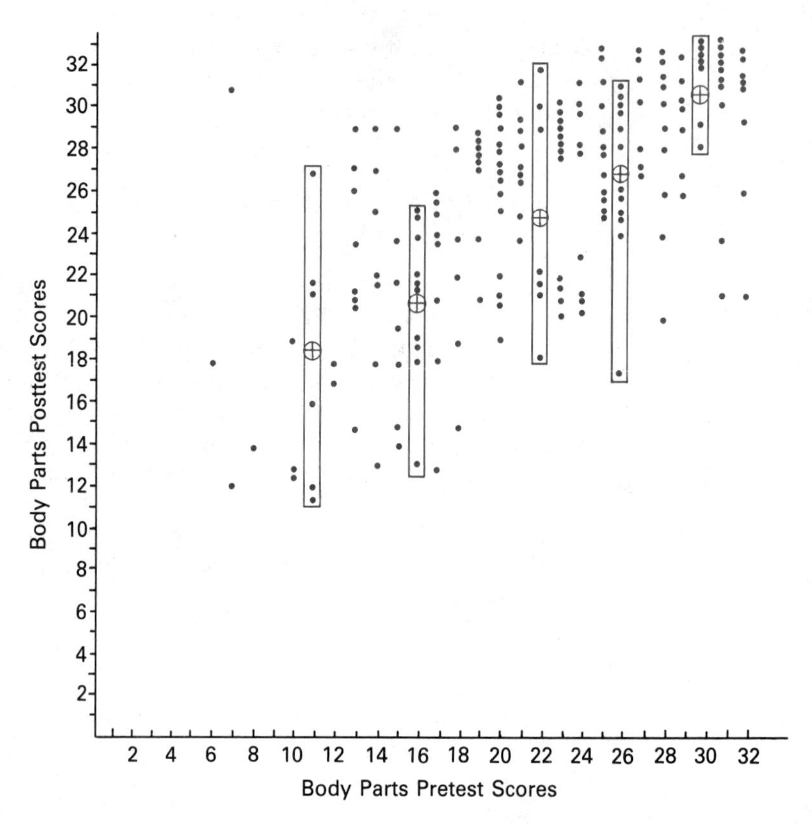

FIGURE 9.2
Scatter diagram of Body Parts pre- and posttest scores from the "Sesame Street" data.

one with a pretest score of, for example, 16, will not have the same posttest score.

To illustrate this point, the posttest scores have been highlighted in separate boxes for individuals who had pretest scores of 11, 16, 22, 26, and 30. Table 9.1 provides the exact posttest scores for each of these individuals. Look at the posttest scores for the 11 persons who had a pretest score of 16. Their posttest scores ranged from a low of 13 to a high of 25. Now look at the pretest score of 26. How many individuals had this score?[3] What was the range of their posttest scores?[4] The distribution of each set of scores in Table 9.1 is called a **conditional distribution,** the distribution of Y scores given a fixed value of X. For predictive purposes it is the mean Y score of each conditional distribution that provides the best predicted score for a given X score. It is also the mean Y score for each conditional distribution that will result in a minimum value for the sum of the squared errors, $\Sigma(Y - Y')^2$. From the conditional

[3]15.
[4]17–30.

TABLE 9.1
Body Parts posttest scores for individuals with selected pretest scores.

Pretest Score (X)	Individual's Posttest Score (Y)	Group Posttest Mean ($\bar{Y}$)
11	11, 12, 16, 21, 22, 27	18.17
16	13, 18, 19, 19, 21, 22, 22, 22, 24, 25, 25	20.91
22	18, 21, 22, 23, 25, 25, 29, 30, 32	25.00
26	17, 24, 25, 25, 26, 26, 27, 27, 28, 28, 29, 30, 30, 30, 30	26.80
30	28, 29, 30, 30, 31, 32, 32, 32, 32, 32	30.80

distributions of Y scores in Table 9.1, what is the best predicted score for an individual who had a pretest score of 16?[5] For a pretest score of 26?[6]

The Y score means of the conditional distributions for pretest scores of 11, 16, 22, 26, and 30 have been identified in Figure 9.2 by a ⊕. What can we note from the plot of these conditional distribution means? The means tend to form a straight line. If the relationship between two variables is linear, we can always expect this outcome. With enough data points in the scatter diagram, the Y score means for the conditional distributions given fixed values of X will *tend* to fall in a straight line. That does not mean they will form an absolutely straight line. Rather, sampling error and measurement error will cause some deviation from a straight line, but the general trend will be a straight line.

This straight-line phenomenon of the conditional distribution means allows us to derive a general, single equation to predict different values of Y for each value of X based on the functional relationship between X and Y. This prediction equation will follow the general equation for a straight line. The straight line defined by the equation is called the best fitting straight line for the data: It gives a Y' value for each X score that will minimize $\Sigma(Y - Y')^2$, the sum of the squared errors of prediction.

The General Equation of a Straight Line

Any straight line can be represented mathematically by the general equation

$$Y = bX + a$$

As indicated in the general linear model, Y and X are variables because they may take on different values. For any specific straight line, the values for b and a will be constants. These values define a specific straight line, such as in the equation $Y = 3X + 2$. Similarly, the equations $Y = 3X - 8$ and

[5]The group posttest mean is 20.91.

[6]26.80.

$Y = -.3X + 1$ both define straight lines that are different. We can find points on these straight lines and plot them on a bivariate diagram. For example, in the first equation, if X has a value of 0, then Y will have a value of 2 (i.e., $Y = 3(0) + 2$). What value of Y would be paired with an X value of 1?[7] In any of the equations, the mathematical expression will determine a value for Y given a value for X. The paired Y score and X score make the equation true when they are substituted into the equation. For each of the three straight-line equations identified, five paired X and Y scores are given that make each equation true. Solve for Y values where question marks appear.

- $Y = 3X + 2$
  If $X = 0$; $Y = 2$
  $X = 1$; $Y = 5$
  $X = -1$; $Y = -1$
  $X = 2$; $Y = 8$
  $X = 1/3$; $Y = ?$
- $Y = 3X - 8$
  If $X = 4$; $Y = 4$
  $X = 2$; $Y = ?$
  $X = 0$; $Y = ?$
  $X = 3$; $Y = 1$
  $X = 4/3$; $Y = -4$
- $Y = -.3X + 1$
  If $X = 0$; $Y = ?$
  $X = 1$; $Y = .7$
  $X = -5$; $Y = ?$
  $X = 2$; $Y = .4$
  $X = -4/3$; $Y = ?$

The data points for the first two equations have been plotted in Figure 9.3. Paired data points are plotted the same way data points were plotted in the scatter diagrams in Chapter 8. Plot the data points for the third equation.[8] As

---

[7] $Y = 5$ when $X = 1$.

[8]

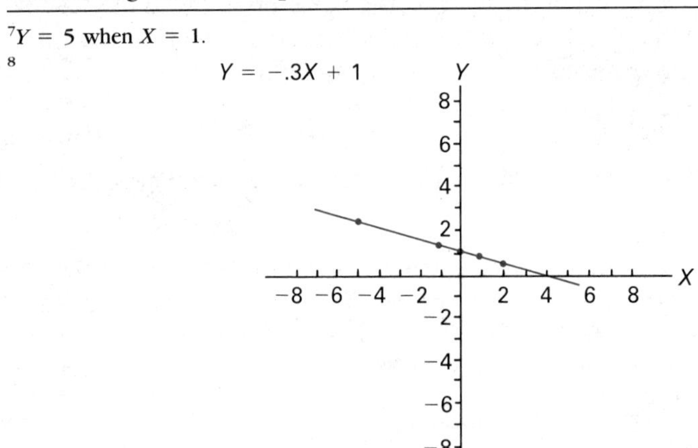

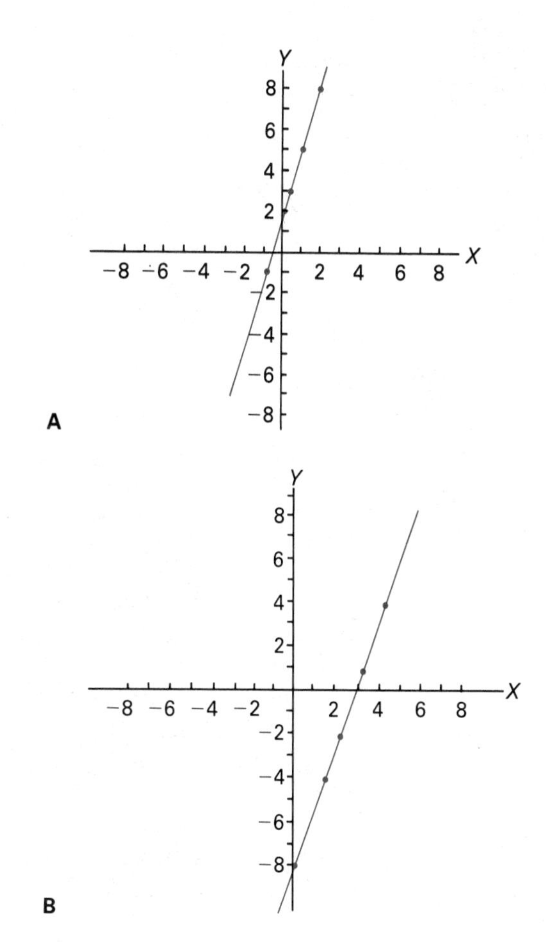

FIGURE 9.3
Graphs of equations for two straight lines. **A**, $Y = 3X + 2$; **B**, $Y = 3X - 8$.

demonstrated in the figure, a straight line can be drawn through the data points for each separate equation. For the equation of a straight line it is important to note that each value given to X determines a value of Y such that the point for the paired X and Y scores will lie on the straight line defined by the equation. Also note that while the form of the equation is the same, $Y = bX + a$, the values for either b or a are different across equations and result in straight lines located in different positions in the bivariate diagrams.

In the general equation for a straight line, b is called the *slope* of the line. The value for b identifies the steepness (slope) of the line and is defined as the amount of change in Y for every unit change in X;

$$b = \frac{\text{change in } Y}{\text{unit change in } X}$$

For the equation $Y = 3X + 2$, the value of b (the slope) is 3. This indicates that for every unit change in X, Y will increase 3 units. Substituting values for X into the equation $Y = 3X + 2$ will illustrate this point. For example, when

FIGURE 9.4
Changes in Y for unit changes in X.

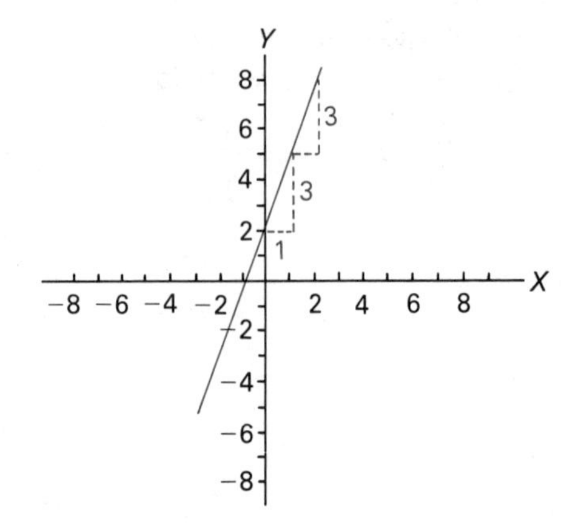

$X = 1, Y = 5$; when $X = 2, Y = 8$; when $X = 3, Y = 11$; and so on. For every unit change in X (when X goes from 0 to 1 or 1 to 2), Y increases 3 units (2 to 5 and 5 to 8). This is illustrated in Figure 9.4. The slopes for the first two equations are the same, $b = 3$ units. The steepness of the line in Figure 9.3B is the same as in Figure 9.3A. The only difference between the two is their locations on the graph. As we will see, the different location is caused by their different values of a. The slope for the third equation does differ from the first two; $b = -.3$. Because the slope is a *negative* .3, the value of Y *decreases* for every unit increase in X. As X goes from a value of 0 to a value of 1, Y decreases from 1 to .7.

The contrast in the equations illustrates an important point about slope and its sign. A positive slope means that the straight line is *increasing* or *ascending* from left to right. These lines are all positively sloped:

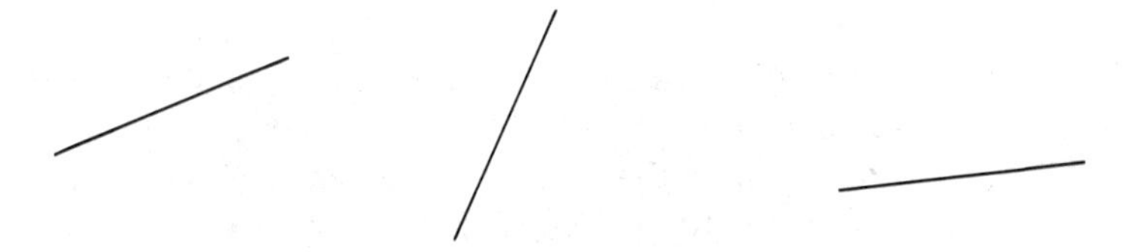

A negative slope indicates that the straight line is *decreasing* or *descending* from left to right. These lines are negatively sloped:

The sign of the slope indicates whether the line is rising (positive) or falling (negative) from left to right. The numerical value of the slope indicates the amount of change in Y per unit change in X. The larger the absolute value of b, the more rapid the rise or fall of the line.

The other constant in the equation for a straight line that determines location is a. In the equation of form $Y = bX + a$, a is the *Y intercept,* the place on the Y axis through which the straight line passes. It is the value of Y when $X = 0$. Examine the plots of the lines for the three straight-line equations. In the first, $Y = 3X + 2$, the Y intercept or value of a is $+2$. The line intercepts the Y axis when $X = 0$ and $Y = 2$. What is the Y intercept for $Y = 3X - 8$?[9] For $Y = -.3X + 1$?[10]

The Linear Regression Equation

As we have seen, the line of best fit for the data points in a scatter diagram follows the general form of the straight-line equation. It serves two purposes: it defines the line of best fit for linearly related variables, and it defines the functional relationship between X and Y. This latter component in the general linear model was represented by the equation

$$Y' = bX + a$$

Note that the only change in the general mathematical equation for a straight line is insertion of the symbol Y', signifying predicted Y scores. The variable b is still the slope of the line and a is the Y intercept. The symbol X represents a predictor variable; it can take on fixed values along the measurement scale for X. Using the equation $Y' = bX + a$, where b and a are determined from a given set of bivariate data, we can find Y' scores for fixed values of X.

The equation of the form $Y' = bX + a$ is commonly called the **linear regression equation.** The predicted scores, Y', based on the regression equation will all lie on the line of best fit defined by the regression equation. Because the predicted scores lie on a straight line, their variability is restricted. They are *regressed* (made closer) as a set to the group mean of the criterion variable relative to the actual raw Y scores. Therefore, the label linear regression equation reflects the regression of predicted scores. Within this prediction context, the slope or b is called the **regression coefficient.** Its magnitude determines the relative amount of regression that will occur. The smaller the regression coefficient, the greater the regression effect.

9.4 CALCULATING THE REGRESSION EQUATION

How do we determine the specific regression equation for a given two-variable data set? Keeping the principle of least squares in mind, values of b and a must

---

[9] $a = -8$.

[10] $a = +1$.

be determined from the data on variables X and Y to minimize the sum of the squared errors of prediction. While this task may seem overwhelming, it is reasonably simple. General formulas for calculating b and a can be developed using familiar information.

Mathematically, the task is to find computational formulas for computing values of b and a that will define a straight line, $Y' = bX + a$, which when used to predict Y scores will minimize $\Sigma(Y - Y')^2$. Formulas have been developed to minimize this quantity. The mathematics involved are not described here. The raw-score formula for calculating b is

$$b = \frac{n\Sigma XY - (\Sigma X)(\Sigma Y)}{n\Sigma X^2 - (\Sigma X)^2} \tag{9.3}$$

Dividing both numerator and denominator by n,

$$b = \frac{\Sigma XY - \dfrac{(\Sigma X)(\Sigma Y)}{n}}{\Sigma X^2 - \dfrac{(\Sigma X)^2}{n}}$$

Do any of these quantities look familiar? They should. The ΣY^2, ΣXY, ΣY, ΣX^2, and ΣX are all statistical quantities from Chapter 8 involving the summation of values for variables X and Y from a bivariate data set. The numerator and denominator of this formula also should look familiar. The numerator is the raw-score formula for computing the sum of the deviation cross-products, Σxy (see Chapter 8, formula 8.4). The denominator is the raw-score formula for computing the sum of squared deviations for variable X, Σx^2. Substituting the deviation summations for their raw-score formulas yields a deviation score formula for computing b:

$$b = \frac{\Sigma xy}{\Sigma x^2} \tag{9.4}$$

Once a value for b is computed, the value of a is calculated using the formula

$$a = \overline{Y} - b\overline{X} \tag{9.5}$$

These formulas for calculating b and a result in values for the line of best fit to the bivariate data based on the least-squares criterion. The distributional characteristics of variables X and Y and their covariability determine the values of b and a, thereby defining the linear regression equation and the functional relationship between X and Y.

Data for examples A and B from Chapter 8 are reproduced here to illustrate these formulas.

Example A			Example B		
Individual	*X*	*Y*	Individual	*X*	*Y*
A	60	16	A	60	6
B	45	13	B	45	8
C	40	12	C	40	12
D	20	8	D	20	13
E	10	6	E	10	16

The summation quantities needed to calculate the regression coefficient, b, and the Y intercept, a, are the same as those needed to calculate the correlation coefficient, r_{xy}. The necessary sums are

Example A

$$\Sigma X = 175 \qquad \Sigma Y = 55$$
$$\Sigma X^2 = 7725 \qquad \Sigma Y^2 = 669$$
$$\Sigma XY = 2245$$

Example B

$$\Sigma X = 175 \qquad \Sigma Y = 55$$
$$\Sigma X^2 = 7725 \qquad \Sigma Y^2 = 669$$
$$\Sigma XY = 1620$$

For the data in example A,

$$b = \frac{n\Sigma XY - (\Sigma X)(\Sigma Y)}{n\Sigma X^2 - (\Sigma X)^2} = \frac{5(2245) - (175)(55)}{5(7725) - 175^2}$$
$$= \frac{1600}{8000}$$
$$= .20$$

and

$$a = \overline{Y} - b\overline{X}$$
$$= 11 - (.20)(35)$$
$$= 11 - 7$$
$$= 4$$

The linear regression equation based on the data in Example A is

$$Y' = .20X + 4$$

Example A is unique in that $r_{xy} = 1.00$. While the correlation is perfect, the regression equation indicates that this does not imply that Y will change one unit for every unit change in X. Rather, the functional relationship indicates that the effect of a one-unit change in X is that Y will increase .20 of a unit. However, because of the perfect relationship, the regression equation

$Y' = .20X + 4$ does result in perfect prediction of Y scores given knowledge of the X scores. To check, substitute an X score value for X in the equation and solve for Y'. Given an X score of 40, what is the predicted value of Y?

$$Y' = .20X + 4$$
$$Y' = .20(40) + 4$$
$$Y' = 8 + 4$$
$$Y' = 12$$

The predicted score of 12 for individual C equals individual C's actual Y score. Check the predicted values given the X scores for individuals A, B, D, and E. Y' should equal Y for each case.

As indicated, this example is unusual, but nicely illustrates the concepts that provide the foundations to linear regression. The formulas for computing b and a certainly worked in this instance. Values found based on the data resulted in an equation that minimized the squared errors of prediction. In this example, $\Sigma(Y - Y')^2 = 0$.

The data in example B produce a slightly different illustration. The correlation between X and Y scores is not perfect, but is exceptionally high ($r_{xy} = -.953$). The relationship is also negative; in example A it was positive. Determine the equation for the best-fitting straight line for the data in example B.[11] Using that equation, which Y scores would you predict for each of the five individuals in example B given their individual X scores?[12] The predicted scores for the five individuals based on the regression equation $Y' = -.191X + 17.685$ are more accurate than for any other possible equation for a straight line. For these data,

$$\Sigma(Y - Y')^2 = (6 - 6.225)^2 + (8 - 9.090)^2 + (12 - 10.045)^2 + (13 - 13.865)^2 + (16 - 15.775)^2 = 5.86$$

---

[11] $$b = \frac{5(1620) - (175)(55)}{5(7725) - 175^2} = \frac{-1525}{8000} = -.191$$

$$a = 11 - (-.191)(35) = 11 + 6.685 = 17.685$$

$$Y' = -.191X + 17.685$$

[12] A: $Y' = (-.191)(60) + 17.685 = 6.225$
B: $Y' = (-.191)(45) + 17.685 = 9.090$
C: $Y' = (-.191)(40) + 17.685 = 10.045$
D: $Y' = (-.191)(20) + 17.685 = 13.865$
E: $Y' = (-.191)(10) + 17.685 = 15.775$

The value 5.86 is thus the smallest possible $\Sigma(Y - Y')^2$ value that can be obtained for these data from a straight-line equation. This is the principle of least squares. Choose any other equation for a straight line and go through the calculations for finding the value of $\Sigma(Y - Y')^2$. Your value of $\Sigma(Y - Y')^2$ will always be larger than 5.86. $Y' = -.191X + 17.685$ is the ***best fitting straight line*** for the set of data points in example B.

Plotting the Regression Equation

Fitting the regression equation to a set of data points involves nothing more than graphing a straight line on a bivariate graph or scatter diagram. We need only two data points on the line to determine the exact location and plot of a straight line. We can obtain these two data points by substituting any two values of X in the regression equation and calculating values of Y'. The plot of the data points for the paired scores X and Y' will lie on the line of best fit defined by the regression equation. One easy data point to obtain is the point on the line when $X = 0$, or the Y intercept, a. The second data point is found by choosing any arbitrary value of X and solving the equation for Y'.

The data points from example B are plotted in Figure 9.5A. Two points are needed to fit the regression line to these data. One point quickly deter-

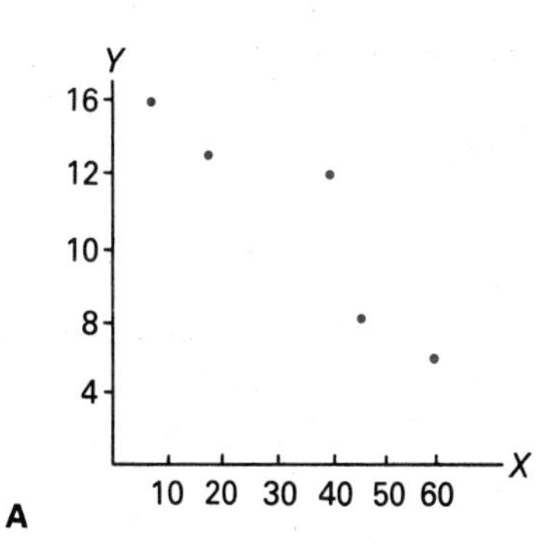

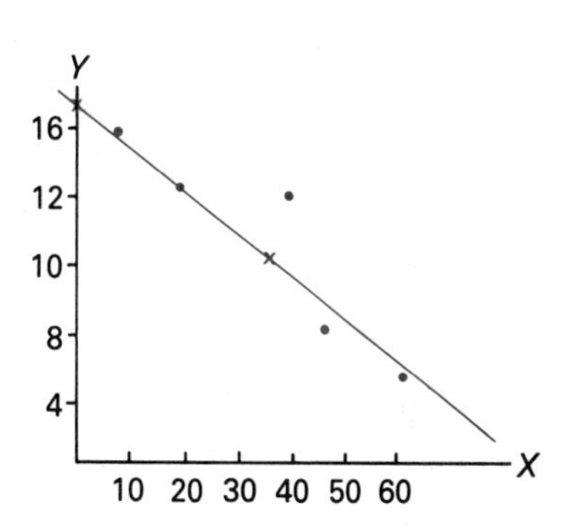

FIGURE 9.5
Fitting the regression line to a set of data points. **A,** example B data; **B,** regression equation $Y' = -.191X + 17.685$.

mined is for $X = 0$ and $Y' = 17.685$. Another point could be that determined by setting $X = 40$. Then $Y' = -.191(40) + 17.865 = 10.045$. The regression line can be fit to the data by drawing a straight line between the points (0, 17.685) and (40, 10.045), as illustrated in Figure 9.5B.

It is important to note that all predicted scores will lie on the straight line defined by the regression equation. Errors of prediction are illustrated by the distance from the regression line of the actual data points. The better the prediction, the closer the actual individual data points "hug" the regression line. Note also the value of b in this example. For these data, Y may be expected to *decrease* .191 units for each unit *increase* in X. If a perfect relationship existed, the observed individual data points would lie exactly on the regression line ($Y' = Y$ in this case). You may want to plot the regression equation for example A to demonstrate this principle.

As a final illustration, let's return to the example given in this chapter examining the predictive relationship between the number of high-school graduates (X) and the number of first-year college enrollees (Y). Hypothetical data for the years 1972–1984 appear in Table 9.2. Two questions can be answered from the data presented. First, what effect on the number of first-year enrollees might we expect as a function of the numbers of high-school graduates based on the pattern of the linear relationship between the two variables for the years 1972–1984? Second, given the regression equation and the number of high-school graduates in 1985 (the number given in Table 9.2 is 34,786), how many first-year students might we expect to enroll in the fall of 1985?

TABLE 9.2
Number of high-school graduates and first-year college enrollees, 1972–1984.

Year	High-School Graduates (X)	First-year college enrollees (Y)
1972	37,577	6543
1973	40,240	6923
1974	42,317	7452
1975	45,381	7981
1976	44,840	7360
1977	43,352	7491
1978	41,111	7018
1979	41,090	7112
1980	39,756	6905
1981	39,142	6896
1982	37,963	6804
1983	36,640	6637
1984	35,377	6118
1985	34,786	?

For 1972–1984 data, $r_{xy} = .948$

Working with such large numbers produces astronomical summations, but bear with the calculations:

$$\Sigma X = 524{,}786 \qquad \Sigma Y = 91{,}240$$

$$\Sigma X^2 = 21{,}297{,}822{,}902 \qquad \Sigma Y^2 = 643{,}092{,}942$$

$$\Sigma XY = 3{,}699{,}856{,}016$$

Substituting these values into formula 9.3,

$$\begin{aligned} b &= \frac{13(3{,}699{,}856{,}016) - (524{,}786)(91{,}240)}{13(21{,}297{,}822{,}902) - (524{,}786)^2} \\ &= \frac{216{,}653{,}568}{1{,}471{,}351{,}943} \\ &= .147 \end{aligned}$$

The straight interpretation of b is that for every high-school graduate, we would expect .147 of a person to enroll at a university. Translated into units of 1000, for every 1000 students graduating, we estimate that 147 or 14.7% will enroll at a university. The exact prediction for a fixed number of graduates would need to be adjusted by the value of a in the regression equation, however. We use formula 9.5 to calculate a. Before applying the formula, $\overline{Y}$ and $\overline{X}$ must be computed.

$$\overline{Y} = \frac{91{,}240}{13} = 7018.46 \qquad \overline{X} = \frac{524{,}786}{13} = 40{,}368.15$$

Then

$$\begin{aligned} a &= 7018.46 - (.147)(40{,}368.15) \\ &= 1084.34 \end{aligned}$$

The regression equation for predicting first-year enrollment based on the past relationship between number of high-school graduates and first-year college enrollment is

$$Y' = .147X + 1084.34$$

Figure 9.6 shows the fit of the regression line to the data. The two points used to draw the line were $X = 37{,}000$, $Y' = 6523$ and $X = 44{,}000$, $Y' = 7552$. The Y intercept is not used because the X and Y axis are not continuous. Portions of the scales are missing to conserve space. Given the regression equation and a 1985 class size of 34,786, we can find the predicted 1985 first-year student enrollment:

$$\begin{aligned} Y' &= .147(34{,}786) + 1084.34 \\ &= 6197.88 \end{aligned}$$

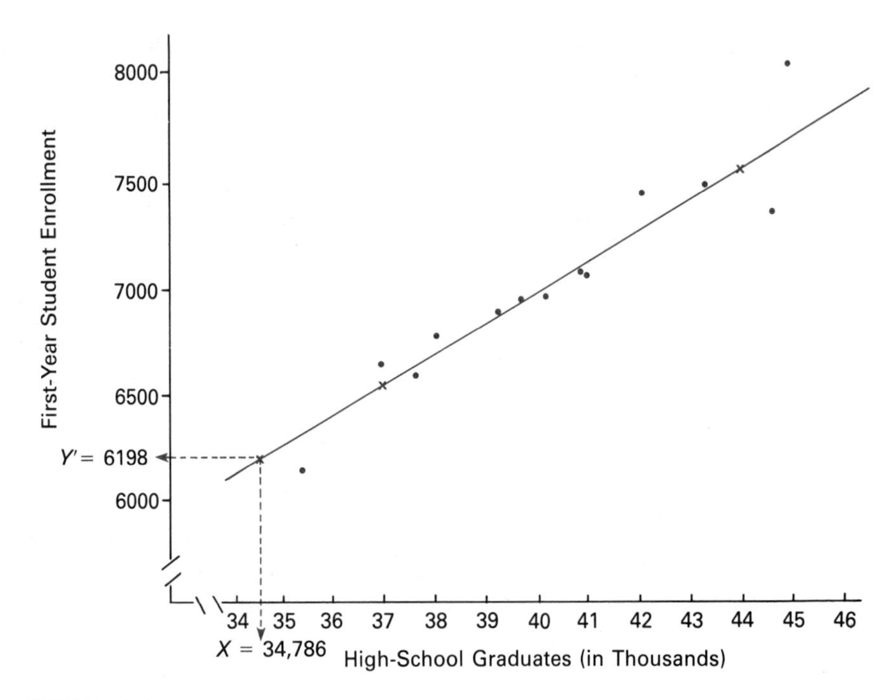

FIGURE 9.6
Regression line for predicting first-year student enrollment.

Based on the past relationship between the two variables, we would expect an enrollment of approximately 6198 freshmen. This predicted value is shown in Figure 9.6.

9.5 HOMOSCEDASTICITY AND ERRORS IN REGRESSION

One basic question not addressed yet in our discussion of regression analysis is the question of accuracy. How accurate is a specific regression equation in estimating scores on the criterion variable? This is a reasonably difficult question to answer because the degree of accuracy depends on the situation, partially on the variability of the criterion variable and partially on the magnitude of r_{xy}. In this section, the degree of regression equation accuracy is explored by examining its complement, the degree of inaccuracy or prediction error. Prediction error implies a lack of fit in a regression analysis. Within the general linear model of section 9.1 a lack of fit between Y and Y' was represented by the model's error component, examined in this section.

Development of the linear-regression equation was based on the least-squares criterion, *minimizing the sum of the squared errors.* Minimizing the *sum* focuses on the errors accumulated across all data points. One of the assumptions made in using the sum of squared errors as a criterion is that each predicted value resulting from the derived regression equation is equally accurate as an estimate of the dependent variable. This is a very important assumption. What does it mean for each predicted score to be equally accurate? Let's examine it in detail.

For any regression equation, a different predicted score, Y', exists for each value of the predictor variable, X. Figure 9.2 and Table 9.1 illustrate that for any given single value of X, there is only one Y' score, but there are many different Y scores for individuals who are equal on X. The many different values of Y form a conditional distribution for a fixed value of X. The question of accuracy for a given Y' score, then, is not one of how well it estimates a single Y score, but rather how well it represents all the many Y scores in the conditional distribution for the fixed value of X. Conceptually, Y' is really an estimate of a mean, the mean Y score for all Y scores in the conditional distribution for a fixed X value. The accuracy of Y' as a representative point for several Y scores is best defined under these conditions in variability terms. The less spread (distance) of Y scores from a Y' score the more accurate the Y' score is as a predicted value for all the Y scores.

How does one measure spread in a distribution? As you know, one statistical index is variance. Important to variance is the concept of squared deviations. Since Y' is an estimate of the mean of the conditional distribution, then taking the sum of squared deviations of Y scores from Y', $\Sigma(Y - Y')^2$, becomes a measure of variability. The smaller the variability of Y scores around Y', the more precise is Y' as an estimate of each Y score in the conditional distribution.

The assumption of equal accuracy of predicted scores can now be defined in variance terms. For all Y' scores to be equally accurate, the variability of Y scores around a predicted value within a conditional distribution must be equal for all predicted values. This assumption of equal conditional distribution variances is called the **assumption of homoscedasticity,** equal variability of Y scores around a predicted score across distributions for fixed values of X. This assumption states that each predicted score is equally accurate.

Note that the assumption of homoscedasticity is just that, an assumption on which prediction is based in using the linear regression equation as a model. In real data, equality of variability is not an absolute, but rather is qualified to imply approximately equal. Sampling error due to a small number of data points for a given value of X will result in conditional distribution variabilities that are definitely not equal in an exact sense. Figure 9.2 vividly illustrates this point. The length of the boxes for the five enclosed sets of data points shows that the spread of Y scores for the conditional distributions when $X = 11, 16, 22, 26$, and 30 is not exactly equal. The real question is whether the variances are *approximately* equal; that is, are they more different than we would expect based on sampling error? Descriptively, Y scores vary most when $X = 8$. There are only two values in this distribution (14 and 31), but the calculated standard deviation is 12.02. The 10 Y scores that form the conditional distribution for an X score of 30 vary least ($s = 1.48$).

Because the variability differs in practice across conditional distributions due to sampling error in our data, it is better to find a single, accumulated index of error variability. As variance and standard deviation indices, the generalized indices of error variability are called the **variance error of estimate** and the **standard error of estimate,** respectively.

9.6 VARIANCE AND STANDARD ERROR OF ESTIMATE

In developing the principle of least squares in section 9.2, we showed that minimizing $\Sigma(Y - Y')^2$ actually minimizes the variance of error scores for a linear equation. If we let $e = Y - Y'$,

$$\Sigma(Y - Y')^2 = \Sigma(e - \bar{e})^2$$

Dividing each of these quantities by $n - 1$ results in the variance of the error scores for all data points. Symbolically, the variance of the error scores is represented by $s^2_{y \cdot x}$ and called the variance error of estimate:

$$s^2_{y \cdot x} = \frac{\Sigma(e - \bar{e})^2}{n - 1} = \frac{\Sigma(Y - Y')^2}{n - 1}$$

We can find the variability of error scores by calculating the variability of actual Y scores from their respective predicted scores. A simpler, mathematically equivalent formula is used

$$s^2_{y \cdot x} = \frac{(Y - Y')^2}{n - 1} = s^2_y(1 - r^2_{xy}) \qquad (9.6)$$

As always, the square root of the variance provides the standard deviation of a set of scores. In this case, the square root of $s^2_{y \cdot x}$ yields the standard deviation of the error scores as an index of prediction error variability. The standard deviation of error scores is called the standard error of estimate and is represented as $s_{y \cdot x}$, the standard error of estimate when predicting Y from X. The computational formula is

$$s_{y \cdot x} = s_y\sqrt{1 - r^2_{xy}} \qquad (9.7)$$

To illustrate the equivalence of formula 9.7 as the standard deviation of error scores, we will use the five data points from example B. For these data, $r_{xy} = -.953$ and the regression equation was $Y' = -.191X + 17.685$. Table 9.3 lists the relevant calculations. Note that the sum of the error scores is zero—$\Sigma(Y - Y') = \Sigma e = 0$—implying that the mean error score will equal

TABLE 9.3
Illustration of formula 9.7.

Individual	X	Y	Y'	$e = Y - Y'$	e^2
A	60	6	6.225	−.225	.051
B	45	8	9.090	−1.090	1.188
C	40	12	10.045	1.955	3.822
D	20	13	13.865	−.865	.748
E	10	16	15.775	.225	.051
				$\Sigma e = 0$	$\Sigma e^2 = 5.860$

zero. The mean error score will in fact be zero for any set of data. To compute the standard deviation of error scores, formula 6.6 is used:

$$s_e = \sqrt{\frac{\Sigma e^2 - \frac{(\Sigma e)^2}{n}}{n - 1}}$$

$$= \sqrt{\frac{5.86 - \frac{0^2}{5}}{4}}$$

$$= \sqrt{1.465}$$

$$= 1.21$$

To use formula 9.7 for calculating $s_{y \cdot x}$, s_y must be calculated first. For the Y scores in example B, $s_y = 4$, and

$$s_{y \cdot x} = 4\sqrt{1 - (-.953)^2}$$

$$= 4\sqrt{.0918}$$

$$= 4(.303)$$

$$= 1.21$$

As demonstrated, $s_{y \cdot x} = s_e$—the standard error of estimate equals the standard deviation of error scores. The variance error of estimate ($s^2_{y \cdot x}$) is applied as an index of the lack of accuracy when regression analysis is employed within the context of inferential statistics. As developed later in section 9.7, it also can be used in identifying the proportion of Y score variability that is *not* functionally related to X. We will return to the variance error of estimate, but for now let us focus our attention on the standard error of estimate. The standard error of estimate is useful primarily when straight prediction is the purpose of our regression analysis. It is discussed within this context.

Interpreting the Standard Error of Estimate

Like the standard deviation, interpretation of the standard error of estimate is situation specific. A value of $s_{y \cdot x}$ must be interpreted within the context of a research situation to evaluate its "goodness." For any interpretation attempt, however, it is helpful to know the potential minimum and maximum values of $s_{y \cdot x}$. As with any standard deviation, the minimum value of $s_{y \cdot x}$ is zero, meaning no variability among error scores. Under what conditions will an $s_{y \cdot x}$ value of 0 occur? Certainly an $s_{y \cdot x}$ of 0 implies perfect prediction. All error scores are 0 resulting in no variability. When does perfect prediction occur? That's right, when $r_{xy} = +1.00$ or -1.00. Substituting these values for r_{xy} into the formula for $s_{y \cdot x}$ quickly shows that $s_{y \cdot x}$ must equal zero.

$$s_{y \cdot x} = s_y\sqrt{1 - r_{xy}^2}$$

If $r_{xy} = \pm 1.00$,

$$s_{y \cdot x} = s_y\sqrt{1 - 1} = s_y\sqrt{0}$$
$$s_{y \cdot x} = 0$$

The minimum value of $s_{y \cdot x}$ is zero. What about its maximum value? When would errors be the greatest? Logically, if $s_{y \cdot x}$ is smallest when prediction is perfect, then $s_{y \cdot x}$ should be the largest when prediction is poorest; when $r_{xy} = .00$. Under this condition there is *no* linear relationship between the criterion and predictor variables. Knowing the value of X for an individual does not help in predicting Y in this instance. What is the maximum value of $s_{y \cdot x}$ when $r_{xy} = .00$? Substituting $r_{xy} = .00$ into formula 9.7, the maximum value for the standard error of estimate is s_y, the standard deviation of Y scores. Under this condition, no variability in Y scores is functionally related to values of X:

$$\begin{aligned} s_{y \cdot x} &= s_y\sqrt{1 - r_{xy}^2} \\ &= s_y\sqrt{1 - 0} \\ &= s_y \end{aligned}$$

Because the interpretation of $s_{y \cdot x}$ will always be relative, the potential range for $s_{y \cdot x}$ within a specific prediction context becomes important information. The magnitude of $s_{y \cdot x}$ can only be evaluated relative to the size of s_y. As the ratio of $s_{y \cdot x}$ to s_y becomes small, the predictive power of the predictor variable X becomes meaningfully large. The relative size of $s_{y \cdot x}$ to s_y does not decrease proportionally in a linear manner as r_{xy} increases, however. In fact, it takes a very large value of r_{xy} before the ratio $s_{y \cdot x}/s_y$ drops below .50 ($s_{y \cdot x}/s_y = .50$ when $r_{xy} = .866$). The graph in Figure 9.7 shows the relationship between the ratio of $s_{y \cdot x}/s_y$ and values of r_{xy}. As illustrated in the figure, $s_{y \cdot x}$ as the index of prediction error variability does not become small, relative to s_y, until values of r_{xy} are into the mid-.90 range.

FIGURE 9.7
Ratio values of $s_{y \cdot x}/s_y$ for values of r_{xy}.

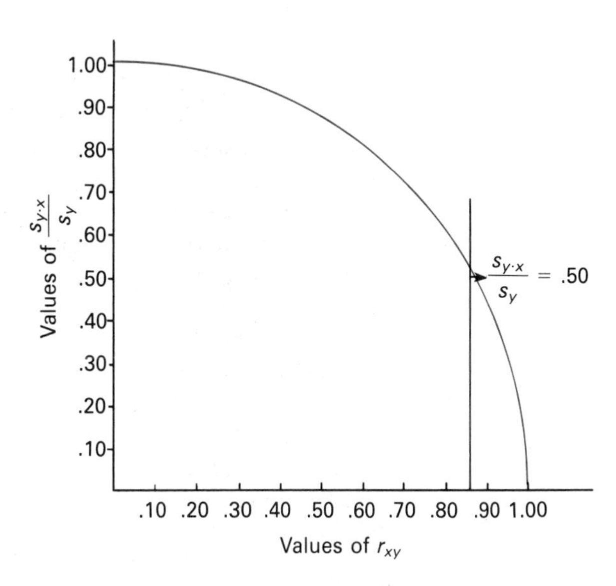

The literal interpretation of $s_{y \cdot x}$ is as the standard deviation of the error scores in the prediction process. It quantifies the extent to which the data points in a scatter diagram tend to vary or distribute themselves around the regression line. If you think of the error scores for a sample as forming a distribution of scores, $s_{y \cdot x}$ is the standard deviation of this distribution. We know from previous information that the mean error score, $\bar{e}$, is 0, so the mean of this error score distribution is zero. If we know the mean of a distribution and the standard deviation, what other information do we need about the distribution to know what the error score data look like?[13] As it turns out, the underlying model for the distribution of error scores is the normal curve. For large samples, error scores resulting from the prediction process tend to be normally distributed with a mean of zero and a standard deviation equal to $s_{y \cdot x}$. Given this distributional information what proportion of the error scores would you expect to be in the range of $\bar{e} \pm 1s_{y \cdot x}$?[14]

The normal curve distributional characteristics of error scores can be used to determine expectations about the magnitude and frequency of prediction errors in our data. As an example, assume you have regression data that produced an equation with a standard error of estimate equal to 3.0. Based on that information, what proportion of the positive and negative error scores would you expect to be five units or larger? In other words, what proportion of the Y scores would you expect to be five units larger or five units smaller than their predicted scores? An error score of $+5$ units can be changed to a z score given that $\bar{e} = 0$ and $s_{y \cdot x} = 3$.

---

[13]The shape of the distribution.

[14]Approximately 68%.

$$z = \frac{5 - 0}{3} = 1.67$$

An error score of −5 units would have a z score equivalent of −1.67. The shaded portion of Figure 9.8 identifies the proportion we want to find. Using Table B.2 to find the shaded areas of the normal curve, we would expect 4.75% of the error scores to be larger than +5 and 4.75% to be smaller than −5 units. This normal curve interpretation of $s_{y \cdot x}$ as the standard deviation of all error scores gives us some feeling for the expected frequencies of large absolute errors in prediction.

While the latter interpretation of $s_{y \cdot x}$ is the strict interpretation based on its definition, a more common use of $s_{y \cdot x}$ is based on the assumption of homoscedasticity. This equal variance assumption across conditional distributions also implies that the conditional distribution variances are not only equal, but that they are all equal to $s_{y \cdot x}$ as a value. In this context, $s_{y \cdot x}$ represents the standard deviation of Y scores around the predicted score, Y', of each conditional distribution for a fixed value of X. With this theoretical assumption, the shape, mean, and standard deviation of each conditional distribution are known. Y scores are assumed to be *normally distributed* around Y' *as the mean* with *variability equal to* $s_{y \cdot x}$. Figure 9.9 illustrates these distributional characteristics.

Under the conditional distribution assumptions, normal-curve theory will allow us to form expectations about predicted scores and their accuracy as summarized by the size of $s_{y \cdot x}$. Remember, a Y' score is derived from a regression equation and it is used as an estimate of several Y scores. It is assumed that for a fixed value of X, Y' is the mean of the conditional distribution of Y scores and that the Y scores are normally distributed around Y' with a variability index of $s_{y \cdot x}$ as the standard deviation. Using normal curve theory, 68% of actual Y scores should be within ±1 standard error of estimate of Y', 95% within $\pm 2s_{y \cdot x}$, and 99.7% within $\pm 3s_{y \cdot x}$ units of Y'. These percentages form intervals within which a given percentage of the Y scores is expected to lie given a specific predicted score.

Using the previously calculated values from the example B data for illustration, $Y' = -.191X + 17.685$ and $s_{y \cdot x} = 1.21$. For individuals who might have a score of 35 on X, a score of 11.0 would be predicted on Y. Since the correlation is not perfect, few persons who have scores of $X = 35$ would also have Y scores of exactly 11. Rather, their Y scores would distribute normally

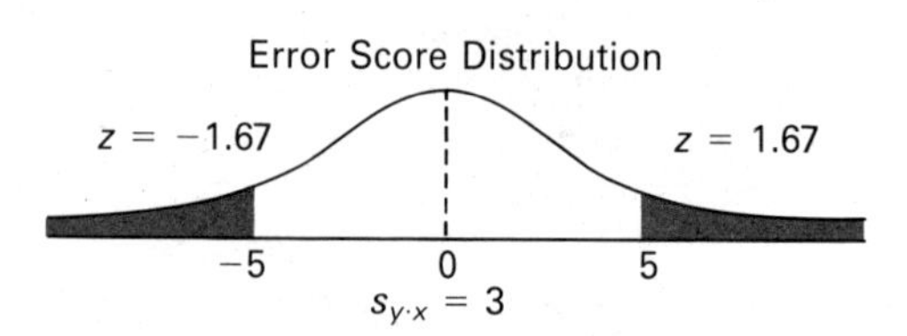

FIGURE 9.8
Areas above and below $e = \pm 5$.

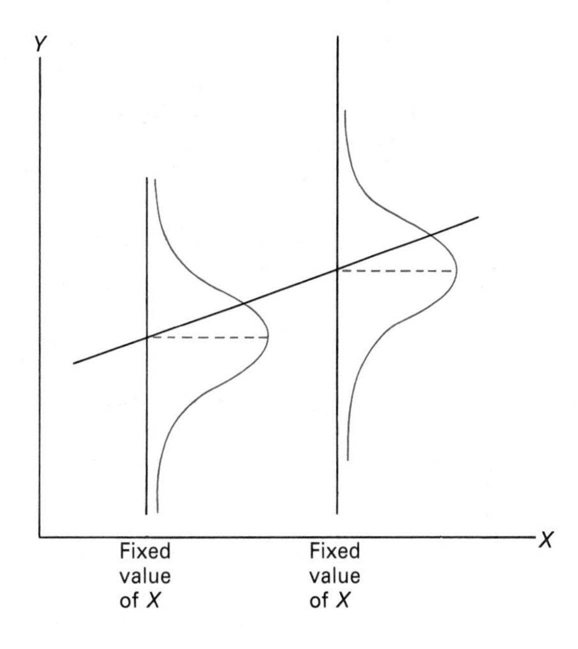

FIGURE 9.9
Normal distribution of conditional Y scores with a mean equal to Y' and standard deviation equal to $s_{y \cdot x}$.

around a score of 11 with a standard deviation of 1.21. Using this information, what would the expected score range be for the middle 68% of those with X scores of 35? The middle 68% of a normal distribution of scores falls within the range of one standard error of estimate above and one standard error of estimate below Y'. In this instance, we would expect 68% of those with X scores of 35 to have actual Y scores between

$$Y' + 1s_{y \cdot x} = 11.0 + 1.21 = 12.21$$

and

$$Y' - 1s_{y \cdot x} = 11.0 - 1.21 = 9.79$$

What are the Y score boundaries that will include the middle 99.7%? The range for this percentage is $\pm 3s_{y \cdot x}$.

$$\begin{aligned} Y' + 3s_{y \cdot x} &= 11.0 + 3(1.21) \\ &= 14.63 \\ Y' - 3s_{y \cdot x} &= 11.0 - 3(1.21) \\ &= 7.37 \end{aligned}$$

For 99.7% of those with X scores of 35, we would expect their observed Y scores to be in the range from 14.63 to 7.37.

The utility of the latter use of $s_{y \cdot x}$ is particularly important when prediction equations are being used for selection or diagnosis. Consider the initial screening and identification of gifted students using teacher ratings as the predictor variable. Student ability as measured by an IQ scale might be the criterion variable. Assume that an IQ of 130 is needed to qualify for the gifted program. Based on data from prior students, the regression equation for predicting student IQ scores (Y) from teacher ratings (X) is $Y' = 1.64X + 50.8$. The standard error of estimate, $s_{y \cdot x}$, is given as 5.72. It also has been decided based on the data that an initial teacher rating of 45 is necessary to be considered further for the program. Given a teacher rating cutoff score of 45, what proportion of the students who had ratings of 45 would you expect to have IQ scores of 130 or greater? The predicted IQ for a teacher rating of 45 is $Y' = 1.64(45) + 50.8 = 124.6$. Using the $s_{y \cdot x}$ value of 5.72, what proportion is expected to have IQs greater than 130?[15]

9.7 REGRESSION AND THE CORRELATION COEFFICIENT

The distinction previously made between correlation and regression was based on the different purposes of each as statistical tools. Correlation is primarily a descriptive index of the magnitude of the relationship between two variables. Regression, while based on the relation between two variables, focuses on the explanatory or predictive powers of this relationship. The adequacy of the linear regression equation as an explanatory prediction model certainly is linked to the magnitude of the linear relationship between the predictor and criterion variables, but the goal in a regression study is entirely different than in a strictly correlational study. Examining the interdependencies between correlational methodology and the regression model will help us understand both procedures.

Pearson *r* and the Regression Equation

All the computational summation quantities necessary to calculate b and a in the regression equation are also necessary to compute r_{xy}. Let's examine the functional relationships between b and r_{xy} first. Consider the deviation score computational formula for b:

$$b = \frac{\Sigma xy}{\Sigma x^2}$$

15

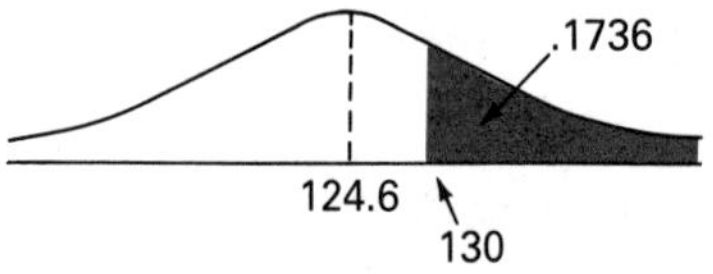

.1736

124.6

130

$$z = \frac{130 - 124.6}{5.72} = .94$$

The proportion of students with teacher ratings of 45 expected to have IQs > 130 is .1736.

Dividing both numerator and denominator by $n - 1$ changes Σxy to the covariance, s_{xy}, and Σx^2 to a variance, s_x^2:

$$b = \frac{\dfrac{\Sigma xy}{(n - 1)}}{\dfrac{\Sigma x^2}{(n - 1)}} = \frac{s_{xy}}{s_x^2}$$

The regression coefficient, b, equals the covariance between X and Y divided by the variance for X. Formula 8.6 provided a similar covariance formula for computing r_{xy}:

$$r_{xy} = \frac{s_{xy}}{s_x s_y}$$

Since both r_{xy} and b can be computed based on s_{xy}, the sign of r_{xy} and b will always be the same. We can algebraically manipulate both sides of this equation to find b in terms of r_{xy}. Multiplying both sides of the equation by s_y/s_x yields

$$r_{xy}\left(\frac{s_y}{s_x}\right) = \frac{s_{xy}}{s_x \cancel{s_y}}\left(\frac{\cancel{s_y}}{s_x}\right) = \frac{s_{xy}}{s_x^2}$$

The regression coefficient b equals s_{xy}/s_x^2, and the expression $r_{xy}(s_y/s_x)$ also equals s_{xy}/s_x^2, therefore,

$$b = r_{xy}\left(\frac{s_y}{s_x}\right) \qquad (9.8)$$

Formula 9.8 identifies the mathematical relationship between b and r_{xy}.

From formula 9.8, you can see that the value of b depends on the size of r_{xy}, but it also depends on the amount of variability in both Y and X scores. Three relationships are important to remember.

1. When $s_y > s_x$, $|b| > |r_{xy}|$. (The absolute value of the regression coefficient will always be larger than the absolute value of the correlation coefficient when there is more variability of scores on the criterion variable than on the predictor variable.)
2. When $s_y < s_x$, $|b| < |r_{xy}|$.
3. When $s_y = s_x$, $b = r_{xy}$.

Figure 9.10 demonstrates how unequal X and Y variances restrict the location of data points in a scatter diagram. As the best-fitting straight line to the data points, the regression line must have a steeper slope when s_y is greater than s_x (Figure 9.10A). When s_y is less than s_x (Figure 9.10B), the slope of the regression line is flatter so the line will fit the arrangement of data points caused by unequal variability.

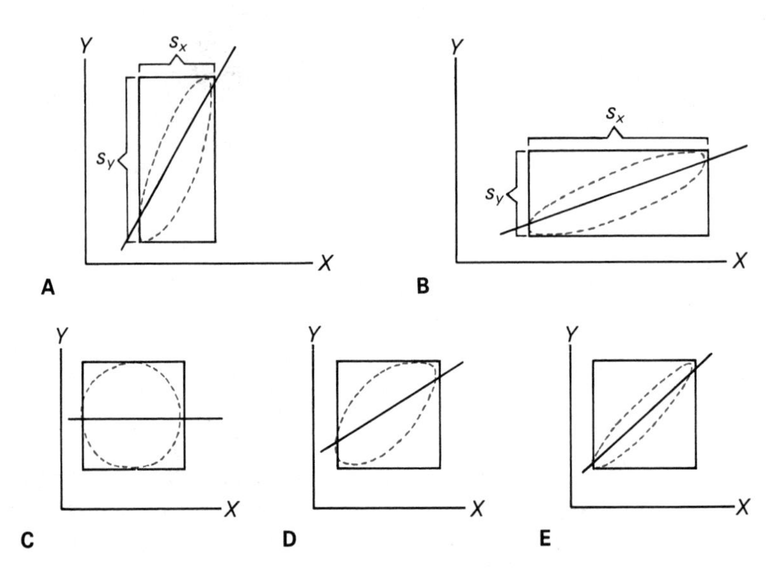

FIGURE 9.10
Effects of s_y, s_x, and r_{xy} on b. A, $s_y > s_x$; B, $s_y < s_x$; C, $r_{xy} = 0$, $s_y = s_x$; D, $r_{xy} = .50$, $s_y = s_x$; E, $r_{xy} = .90$, $s_y = s_x$.

The effect of r_{xy} on b is direct. The magnitude of r_{xy} determines the extent to which the data points cluster together. When $r_{xy} = .00$, the data form a circular or rectangular array of points and the best-fitting straight line is one with zero slope, a horizontal line parallel to the X axis. As r_{xy} increases in magnitude, the slope increases as data points cluster closer in the shape of an ellipse. Figures 9.10C, D, and E illustrate the effect of r_{xy} on b when $s_y = s_x$.

The regression coefficient and correlation coefficient for the example B data illustrate the relationship between b and r_{xy}. For these data, $r_{xy} = -.953$, $s_y = 4$, $s_x = 20$, and the calculated regression equation was $Y' = -.191X + 17.685$. The value for b of $-.191$ was calculated from the raw score formula 9.3. The same value would be obtained using formula 9.8.

$$b = r_{xy}\left(\frac{s_y}{s_x}\right) = (-.953)\left(\frac{4}{20}\right)$$
$$= -.191$$

In this example, the criterion score variability, $s_y = 4$, is less than the predictor score variability, $s_x = 20$, resulting in a value of b whose absolute value is less than the absolute value of r_{xy} (i.e., $.191 < .953$).

As with b, the Y intercept may be expressed as a function of r_{xy}. Substituting for the value of b in the computational formula for a,

$$a = \overline{Y} - b\overline{X}$$
$$= \overline{Y} - r_{xy}\left(\frac{s_y}{s_x}\right)\overline{X} \tag{9.9}$$

The linear regression equation may now be rewritten as a function of the means, standard deviations, and correlation coefficient for X and Y.

$$Y' = \overbrace{r_{xy}\left(\frac{s_y}{s_x}\right)}^{b}X + \overbrace{\overline{Y} - r_{xy}\left(\frac{s_y}{s_x}\right)\overline{X}}^{a} \tag{9.10}$$

Formula 9.10 is often used as the form for the linear regression equation. Inserting r_{xy}, s_y, s_x, $\overline{Y}$, and $\overline{X}$ into formula 9.10 will produce the same results as calculating b and a by formulas 9.3 and 9.5, respectively. Whichever formula is used, however, they all produce the same slope and Y intercept values that define the best-fitting straight line to the bivariate data.

Formula 9.10 can illustrate certain conceptual points concerning regression. For example, if no linear relationship exists between X and Y, then knowing values of X will not help in predicting values of Y. If we substitute a value of .00 for r_{xy} in formula 9.10,

$$Y' = (0)\left(\frac{s_y}{s_x}\right)X + \overline{Y} - (0)\left(\frac{s_y}{s_x}\right)\overline{X}$$
$$Y' = \overline{Y}$$

When $r_{xy} = .00$, the regression line has zero slope ($b = .00$) and every value of X results in the same predicted value, $\overline{Y}$. The regression line Y intercept is at the Y score mean ($a = \overline{Y}$). Knowing values of X does not improve predictions of their Y values in this case. The sum of the squared errors of prediction equals the sum of the squared deviations around $\overline{Y}$ when $r_{xy} = .00$.

$$\Sigma(Y - Y')^2 = \Sigma(Y - \overline{Y})^2$$

This implies that $s^2_{y \cdot x} = s^2_y$.

As r_{xy} varies in magnitude, its regression effect in prediction can be demonstrated. Formula 9.10 can be rewritten in deviation score terms so that

$$Y' - \overline{Y} = r_{xy}\left(\frac{s_y}{s_x}\right)(X - \overline{X})$$

The deviation scores can then be standardized (changed to z scores).

$$\frac{(Y' - \overline{Y})}{s_y} = r_{xy}\frac{(X - \overline{X})}{s_x}$$

The linear regression equation in standard score form becomes

$$z'_y = r_{xy}z_x \tag{9.11}$$

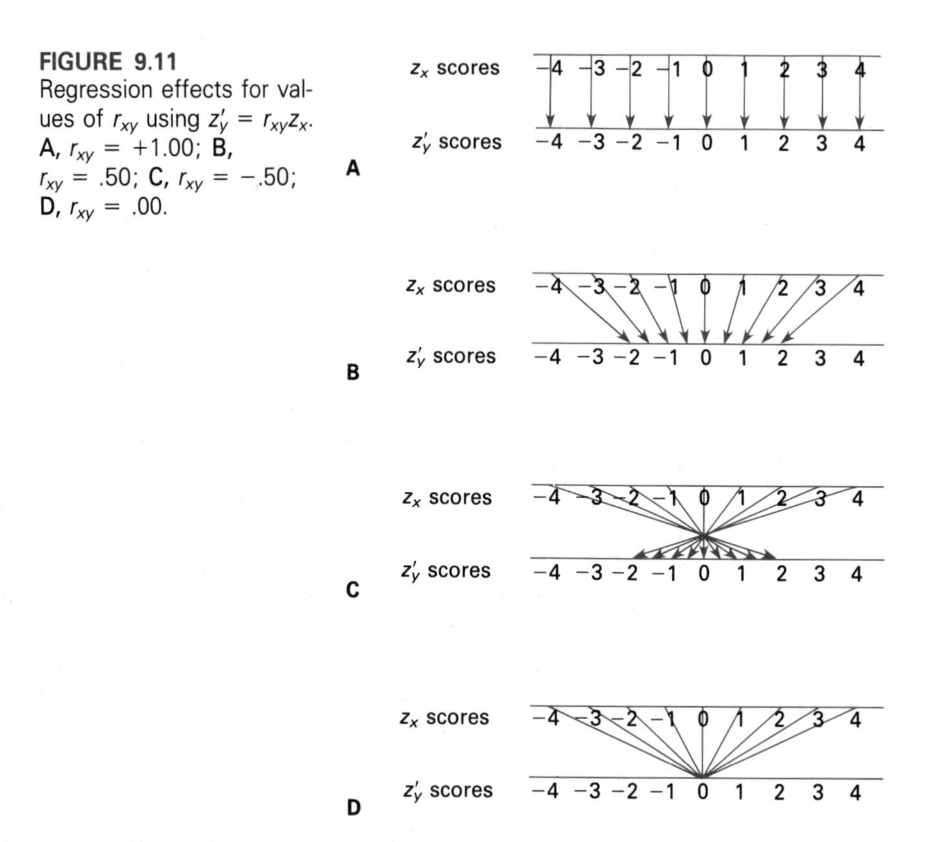

FIGURE 9.11
Regression effects for values of r_{xy} using $z'_y = r_{xy}z_x$. A, $r_{xy} = +1.00$; B, $r_{xy} = .50$; C, $r_{xy} = -.50$; D, $r_{xy} = .00$.

The regression phenomenon of predicted scores based on the magnitude of r_{xy} is most clearly illustrated by the standard score regression equation. Figure 9.11 illustrates the regression effect. When $r_{xy} = \pm 1.00$, there is no regression effect, z_x scores are matched perfectly with z_y scores, and prediction is perfect (See Figure 9.11A). When r_{xy} is less than perfect, predicted scores will be closer to the mean of the Y distribution than the respective X scores are to $\overline{X}$. The lower the r_{xy} value, the closer the predicted scores are to $\overline{Y}$. When $r_{xy} = .00$, all predicted scores regress to equal $\overline{Y}$ or $\overline{z}_y = 0$. This regression phenomenon has been labeled *regression to the mean.* The lower the value of r_{xy}, the more predicted scores regress to the mean of the Y score distribution. The regression effects for r_{xy} values of .50, $-.50$, and .00 are given in Figures 9.11B, C, and D.

Regression Concepts in the Interpretation of r_{xy}

The general linear model discussed in section 9.2 indicates that each individual's obtained Y score has two components, one part that is predictable, Y', and one that is not, the error component $Y - Y'$. In equation form:

$$Y = Y' + (Y - Y')$$
$$= Y' + e$$

Over individuals in a group, the set of predicted scores and error scores are independent of each other ($r_{y'e} = .00$). The independence of Y' scores and error scores also means that their variances are independent, so we can express the variance of the Y scores as the sum of the two independent variances for predicted and error scores.

$$s_y^2 = s_{y'}^2 + s_e^2 \tag{9.12}$$

The relationship in formula 9.12 is important to conceptualizing correlation and prediction in terms of variance components. What is s_y^2? It is a summary index of the amount of individual differences for a group on variable Y. Part of this variability of Y scores is predictable due to knowledge of X for individuals in the group. The predictable portion of s_y^2 is represented by $s_{y'}^2$. This value is the calculated variance of predicted Y scores for all individuals based on the regression equation $Y' = bX + a$. Based on discussion of the regression of Y' scores to the mean, the variance of predicted scores will always be less than the variance of Y scores except when $r_{xy} = \pm 1.00$. The remaining portion of Y score variance is due to individual differences; it is not predictable based on knowledge of X. Figure 9.12 illustrates the relationship of $s_{y'}^2$ to s_y^2. Remember that predicted scores always lie on the regression line. Because data points will lie above and below the regression line for all values of $r_{xy} \neq \pm 1.00$, the variability of these data points will be greater than the variability of the predicted points on the regression line. While not illustrated, s_e^2 indicates the extent to which the data points deviate from the predicted points on the regression line, or variability of the errors in prediction.

The variance of error scores should be familiar to you. It was defined in section 9.6 as the variance error of estimate, $s_{y \cdot x}^2$. Computationally,

$$s_{y \cdot x}^2 = s_y^2(1 - r_{xy}^2)$$

Substituting the right value for s_e^2 in formula 9.12 allows us to solve for r_{xy}^2 in terms of predicted and actual Y score variance.

$$s_y^2 = s_{y'}^2 + s_y^2(1 - r_{xy}^2)$$
$$s_y^2 - s_y^2(1 - r_{xy}^2) = s_{y'}^2$$
$$s_y^2[1 - (1 - r_{xy}^2)] = s_{y'}^2$$
$$s_y^2(r_{xy}^2) = s_{y'}^2$$
$$r_{xy}^2 = \frac{s_{y'}^2}{s_y^2} \tag{9.13}$$

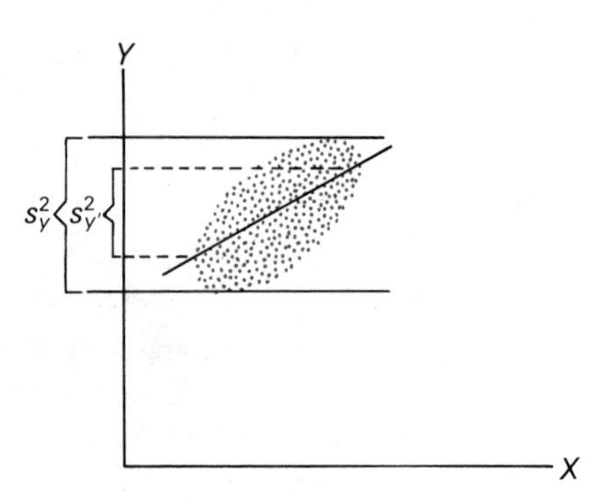

FIGURE 9.12
The relationship between predicted score variance and total Y score variance.

The relationship in formula 9.13 is important conceptually to interpreting the correlation coefficient within a regression model. In section 8.5, r_{xy}^2 was described as the proportion of variance shared between variables Y and X. In regression terms, r_{xy}^2 represents that portion of Y score variance that is predictable or accounted for by X. The value, r_{xy}^2, is called the **coefficient of determination.**

Similarly, manipulation of formula 9.13 produces the following relationship:

$$(1 - r_{xy}^2) = \frac{s_{y \cdot x}^2}{s_y^2} \tag{9.14}$$

The proportion of Y score variance that is unpredictable error variance is represented by $1 - r_{xy}^2$. This value is called the **coefficient of nondetermination.** It is the proportion of Y score variance *not* determined or predictable from knowledge about scores on X.

The square root of the coefficient of nondetermination produces another coefficient, discussed in section 9.6, but not formally named.

$$\sqrt{1 - r_{xy}^2} = \frac{s_{y \cdot x}}{s_y}$$

Does this look familiar? We had used the ratio of the standard error of estimate to the standard deviation of Y scores in section 9.6 as an index of the accuracy of prediction for values of r_{xy}. The relationship between values of r_{xy} and the ratio of $s_{y \cdot x}/s_y$ was shown in Figure 9.7. This ratio is called the **coefficient of alienation** and is represented by the letter k.

$$k = \sqrt{1 - r_{xy}^2} = \frac{s_{y \cdot x}}{s_y} \tag{9.15}$$

Again, data from example B can help illustrate these relationships. Relevant statistical indices previously calculated include

$$\begin{array}{lll} \overline{X} = 35 & \overline{Y} = 11 & r_{xy} = -.953 \\ s_x = 20 & s_y = 4 & s_{x \cdot y} = 1.21 \\ s_x^2 = 400 & s_y^2 = 16 & Y' = -.191X + 17.685 \end{array}$$

Based on the regression equation, the predicted scores for each of the five individuals were A = 6.225; B = 9.090; C = 10.045; D = 13.865; E = 15.775. The variance of these predicted scores is calculated using the following summations of these scores.

$$\Sigma Y' = 55.00 \qquad \Sigma(Y')^2 = 663.37$$

$$s_{y'}^2 = \frac{663.37 - \dfrac{55^2}{5}}{4}$$

$$= 14.59$$

The error variance is $s_{y \cdot x}^2 = 1.21^2 = 1.46$. Based on formula 9.12, the variance of Y scores, s_y^2, should equal $s_{y'}^2 + s_{y \cdot x}^2$ within rounding error.

$$s_y^2 = 16$$

$$s_{y'}^2 + s_{y \cdot x}^2 = 14.59 + 1.46 = 16.05$$

From formula 9.13, the squared correlation coefficient should equal $s_{y'}^2/s_y^2$.

$$r_{xy}^2 = .91$$

$$\frac{s_{y'}^2}{s_y^2} = \frac{14.59}{16} = .91$$

Of the variability in Y scores, 91% is predictable from a knowledge of X scores, and 9% is nonestimable error variance. The coefficient of alienation is

$$k = \frac{s_{y \cdot x}}{s_y}$$

$$= \frac{1.21}{4} = .30$$

or

$$k = \sqrt{1 - r_{xy}^2}$$

$$= \sqrt{1 - (-.953)^2} = .30$$

The ratio of the standard error of estimate to the standard deviation of Y is still quite large even with $r_{xy} = -.953$.

9.8 A FINAL REGRESSION EXAMPLE

The "Sesame Street" Body Parts pre- and posttest data were used earlier in the introduction of the prediction process, but the actual regression equation was never calculated or fit to the scatter diagram in Figure 9.2. These data will be used as a final example of the concepts covered in this chapter. The needed summation quantities and summary statistics for the two variables on the 240 individuals follow.

Body Parts Pretest (X)	**Body Parts Posttest (Y)**
$\Sigma X = 5136$	$\Sigma Y = 6063$
$\Sigma X^2 = 119{,}672$	$\Sigma Y^2 = 160{,}167$

$$\Sigma XY = 135{,}125$$

$\overline{X} = 21.400$	$\overline{Y} = 25.263$
$s_x = 6.391$	$s_y = 5.412$

$$r_{xy} = .6504$$

The regression coefficient and Y intercept may be calculated using formulas 9.3 and 9.5, respectively.

$$b = \frac{n\Sigma XY - (\Sigma X)(\Sigma Y)}{n\Sigma X^2 - (\Sigma X)^2} = \frac{240(135{,}125) - (5136)(6063)}{240(119{,}672) - 5136^2}$$

$$= \frac{1{,}290{,}432}{2{,}342{,}784}$$

$$= .551$$

$$a = \overline{Y} - b\overline{X} = 25.263 - .551(21.4)$$

$$= 13.472$$

The regression equation is $Y' = .551X + 13.472$. Why is the regression coefficient less than r_{xy}?[16] Use formula 9.12 to calculate the regression equation. Do you get the same equation?[17] If r_{xy}, $\overline{X}$, $\overline{Y}$, s_x, and s_y are given, formula 9.12 is an easier formula to use. Based on the regression equation, what are the predicted scores and error scores for Individuals 75 and 155 in Appendix A?[18]

---

[16]Because s_y is less than s_x; $s_y = 5.412$ and $s_x = 6.391$.

[17]The calculations are

$$Y' = r_{xy}\left(\frac{s_y}{s_x}\right)X + \overline{Y} - r_{xy}\left(\frac{s_y}{s_x}\right)\overline{X}$$

$$= .6504\left(\frac{5.412}{6.391}\right)X + 25.263 - .6504\left(\frac{5.412}{6.391}\right)21.4$$

$$= .551X + 25.263 - .551(21.4)$$

$$= .551X + 13.472$$

[18]For Individual 75, the Body Parts pretest score was 19 and the posttest score was 28.

$$Y' = .551(19) + 13.472 = 23.941$$

$$\text{Error} = Y - Y' = 28 - 23.941 = 4.059$$

For Individual 155, $X = 26$ and $Y = 17$.

$$Y' = .551(26) + 13.472 = 27.798$$

$$\text{Error} = Y - Y' = 17 - 27.798 = -10.798$$

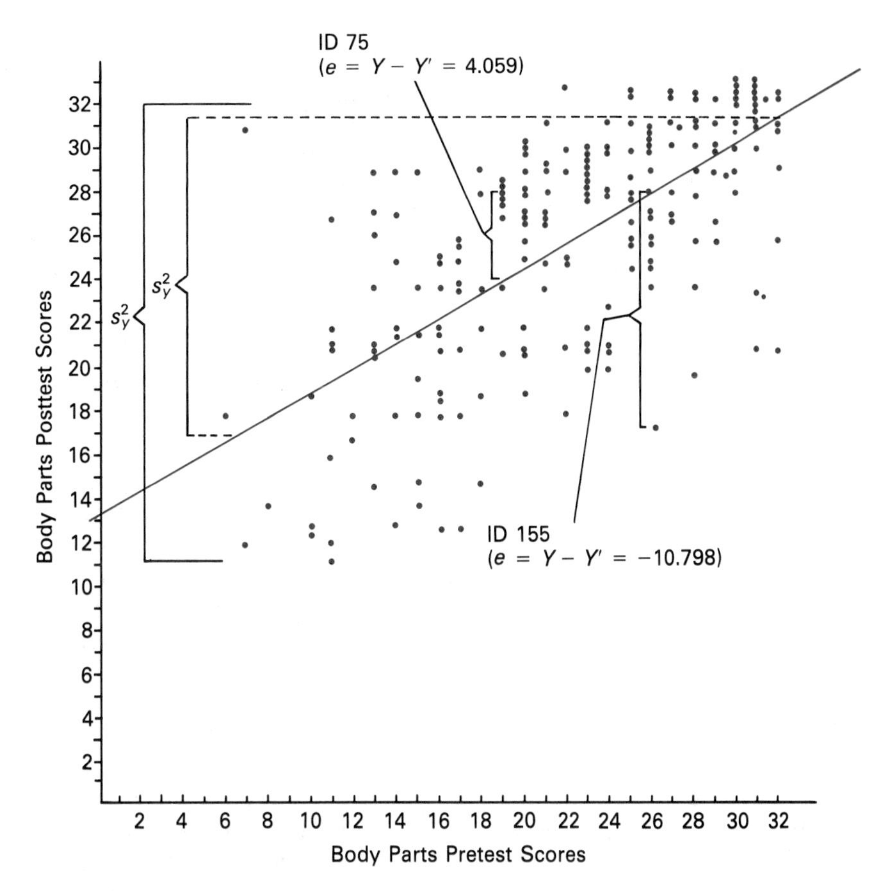

FIGURE 9.13
Scatter diagram of Body Parts pre- and posttest scores from the "Sesame Street" data.

The regression line is fit to the data in Figure 9.13 using the points (0, 13.472) and (32, 31.104). The data points for Individuals 75 and 155 have been marked. The distance of these data points from the regression line are equal to the error scores you calculated. Note that the predicted scores range from a low of 16.778 (when $X = 6$) to a high of 31.104 (when $X = 32$). The range for the actual Y scores is from a low of 11 to a high of 32. The calculated variances of predicted scores and actual Y scores for all 240 individuals confirm the discrepancy in the variability of predicted and actual scores, $s_y^2 = 29.29$ and $s_{y'}^2 = 12.40$. Using formula 9.13, the ratio of these two variances should produce r_{xy}^2.

$$\frac{s_{y'}^2}{s_y^2} = \frac{12.40}{29.29} = .423$$

The square root of .423 equals .6504, exactly the calculated value of r_{xy}.

Theoretically, the error scores should be normally distributed. They will have a mean of 0 and a standard deviation equal to $s_{y \cdot x}$, the standard error of estimate. Error scores for each of the 240 individuals were computed. A histogram of these error scores is presented in Figure 9.14. The distribution of errors tends to be slightly negatively skewed. This is due to the negative skewness of the *X* and *Y* distributions caused by the apparent ceiling on the test. The mean error score was zero as it always will be. The standard deviation of error scores, s_e, was 4.11. The calculated s_e from actual error scores should be equal to $s_{y \cdot x}$, calculated from formula 9.7.

$$\begin{aligned} s_{y \cdot x} &= s_y \sqrt{1 - r_y^2} \\ &= 5.412\sqrt{1 - .6504^2} \\ &= 4.11 \end{aligned}$$

The standard error of estimate relative to s_y is very large, and $k = .76$.

One additional type of question can be answered from our discussion of regression: "What proportion of persons would be expected to have posttest scores less than *Y*, greater than *Y*, or within a specific range of *Y* scores given a specific value on *X*?" For example, what proportion of the "Sesame Street" children who had pretest scores of 22 would you expect to have scores lower than 22 on the posttest?[19]

9.9 SUMMARY

Linear regression procedures based on the general linear model have been widely used as a generalized data-analysis technique over the past several years. Our discussion of linear regression was restricted to the simplest case, that involving only two variables in which one variable is viewed as a predictor variable and the other as the criterion variable. Applying the linear regression model helps describe the functional relationship between the two variables. The linear regression model has other uses in inferential statistics. The model also can be expanded to consider several independent variables. **Multiple regression** procedures are employed in the latter case. Coverage of the more complex regression procedures may be found in Cohen and Cohen (1983) and Pedhazur (1982).

---

[19] Y' for an *X* score of 22 is $.551(22) + 13.472 = 25.59$; the z score equivalent for a *Y* score of 22 in the conditional distribution with a mean of 25.59 and a standard deviation of $s_{y \cdot x} = 4.11$ is

$$z = \frac{22 - 25.59}{4.11} = -.87$$

The proportion of the normal curve to the left of $z = -.87$ is the proportion of "Sesame Street" students expected to have posttest scores of 22 or lower given that their pretest score was 22. Using the normal curve table in Appendix B.2 reveals that approximately 19% of the population in a normal distribution is expected to be located at or below a z score of $-.87$.

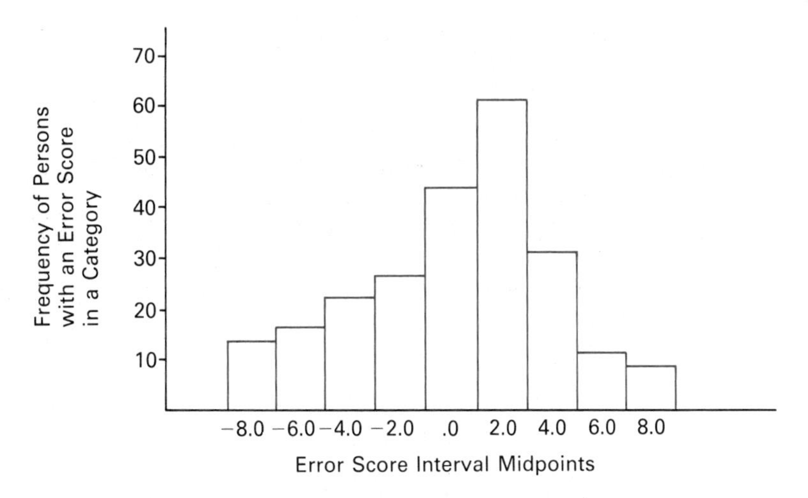

FIGURE 9.14
Histogram of error scores when predicting Body Parts posttest scores from pretest scores.

The simple linear regression equation was defined by the equation for a straight line:

$$Y' = bX + a$$

where b and a are constants whose values are calculated from data collected on variables X and Y. The constant b is the slope of the line when fit to the X, Y scatter plot and as the regression coefficient represents the change in Y expected as a function of a unit change in X. The constant a is the Y intercept and is a scale adjustment factor. The principle of least squares criteria were used in developing the computational formulas for b and a.

$$b = \frac{\Sigma xy}{\Sigma x^2} \quad \text{or} \quad b = \frac{n\Sigma XY - (\Sigma X)(\Sigma Y)}{n\Sigma X^2 - (\Sigma X)^2}$$

$$a = \overline{Y} - b\overline{X}$$

Application of these formulas will result in values for b and a that define a functional linear relationship between X and Y that minimizes the sum of the squared errors in the model. The identical linear regression equation also can be derived using the following formula:

$$Y' = r_{xy}\left(\frac{s_y}{s_x}\right)X + \overline{Y} - r_{xy}\left(\frac{s_y}{s_x}\right)\overline{X}$$

When the correlation between the predictor and criterion is less than perfect, we can expect error in prediction. The extent of error is reflected in an index termed the standard error of estimate, $s_{y \cdot x}$, and is inversely related to the correlation coefficient, or

$$s_{y \cdot x} = s_y \sqrt{1 - r_{xy}^2}$$

As the standard deviation of the errors in prediction, $s_{y \cdot x}$ is useful when expanded about predicted scores, for example, $Y' \pm 1s_{y \cdot x}$, $Y' \pm 2s_{y \cdot x}$, and $Y' \pm 3s_{y \cdot x}$. This framework provides insight about the likely variability of actual observed Y scores about the predicted score.

Finally, this chapter used regression concepts to aid understanding of r_{xy}^2, the proportion of variance in common between X and Y. Using the relation $s_y^2 = s_{y'}^2 + s_{y \cdot x}^2$ we find that

$$r_{xy}^2 = \frac{s_{y'}^2}{s_y^2} \quad \text{and} \quad (1 - r_{xy}^2) = \frac{s_{y \cdot x}^2}{s_y^2}$$

Thus, in the context of the Y variable r_{xy}^2 represents the ratio of the variance that can be accounted for ($s_{y'}^2$) in the relation between X and Y to the total variance observed for Y.

PROBLEMS

Problem Set A

1. An educational program uses a composite ability test score to place students into instructional groups. A common teacher-made test evaluates the extent of student learning. If simple linear regression were used to analyze data from a study attempting to identify the functional relationship between the two variables, which would be classified as the criterion variable and which as the predictor variable?

2. A study investigates the activity level in children as a function of the amount of sugar in the diet per the child's weight. Which variable would be classified as the predictor and which as the criterion variable?

3. Given the regression equation $Y' = .7X + 1.2$, what is the estimated change in Y scores as X changes from 8 to 9? As X changes from 5 to 14? Demonstrate these estimated changes by computing the difference in Y' scores for the changes in X and relating these changes to the value of the regression coefficient.

4. Given the regression equation $Y' = 1.5X - 1$, plot the predicted Y scores when X equals 0, 1, 4, 6, and 10. Assume that two individuals had X scores of 10, but the first had an actual Y score of 12 and the second an actual Y score of 15 in the data. What would the value of the error component in the general linear model be for each of these individuals? Plot each individual's Y score on the graph of Y' scores and demonstrate that each individual's actual Y score is made up of two components, a Y' score and an error score (i.e., $Y = 1.5X - 1 + e_i$).

5. Plot the regression lines defined by (a) $Y' = -2X + 2$ and (b) $Y' = -.8X - 2.5$.

6. Calculate estimates of the regression coefficient and the Y intercept based on the functional relationship between X and Y in the following set of sample data, then answer questions a–g.

Individual	A	B	C	D	E	F	G	H
X Score	5	2	11	5	7	6	6	8
Y Score	4	4	12	6	9	10	7	10

a. What is the change in Y estimated for every unit change in X?
b. Write the regression equation.
c. Plot the raw score data for the eight individuals and fit the regression line to the data.
d. Calculate the error score in the general linear model for each individual, then find the mean ($\bar{e}$), variance (s_e^2) and standard deviation (s_e) of the calculated error scores.
e. Calculate the variance error of estimate ($s_{y \cdot x}^2$) and the standard error of estimate ($s_{y \cdot x}$) using formulas 9.6 and 9.7, respectively. Do these values equal the values for s_e^2 and s_e calculated in part d within rounding error?
f. What proportion of Y score variance is accounted for by knowledge of X values?
g. You had to compute the predicted score for each individual in part d. Demonstrate that the variance of the predicted scores equals $r_{xy}^2(s_y^2)$ within rounding error.

7. The means and standard deviations for X and Y are given below along with the value of r_{xy}. Calculate the estimates of b and a from this information and write the equation for the best-fitting straight line to the sample data summarized by these statistical values.

$$\bar{X} = 53 \qquad s_x = 12$$
$$\bar{Y} = 24 \qquad s_y = 6$$
$$r_{xy} = -.80$$

8. The variability of a set of scores may be increased or decreased by transforming the scores to a difference scale of measurement (see Chapter 7). In Chapter 8, this type of transformation did not change the value of the correlation coefficient, r_{xy}. What effect does increased or decreased variability due to scale changes in X or Y have on the regression coefficient? Demonstrate these effects by calculating b using $s_y = 3$ (a decrease) in problem 7, $s_y = 9$ (an increase), $s_x = 9$, (a decrease), and finally, $s_x = 15$, keeping all other values in the problem constant.

9. Performance scores on the midterm (X) and final examination (Y) were collected for 250 students enrolled in an introductory statistics class. The intent was to develop a regression equation to examine the predictive relationship between the two sets of scores. Following is a summary of the data collected.

$$\bar{X} = 40 \qquad \bar{Y} = 52$$
$$s_x = 6 \qquad s_y = 8$$
$$r_{xy} = .70$$

a. Compute the regression equation that identifies the functional relationship between final and midterm exam scores.
b. What score would you predict on the final for individuals who received a score of 43 on the midterm?
c. Compute the standard error of estimate.
d. Assuming that the assumption of homoscedasticity holds and that actual Y scores are normally distributed around a predicted score, what proportion of individuals who received a score of 43 on the midterm would you theoretically expect to have scores of 56 or higher on the final?
e. Given the same assumptions as in part d, what proportion of individuals who were one standard deviation unit above the mean on the midterm would you theoretically expect to score at the mean or lower on the final?
f. Using the index of the proportion of final exam score variability accounted for by knowledge of midterm scores, compute the variance of both predicted and error scores.

Problem Set B

1. The predictive validity of the Nelson-Denny Reading Test using first-year college students' grade point average (GPA) as the criterion variable was investigated recently by Wood (1982). In addition to the Nelson-Denny test, the potential validity of other possibly important predictor variables was examined. As in several other studies, Wood found that of the predictor variables investigated, high school GPA was the best predictor of GPA at the end of the first year at college. While Wood collected data on over 1500 cases, we have simulated high school and first-year college students' GPAs for only 10 students in Table 9.4.

a. Based on this limited data set, compute the regression equation for predicting college GPA from high school GPA.
b. Estimate the proportion of high school seniors with GPAs of 3.00 that you would expect to have first-year college GPAs of 3.00 or more.
c. What proportion of these same high school seniors (GPA = 3.00) would you

TABLE 9.4
Simulated GPAs for 10 students.

Individual	High School GPA	First-Year College GPA
1	3.72	2.81
2	2.47	2.52
3	2.98	2.13
4	3.23	2.39
5	3.42	2.36
6	2.76	2.83
7	3.00	2.85
8	2.44	2.37
9	3.66	2.88
10	3.92	3.31

estimate to have first-year college GPAs of 2.00 or less based on the data in this problem?

d. What proportion of college GPA variability is predictable from knowledge of high school GPA?

e. Compute the variance of predicted and error scores for these data.

2. A study by Miskel, Glasnapp, and Hatley (1975) investigated the level of job satisfaction for educators as a function of perceived levels of work motivation and work incentives. One of the subscales used as a predictor variable was tolerance for work pressure. Summary data for this subscale and the criterion variable (job satisfaction) are given here for 516 secondary male teachers. Calculate the regression equation defining the functional relationship between the two variables. What magnitude of change in the level of job satisfaction would you expect for a five-unit change in the tolerance for work pressure based on these data?

Tolerance for Work Pressure (X)	Job Satisfaction (Y)
$\overline{X} = 26.3$	$\overline{Y} = 20.5$
$s_x = 3.3$	$s_y = 4.0$

$$r_{xy} = .36$$

3. An article by Karweit and Slavin (1981) is just one of many studies of the relationship between time spent in instruction/learning and achievement gains. While this study focused on the outcomes obtained by employing different measurement and modeling choices in the study of time and learning, it still illustrates the use of regression in trying to understand the functional relationship between time spent and achievement level. One time variable used as a predictor variable was *engaged* rate, the ratio of the average total number of minutes actually spent on-task to the average total instructional time per day. The Comprehensive Test of Basic Skills (CTBS) was one measure of achievement used. We have simulated data on the two variables for six individuals.

a. What is the estimated change in CTBS scores as a function of a 1⁄10 (.1) unit change in engagement rate?

b. What proportion of CTBS variability is functionally related to engagement rate scores?

c. For individuals whose CTBS predicted score would be 64 based on the regression equation, what increase in the proportion of on-task instructional time would you suggest to increase their predicted CTBS score to 68?

Individual	Engagement Rate	CTBS
1	.62	58
2	.85	75
3	.54	60
4	.68	71
5	.73	62
6	.75	77

Problem Set C

1. Peabody mental age (PMA) scores should be related to performance on the pretest measures. Based on the following information on PMA scores (X) and the Numbers pretest (Y), estimate the change in Numbers pretest performance expected for every change of 10 units in PMA scores.

$$\overline{X} = 46.78 \qquad \overline{Y} = 20.90$$
$$s_x = 15.99 \qquad s_y = 10.69$$
$$r_{xy} = .623$$

2. Calculate the regression equation for predicting Numbers pretest scores from PMA scores. What Numbers pretest score would you predict for individuals whose PMA score was 40? What would you estimate as the variability of errors in prediction for these individuals given that all assumptions are met?

3. Identify individuals in the "Sesame Street" data base with PMA scores of 40. Based on their actual Numbers pretest scores, compute the value of the error component in the general linear model for each individual.

4. While not true, assume for the purposes of the present problem that the category codes for the viewing quartile variable in the "Sesame Street" data base are at equal intervals so that viewing quartile may be used as a legitimate variable in a simple regression analysis. Record the scores on (a) viewing quartile, (b) Letters posttest, and (c) Relational Terms posttest for the following random sample of 20 individuals from the data base. From the recorded data, determine the separate regression equations that identify the functional relationships between each of the posttest scores as criterion variables and viewing quartile as the predictor variable.

004	055	077	171
009	056	100	193
024	060	128	196
035	066	139	206
051	068	162	207

5. On which posttest would you estimate a greater change if an individual were to move from one viewing category to the next most frequent viewing category (a one-unit change in X)? Compare statistical information on the two posttest variables to determine why one would result in an estimate of greater change per unit change in viewing quartile than the other would.

6. For which posttest is knowledge of the viewing category variable more beneficial in explaining variability among these individuals?

10

Inferential Statistics and Probability

Chapter 1 drew a distinction between two types of statistical content, descriptive and inferential. The two are often treated as if they were independent of each other; they are not. The content of descriptive statistics is a prerequisite to the study of inferential statistics. The foundation, the nuts and bolts, the elements of focus in the inferential process are the indices of descriptive statistics. Because 80%–90% of statistical content focuses on inferential techniques, the relative importance of descriptive statistics is often minimized. It should not be. For either sample or population data, the distribution shape, the "average" value, the amount of spread, and the magnitude of relationships among variables are all important descriptive indicators of the data. Remember, these descriptive indices do not merely summarize a set of numbers, they describe a behavior, trait, or characteristic of interest. The indices of descriptive statistics are important by themselves as a content domain. They become doubly important because of their role as indices of focus in the inferential statistics process.

10.1 THE PROCESS OF STATISTICAL INFERENCE

Inferential statistics and the statistical inference process were introduced in Chapter 1 and discussed in more detail in Chapter 3. These prior discussions looked at inference within the context of the more general research process. The fundamental goal is very simple: Based on information obtained from a sample of elements, draw conclusions about a population by inferring that what is observed in the sample reflects what we might expect to be true in the population.

As Figure 10.1 indicates, the general process is cyclic. We start with an identifiable characteristic of interest, a defined population of potential sampling units, and the statistical hypotheses to be tested or the population parameters to be estimated. We select a sampling plan that we follow to obtain a sample. The characteristic of interest (the dependent variable) is measured. These measurements are analyzed using statistical procedures. Then, inferences are made about the population based on the results of the statistical analysis. Interpretations are made and conclusions are drawn about the characteristic of interest for the population based on the results of the statistical inferences. Adequate sampling, measurement, and statistical analysis are all important in determining the validity of interpretations and conclusions, but the theory of statistical inference provides the framework for the entire generalization process.

Statistical inference is a process that does not change. Specific statistical

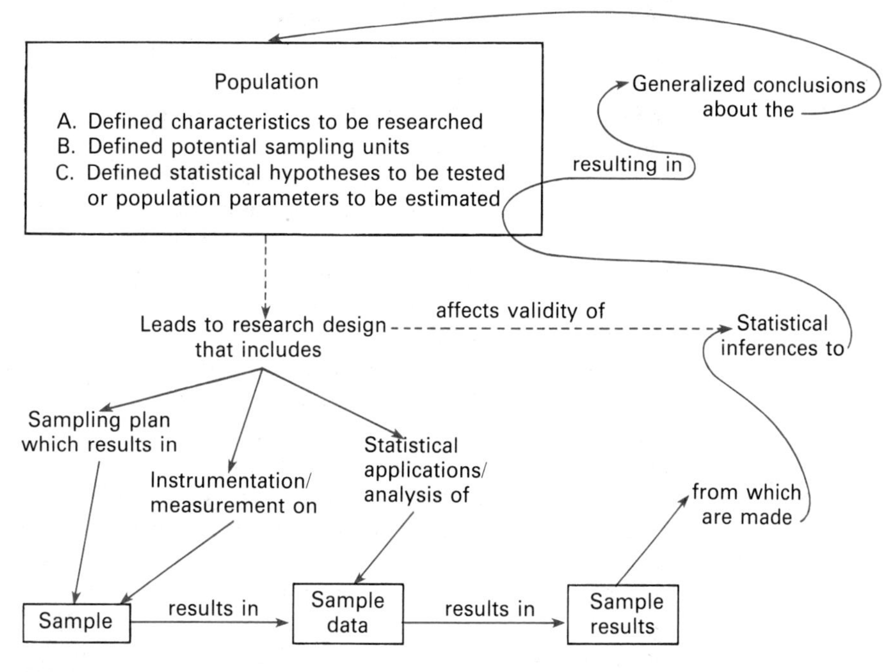

FIGURE 10.1
The research process.

methods or techniques applied within the process will change across research situations, but the theory underlying statistical inference remains the same. This and the following two chapters introduce the foundations of statistical inference. These chapters will examine the basic concepts on which statistical inference is built. Understanding these concepts is crucial to understanding application of statistical procedures. After these three chapters, the majority of statistical content in this text and in more advanced texts emphasizes the many statistical techniques available for use as tools within the inferential process.

Statistical analysis is necessary to most research. Specific statistical techniques for performing different data analyses comprise the bulk of statistical content, but the techniques are nothing more than tools to be applied when appropriate within the more general research process.

10.2 PROBABILITY'S ROLE IN STATISTICAL INFERENCE

Statistical inference assumes that sample data are being analyzed. When we deal with sample data and not data from the entire population, any statement about the population is somewhat risky. The only way to have exact information about the population is to measure every element in the population. Once we deviate from measuring the entire population, we can no longer be 100% certain that our information accurately reflects the population characteristic(s). Some amount of sampling error always exists, producing some degree of uncertainty about any conclusions.

Chapter 3 stressed the importance of sampling in the research process, primarily because the adequacy of statistical inferences depends on how well the sample represents the population. If we are uncertain about how representative our sample is, then we should be equally uncertain about inferring any results to a specific population. As you realize from our discussions of sampling, it is impossible to know exactly how representative or unrepresentative any one sample is of the population. We can, however, employ a sampling plan that results in known sampling errors. Based on the known sampling error characteristics, we can quantify the uncertainty about sample representativeness in a probability statement. The probability statement about the likelihood of a particular sample event occurring is a confidence measure used by the researcher when making particular inferences based on sample information.

"What is the risk involved?" "How uncertain are we?" "How likely is it that a particular event will happen?" These questions of probability are all pertinent to the statistical inference process. Probability is a basis for making decisions about the kinds of inferences to be made. Statistical inference is not an exact science. It does not result in certainties. Probability allows us to make decisions with predetermined levels of confidence. Statistical inferences are made based on the answer to the question "How likely is the occurrence of sample events, like the one observed, when the sample is assumed to be representative of the population?" "How likely is the occurrence" translates into "What is the probability?" If the probability is high based on expectations from sampling error theory, then the population is inferred to have the characteristic described by the sample. If the probability is low, the inference is that the sample is not representative of the defined population and therefore the population does not have the characteristic described by the sample.

While probability levels provide the basis for all statistical inference decisions, we do not expect you to become an expert in the mathematics of probability theory. You should, however, have an intuitive understanding of probability and the logic of probability thinking as it applies to the statistical inference process.

10.3 PROBABILITY: DEFINITION AND RULES

A **probability** is a mathematical quantity expressed in the form of a proportion or decimal between zero and one, inclusive. It summarizes the relative frequency with which something will happen over a large number of repeated trials. The "something" to which a probability value is attached usually is labeled an **event.** An event is any observable happening or outcome, such as an individual having a car accident, being selected as an element from a population, passing a test, winning a state lottery, or changing one's behavior. In one form, probability may be defined as the frequency with which an event will occur in a population relative to the total number of possible events in the population.

In random sampling, for example, an event is the selection of a single individual. If a finite population contains 500 individuals, what is the probability that individual 322 will be selected at random if only one person is selected? Individual 322 occurs only once in the population. The total number of possible events in the population is 500. The probability of selecting individual 322 at random from the 500 possible people is 1/500 or, expressed as a relative proportion, .002.

Probabilities often are interpreted with reference to repeated samplings, trials, or experiments. How often would you expect the event to occur over repeated trials (or, in common probability language, "in the long run")? The probability gives the relative frequency with which an event occurs in the population and forms a basis for our expectation of how often we might observe the event in practice. A frequently used illustration is the two-event population resulting from a single flip of a balanced coin. The two possible population events are "heads" and "tails." For any one flip of the coin, the frequency with which heads could occur is once. As a relative proportion, the probability that heads will occur on any one flip of a coin is one half or .50. The same probability holds for the occurrence of tails. These probabilities allow us to form expectations about the frequency with which we would observe heads or tails in repeated flips of the coin. In the long run or for a very large number of coin flips, we would expect approximately half of the flips to end up heads and half to end up tails.

The long-term expectation based on the probability for an event is important for statistical inference: It will allow us to anticipate the likely frequency of certain outcomes. Based on the notion of probability and expected outcomes, we will be able to form a hypothetical frequency distribution of anticipated outcomes in the population without observing or counting every event.

Probability Rules

For our applications, we need several simple rules of probability. Some, included in the discussion and definition of probability, are merely common sense.

- Rule 1 Probability values range from zero to one, inclusive.
- Rule 2 The sum of the probabilities of all possible events in a population must equal one.

- Rule 3 The probability that any one of a set of equally likely events will occur is one divided by the total number of events in the population (e.g., the random selection of individual 322 from a population of 500).
- Rule 4 If multiple events of the same kind exist in the population, the probability that the event will occur equals the frequency of the event in the population divided by the total number of possible events (e.g., if there are four males in a group of 52 teachers, the probability of selecting a male at random from the group is 4/52 = 1/13 or .077).

Two additional probability rules—the addition rule and multiplication rule—form the basis for solving most problems in probability.

- Rule 5 (Addition Rule) For a group of mutually exclusive events, the probability that any one of the events will occur equals the sum of probabilities of the separate events.

The key to Rule 5 is the phrase *mutually exclusive events.* Two events are mutually exclusive if the occurrence of one event precludes occurrence of the other event on the same trial. This rule is appropriately applied to any either/or type of probability problem. For example, what is the probability, for any one flip of a coin, that *either* a tail *or* a head will occur? The events "tails" and "heads" are mutually exclusive: If one happens on any one flip, the other cannot happen. If the probability is the sum of the probabilities of the separate events, then the probability that either tails or heads will occur is equal to the probability of tails (.50) plus the probability of heads (.50) or 1.00. The example is simple, but illustrates the rule. Let's look at a more complex example that applies several of the rules. Assume we are selecting one individual at random from the "Sesame Street" data base ($n = 240$). What is the probability that the randomly selected individual would be from either Site I ($n = 60$), Site IV ($n = 43$), or Site V ($n = 18$)? The separate events of interest are mutually exclusive, so that a person selected from Site I cannot be from Sites IV or V, too. Using Rule 4, the probability for selection from

$$\text{Site I} = 60 \div 240 = .250$$

$$\text{Site IV} = 43 \div 240 = .179$$

$$\text{Site V} = 18 \div 240 = .075$$

Applying Rule 5, the probability that the individual randomly selected would be from one of the three sites is $.250 + .179 + .075 = .504$. The probability of .504 means that over repeated samplings (assuming with replacement) in the long run, we expect approximately 50.4% of the individuals selected would be from one of the three sites.

■ Rule 6 (Multiplication Rule) The probability that a combination of independent events will occur over trials is the product of the separate probabilities of the events.

Rule 5 specifies the probability of one of many mutually exclusive events occurring for any one given trial. Rule 6 specifies the probability for a series of outcomes over more than one trial. It applies to the *and* condition in a probability problem. What is the probability of getting a head on the first flip of a coin *and* getting a head on the second flip? Application of Rule 6 requires that the trials be independent of each other, meaning outcomes on one trial do not affect the outcomes on any other trials. What is the probability of getting two heads on two successive flips of a coin? The two flips are independent of each other, so the probability of a head on either flip is .50. The probability of two heads on two successive flips is (.50)(.50) = .25. The probability of .25 for the outcome of two heads also can be obtained by looking at the relative frequency of the outcome in the total population of outcomes. The population outcomes are easily identified for two flips of a coin: (H, H), (H, T), (T, H), or (T, T). From the basic definition of probability, the probability of the event (H, H) equals its frequency in the population divided by the number of possible events in the population; for example, the probability of (H, H) = 1/4 = .25. As another example, what is the probability of randomly selecting an individual from Site I on the *first* selection *and* someone from Site V on the *second* selection (assuming random sampling with replacement)?[1] Our expectation based on the probability of .019 is that samples of two persons chosen at random with replacement *in the identified order* would occur approximately 19 times out of 1000 samples.

Typical probability problems involve outcomes over a series of trials rather than on just one trial. When a series of outcomes is of interest, both the different combinations of potential outcomes that could occur and the different orders in which they could occur become relevant to calculating the probability of any one event. In the two-trial coin-flipping example, there were three different combinations of outcomes: two heads, one head and one tail, or two tails. When outcome order is considered, the population consists of four possible outcomes. The combination of one head and one tail could occur twice with each outcome defined by a different order of occurrence: (H, T) and (T, H). When looking at a series of potential outcomes, we must consider both the number of different combinations and the different orders of occurrence to define the number of population events. One way to calculate the probability that a particular sequence or combination of events will occur is to list or count the number of possible population outcomes and the frequency

---

[1]*p*(Site I) = .250
p(Site V) = .075
p(Site I on first selection, Site V on second) = (.250)(.075) = .019

with which the specific sequence or combination of interest occurs in the population. The ratio of the latter frequency to the total number of possible population outcomes is the probability. It is usually inconvenient or practically impossible to list all possible outcomes, however. It was easy to list all potential outcomes for the two-trial coin-flipping example because the number of potential outcomes was limited. This would not be true for the other example, selecting two persons from the "Sesame Street" data—the first from Site I and the second from Site V. Do you see the problem in listing all the combinations and sequences of potential events in the population? We can form 57,360 different ordered pairs from the 240 persons in the "Sesame Street" data base. While it would not be impossible to list all of them, it certainly would be time consuming! Listing and physically counting individual events fortunately is not necessary. We can use mathematical rules to calculate the number of possible combinations or sequences for a defined population of events.

10.4 COMBINATIONS AND PERMUTATIONS

A **combination** is defined as an unordered group of r objects. The key word in the definition is *unordered.* Any grouping of r objects, no matter what its pattern or arrangement, satisfies the definition of a combination. In contrast, the arrangement or ordered pattern of objects is the key element in the definition of a **permutation.** A permutation of r objects is defined as the combination of r objects sequenced in a particular pattern. Suppose a teacher is asked to select three students from a class of 18 to help in the cafeteria during lunch hour. How many different groups of three students could the teacher potentially select from the class of 18? This is a combinations problem. How many different combinations of three students can be selected from 18? Ordering within the group selected is not important. If the teacher had been requested to select three students and then rank order them in terms of preference for working in the lunch room, the number of different ordered sequences of three students would be a permutation problem. Whether the objects are to be selected and ranked (permutations) as opposed to just selected (combinations) affects the number of distinct events in the population of possible outcomes. Assume that Ed, Sue, and Jo were selected to work in the lunch room. As a group, they constitute a single combination; however, we can form six different permutations as possible outcomes: (Ed, Sue, Jo), (Sue, Jo, Ed), (Jo, Sue, Ed), (Ed, Jo, Sue), (Sue, Ed, Jo), or (Jo, Ed, Sue).

The mathematical formulas for determining the number of possible combinations or permutations employ a special type of notation, **factorial notation.** Factorial notation is symbolically represented by $n!$ (read n factorial) which is the product of all of the integers from n to 1. For example, $3! = 3 \times 2 \times 1 = 6$. The computed value of $n!$ represents the number of different ways n objects can be arranged in order. The number of different orderings or sequences of Ed, Sue, and Jo is six or 3!. As n increases in size, the number of possible different orderings becomes extremely large (e.g., $7! = 5040$). There are 5040 different ways to arrange seven objects. In factorial notation, one special case is the occurrence of 0! (zero factorial). It is customary to let 0! equal one ($0! = 1$).

As a product, $n!$ gives the number of different permutations (arrangements) of n objects taken n at a time. There are times when the entire set of n objects is not desired, but some subset of r objects is selected from the n possible objects. To calculate the number of permutations of n objects taken r at a time, the following formula is used.

$$P\binom{n}{r} = \frac{n!}{(n-r)!} \tag{10.1}$$

$P\binom{n}{r}$ represents the number of distinct arrangements of r objects that could be sampled from a population of n objects. How many different arrangements are there for three objects selected from a population of five objects?

$$\begin{aligned} P\binom{5}{3} &= \frac{5!}{(5-3)!} = \frac{5!}{2!} = \frac{5 \times 4 \times 3 \times 2 \times 1}{2 \times 1} \\ &= \frac{120}{2} \\ &= 60 \end{aligned}$$

Referring back to the teacher selection of three students ordered by preference, how many possible different rankings of 3 students from the class of 18 could the teacher have made?[2] The population of ordered pairs from the "Sesame Street" example used earlier was a permutation problem. How many different ordered pairs could one make when selecting two individuals from 240?

$$P\binom{240}{2} = \frac{240!}{238!} = \frac{240(239)(238!)}{238!} = (240)(239) = 57{,}360$$

When computing the possible number of combinations, ordering is not important. Any combination of r objects is counted only once. The formula for computing the number of combinations of r objects sampled from a population of n objects is

$$C\binom{n}{r} = \frac{n!}{r!(n-r)!} \tag{10.2}$$

---

[2] $$\begin{aligned} P\binom{18}{3} &= \frac{18!}{(18-3)!} = \frac{18!}{15!} = \frac{18 \times 17 \times 16 \times 15!}{15!} \\ &= 18 \times 17 \times 16 \\ &= 4896 \end{aligned}$$

The number of different combinations of three objects taken from a population of five objects is

$$C\binom{5}{3} = \frac{5!}{3!(5-3)!} = \frac{5 \times 4 \times \cancel{3!}}{\cancel{3!}(2 \times 1)}$$
$$= \frac{20}{2} = 10$$

There were 60 different permutations for the same problem, but only 10 different combinations. How many different combinations of 3 students could we select from a class of 18 students?[3]

The use of formulas 10.1 and 10.2 can save time and effort in computing certain probabilities. To illustrate, let's look at a marketing strategy used by some businesses to attract customers This strategy uses a game of chance or contest that allows the customer the possibility of winning a free prize. One such contest might include nine covered spaces on a card. By erasing two spaces, the customer can win if the objects in the two erased spaces match. Assuming that the objects have been placed on the card at random and that there is one and only one winning combination on each card, what is the probability of winning a free prize? There is only one winning combination on the card and the order of erasures makes no difference. Because of the random arrangement of objects on the card, all combinations of two objects are equally likely to occur. The probability of selecting the one winning combination equals one divided by the total number of possible combinations of two objects selected from nine objects.

$$C\binom{9}{2} = \frac{9!}{2!(9-2)!} = \frac{9 \times 8}{2} = 36$$

The probability of winning is thus 1/36 or .028. What would the probability of winning be if there were 15 spaces, but we needed to match 3 objects?[4]

Assume the contest had four columns of 10 spaces each. Each space in a column contains one of the numbers 0 through 9, which have been randomized so they do not occur in sequence. The contest is to erase one space in each of the four columns. If the sequence of numbers you erase in columns 1–4 match a random four-digit number at the top of the card, you win. What is

---

[3] $C\binom{18}{3} = \frac{18!}{3!(18-3)!} = \frac{18!}{3!15!} = \frac{18 \times 17 \times 16 \times \cancel{15!}}{3 \times 2 \times 1 \times \cancel{15!}}$

$$= \frac{4896}{6} = 816$$

[4] The total number of possible combinations is

$$C\binom{15}{3} = \frac{15!}{3!(15-3)!} = \frac{15 \times 14 \times 13 \times \cancel{12!}}{3 \times 2 \times 1 \times \cancel{12!}} = 455$$

The probability is 1/455 or .0022 of winning on any one card.

the probability that any one customer will win? The sequence of digits across the four columns is important in this problem, so we are interested in the total number of possible permutations of 4 numbers selected from the population of the 10 numbers 0–9. There is only one winning sequence, so the probability is 1 divided by the number of permutations of 10 numbers taken 4 at a time.

$$P\binom{10}{4} = \frac{10!}{(10-4)!} = \frac{10 \times 9 \times 8 \times 7 \times \not{6}!}{\not{6}!}$$
$$= 5040$$

The probability of winning is 1/5040 or .0002.

10.5 THE BINOMIAL DISTRIBUTION: AN ILLUSTRATION

Probability models allow us to make decisions in statistical inference. We use these models to approximate what we might expect to happen in reality. Based on an appropriate probability model, we can identify the relative frequency with which certain events are expected to occur. Probabilities identify long-run expectancies of the frequency with which an event will occur over a very large number of trials. In the short run over a limited number of trials, however, the actual number of outcomes of a particular type tends to follow a random model in deviating from the long-run expectancies. For example, in answering 10 true/false test items by random guessing, everyone would not get exactly five items correct. Chance or luck determines how many items an individual answers correctly purely by guessing. This chance or luck is really randomness at work. Particularly important to statistical inference is the expected occurrence of events under such a random model. Theoretical frequency distributions of outcomes can be generated based on the mathematical models of probability and randomness. These theoretical frequency distributions are models we use to form expectations about reality. These expectations are in turn used to make decisions in statistical inference.

One such theoretical frequency distribution is called the **binomial distribution.** A binomial distribution can be used to identify the expected relative frequency of dichotomous outcomes for a fixed number of independent trials. The occurrence of any event for which the outcome may be viewed as dichotomous (e.g. success/failure, correct/incorrect, selected/not selected) is termed a binomial trial. A binomial distribution is formed by determining the relative frequencies with which the success outcome will occur over a fixed number of independent binomial trials in which the probability of success is constant for each trial. A relative frequency really is a probability, so a binomial distribution may be viewed as a distribution of probabilities. A binomial distribution describes the probabilities associated with each of r successes for a fixed number of n binomial trials.

Consider the example of random guessing in answering six true/false test items. Each item represents a binomial trial in which answering the item results in a dichotomous outcome: Either the answer is correct or it is incorrect. The probability of success—guessing the answer correctly—for each item is .5. By guessing, the response to any one item is a random response and the responses over all six binomial trials (items) will follow the laws of randomness. Assuming randomness of response, what is the probability of getting zero items correct? One item correct? Two items correct? Three, four, five, or six items correct? These probabilities as a group form a binomial distribution. The binomial distribution formed describes the probabilities associated with the occurrence of each of r successes given n binomial trials in which r takes on integer values from 0 to n, inclusive.

What is the probability of correctly answering all six items by chance, that is, of having six successes? Given that the probability of guessing any one answer correctly is .5, how often would we expect a sequence of six right answers? This probability is easily calculated. The events are independent, so we can use the multiplication rule of probability (Rule 6). The probability of six correct answers equals $1/2 \times 1/2 \times 1/2 \times 1/2 \times 1/2 \times 1/2 = (1/2)^6 = 1/64 = .016$. The long-run frequency with which we would expect this event to happen at random is less than twice out of 100 cases. What is the probability of getting 5, 4, 3, 2, or 1 item(s) correct by chance (at random)? These probabilities are not as easily calculated because more than one combination of correctly answered items out of the six could produce the required number of successes. Not only do we need to consider the probabilities of success or failure across the independent trials, we also need to consider the number of different combinations that could result in the number of desired successes. Fortunately, a general mathematical equation can help us find the probability of r successes in n independent binomial trials. Using the equation to compute the probabilities for every integer value of r between 0 and n, inclusive, results in a **binomial probability distribution.** The general form of the equation for generating the binomial probability distribution is

$$P(r \text{ successes}) = C\binom{n}{r}p^r q^{n-r} \tag{10.3}$$

where

$P(r \text{ successes})$ = the probability of r successes in n trials

n = the number of binomial trials in a series

r = the number of successful outcomes

$C\binom{n}{r} = \dfrac{n!}{r!(n-r)}$ = the number of *combinations* of n things taken r at a time

p = the probability of a success on any one of the n trials

$q = 1 - p$ or the probability of a failure on any one of the n trials

In the equation $C\binom{n}{r}$ is the number of different combinations of n trials that could result in r successes. The product $p^r q^{n-r}$ is the probability of occurrence for any one combination of r successes and $n - r$ failures. Multiplying $C\binom{n}{r}$ times $p^r q^{n-r}$ is the same as applying the additive property of probabilities (Rule 5). The complete equation gives the probability that *either* combination 1, 2, or 3, for example, will occur.

Table 10.1 shows the results of applying formula 10.3 to the example of six true/false (T/F) items. Although the symbols are different, r and C are equivalent to scores (X) and frequencies (f) in a frequency distribution. For example, at an r value of 4 there are 15 observations. Note also the distribution of probabilities across values of r. The most probable number of items correct is three, or half the items. Based on a .5 probability of guessing the correct answer to any one T/F item, the long-run expectation is that half the items would be answered correctly. Over a limited number of trials, however, getting half the items correct is not a 100% certainty. Chance or randomness enters into the response pattern, creating errors that deviate from our long-run expectations. For dichotomous outcomes, the binomial distribution is a model for describing the probabilities of a particular deviation or outcome occurring under a random response pattern. For example, assuming that responses are made at random, how likely is it that an individual would only get one of six T/F items correct? Responding at random, it is *possible* to only get one T/F item correct out of six items, but this event would occur less than 10% of the time—that is, $p(1 \text{ success}) = .094$—in the population of random responses.

Let's change the example slightly using four-option multiple choice (MC) items instead of T/F items. If each option is equally likely, the probability of guessing the correct answer to any one MC item is 1/4. The probability of an incorrect response ($q = 1 - p$) is 3/4. Table 10.2 displays the binomial probabilities associated with getting r correct answers by chance in a 20-item test. We can use the additive rule of probability to find other probabilities from the distribution in Table 10.2. For example, what is the probability that someone who is guessing would get 10 or more MC items correct? It is the sum of the

TABLE 10.1
Binomial probability distribution of random correct answers to six true/false items.

r	$C\binom{n}{r}$	p^r	q^{n-r}	Probability
6	1	$(1/2)^6$	$(1/2)^0$	1/64 = .016
5	6	$(1/2)^5$	$(1/2)^1$	6(1/32)(1/2) = .094
4	15	$(1/2)^4$	$(1/2)^2$	15(1/16)(1/4) = .234
3	20	$(1/2)^3$	$(1/2)^3$	20(1/8)(1/8) = .312
2	15	$(1/2)^2$	$(1/2)^4$	15(1/4)(1/16) = .234
1	6	$(1/2)^1$	$(1/2)^5$	6(1/2)(1/32) = .094
0	1	$(1/2)^0$	$(1/2)^6$	1/64 = .016

TABLE 10.2
Binomial probability distribution for random guessing on 20 four-option MC items.

r	$C\binom{n}{r}$	p^r	q^{n-r}	Probability
20	1	$(1/4)^{20}$	$(3/4)^{0}$	.0000
19	20	$(1/4)^{19}$	$(3/4)^{1}$	.0000
18	190	$(1/4)^{18}$	$(3/4)^{2}$	.0000
17	1140	$(1/4)^{17}$	$(3/4)^{3}$	.0000
16	4845	$(1/4)^{16}$	$(3/4)^{4}$	.0000
15	15504	$(1/4)^{15}$	$(3/4)^{5}$	.0000
14	38760	$(1/4)^{14}$	$(3/4)^{6}$	.0000
13	77520	$(1/4)^{13}$	$(3/4)^{7}$	.0002
12	125970	$(1/4)^{12}$	$(3/4)^{8}$	.0008
11	167960	$(1/4)^{11}$	$(3/4)^{9}$	.0030
10	184756	$(1/4)^{10}$	$(3/4)^{10}$	.0099
9	167960	$(1/4)^{9}$	$(3/4)^{11}$	.0271
8	125970	$(1/4)^{8}$	$(3/4)^{12}$	.0609
7	77520	$(1/4)^{7}$	$(3/4)^{13}$	.1124
6	38760	$(1/4)^{6}$	$(3/4)^{14}$	.1686
5	15504	$(1/4)^{5}$	$(3/4)^{15}$	.2023
4	4845	$(1/4)^{4}$	$(3/4)^{16}$	.1897
3	1140	$(1/4)^{3}$	$(3/4)^{17}$	.1339
2	190	$(1/4)^{2}$	$(3/4)^{18}$	.0669
1	20	$(1/4)^{1}$	$(3/4)^{19}$	.0211
0	1	$(1/4)^{0}$	$(3/4)^{20}$	.0032

probabilities for the mutually exclusive events $r = 10$, $r = 11$, . . . $r = 20$. The probability of getting 10 or more four-option *MC* items correct by guessing is .0099 + .0030 + .0008 + .0002 = .0139. What is the probability of correctly guessing 2 or fewer items out of 20 *MC* items?[5]

The binomial probability distribution may be thought of as a *theoretical* probability distribution—a mathematical model for describing the probabilities of random events with dichotomous outcomes occurring. It describes what we should expect for the occurrence of different outcomes under an assumed model of randomness. Randomness and the probabilities associated with random events occurring are the basis for our expectations. In conducting research, a theoretical probability distribution forms a criterion against which sample results are compared to see if we would expect them under a model of randomness or if the results deviate so much from our expectations that we might conclude something other than randomness is operating. We will spend the next two chapters expanding this latter idea, but here we present an example of how we might use a binomial probability distribution to conduct a research experiment.

[5] $p(r = 2 \text{ or } r = 1 \text{ or } r = 0) = .0669 + .0211 + .0032$

$= .0912$

The psychology of consumer preference is often used as a mechanism in advertising campaigns. Some commercials show how individuals prefer one product over others in a direct-comparison test. Assume we want to see if consumers prefer Cola A to Colas B and C. We set up an experiment in which a randomly selected sample of 15 individuals will participate; we will consider each individual a trial. We will randomly arrange the three colas for each trial on a table. Each participant will taste the three colas in the random arrangement and select the one that they prefer. A trial is scored as a success if Cola A is preferred and as a failure if Colas B or C are chosen. How would we assess the outcome of this experiment if 10 or more of the 15 subjects preferred Cola A?

In this example, tne binomial probability distribution can serve as our theoretical model of expected outcomes if a preferred cola is really selected at random. A model of random selection implies no difference in the preferences for a particular cola: Each is equally likely to be chosen. On any one trial, the probability of choosing Cola A if the choice is random is 1/3. The probability of not choosing Cola A (choosing Colas B or C) is 2/3. How likely is it that 10 or more successes would occur by chance under the conditions of this experiment? The theoretical binomial distribution of probabilities used as a basis for answering this question is given in Table 10.3. The probability that 10 or more people would select Cola A by chance is .0085, the sum of the probabilities for values of r = 10, 11, 12, 13, 14, and 15. The expected probability based on the theoretical binomial probability distribution suggests that 10 or more persons

TABLE 10.3
Binomial probability distribution for the chance selection of one of three soft drinks over 15 trials.

r	$C\binom{n}{r}$	p^r	q^{n-r}	Probability
15	1	$(1/3)^{15}$	$(2/3)^{0}$	.0000
14	15	$(1/3)^{14}$	$(2/3)^{1}$	.0000
13	105	$(1/3)^{13}$	$(2/3)^{2}$	.0000
12	455	$(1/3)^{12}$	$(2/3)^{3}$	.0003
11	1365	$(1/3)^{11}$	$(2/3)^{4}$	.0015
10	3003	$(1/3)^{10}$	$(2/3)^{5}$	.0067
9	5005	$(1/3)^{9}$	$(2/3)^{6}$	.0223
8	6435	$(1/3)^{8}$	$(2/3)^{7}$	.0574
7	6435	$(1/3)^{7}$	$(2/3)^{8}$	.1148
6	5005	$(1/3)^{6}$	$(2/3)^{9}$	.1786
5	3003	$(1/3)^{5}$	$(2/3)^{10}$	.2143
4	1365	$(1/3)^{4}$	$(2/3)^{11}$	.1948
3	455	$(1/3)^{3}$	$(2/3)^{12}$	.1299
2	105	$(1/3)^{2}$	$(2/3)^{13}$	.0599
1	15	$(1/3)^{1}$	$(2/3)^{14}$	.0171
0	1	$(1/3)^{0}$	$(2/3)^{15}$	.0023

out of 15 selecting Cola A by chance is highly unlikely. (This outcome would be expected to occur less than 1% of the time under a random model of outcomes.) Because the outcome of our experiment is so unlikely to occur by chance, the evidence suggests that Cola A is preferable to Colas B and C.

10.6 THE BINOMIAL DISTRIBUTION: PROPERTIES AND THE NORMAL APPROXIMATION

The binomial distribution is a discrete probability distribution. The scale values, r successes in n trials, do not represent some underlying continuum. Even as a discrete distribution, the binomial distribution can be graphed using the techniques in Chapter 4. Figure 10.2 plots the binomial distributions from

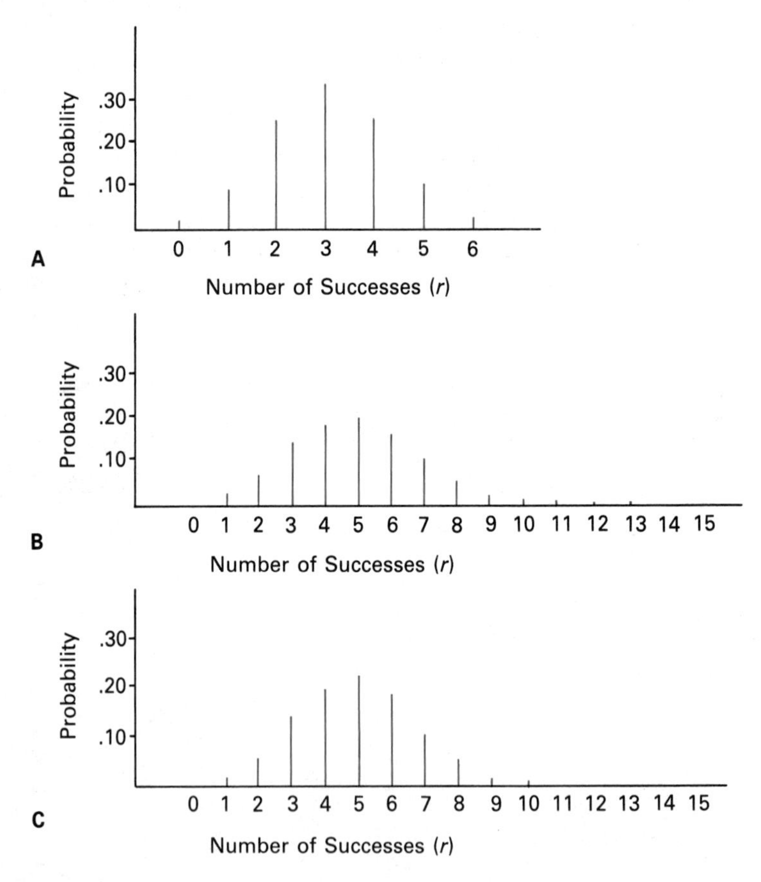

FIGURE 10.2
Binomial distribution graphs. **A,** $n = 6$, $p = 1/2$; **B,** $n = 20$, $p = 1/4$; **C,** $n = 15$, $p = 1/3$.

Tables 10.1, 10.2, and 10.3. Note that when $p = .5$, the distribution will be exactly symmetrical (see Figure 10.2A). When p is less than .5 (as in Figures 10.2B and C), the binomial distribution will tend to have a slight positive skew. Negatively skewed binomial distributions will occur for values of $p > .5$. The degree of skewness in binomial distributions is affected both by n and p. As n increases (more trials are observed), the binomial distribution tends to be less skewed and more symmetrical for all values of p. We can see this in Figures 10.2B and C, which have a slight skew. Because of the relatively large number of trials, however, they tend to approach symmetry. A more apparent skew is evident in Figures 10.3A and B in which $p = 1/4$, but $n = 10$ and 5, respectively. The value of p affects the shape of the distribution in that the closer p is to .5, the less skewed and more symmetrical the distribution becomes for any given number of trials.

As you can see from formula 10.3, computing binomial probabilities can be very laborious, particularly if they need to be calculated for an entire distribution in which the number of trials is large. Tables do exist for a full range of p values and a large number of trials, but complete sources tend to be voluminous and not easily accessible (e.g., *Tables of the Binomial Probability Distri-*

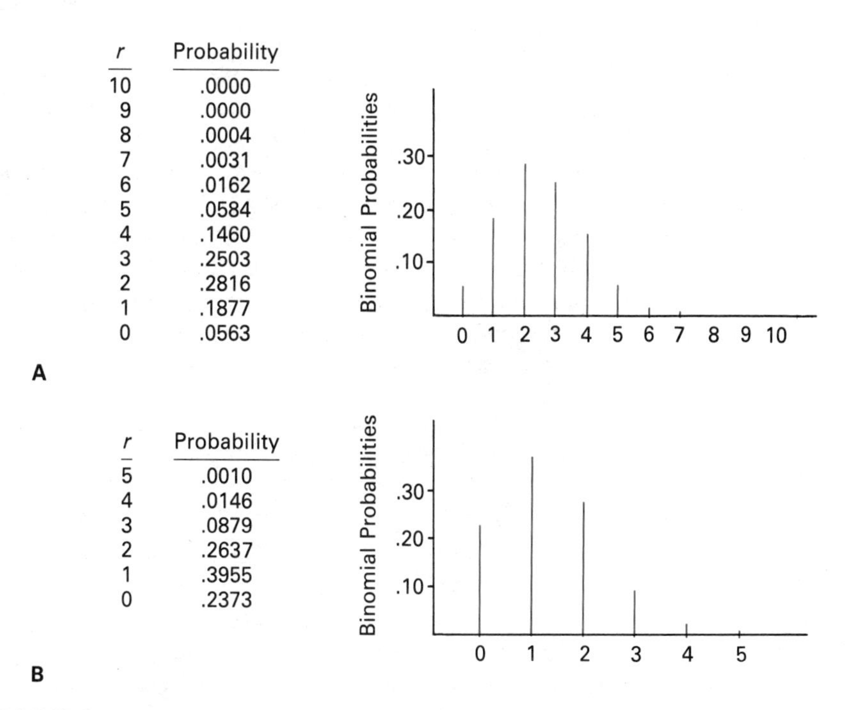

A

r	Probability
10	.0000
9	.0000
8	.0004
7	.0031
6	.0162
5	.0584
4	.1460
3	.2503
2	.2816
1	.1877
0	.0563

B

r	Probability
5	.0010
4	.0146
3	.0879
2	.2637
1	.3955
0	.2373

FIGURE 10.3
Binomial probability distributions for trials of 10 and 5 when $p = 1/4$. **A**, $n = 10$, $p = 1/4$; **B**, $n = 5$, $p = 1/4$.

bution, U.S. Government Printing Office, 1949, 195 pages). Computer programs also are readily available for generating binomial probabilities. However, these resources are often inconvenient. Fortunately, the problems in generating the binomial probabilities can be avoided through use of the normal distribution in estimating the probability values in a binomial distribution. Even though the binomial distribution is discrete, we can use its tendency to be symmetrical and normal for a large number of trials (and/or values of p close to or equal to .5) to our advantage. We can treat the binomial as if it were a continuous distribution to estimate the discrete binomial probabilities. When the product of np or nq, whichever is smaller, is greater than or equal to 5, a binomial distribution tends to resemble a normal distribution. When $p = q = .5$, the binomial distribution for 10 or more trials ($n \geq 10$) would be approximately normally distributed. If $p = 1/4$ and $q = 3/4$, a minimum of 20 trials would be needed for an acceptable approximation of binomial probabilities from normal curve values.

Both a binomial distribution and the normal curve may be thought of as proportional frequency distributions. We can determine and interpret the relative frequency with which values of r or greater will occur in a binomial distribution the same as the relative frequency with which values of z or greater will occur in a normal curve. Approximating binomial probabilities involves treating the binomial distribution as a continuous distribution, changing the interval represented by r to z scores, and looking up the appropriate tabled proportion of the normal curve associated with the obtained z scores. The resulting interpretation is the same. For example, the area of the normal curve to the right of a value of z is *the proportion of scores expected to have z scores greater than the one identified.* The proportions in Table B.2 in Appendix B are relative frequencies. They can be viewed as probabilities that a score drawn at random from a normal distribution would have a z score greater than the one identified. If the frequencies of r values in a binomial distribution tend to be normally distributed, then the probability of obtaining r successes or more is the same as the proportion of the normal curve to the right of the z score equivalent of r.

How do we change a value of r in a binomial distribution to its z score equivalent? The standard z score formula is

$$z = \frac{X - \overline{X}}{s_x}$$

The formula requires the mean and standard deviation of the binomial distribution. The mean for any binomial distribution is calculated by the product np. This mean is the long-run expected value for the number of successes. In the short run, the actual number of successes observed will distribute themselves randomly around the mean, np. For example, there were 20 trials (*MC* items) for the distribution in Table 10.2. The value of p was 1/4 or .25. The mean of the binomial distribution in Table 10.2 is (20)(1/4), an r value of 5. The product npq is the variance of any binomial distribution. The standard deviation is the square root of npq. Using these products for the mean and standard deviation of a binomial distribution, the z score formula for converting any value of r is

$$z_r = \frac{r - np}{\sqrt{npq}} \quad (10.4)$$

In this formula the value of r is a discrete number, but in the z score transformation we are treating it as a continuous value. To overcome this discontinuity in measurement, we treat each event (r) as having real limits of $r \pm .5$. To determine the probability that 7 successes would occur in a binomial distribution using the normal curve approximation, we would need to calculate the area of the normal curve between the values 6.5 to 7.5. If we wanted to determine the probability of 7 or more successes, we would find the area of the normal curve to the right of the z_r score 6.5. It is essential to use the exact score limits for r when we use the continuously distributed normal curve to approximate the probability values in the discrete binomial distribution.

The binomial distribution given in Table 10.3 for $n = 15$ and $p = 1/3$ meets the minimum criteria for using the normal-curve approximation, $np = 5$ and $nq = 10$. Using normal-curve values to obtain an approximate probability for 3 successes occurring requires determining the proportion of the normal curve between the z_r score equivalents for 2.5 and 3.5. The mean of this binomial distribution is $np = 15(1/3) = 5$ and the standard deviation is

$$\sqrt{npq} = \sqrt{15\left(\frac{1}{3}\right)\left(\frac{2}{3}\right)} = \sqrt{3.333} = 1.826$$

The z score equivalents for 2.5 and 3.5 are

$$z_{2.5} = \frac{2.5 - 5}{1.826} = -1.369 \qquad z_{3.5} = \frac{3.5 - 5}{1.826} = -.821$$

The proportion of the area between z scores of -1.369 and $-.821$ from Table B.2 is .1208. As an approximate value, .1208 is acceptably close to the true binomial probability of .1299. Figure 10.4 illustrates the fit of the normal curve and the area calculated. From this illustration, we see that the normal curve can serve as a probability distribution. As another example, let's compare the probability using the normal approximation to the binomial probability for $r \geq 7$ from Table 10.3. We can obtain the binomial probability from the sum of appropriate probabilities in Table 10.3.

$$P(r \geq 7) = .1148 + .0574 + .0223 + .0067 + .0015 + .0003$$
$$= .2030$$

The normal approximation is the proportional area of the curve to the right of the z_r score 6.5

$$z_r = \frac{6.5 - 5}{1.826} = .821$$

The area from Table B.2 is .2061, again reasonably close to .2030, the exact binomial probability.

As mentioned previously, for larger numbers of trials and/or p values closer to .50, the normal approximations come very close to the actual binomial probabilities. Assume a binomial distribution with 100 trials and $p = .5$. What is the probability that 60 successes will occur by chance? The binomial probability is

$$P(r = 60) = C\binom{100}{60}\left(\frac{1}{2}\right)^{60}\left(\frac{1}{2}\right)^{40}$$
$$= .0108$$

The z scores for $r = 59.5$ and 60.5 are

$$z_{59.5} = \frac{59.5 - 50}{\sqrt{100\left(\frac{1}{2}\right)\left(\frac{1}{2}\right)}} = 1.90$$

$$z_{60.5} = \frac{60.5 - 50}{5} = 2.10$$

The proportional area of the normal curve between $z = 1.90$ and $z = 2.10$ is $.4821 - .4713 = .0108$. The probabilities are identical from both calculations. If you try to compute the binomial probability by hand for a problem with this many trials, you can see what an enormous task it is. Using the normal approximation is much easier and just as accurate.

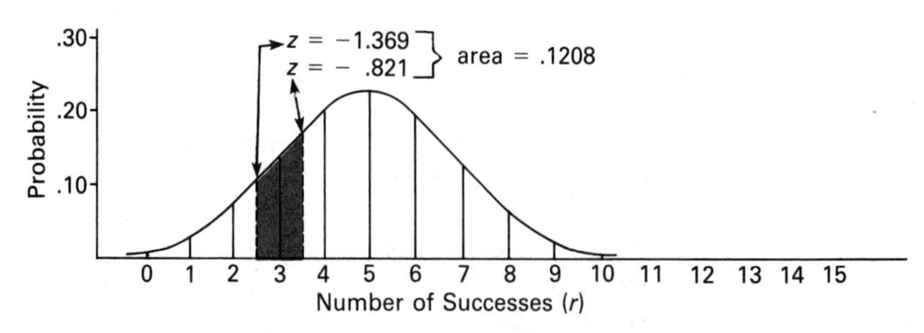

FIGURE 10.4
Fit of the normal curve to a binomial distribution in which $n = 15$, $p = 1/3$.

As a final illustration, calculate the probability for 10 successes in 30 trials by chance if the probability of success on any one trial is .2. Use both the binomial equation and the normal approximation.[6]

10.7 SUMMARY

The rules and concepts of probability were introduced as a foundation for understanding the process of statistical reasoning in the inferential process. Probability was defined as the relative frequency with which random events would occur over the long run. Theoretical frequency distributions of outcomes for specific random events can be generated based on the mathematical models of probability and randomness. These theoretical frequency distributions provide models for forming expectations about what will occur in reality. The expectations are in turn used to make decisions in statistical inference.

The binomial distribution was one such theoretical distribution covered. We can use it to identify the relative frequency with which dichotomous (success/failure) outcomes are expected to occur for a fixed number of independent trials. A binomial distribution describes the probabilities associated with each of r successes for a fixed number of n independent binomial trials. The general mathematical equation that defines the probabilities for every integer value of r between 0 and n is

$$P(r \text{ successes}) = C\binom{n}{r}p^r q^{n-r}$$

This equation is the general form for generating a **binomial probability distribution.** The binomial distribution may be thought of as a theoretical probability distribution—a mathematical model for describing the probabilities of random events with dichotomous outcomes occurring.

The binomial probability distribution is a discrete distribution in that values of r are fixed points. However, any binomial distribution is approximately normally distributed with a mean of np and standard deviation equal to

---

[6]*Binomial probability:*

$$\begin{aligned} P(r = 10) &= C\binom{30}{10}(.2)^{10}(.8)^{20} \\ &= (30{,}045{,}015)(.000000102)(.011529215) \\ &= .0355 \end{aligned}$$

Normal approximation:

$$z_{9.5} = \frac{9.5 - 30(.2)}{\sqrt{30(.2)(.8)}} = \frac{3.5}{2.191} = 1.597$$

$$z_{10.5} = \frac{10.5 - 6}{2.191} = 2.054$$

Area between $z = 2.054$ and $z = 1.597$ is .0346.

The difference in the two probabilities is .0009. Even with a p value quite discrepant from .50 ($p = .2$), the normal curve approximation of the binomial probability is very close.

the square root of npq when the products np and nq are both greater than or equal to 5. Under these conditions, we can use the continuous unit normal curve as a probability distribution to approximate the probabilities in any binomial distribution. The formula to find the equivalent z score for a value of r in the binomial distribution was

$$z_r = \frac{r - np}{\sqrt{npq}}$$

where the value of r is corrected for the continuous nature of the z score scale by using the lower real limit $(r - .5)$ or the upper real limit $(r + .5)$ for the exact value of r. The appropriate probability for z_r is found in the table of proportional areas for the unit normal curve (Table B.2).

PROBLEMS

Problem Set A

1. A sack of jelly beans contains the following quantities of different flavors:

12 coconut	14 grape
13 lemon	5 banana
22 cherry	16 chocolate
8 orange	10 licorice

When selecting single candies at random from the sack, what are the probabilities of selecting each of the following flavors?

a. a coconut?
b. a banana?
c. a lemon *or* an orange?
d. a licorice *or* a grape *or* a cherry?
e. any kind but a chocolate?
f. on two consecutive trials with replacement, two licorice?
g. on three consecutive trials with replacement, one cherry, one orange, *and* one coconut?

2. When flipping a coin for three consecutive trials, list the population of potential outcomes that could occur. What is the probability that

a. three tails would occur?
b. two heads and one tail would occur in any order?
c. two heads would occur on the first two trials and a tail on the third?
d. at least one head would occur on three trials?

3. Calculate (a) $C\binom{6}{2}$, (b) $C\binom{15}{8}$, (c) $P\binom{6}{2}$, (d) $P\binom{15}{8}$, (e) $C\binom{500}{4}$, (f) $P\binom{500}{4}$.

4. For the problem in section 10.4 requiring that symbols match on erased spaces to win, what is the probability of winning if there were

a. 5 spaces, but you need to match 2 objects?
b. 8 spaces, but you need to match 4 spaces?
c. 20 spaces, but you need to match 4 spaces?

5. Given a six-trial binomial distribution in which the probability of a success is 1/10, what is the probability of
a. exactly three successes?
b. zero or one success?
c. four or more successes?

6. Assume that Paul has contracted to use a computer dating service. He knows that his personality, background, interests, and so on should make him initially compatible with 40% of the general population. If the computer dating service is doing nothing more than making random assignments and its clientele is representative of the general population, what is the probability that out of 10 independent uses of the dating service that
a. Paul would find no one with whom he was initially compatible?
b. Paul would be matched with *at least one* person with whom he would be initially compatible?

7. In problem 6, probabilities are based on random assignment. Paul could test to see if the computer dating service were using a more valid matching procedure than random assignment. He might set as a criterion that if r or more successes occur with a probability of less than .15 under the random assignment model, he would conclude that something other than random assignment was being used successfully. For the 10 trials, how many successes would meet the probability criterion, $P(r \text{ or more successes}) \leq .15$?

8. Could you legitimately use the normal approximation to the binomial to solve problem 7? Why?

9. Assume 20 trials in problem 7 with $p = .40$ or 2/5. Use the normal approximation to estimate
a. $P(r \geq 10)$
b. $P(r = 3)$
c. $P(r \leq 6)$
d. $P(4 \leq r \leq 8)$

Problem Set B

The question of cultural bias in test instruments has received considerable attention as a critical issue in educational and psychological measurement over the past few years. One recent proposal for assessing cultural bias in multiple-choice test items focused on the use of information in examinees' responses to the incorrect options of a multiple-choice item, called distractors or foils (Veale and Foreman, 1983). They reasoned that

> when one group is attracted to a particular foil and the other groups are drawn to other foils, cultural bias may be present in the item. This response configuration will likely occur when (1) a foil is actually the correct response for a particular cultural group or (2) a foil, although clearly incorrect for all groups, contains culture-specific stimuli which attract (or repel) members of some group. (p. 249)

To provide evidence that cultural bias is affecting responses to an item, the pattern of the proportion of individuals in different cultural groups responding to each option is assessed. Veale and Foreman reasoned that for individuals in each cultural group who miss an item, the proportion selecting each foil should be approximately the same for each group. If the pattern of selection across the foils is different for the separate cultural groups and there are differences in the proportion of individuals getting the item correct, then there is evidence that the item is biased: One of the foils is stronger in attracting individuals in the lower-performing group. This foil has a tendency to pull or attract more individuals than would be expected.

The model Veale and Foreman proposed as a basis for identifying culturally biased items is called the overpull probability model. Developing and applying the model requires greater knowledge of probability and statistical inference than is appropriate for coverage of the topics in this chapter. However, the basic data in terms of the frequencies and proportions of individuals from different cultural groups that respond to different multiple choice foils in an item can be used to illustrate some of the probability concepts of the present chapter. Table 10.4 is an example Veale and Foreman gave to illustrate how differential response patterns across foils might look in a biased item. Based on these data, respond to the following:

1. What is the probability that an individual chosen at random from Group A answered the item correctly?
2. What is the probability that a randomly selected person from Group B responded incorrectly to the item?
3. *Given that an individual from Group A missed the item,* what is the probability that the person responded to foil B? foil C? foil D?
4. What are the similar probabilities for each foil for individuals from Group B *given that they missed the item?*
5. Assess the differential probabilities of selecting a foil for the two groups. Does there appear to be a bias in the response pattern? Identify which foil appears to be contributing to the bias.

TABLE 10.4
Differential response patterns for a biased item.

Cultural Group	Sample Size	Proportion Getting Item Correct	Foils		
			B	C	D
Cultural Minority (Group A)	98	.45	6*	14	29
Other (Group B)	412	.78	17	42	33

*frequency responding to a specific foil

6. Assuming that selecting an individual at random who got an item correct is considered a success,
   a. what is the probability of selecting four individuals from Group A in which at least three are identified as successful?
   b. what is the same probability if selecting four individuals from Group B?

7. Under the same sampling conditions as in Problem 6,
   a. what is the probability of selecting 30 individuals from Group A in which 10 or fewer were successful (answered the item correctly)?
   b. what is the same probability if selecting 30 individuals from Group B?

Problem Set C

1. Table 10.5 is a frequency distribution of scores on the "Sesame Street" test. If we are randomly selecting one individual from the "Sesame Street" data base, what is the probability that the individual would have a score
   a. of 10?
   b. less than or equal to 4?
   c. greater than 7?
   d. of 5, 6, or 7?

2. The numbers of males and females in each of the school attendance categories from the "Sesame Street" data are given in Table 10.6. What is the probability of selecting an individual at random from the data base who
   a. is a male?
   b. is a female?
   c. attended Headstart?
   d. did not attend preschool?
   e. attended Headstart *or* a day-care center?

TABLE 10.5
Frequency distribution of scores from the "Sesame Street" Test.

Score Value	Frequency (f)
10	76
9	26
8	24
7	16
6	12
5	20
4	14
3	14
2	14
1	12
Missing Data	12
TOTAL	240

TABLE 10.6
School attendance category by sex.

	School Attendance Category					
Sex	None	Kinder-garten	Nursery School	Headstart	Day-Care Center	Other
Female	59	2	14	25	11	3
Male	62	2	17	26	15	2

f. is a male *and* attended kindergarten?
g. is a female *and* attended no preschool?

3. Assume that the random selection of an individual who attended Headstart is a success. What is the probability of a success on any one trial? Assuming replacement after each selection so that the probability of success is constant over individual samplings, generate the binomial probability distribution for each of r successes over six independent random selections. What is the probability that the final sample of six individuals will contain *five or six* persons who attended Headstart? (Round the probabilities to the nearest hundredth in your calculations.)

4. If you were to randomly sample 50 individuals from the "Sesame Street" data base under the conditions of problem 3, what is the probability that
a. exactly 10 of the individuals selected attended Headstart?
b. 40 or more individuals selected attended Headstart?
c. fewer than five individuals selected attended Headstart?

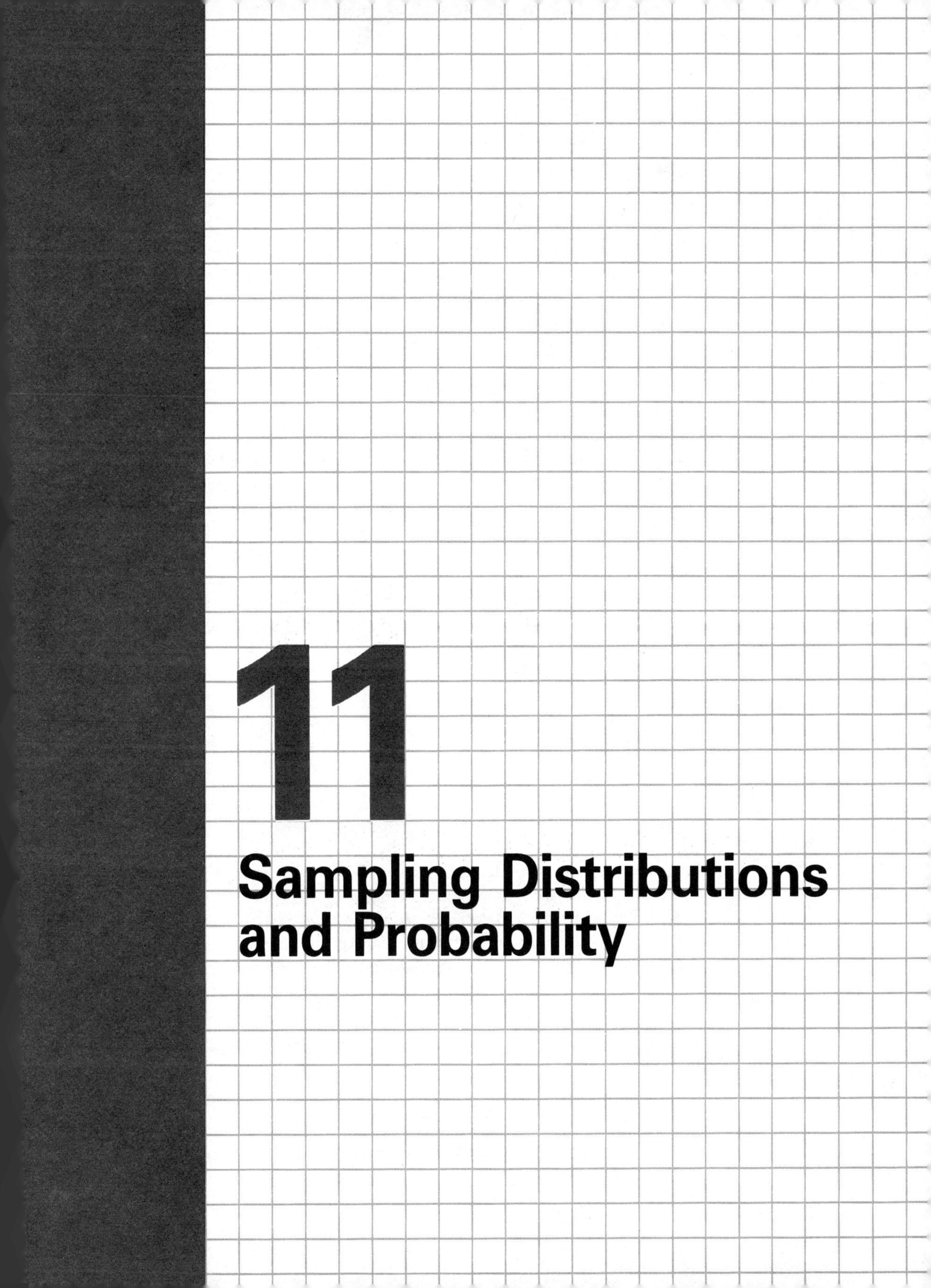

11

Sampling Distributions and Probability

Randomness is vital to much statistical reasoning. Conceptually, it implies some apparently conflicting notions. On the one hand, individual random events are by definition unpredictable; their occurrence is unsystematic. The same holds true for any permutation or combination of a group of random events: The occurrence of any specific set is unsystematic. In contrast, while the occurrence of any one event is unpredictable, the *frequency* with which the individual events will occur over repeated trials is predictable. The long-run frequency with which random events will occur follows a lawful, predictable distribution of outcomes. One task a theoretical statistician has is to determine the mathematical functions defining these long-run frequency distributions for specific types of outcomes. These long-run frequency distributions are termed theoretical probability distributions; they describe the lawful nature of randomness over repeated trials in terms of the probability with which particular events will happen. These theoretical probabilities allow us to make decisions about outcomes for any sample event or series of sample events. If an event is assumed to be random, how likely is it that the specific outcome observed would occur by chance? The outcome of a specific random event is unpredictable, but the lawful nature of randomness as described by a theoretical probability distribution allows us to answer the latter question.

11.1 PROBABILITY DISTRIBUTIONS

For the purposes of this text, we will think of a probability distribution as a theoretical relative frequency distribution. It describes the relative frequency with which events are actually expected to occur using a proportion or percentage. We have already introduced two theoretical probability distributions, one discrete (the binomial distribution) and one continuous (the normal distribution). Since we know the mathematical functions defining each of these distributions, we can generate the probabilities of a particular event occurring without having to perform repeated trials of the events to see what happens. These familiar functions developed by a theoretical statistician are the binomial equation,

$$P(r \text{ successes}) = C\binom{n}{r}p^r q^{n-r}$$

and the equation for the normal curve,

$$y = \frac{1}{\sigma\sqrt{2\pi}}e^{-(X-\mu)^2/2\sigma^2} = \frac{1}{\sqrt{2\pi}}e^{-z^2/2}$$

Actually, the binomial distribution is not a single distribution, but rather a family of distributions defined by the same parameters. Its shape is not constant. The binomial equation results in a different distribution for different values of n and p. The two parameters, n and p, affect the shape of the binomial distribution. Therefore, there is no single standardized binomial distribution, and the probability for a particular outcome occurring changes depending on the values of n and p.

In contrast to the family of binomial distributions, the family of normal curves has a constant shape for all values of μ and σ. The latter distributions also may be transformed (by changing raw scores to z scores) to a single, standardized distribution. Labeled the unit normal curve, this standardized distribution has a constant mean, $\mu_z = 0$, a constant variance and standard deviation, $\sigma_z^2 = \sigma_z = 1$, and a mathematically predictable, constant shape. The area under the unit normal curve defined by z score intervals also is constant. This area can be translated as a probability value implying that once raw scores are changed to z scores in a distribution (and if the distribution is approximately normally distributed), the probability of equivalent z scores occurring across situations is constant. All normal curves have only one probability distribution, and the probabilities are tabulated, as in Table B.2. The constancy of the unit normal distribution probabilities and their applicability to normally or approximately normally distributed values make the normal distribution effective. As shown in Chapter 10, applying the normal approximation to the binomial for large values of n is only one example of the normal distribution's utility.

We use theoretical probability distributions to describe our long-run expectations for particular sample results from individual experiments. Can we reasonably conclude that the sample result of a specific experiment occurred randomly? Statistical inferences are made based on the answer to this question. We can use a theoretical probability distribution to answer it.

The kind of theoretical probability distribution we use depends on the specific research situation and the assumptions we can legitimately make. If the outcomes are dichotomous with constant probability of occurrence over trials, then the theoretical binomial probability distribution is appropriate. For the same situation, but with np and nq greater than 5, we may want to use the normal curve as the approximate probability distribution. If statistical values of interest are continuous and normally distributed, we should use the normal curve as the theoretical probability distribution. Other probability distributions will be discussed in later chapters. (You may have heard of some of them; the t distribution, the chi-square distribution, and the F distribution are all probability distributions used in testing statistical hypotheses.)

Several examples in the last chapter used the binomial distribution as a probability distribution. The soft drink comparison illustrated its application in a possible research study. While the binomial distribution is useful as a probability distribution, it is limited to dichotomous outcomes. The normal distribution, however, has exceptionally widespread application, not only to continuously distributed outcomes, but, as we have seen, as a good approximation under certain conditions to the probabilities of dichotomous outcomes. Thinking of the normal distribution not just as a potential raw score distribution as in Chapter 7, but as a probability distribution, indicates the probability or relative frequency with which we expect certain events to occur.

Probability questions can be written symbolically as probability statements. Assuming an approximately normally distributed set of scores, what is the probability that an individual drawn at random would have a raw score with a z score equivalent between .75 and 1.35? The probability statement is

$$P(.75 \leq z \leq 1.35) = \underline{\hspace{3cm}}$$

The unit normal curve as a probability distribution provides the expected proportion of individuals with z scores of .75–1.35. The expected proportional frequency is the probability that the event will occur by chance (the individual was drawn at random). What is the probability (use Table B.2)?[1]

As another example, assume we want to draw a random sample of 32 cases from the "Sesame Street" data. Assume sampling with replacement. What is the probability that our sample will contain 12 or more students from Site I? This problem is really a binomial problem, but we will use the normal approximation to find the probability. Each individual selected constitutes a trial. For any one trial, success is selecting a student from Site I. When random sampling with replacement, the probability of selecting a student from Site I on any one of the 32 trials is 60/240 = 1/4. The probability statement may be written

$$P\left(r \geq 12; \quad n = 32, p = \frac{1}{4}\right)$$

meaning the probability that r is greater than or equal to 12 *given that* $n = 32$ and $p = 1/4$. What is the probability value using the normal approximation?[2] When selecting a random sample of 32 individuals from the 240 "Sesame Street" students, we would expect less than 8% of the samples to have 12 or more students from Site I. The normal curve provides the relative frequency (probability) with which we would expect this result. The actual binomial probability values for selecting a random sample of 32 containing r students from Site I appears in Table 11.1. As you can see, the sum of the individual binomial probabilities for $r = 12, r = 13, \ldots r = 32$ is .0804. Once again, with this many trials ($n = 32$) the normal-curve approximation is accurate and much easier to compute.

---

[1] $p(.75 \leq z \leq 1.35) = .1381.$

[2] $np = (32)\left(\frac{1}{4}\right) = 8$

$$npq = 32\left(\frac{1}{4}\right)\left(\frac{3}{4}\right) = 6$$

$$z = \frac{11.5 - 8}{\sqrt{6}} = 1.43$$

$p(z \geq 1.43) = .0764.$

TABLE 11.1
Binomial probabilities for $n = 32$ and $p = 1/4$

r	Probability	r	Probability	r	Probability
32	.0000	21	.0000	10	.1097
31	.0000	20	.0000	9	.1431
30	.0000	19	.0000	8	.1610
29	.0000	18	.0001	7	.1546
28	.0000	17	.0004	6	.1249
27	.0000	16	.0014	5	.0832
26	.0000	15	.0040	4	.0446
25	.0000	14	.0099	3	.0185
24	.0000	13	.0219	2	.0055
23	.0000	12	.0427	1	.0011
22	.0000	11	.0732	0	.0001

11.2 SAMPLING DISTRIBUTIONS

Inferential research involves selecting a sample, measuring sampling units selected, statistically analyzing sample data, and then making inferences based on sample results as they relate to hypotheses about population parameters. We make inferences from a single event, a single experiment, a single set of sample results. Assuming a model of randomness, we are usually interested in determining how probable a result is by chance for the single event or sample. To make this determination, we need to know something about how the sample outcomes are distributed in general so that we will have some basis for forming expectations about the frequency of certain sample events. Does it sound like we need a theoretical probability distribution? Eventually yes, but first let's examine the basic frequency distribution of sample events.

Using the binomial distribution, previous examples have tried to find the number of successes resulting from a fixed number of sample trials. Dichotomous outcomes are only one possible type of many potential sample outcomes of interest. In fact, most studies examine sample results using measurement of continuous variables. Instead of the number of successes in a sample, we might be interested in the sample mean for a variable, the sample variance, or a sample correlation coefficient. We might want to generalize from sample statistics to a population using statistical inference. To make the inference, we will need information on how the sample statistic is distributed over repeated samplings.

The frequency distribution of a sample statistic over repeated samples of the same size is a **sampling distribution.** Unlike a population frequency distribution or sample frequency distribution, which are formed using raw score frequencies for individual elements or sampling units in the population or sample, a sampling distribution uses a sample summary statistic such as the sample mean as the raw score in forming the frequency distribution. Conceptually, a sampling distribution could be formed by graphing the frequency of different values of the sample statistic for an infinite (extremely large) number of samples. Many different sampling distributions are possible: a sampling distribution of sample means, a sampling distribution of sample modes, one for sample variances or one for sample correlation coefficients, for example. To generate a sampling distribution empirically, such as a sampling distribution of means, we would repeatedly select all possible independent samples of a particular size and calculate each sample's mean for the variable of interest. The frequency distribution of all these sample means would form a sampling distribution.

To illustrate the concept of sampling distribution, we have drawn 100 independent random samples of 12 from the "Sesame Street" data and computed each sample's mean for the Numbers posttest variable (see Table 11.2).

TABLE 11.2
Numbers posttest means for 100 random samples ($n = 12$).

Sample Number	Mean	Sample Number	Mean	Sample Number	Mean	Sample Number	Mean
1	32.17	26	32.33	51	26.00	76	27.50
2	33.42	27	34.00	52	24.17	77	31.83
3	31.08	28	26.25	53	26.42	78	32.58
4	25.75	29	36.25	54	30.42	79	30.00
5	38.08	30	29.50	55	31.42	80	24.75
6	29.00	31	29.67	56	30.08	81	33.25
7	28.17	32	30.08	57	25.67	82	33.50
8	24.92	33	32.92	58	35.92	83	30.00
9	29.17	34	25.50	59	33.83	84	29.91
10	31.00	35	26.67	60	32.00	85	31.50
11	31.92	36	35.58	61	36.42	86	35.92
12	26.08	37	31.08	62	29.83	87	35.75
13	32.17	38	30.00	63	30.42	88	28.00
14	28.42	39	26.50	64	25.33	89	26.25
15	28.75	40	30.83	65	28.08	90	27.08
16	28.83	41	34.17	66	24.75	91	26.08
17	33.17	42	28.17	67	31.75	92	30.58
18	32.25	43	31.83	68	35.17	93	35.17
19	28.25	44	29.75	69	30.33	94	35.17
20	34.83	45	23.33	70	31.17	95	32.25
21	28.83	46	32.50	71	31.83	96	36.50
22	31.83	47	28.50	72	27.67	97	35.33
23	32.17	48	34.67	73	28.08	98	39.08
24	30.00	49	31.83	74	27.00	99	24.25
25	26.67	50	25.92	75	27.67	100	27.17

TABLE 11.3
Sampling frequency distribution of Numbers posttest means for 100 random samples (n = 12).

Interval	f
38.01–40.00	2
36.01–38.00	3
34.01–36.00	11
32.01–34.00	15
30.01–32.00	21
28.01–30.00	21
26.01–28.00	15
24.01–26.00	11
22.01–24.00	1

The means in Table 11.2 appear as a frequency distribution in Table 11.3. Figure 11.1 plots a frequency polygon of the distribution of these sample means. The 100 random samples represent a reasonable number of repeated samplings to indicate how Numbers posttest mean scores from random samples of size 12 will distribute themselves. The data in Figure 11.1 represent a subset of all the samples of size 12 needed to generate an empirical sampling distribution. However, we can use these 100 sample means to get some idea about a sampling distribution in terms of the frequency with which particular sample means occur. The miniature sampling distribution represented tells us what to expect in terms of sample results relative to the mean when randomness is operating. Using the tables and Figure 11.1, our sample means ranged from a low of 23.33 for sample 45 to a high of 39.08 for sample 98 and were unimodal in shape with the greatest frequency of sample means in the middle intervals.

We can generate a miniature sampling distribution for any sample statistic other than the mean. To illustrate, we also calculated the variances for each of the 100 samples in Table 11.2. The frequency distribution of these 100

FIGURE 11.1
Frequency polygon of the sampling distribution of Numbers posttest means for 100 random sample means (n = 12).

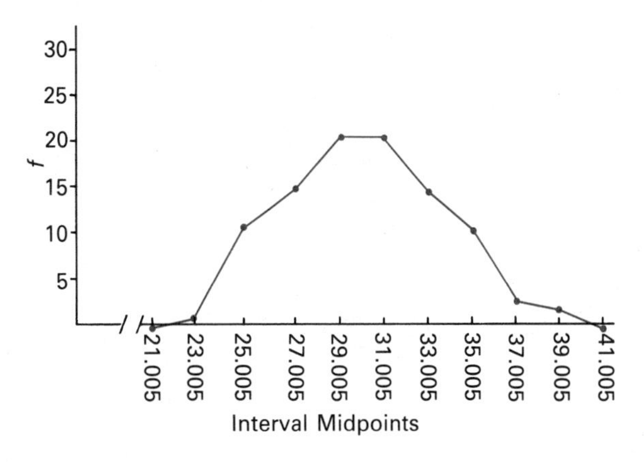

TABLE 11.4
Sampling frequency distribution of Numbers posttest variances for 100 random samples (n = 12).

Interval Limits	*f*
310.00–339.99	1
280.00–309.99	2
250.00–279.99	3
220.00–249.99	10
190.00–219.99	13
160.00–189.99	21
130.00–159.99	25
100.00–129.99	18
70.00– 99.99	5
40.00– 69.99	2

sample variances appears in Table 11.4. This frequency distribution represents a subsampling of the complete sampling distribution of variances for random samples of size 12. The frequency polygon of the 100 variances is graphed in Figure 11.2. Note that the variance distribution tends to be positively skewed while the distribution for the means tends to be symmetrical.

As with probability distributions, the utility of the sampling distribution concept is based on randomness. Under an assumed model of random sampling, how might we expect sampling error to affect the distribution of sample means, variances, or some other sample statistic? When we select samples at random, the magnitude of a sample statistic will exhibit a pattern of randomness individually across repeated samplings, but as a group, the values of a sample statistic will occur with a predictable frequency over these repeated samplings. The sampling distribution describes the frequency of different values of a sample statistic resulting from random sampling error.

Fortunately, we need not empirically generate a sampling distribution subset every time we do a study. We depend on the theoretical statisticians to

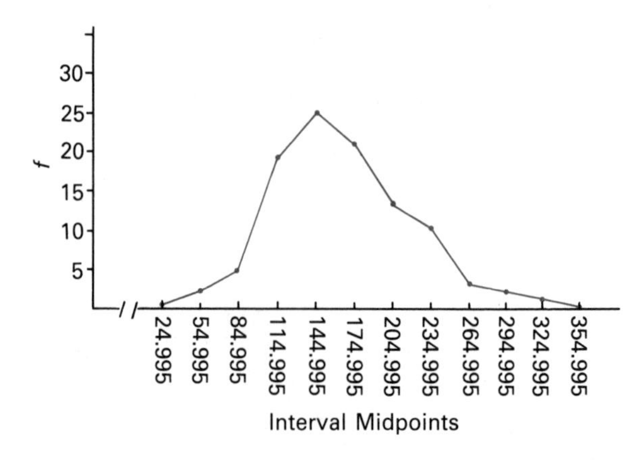

FIGURE 11.2
Frequency polygon of the sampling distribution of Numbers posttest variances for 100 random sample variances (n = 12).

develop the mathematical basis for the characteristics of sampling distributions for various sample statistics. If we know three things about most frequency distributions—the shape, central tendency, and amount of variability—we have the necessary information to describe the data. The knowledge developed in theoretical statistics will tell us the shape, mean, and standard deviation of the sampling distributions of interest. Given these three descriptors, we can determine the theoretical sampling distribution for a statistic and the relative frequency with which we can expect sample events to occur without having to draw repeated samples and actually construct an empirical distribution.

11.3 THE CENTRAL LIMIT THEOREM

The samples in Table 11.2 were drawn at random from a population of individuals whose mean Numbers posttest score was 30.43 ($\mu = 30.43$) under the assumption that the "Sesame Street" data base represents a finite population. The standard deviation, σ, of Numbers posttest scores in the finite population of "Sesame Street" students is 12.60. By drawing random samples of size 12 from this population of Numbers posttest scores, computing their means, and plotting them, we can begin to mirror the characteristics of an *expected* distribution of sample means. The empirically generated data for 100 samples in Tables 11.2 and 11.3 and Figure 11.1 help us envision a sampling distribution of means. Based on the distributional characteristics of the empirically generated sample data, the sampling distribution of means tends to be symmetrical in shape. In Figure 11.1, the mean of the sampling distribution (the mean of the 100 sample means) is 30.36, approximately equal to the population mean ($\mu = 30.43$), and the standard deviation of the sampling distribution (of the 100 sample means) is 3.50, considerably smaller than the population raw score standard deviation of Numbers posttest scores ($\sigma = 12.60$). These characteristics of our empirically generated distribution of means are expected and predictable from several separate mathematical theorems. One such theorem, the one on which we will base our discussion, is the **central limit theorem.** The central limit theorem says that regardless of the shape of the population distribution of scores, if a population has a finite variance σ^2 and mean μ, then the sampling distribution of means for random samples of size n approximates a normal distribution with a mean equal to the population mean, μ, and a variance equal to the population variance divided by the sample size, σ^2/n. The shape of the sampling distribution tends to approximate the normal distribution more closely as n increases. The central limit theorem identifies the three characteristics necessary to describe fully the expected distribution of sample means resulting from random samples of size n.

- Shape: It approaches normality as sample size increases and for a very large n is approximately normal.

- Measure of central tendency: The mean of a sampling distribution of means ($\mu_{\bar{x}}$) for samples of any size n equals the population mean (μ).

$$\text{mean} = \mu_{\bar{x}} = \mu$$

- Measure of variability: The variance of the sampling distribution of means ($\sigma^2_{\bar{x}}$) equals the population variance (σ^2) divided by n.

$$\text{variance} = \sigma^2_{\bar{x}} = \frac{\sigma^2}{n}$$

$$\text{standard deviation} = \sigma_{\bar{x}} = \frac{\sigma}{\sqrt{n}} \quad (11.1)$$

The standard deviation of the sampling distribution of means (formula 11.1) is called the **standard error of the mean.** It represents the amount of variability we can expect in the sample means due to random sampling error. Symbolically, the standard error of the mean is represented by $\sigma_{\bar{x}}$. The central limit theorem allows us to conceptualize the characteristics of the sampling distribution of means using population parameters. We need only know the population mean (μ) and the population variance (σ^2) or standard deviation (σ) to know the expected characteristics of the theoretical sampling distribution of means for constant sample size.

While the central limit theorem applies only to samples from infinite populations and complete sampling distributions, the data in Table 11.2 of the 100 sample means can illustrate central limit theorem properties. The calculated standard deviation of the 100 sample means in Table 11.2 is 3.50. This value can be compared to the value of the standard error of the means calculated by formula 11.1. The population standard deviation of Numbers posttest scores for the 238 "Sesame Street" participants was 12.60. For random samples of size 12, the central limit theorem indicates that the standard deviation of the sampling distribution of means should be

$$\sigma_{\bar{x}} = \frac{\sigma}{\sqrt{n}} = \frac{12.60}{\sqrt{12}}$$
$$= 3.64$$

The empirically derived standard deviation of 3.50 for the 100 sample means is close to the value of 3.64 calculated by the central limit theorem. If we draw more samples than 100 to form our empirical sampling distribution, the calculated standard deviation of the sample means should be even closer to the theoretical central limit theorem value of 3.64.

The central limit theorem makes several subtle points. The identified characteristics of the sampling distribution depend on random sampling. Remarkably, however, the population distribution of the variable of interest need not be normal. Even if sampling occurs from a nonnormal population distribution, the resulting sampling distribution of sample means will approach normality as the sample size increases. Again, the normal approximation is closer for larger sample sizes. If the population set of scores is normally distributed, the sampling distribution of means will be of normal shape. Each sample is assumed to be representative of the population because of random sampling. Each sample mean for the variable of interest is therefore an estimate of the population mean for that characteristic. Deviant samples will occur by chance due to random sampling error. The sampling distribution of means describes how much we can expect the sample means to deviate as estimates of the population mean. The sample means will distribute themselves with approximate normal-curve frequency around the population mean, which is also the sampling distribution mean. Theoretically, half the sample means will be below the value of μ and half will be above μ. How much will the sample means deviate as estimates of μ? That is given by the standard error of the mean, $\sigma_{\bar{x}}$. As we might expect, the sample means are summary average indices and do not exhibit nearly as much variability as do the raw scores in the population, $\sigma_{\bar{x}} \leq \sigma$. In fact, formula 11.1 indicates that the relationship between $\sigma_{\bar{x}}$ and σ is a direct function of the sample size. By increasing the size of the sample drawn, the variability of sample means from μ in the sampling distribution decreases dramatically. By increasing n, each sample drawn at random has a higher probability of being more representative and the sample means will tend not to deviate as far from the population mean value. How much would you expect sample means to vary if random samples of size 25 had been drawn instead of size 12 for the Numbers posttest example?[3] What if n had been of size 64?[4] Increasing sample size increases the precision of each sample mean as an estimate of the population parameter. This is sometimes called the **law of large numbers.** It should be noted that the relationship between changes in n and the size of $\sigma_{\bar{x}}$ is not multiplicative. Doubling the sample size does not cut $\sigma_{\bar{x}}$ in half. The reduction in size of $\sigma_{\bar{x}}$ is a function of $n^{1/2}$, not n. (Note: The notation $^{1/2}$ is used to represent $\sqrt{\ }$ in text.)

11.4 USING SAMPLING DISTRIBUTIONS TO MAKE INFERENCES

The sampling distribution of means is a theoretical distribution whose characteristics several independent mathematical theorems have proven. We have chosen to focus on one form of the central limit theorem to tell us what we can legitimately expect about the distribution of sample means if we draw a series of random samples of a particular size. For example, suppose we were drawing

---

[3]When $n = 25$; $\sigma_{\bar{x}} = \dfrac{12.60}{\sqrt{25}} = 2.52.$

[4]When $n = 64$; $\sigma_{\bar{x}} = \dfrac{12.60}{\sqrt{64}} = 1.58$

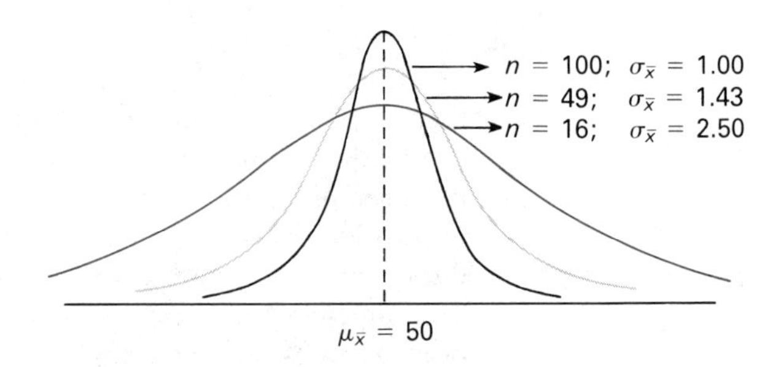

FIGURE 11.3
Distributions of various-sized random samples.

random samples of 16 from a population with $\mu = 50$ and $\sigma = 10$ for the variable of interest. How would the theoretical sampling distribution of means appear?[5] If we had selected samples of 49 or 100, the sampling distribution would approximate a normal curve more closely, would still be centered around a mean of 50, but would be less variable (see Figure 11.3).

Statistical inference uses information from only a single sample. The sample of size n is drawn using a sampling plan that produces a sample representative of the population. If we can assume that the sampling plan yielded a representative sample that approximates randomness, then we can use the characteristics of the sampling distribution of means as described by the central limit theorem as a model for expected outcomes. In practice, the sample statistic of interest (e.g., $\bar{X}$) plus knowledge about the theoretical sampling distribution are the only information available to make inferences about a population characteristic or to estimate the value of the population parameter. The actual population parameter of interest will be unknown. Rather, the sampling distribution becomes a *hypothetical sampling distribution for an assumed population characteristic.* The logic of the process follows a form of "*if-then*" thinking. *If* the population parameter is such and such, *then* we can predict characteristics of the sampling distribution for the sample statistic. If the sampling distribution really is distributed as expected given the assumed population parameter, then we would expect certain random sample outcomes to occur more often than certain other outcomes. A sample statistic is computed from sample data as an estimate of the true population parameter and is used to confirm or refute the initial assumption about the population parameter value. If the calculated sample statistic is a likely outcome expected in the hypothetical sampling distribution, then the sample results provide evidence permitting an inference supporting the initially assumed population characteristic. In contrast, if the calculated sample statistic is *not* a likely out-

---

[5]For random samples of size 16 from a population with $\mu = 50$ and $\sigma = 10$, the sampling distribution of means would be approximately normal with a mean of 50 and a standard deviation of

$$\sigma_{\bar{X}} = \frac{10}{\sqrt{16}}$$
$$= 2.5$$

come in the hypothetical sampling distribution, then we would infer from the sample evidence that the initial assumed population characteristic was not correct. Based on the hypothetical sampling distribution expected *if the assumed population characteristics are true,* statistical inference uses sample results to confirm or refute the initial assumptions about the population characteristics.

Statistical inference is a logical process. It starts with an assumption about a population parameter. A hypothetical sampling distribution is generated which has specific characteristics *if the assumed population parameter is true.* The results from statistical analysis of sample data are compared to the hypothetical sampling distribution to see if they are expected or unexpected outcomes. Inferences are made about the validity of the initial assumed population parameter value based on whether the sample results fit the expected outcomes defined by the hypothetical sampling distribution. Figure 11.4 summarizes the logic of statistical inference. You should be able to see that the important link in inference is the use of the hypothetical sampling distribution as a reference for expected sample outcomes under a given, assumed true, population parameter value.

To this point, we have addressed the role of sampling distributions in the statistical inference process at a general, conceptual level. Chapter 12 introduces the practical applications of actually making statistical inferences. Before we reach that point, however, other fundamental understandings about the role and use of sampling distributions are prerequisites. When we talk about a

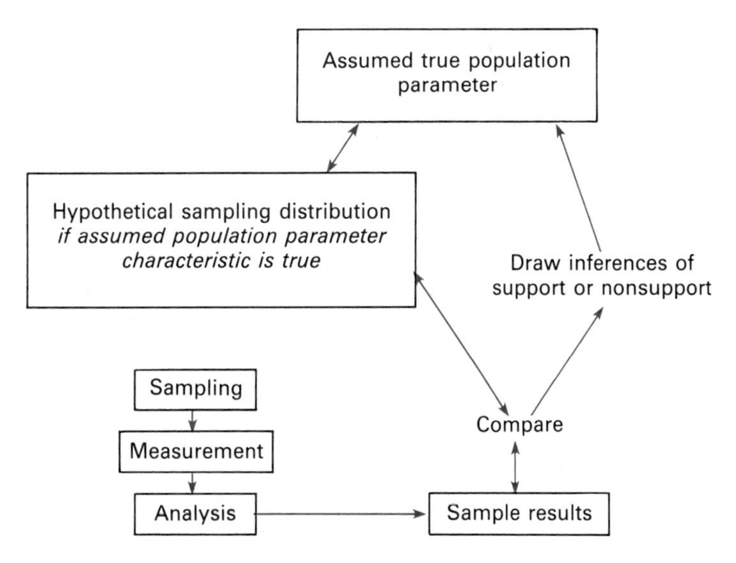

FIGURE 11.4
Role of the hypothetical sampling distribution in statistical inference.

sampling distribution, we need to identify a sample statistic. There is not just one sampling distribution. There are different sampling distributions for various specific sample statistics, such as means, variances, and so on. In the inferential process, the type of sampling distribution of interest must be compatible with the population parameter characteristic of interest. Similarly, the sample statistic result must correspond to the population parameter of interest. If μ is the population parameter of interest, then the sampling distribution of sample means is the appropriate hypothetical sampling distribution and the sample mean would be the result of interest. We use the value of the sample statistic as an estimate of the corresponding population parameter. How much the sample estimate deviates from the assumed value of the population parameter (as determined by the hypothetical sampling distribution characteristics) becomes the important criterion for making decisions in statistical inference.

11.5 POINT ESTIMATION

Statistical inference uses a sampling plan to select a sample representative of some population. As a representative sample, the statistical indices calculated from the sample data also should be representative of their corresponding population parameters. In other words, sample statistics from representative samples should be good estimators of corresponding population parameter values for populations the sample represents. The sample statistic value acts as an estimator of a population parameter that, with the help of a hypothetical sampling distribution, allows us to draw inferences about a specific population parameter value.

To be a good estimator of a population parameter, a sample statistic should meet at least two primary criteria. It should be unbiased and consistent. Other criteria exist (e.g., efficiency and sufficiency), but unbiasedness and consistency are the most important.

Unbiasedness

A sample statistic is said to be an **unbiased** estimator of a population parameter if the mean of the sampling distribution of the statistic equals the value of the parameter it is estimating. A sample mean ($\overline{X}$), regardless of the nature of the population from which the sample is drawn, is an unbiased estimator of the population mean (μ). The mean of the sampling distribution of means ($\mu_{\overline{x}}$) equals the value of the population mean from which the samples were drawn. If samples are drawn randomly from any symmetrical, unimodal population, both the median and the mode also are unbiased estimators of the population mean. Over an infinite (very large) number of random samples, the mean of the sample medians and the mean of the sample modes equal the population mean. If the population distribution is skewed, the sample mean is an unbiased estimator of the population mean, but both the median and the mode would be biased estimators of the population mean. When sampling from negatively skewed population distributions, they would tend to overestimate the value of the population mean; in positively skewed population distributions, they would underestimate it.

In mathematical terms, unbiasedness is defined by expected value notation. The expected value of a statistic (T) is denoted by $E(T)$, the long-run expectation or the average (mean) value of T over an infinite series of repeated trials, samples, or experiments. The concept of an expected value should be familiar from the discussion of probability and the binomial distribution in Chapter 10. The expected long-run proportion of specific outcomes was equal to p, the probability that the outcome would occur on any one trial. In the binomial distribution over an infinite number of experiments with n trials, the expected number of successes would equal np. Using expected value terminology, a statistic (T) is an unbiased estimator of a population parameter if

$$E(T) = \text{the value of the population parameter}$$

The sample mean is an unbiased estimator of the population mean if samples are drawn at random, because $E(\overline{X}) = \mu$.

In Chapter 6 the variance was defined as the average squared deviation of raw scores from the mean, but the formula divides the sum of squared deviation scores by $n - 1$ rather than n. Mathematically, it is only when Σx^2 is divided by $n - 1$ that the expected value of s_x^2 equals σ_x^2. Using the formula $s_x^2 = \Sigma x^2/(n - 1)$ results in a sample variance index that is an unbiased estimator of the population variance:

$$E(s_x^2) = \sigma_x^2 \quad \text{when } s_x^2 = \frac{\Sigma x^2}{n - 1}$$

Using n as the divisor in the formula would have resulted in a negative bias when using the sample variance as an estimate of σ_x^2. Dividing by $n - 1$ corrects the bias. While not going into detail, the sample standard deviation (s_x) is *not* an unbiased estimator of the population standard deviation. For this reason, inferences about the variability of scores in a population will generally involve variances and not standard deviations.

Consistency

A sample statistic is a **consistent** estimator if it tends to get closer and closer to the value of the parameter it is estimating as sample size increases. This principle is the **law of large numbers.** We can think of consistency as the increased precision with which the statistic estimates the parameter as the sample size increases. Any statistic for which a standard error of the sampling distribution decreases with increasing values of n would be a consistent estimator. The sample mean is a prime example of a consistent estimator. The size of the standard error of the mean $\sigma_{\overline{X}}$ is linked directly to sample size. As n increases, $\sigma_{\overline{X}}$ decreases, implying that the sample means bunch closer together around μ in the hypothetical sampling distribution. If increasing sample size results in smaller deviations of sample means from the population mean, the sample means are more accurate estimators.

Point Estimates

The two criteria for defining a good estimator refer primarily to the use of a sample statistic as a **point estimate** of a population parameter. Point estimation is the degree of correspondence between a sample value as a single point and a single population point on some scale. How do we judge the accuracy of a sample statistic as an estimate of a population parameter? Suppose we have a random sample of size 25 with a mean of 30. Our best point estmate of the population mean from which the sample was drawn is 30, but how likely is it that the mean from a single sample is exactly equal to the population mean? From your knowledge of sampling distributions, you should know that this outcome isn't very likely. Look at the 100 random sample means in Table 11.2. How many of the 100 sample means are exactly equal to the population mean of 30.43? None are exactly equal. Many of them come very close as estimates, but none are exact estimates. The precision expected from any one sample mean as a point estimate of the population mean is best summarized by the standard error of the mean. It indicates the extent to which we should expect individual sample means to deviate from the true population mean. The smaller the standard error, the more confident we are that any one sample mean is a reasonable estimate of the population mean. Would you be more confident that our sample mean of $\overline{X} = 30$ is closer to the population mean sampling from a population with $\sigma_x^2 = 100$ or from one in which $\sigma_x^2 = 256$?[6]

In fact, occurrence of the random sample mean 30 is possible as a result of sampling from an infinite number of distinct populations (i.e., populations with different values of μ and σ_x^2). Because of this possibility, we can never be 100% certain that a specific value of the population mean is the one being estimated by the sample mean. Rather, statistical inference usually asks a comparative question, such as, "Given a choice between two different population

---

[6]The size of the population variance affects the amount of variability to be expected in the sample means.

$$\text{When } \sigma^2 = 100, \quad \sigma_{\overline{X}} = \frac{10}{\sqrt{25}} = 2.00$$

$$\text{When } \sigma^2 = 256, \quad \sigma_{\overline{X}} = \frac{16}{\sqrt{25}} = 3.2$$

We would have more confidence that $\overline{X} = 30$ is a closer estimate of μ for the population in which $\sigma^2 = 100$. The sampling distribution of means has less variability.

means, μ_1 or μ_2, from which population is it more likely our sample was randomly selected based on the value of the sample mean as an estimate of population mean?" Phrased in another way, "For which of the populations is the sample mean more likely to occur as a point estimate of the population mean resulting from random sampling?"

To answer these questions we look at the standardized distance of the sample mean from a particular assumed value of the population mean. The closer they are, the more likely the sample mean is to be a point estimator of the population mean resulting from random sampling from a population with that mean. How do we standardize the distance between the two means to give results with constant interpretative meaning across situations? You guessed it, z scores come to the rescue again. The central limit theorem indicates that if a sample of size n is randomly selected from a population with a mean of μ and variance of σ_x^2, the sampling distribution of sample means will be approximately normal with a mean of μ and a standard deviation of $\sigma_x/n^{1/2}$. We can use these characteristics to change a sample mean to a z score by referring to the mean of the population from which it is assumed to have been a random sample. The formula from Chapter 7 for converting a raw score to a z score was

$$z = \frac{X - \overline{X}}{s_x}$$

The distance of the raw score from the mean of the distribution is divided by the standard deviation of the raw scores. In the sampling distribution of means, the raw scores are sample means. The mean of the distribution is μ and the standard deviation of the scores in the sampling distribution is $\sigma_{\overline{x}}$, the standard error of the mean. To transform a sample mean to a z score in a sampling distribution, we substitute values in the z score formula:

$$z_{\overline{x}} = \frac{\overline{X} - \mu}{\sigma_{\overline{x}}} \quad \text{where} \quad \sigma_{\overline{x}} = \frac{\sigma}{\sqrt{n}} \tag{11.2}$$

$$= \frac{\overline{X} - \mu}{\frac{\sigma}{\sqrt{n}}}$$

Interpreting $z_{\overline{x}}$ is the same as for any z score: It is the number of standard deviation (standard error in this case) units the sample mean is from the population mean in the sampling distribution of means. Because the sampling distribution of means is approximately normally distributed, we can interpret $z_{\overline{x}}$ as we would any z score in a normal distribution (e.g., 68% of the sample $\overline{X}$s should fall within the interval range $1z_{\overline{x}}$ above and below μ).

Suppose we are sampling from a population with a mean of 50 and standard deviation of 10. We obtain a sample mean of 55 from a random sample of size 25. To find out how far apart the means are in standard error units, we use formula 11.2:

$$z_{\bar{X}} = \frac{55 - 50}{\frac{10}{\sqrt{25}}} = \frac{5}{2} = 2.50$$

The sample mean of 55 is 2.50 standard error units from the population mean. Based on our knowledge of the unit normal curve, we would not expect a random sample taken from this population to have a mean this discrepant too often. A z score of 2.5 or greater will not occur very often in a normal distribution (From Table B.2, less than 1% of the time).

As another example, assume we have a random sample of size 16 with a mean of 41. From which of the following two discrepant populations is it more likely that this sample was randomly selected? (a) *population A*—$\mu = 45$, $\sigma^2 = 64$; or (b) *population B*—$\mu = 34$, $\sigma^2 = 196$. If we assume the sample was selected from population A, it is two standard error units below the population mean in the sampling distribution:

$$z_{\bar{X}} = \frac{41 - 45}{\frac{8}{\sqrt{16}}} = -2.00$$

If we assume the sample was selected from population B, how many standard error units is $\bar{X} = 41$ from the population mean?[7] Based on the standardized distances of our point estimate from the means of population A ($\mu = 45$) and B ($\mu = 34$), respectively, the random sample is equally likely to have been selected from either population; the sample mean is equidistant from both population means.

11.6 INTERVAL ESTIMATION

Point estimation of a parameter from a single sample statistic may not seem very practical after the discussion in the last section. The potential for error in drawing a definitive conclusion about the specific value of a population parameter may seem overwhelming. Probably the most important consideration to increase one's confidence when using a sample statistic value as a point estimate is sample size. By increasing sample size, an estimator will become more accurate and precise *if it is a consistent estimator.* By definition of consist-

---

[7] $$z = \frac{41 - 34}{\frac{14}{\sqrt{16}}} = \frac{7}{3.5} = 2.00$$

$\bar{X} = 41$ is 2.00 standard error units above μ.

ency, increased sample size should give us more confidence that the sample value is closer to the true population parameter value.

In addition to point estimation, another method of estimation yields information about the population parameter from sample data and is intuitively more comfortable. This alternative method of estimation is called **interval estimation.** Rather than using a single value as a direct estimate of a parameter, interval estimation establishes a range or interval to which we can attach a level of confidence that the range contains the parameter. Establishing a range for the population mean based on sample data with an upper and lower bound intuitively provides more information about the mean, but in terms of accuracy of the estimate, interval estimation and point estimation are identical. Both are based on the same information, but the information is conceptualized and presented differently.

The reasoning behind interval estimation directly follows the concepts of point estimation. When we view the sample mean as a point estimator of the population mean, the central limit theorem describes the expected distribution of all potential point estimates of a specific population mean that could result from random sampling. Figure 11.5 describes our sampling distribution expectations for random sample means. Normal-curve theory indicates that approximately 68% of our sample point estimates will be within the interval range between $\mu - 1\sigma_{\bar{x}}$ and $\mu + 1\sigma_{\bar{x}}$ in the sampling distribution. About 95% of the sample means (point estimates) are expected to fall within the interval range between $\mu - 1.96\sigma_{\bar{x}}$ and $\mu + 1.96\sigma_{\bar{x}}$. Similarly, 99% of the sample means are expected to lie within the interval range between $\mu - 2.575\sigma_{\bar{x}}$ and $\mu + 2.575\sigma_{\bar{x}}$. These normal-curve intervals are intervals around *the population mean* in the sampling distribution of means. They give us some idea about the expected precision of point estimates as a group in terms of how much we can expect them to vary and the frequency we can expect for different point estimates. These intervals are formed around the population parameter, are constant for any one parameter, and contain several point estimates as elements.

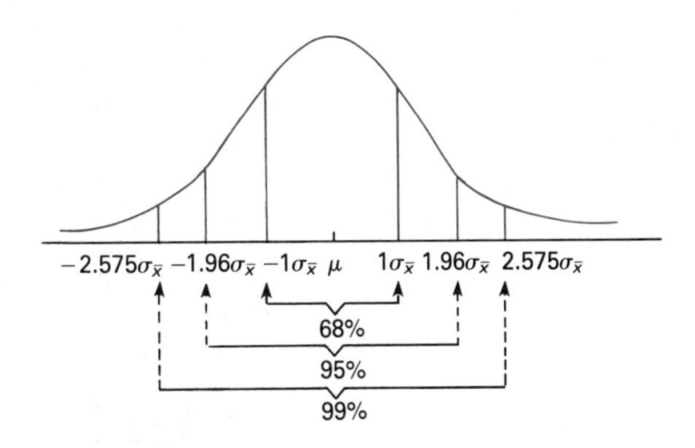

FIGURE 11.5
Proportion of sample means expected within fixed intervals around the population mean in the sampling distribution of means.

In contrast to the preceding intervals in which the population mean is the center, interval estimation procedures form interval ranges *around* the sample mean. The logic behind interval estimation is as follows. Using point estimation, we know that 95% of the random samples of size n will produce a sample mean that lies within a range of $1.96\sigma_{\bar{X}}$ units on either side of the population mean. If 95% of the sample means lie within a distance of $1.96\sigma_{\bar{X}}$ units from the population mean, then it follows that an interval formed by adding and subtracting $1.96\sigma_{\bar{X}}$ units to any of the sample means should contain the population mean 95% of the time. This latter logic is the basis for interval estimation. An interval is formed around a single sample mean with the confidence that, in the latter example, 95% of the intervals so formed (by adding $\pm 1.96\sigma_{\bar{X}}$ to each of the $\bar{X}$s) across repeated random samplings would contain the actual value of μ. In contrast to the intervals around the population mean illustrated in Figure 11.5, Figure 11.6 illustrates that for any of the 95% of the sample means within $1.96\sigma_{\bar{X}}$ units of the population mean, the boundaries of the intervals formed around the sample means will include the population mean (note as examples $\bar{X}_2$ and $\bar{X}_3$ in Figure 11.6). Only intervals formed around the 5% of the sample means beyond $1.96\sigma_{\bar{X}}$ units from the population mean will not contain the value of the population mean within the interval limits (Note $\bar{X}_1$ and $\bar{X}_4$ in Figure 11.6).

The interval formed around a given sample mean in interval estimation is called a **confidence interval.** Because the sampling distribution of means is approximately normally distributed, we use the concepts and values from nor-

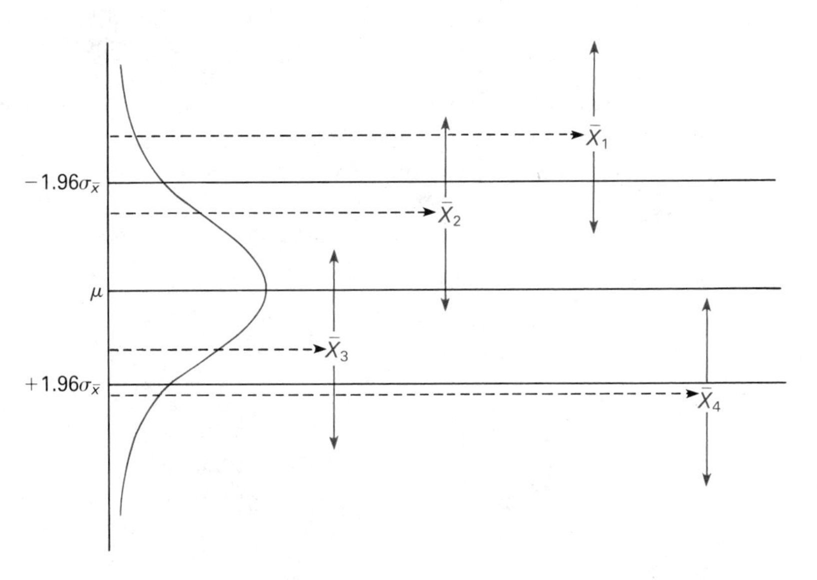

FIGURE 11.6
Confidence intervals around sample means.

mal curve theory to construct the limits of a confidence interval. For a given confidence level, say 95%, we determine the z score value from the table of the unit normal curve such that the middle 95% of the normal-curve distribution lies between $+z$ and $-z$. For the 95% confidence level, $z = 1.96$. If we wanted a confidence level of 90%, what would z equal?[8] Standard notation for confidence intervals is to let the confidence level equal the quantity $1 - \alpha$. The reason for this notation will become apparent in subsequent chapters. If $\alpha = .05$, the confidence level is .95 or 95%. The value of z corresponding to the $1 - \alpha$ confidence level is $z_{1-\alpha/2}$ or the z score equivalent for the $1 - \alpha/2$ percentile. If $1 - \alpha = .95$, then $\alpha = .05$ and the necessary z score is $z_{1-.05/2} = z_{.975}$. The z score value needed to establish the 95% confidence interval is $z = 1.96$, the z score value below which .975 of the distribution lies and above which .025 of the distribution lies. The upper and lower limits of the confidence interval are found using the formulas

$$\text{upper limit} = \overline{X} + (z_{1-\alpha/2})\sigma_{\overline{x}}$$
$$\text{lower limit} = \overline{X} - (z_{1-\alpha/2})\sigma_{\overline{x}}$$

The general form of the confidence interval is often written as an inequality identifying the confidence level, $1 - \alpha$, and the upper and lower limits of the interval.

$$C(\overline{X} - z_{1-\alpha/2}\sigma_{\overline{x}} \leq \mu \leq \overline{X} + z_{1-\alpha/2}\sigma_{\overline{x}}) = 1 - \alpha \qquad (11.3)$$

The concept of interval estimation can best be illustrated by considering the data in Table 11.2, which provides the means for 100 different random samples from the population of "Sesame Street" Numbers posttest scores. The population mean and variance are $\mu = 30.43$ and $\sigma_x^2 = 158.76$. Each mean value is a point estimate of $\mu = 30.43$. We also can construct a confidence interval around each of the 100 different sample means. As an example, let's construct the 90% confidence interval around the second sample mean, $\overline{X} = 33.42$. We will need the following values to use formula 11.3:

$$1 - \alpha = .90; \quad \alpha = .10$$
$$z_{1-.10/2} = z_{.95} = 1.645$$
$$\sigma_{\overline{x}} = \frac{\sqrt{158.76}}{\sqrt{12}} = 3.64$$
$$C[33.42 - 1.645(3.64) \leq \mu \leq 33.42 + 1.645(3.64)] = .90$$
$$C(27.43 \leq \mu \leq 39.41) = .90$$

As an interval estimate of μ based on the mean for random sample 2, the interval limits are 27.43–39.41. This interval does contain or capture the value of μ (30.43), but then we would expect 90% of the intervals formed around

[8] $z = 1.645$.

TABLE 11.5
Upper and lower limits for 90% confidence interval.

Sample	$\overline{X}$	Lower Limit	Upper Limit
2	33.42	27.43	39.41
14	28.42	22.43	34.41
29	36.25	30.26	42.24
49	31.83	25.84	37.82
66	24.75	18.76	30.74
89	26.25	20.26	32.24
98	39.08	33.09	45.07

the means for different random samples to contain the value of μ. We would expect that approximately 10% of the intervals formed from the sample means in Table 11.2 would not contain the value of $\mu = 30.43$ within the interval range. The upper and lower limits to the 90% confidence intervals for samples 14 ($\overline{X}_{14} = 28.42$), 29, 49, 66, 89, and 98 have been computed and are listed in Table 11.5. Confirm the computation of the limits for sample 14.[9] What would the limits be for sample 14 if the confidence level were changed to 80%?[10]

Note that the confidence interval upper and lower limits change across samples. The sample mean is different for each sample as a function of random sampling and results in a unique interval estimate of the population mean. Because the upper and lower limit interval values vary across repeated samplings, their actual value is not important in an absolute sense. Too often the interpretation of a single confidence interval is that the probability equals the confidence level that the population mean is between the upper and lower limit values of the confidence interval. *This interpretation is incorrect.* For example, if the confidence interval is $C(27.43 \leq \mu \leq 39.41) = .90$, the probability is *not* 90% that μ actually lies somewhere between 27.43 and 39.41. The values 27.43 and 39.41 are fixed only for one sample mean. They have no meaning as absolute limits on μ. The correct interpretation reflects the repeated sampling concept of the sampling distribution: *Over repeated random samplings of size n, the long-run occurrence (probability) is that $(1 - \alpha)$% of all the confidence intervals that we could construct around the sample means will contain the population mean.* The probability interpretation is not one about a single interval, but refers to the intervals constructed around all possible sample means from repeated random samplings of size n. For any

---

[9] $C[28.42 - (1.645)(3.64) \leq \mu \leq 28.42 + (1.645)(3.64)] = .90$

$C(22.43 \leq \mu \leq 34.41) = .90$

[10] If $1 - \alpha = .80$, then $\alpha = .20$; $z_{1-\alpha/2} = 1.28$; $\sigma_{\overline{X}} = 3.64$

$C[28.42 - (1.28)(3.64) \leq \mu \leq 28.42 + (1.28)(3.64)] = .80$

$C(23.76 \leq \mu \leq 33.08) = .80.$

single interval, you are $(1 - \alpha)\%$ confident that the interval contains the population mean within its range. The population mean will not be contained within the interval limits for $\alpha\%$ of the confidence intervals constructed. Figure 11.7 displays the confidence intervals calculated for the seven samples selected from Table 11.2 as an illustration of the repeated sampling concept. If we were to display 90% confidence intervals for all 100 random sample means in Table 11.2, our expectation would be that approximately 90 of the intervals would contain the value 30.43 ($\mu = 30.43$) and 10 intervals would not. Note that six of the seven confidence intervals in Figure 11.7 bound the value of $\mu = 30.43$. The confidence interval for $\overline{X} = 39.08$ (sample 98) does not.

Correct interpretation of confidence intervals often obscures their main purpose, serving as **interval estimates** of the population mean. As interval estimates, the limits indicate the scale vicinity in which the population mean might occur. This is the intuitively comforting aspect of interval estimation. Probably equally or more important, however, is interval estimate information on the magnitude of the range or distance of the interval. Interval size relates directly to estimate precision: The smaller the interval range, the more precise the interval estimate of the population mean. The same two factors affect the precision of point estimates—sample size (n) and the amount of variability in the population (σ^2). Both n and σ^2 directly affect the size of the standard error

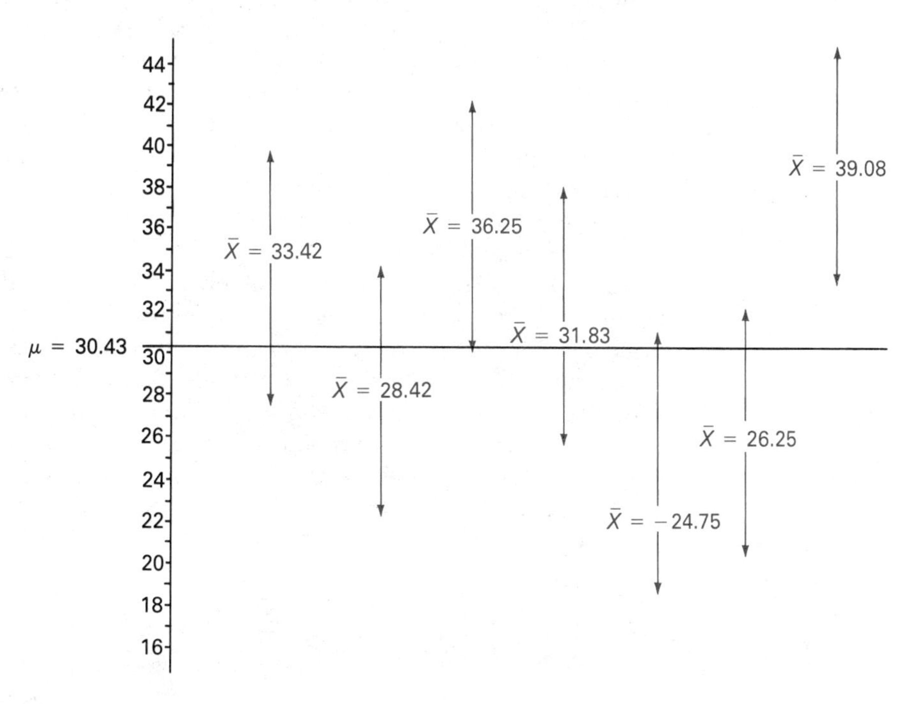

FIGURE 11.7
Ninety percent confidence intervals for seven random sample means.

($\sigma_{\bar{X}} = \sigma / n^{1/2}$), a direct indicator of the precision of an estimator. Increasing sample size improves estimator precision ($\sigma_{\bar{X}}$ is smaller). Less variability in the population set of scores means that $\sigma_{\bar{X}}$ will be smaller and the estimation will be more precise. In addition to these two factors, the confidence level also will affect the size of the interval in interval estimation. Logically, as the confidence level $1 - \alpha$ is increased, and n is fixed, the size of the interval increases. The larger interval range is a necessary result if a greater percentage of the intervals are to contain the population mean.

11.7 OTHER SAMPLING DISTRIBUTIONS

In introducing sampling distributions we stated that they exist for almost any statistical index that we can calculate from sample data. We have focused on the sampling distribution of means in our discussions because it is relatively easy to understand and because it is among the most important sampling distributions in statistical inference. The sampling distribution concept itself does not change for other statistics. The sampling distributions of other statistical indices occur as a result of repeated sampling over an infinite number of trials. Table 11.4 illustrated a subset of a sampling distribution of variances empirically generated for 100 random samples (displayed as a frequency polygon in Figure 11.2). We could have generated the same type of sampling distribution for sample modes or for semi-interquartile ranges or correlation coefficients between two variables. Some combine values from more than one sample. For example, the sampling distribution for the difference between two independent sample means, $\bar{X}_1 - \bar{X}_2$, will be one important sampling distribution. We will discuss these in later chapters.

As the concept of a sampling distribution does not change, neither does its role in statistical inference. (You may want to review this role as illustrated in Figure 11.3.) The appropriate sampling distribution usually depends on the population parameters about which we want to make inferences. If we want to make inferences about the difference between two population means, for example, mean differences on some variable between females and males, the appropriate sampling distribution will be the sampling distribution of mean differences between two independent samples. The role of any sampling distribution is to describe our expectations of the frequency with which sample results occur when samples are selected randomly from a population with assumed characteristics.

11.8 DEFINING EXPECTATIONS: SAMPLING DISTRIBUTIONS AS PROBABILITY DISTRIBUTIONS

The content of this section should come as no surprise. Throughout the discussions in this chapter, we have continually used statistical inference as a framework for understanding sampling distributions. If there is one point that should be clear from your reading, it is that we can use a sampling distribution to describe the *expected* relative frequency of occurrence of sample outcomes over repeated random samplings. In this manner, a sampling distribution acts as a probability distribution. This function is particularly important when the shape of the sampling distribution approximates the shape of a known probability distribution such as the normal distribution. When this occurs, we can compare values in the sampling distribution to the known probability distribution. We can then define our expectations about sample outcomes in probability terms. Or better yet, we can attach a probability value from the distribution to the expected occurrence of a given sample outcome. We then use the probability value to make statistical inferences.

The basic process is to translate or convert the outcome expected from a sampling distribution to a statement of probability. The sampling distribution of means illustrates the process. Suppose we select a random sample of 25 from a population with $\mu = 60$ and $\sigma_x = 15$. What is the probability that the sample mean will be equal to or greater than 64? The probability statement can be written as follows:

$$P(\overline{X} \geq 64; \quad \mu = 60, \quad \sigma = 15, \quad n = 25)$$

Because the sampling distribution of means is approximately normally distributed, the unit normal distribution is the appropriate probability distribution. The proportion of the normal curve area to the right of the z score equivalent of $\overline{X} = 64$ is the probability needed. We can use formula 11.2 to find the z score equivalent of $\overline{X} = 64$ in the sampling distribution of means when random sampling from an assumed population with $\mu = 60$ and $\sigma_x = 15$.

$$z_{\overline{x}} = \frac{64 - 60}{\frac{15}{\sqrt{25}}} = \frac{4}{3} = 1.33$$

The probability statement converts to a normal curve probability.

$$P(z \geq 1.33) = .0918$$

The value of .0918 was obtained from the areas in the normal curve in Table B.2. The probability associated with the occurrence of a sample mean of 64 or greater when drawing random samples of size 25 from a population with a mean of 60 and standard deviation of 15 is about 9%. Figure 11.8 illustrates the correspondence between the sampling distribution and the normal probability distribution. What is the probability of drawing a random sample of size 100 with a mean equal to or greater than 62 from the same population ($\mu = 60$ and $\sigma_x = 15$)?[11] Surprised? The probability of an $\overline{X} \geq 62$ for samples of size

---

[11] $z_{\overline{x}=62} = \frac{62 - 60}{\frac{15}{\sqrt{100}}} = \frac{2}{1.5} = 1.33$

$p(\overline{X} \geq 62; \quad \mu = 60, \sigma^2 = 225, n = 100) = .0918$

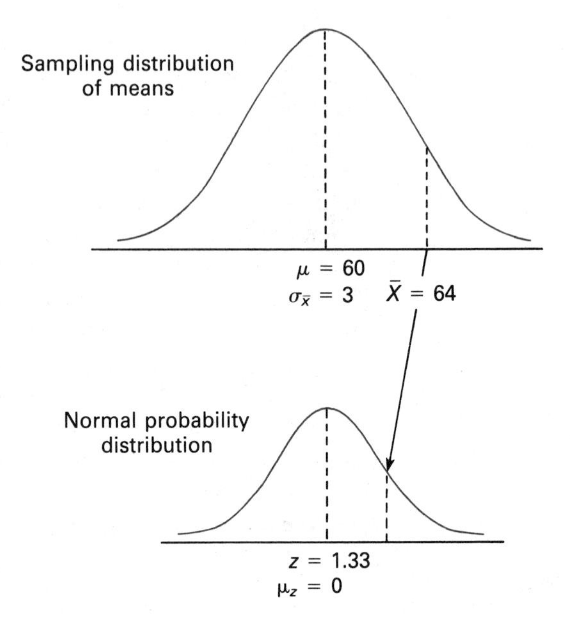

FIGURE 11.8
Correspondence between the sampling distribution of means and the unit normal probability distribution.

100 is the same as for an $\bar{X} \geq 64$ for samples of size 25. The increased sample size decreased the variability of sample means. A sample mean of 62 or greater when $n = 100$ is not likely to happen with any greater frequency than a sample mean of 64 or greater when $n = 25$.

11.9 SUMMARY

This chapter emphasized the importance of the sampling distribution and probability in statistical inference. The information necessary to understand the process of hypothesis testing (to be covered in the next chapter) was stressed. Probability and sampling distributions and their characteristics are not important just to testing hypotheses in statistical inference, however. Statistical inference is based on parameter estimation, both within hypothesis testing and as the possible sole purpose of a study. Sampling distribution characteristics describe theoretical models to indicate a sample statistic's precision in estimating a population parameter of interest. The chapter discussed both point estimation and interval estimation. Both types should be kept in mind in the next few chapters in which sample statistical values help us estimate population parameters and test hypotheses about these parameters based on the sample data.

PROBLEMS

Problem Set A

1. Compute
   a. $P(\overline{X} \leq 38; \mu = 40, \sigma = 10, n = 16)$
   b. $P(\overline{X} \leq 38; \mu = 40, \sigma = 10, n = 49)$
   c. $P(\overline{X} \leq 38; \mu = 40, \sigma = 10, n = 100)$
   d. $P(\overline{X} \leq 38; \mu = 40, \sigma = 10, n = 625)$

2. Compute
   a. $P(\overline{X} \leq 38; \mu = 40, \sigma = 5, n = 25)$
   b. $P(\overline{X} \leq 38; \mu = 40, \sigma = 15, n = 25)$
   c. $P(\overline{X} \leq 38; \mu = 40, \sigma = 25, n = 25)$

3. Compute
   a. $P(92 \leq \overline{X} \leq 98; \mu = 100, \sigma = 10, n = 25)$
   b. $P(\overline{X} \geq 98; \mu = 100, \sigma = 10, n = 25)$
   c. $P(\overline{X} \leq 96 \text{ or } \overline{X} \geq 104; \mu = 100, \sigma = 10, n = 25)$
   d. $P(95.6 \leq \overline{X} \leq 102.3; \mu = 100, \sigma = 10, n = 25)$

4. When random sampling from each of the following populations, from which would it be most probable that a sample of size 16 would produce a mean of 46 or less? (a) population A—$\mu = 50, \sigma = 2$; (b) population B—$\mu = 53, \sigma = 6$; (c) population C—$\mu = 60, \sigma = 16$; (d) population D—$\mu = 56, \sigma = 12$.

5. What is the minimum sample size to ensure that a sample mean of 522 or greater would occur with a probability of .10 or less given random sampling from a population with $\mu = 500$ and $\sigma = 100$?

6. Compute the 95% and 90% confidence intervals for a sample mean of 12 using a random sample of size 25 from a population with a variance of 36.

7. Find the 80% confidence interval limits for the following: $\overline{X} = 122.4, \sigma = 6.32, n = 16$.

8. Find the 99% confidence interval limits for the following: $\overline{X} = 70, \sigma = 7, n = 25$.

9. What can you infer about μ given the confidence interval in problem 8?

Problem Set B

I. Newborn behavioral effects of lower-level alcohol use during pregnancy were examined in 417 infants by Streissguth, Barr, and Martin (1983). The researchers determined alcohol use by self-report during the fifth month of pregnancy. Factors of the Brazelton Neonatal Behavioral Assessment Scale were used to define specific behavioral responses to be analyzed. Two behavioral response outcomes, habituation and low arousal, were found to be related to levels of alcohol use during pregnancy. Habituation scores indicate the newborn's ability to habituate readily to repetitive stimulation and thus to "tune out" different environmental stimuli after a brief response. Habituation was thought to be a protective mechanism against the many novel stimuli with which

TABLE 11.6
Mean habituation and low arousal scores of infants for different levels of maternal alcohol use (AA) during midpregnancy.

	Habituation		Low Arousal	
AA Interval	$\overline{X}$	n	$\overline{Y}$	n
0	17.0	88	9.3	74
.01–.10	16.0	103	8.8	102
.11–.99	15.9	119	9.2	107
1.00–1.99	15.0	20	10.7	23
2.00 or more	13.5	6	12.0	4
TOTAL	16.1	336	9.2	310

the newborn must contend. Low arousal scores indicate the extent to which newborns are easy to console, frequently self-quiet, and seldom very upset or excited.

Mean habituation and low arousal scores from the Streissguth study appear in Table 11.6 for arbitrarily measured categories of alcohol consumption (AA). Additional data indicated that the mean reported AA score for all mothers in the study at midpregnancy was .28 ounces with a standard deviation of .63. The mean reported AA score for the mothers who reported using any alcohol in midpregnancy was .38 ounces with a standard deviation of .71. This latter mean was estimated to be roughly equivalent to two-thirds of a drink of liquor or three-fourths of a drink of beer or wine per day or approximately five drinks per week on the average. Answer the following questions.

1. Assuming that alcohol consumption scores are normally distributed for individuals who reported that they consumed some alcohol, what proportion of the population represented by this sample who report consuming some alcohol would you estimate to consume an average of 1.00 or more ounces of alcohol per day?

2. Assume that the mean daily alcohol consumption for the population of adult women in the United States is .35 ounces with σ equal to .63. If the current sample were a random sample from this population, what is the probability that the sample mean ($\overline{X} = .28$) or one smaller would have occurred solely due to random sampling error? (Use the total sample size of 336.)

3. Compute the 90% confidence interval limits for the sample mean in problem 2. Does the identified interval contain $\mu = .35$? Based on the confidence intervals, would you conclude that the current sample is a probable or improbable random sample from a population with a mean of .35?

4. When examining the effect of reported alcohol consumption on infant habituation behavior, the mean habituation score reported for the nonalcohol-consuming group should be the expected level for all groups if alcohol consumption has no effect. Using this group mean as the population value ($\mu = 17.0$), compute the probability that each of the other group means (some mean value less than or equal to the reported group mean) would occur purely as a result of random sampling from this

population. Use the group sample size indicated in Table 11.6 and assume that the population habituation score standard deviation is 4.0.

5. Perform the same analysis as in problem 4 for the low arousal means using 9.3 as the population value, but calculate the probability that the mean would be greater than or equal to the one reported. Assume the standard deviation is 4.0. Note that the sample size for each alcohol consumption group differs slightly from that used to calculate the habituation means.

6. Use a criterion of a calculated probability equal to or less than .025 to conclude that a specific group mean is too far from the nonconsuming group mean to be due only to random sampling (and therefore that the alcohol consumption must have affected the behavior level). Which, if any, alcohol level groups meet this criterion for either habituation or low arousal scores? Based on these results, what conclusion might you reach about the reported level of alcohol consumption during midpregnancy and its effects on habituation or low arousal behavior level in infants?

II. Chapter 6, Problem Set B provided simulated data on achievement outcomes for two separate groups, one instructed using a mastery learning strategy and the other as a nonmastery strategy. Assuming that the nonmastery group represents the population of achievement outcomes expected under regular instructional practices, the mean is 42.9 and the standard deviation is 5. Further, if we assume that mastery learning does not affect achievement any more or less than regular instruction, then the distribution of sample achievement means for random samples of individuals instructed under a mastery strategy should follow that predicted by the central limit theorem.

1. What would the characteristics of the sampling distribution of means be under these conditions given the sample size for the data in Chapter 6?

2. What is the probability that the mastery sample in Chapter 6 would have a mean achievement value as large or larger than it would if it were indeed a random sample from the population described and the mastery instructional strategy had no effect on the mean level of achievement?

3. Compute the 99% confidence interval limits using the mastery group sample mean. Given these limits, what might you conclude about mastery learning achievement outcomes relative to the population mean of 42.9, which was assumed to describe the average outcome expected?

Problem Set C

1. The Numbers posttest sample means in Table 11.2 represent a sampling distribution of means for samples of 12 using the "Sesame Street" data as the population ($\mu = 30.43$ and $\sigma = 12.60$). Using the central limit theorem, what proportion of the 100 sample means would you theoretically expect to be
   a. greater than 34.005?
   b. between 26.005 and 30.005?
   c. less than 26.005?

2. The theoretical expectations are given in the answers to problem 1. Calculate the exact proportion (number) from Table 11.3 of the 100 sample means that actually were

a. greater than 34.005.
b. between 26.005 and 30.005.
c. less than 26.005.
Compare the actual proportions obtained with the expected proportions calculated in problem 1.

3. Demonstrate the effect of increased sample size on the precision of the sample mean as a point of estimate of the population mean. Again, use the Numbers posttest scores with $\mu = 30.43$ and $\sigma = 12.60$ as the population. Draw a random sample of size 5 and calculate its mean, then draw an additional random sample of size 18. Calculate the mean for each sample. As the sample size increases, the sample means should tend to be more accurate estimates of the population mean, $\mu = 30.43$. Is this the case with your sample means or do you have a low-probable occurring sample mean due to random sampling error?

4. Calculate the standard error of the mean for each sample in problem 3. Compute the probability of occurrence of a sample mean as large or larger (if above μ) or as small or smaller (if below μ) due to random sampling for each sample mean obtained in problem 3. Are any of your sample means low-probable occurring events if a probability of .05 or less is the criterion used?

5. Compute the 90% confidence intervals for each of the sample means obtained in problem 3. Does the value of $\mu = 30.43$ lie within the limits for all your confidence intervals? It should for 90% of the intervals constructed. If you have an interval that does not contain $\mu = 30.43$, the sample mean used to construct the interval is a low-probable occurring event. These results should agree with your results in problem 4.

12

Hypothesis Testing

A process is a series of activities performed to move from a starting point to an end point. The activities defining a process need not be sequential or hierarchical, but they often are. Throughout the text, we have stressed that statistical application is one of several activities in the general research process. Statistical manipulation of data, while a component of the research process, does not exist in a vacuum. As an activity, it both affects and is affected by other activities in the research process. Because this is a statistics text and not a research methods book, the last two chapters focused on another process within the more general research process—statistical inference. Statistical inference is both a process, a series of activities or steps, and a way of thinking. It dictates the logic for one form of statistical reasoning. Chapters 10 and 11 discussed two basic ingredients necessary to understanding statistical inference—probability and sampling distributions. This chapter employs probability and sampling distributions to examine a fundamental statistical inference process: hypothesis testing.

A hierarchical relationship exists between the research process and the statistical inference process. The research process is all-encompassing and provides the framework for general research activities. Statistical inference imposes a logical reasoning process on the research process, resulting in inferences about population characteristics based on sample data. Statistical reasoning and procedures for making inferences about populations from sample data are defined by confidence intervals and the hypothesis-testing process. By emphasizing hypothesis testing within the statistical inference process, we will narrow the focus to specific statistical techniques and their applications. **Hypothesis testing** is applying statistical techniques and statistical reasoning to the inferential process.

A general sequence of steps applies to all hypothesis-testing situations. The hypotheses will change, the statistical techniques applied will change, but the process will not change. This chapter introduces the reasoning behind hypothesis testing, focusing on the fundamental roles of probability and sampling distributions. Hypothesis testing follows this process:

1. Identify the specific statistical hypothesis about the population parameter of interest.
2. Calculate the corresponding statistical index from sample data necessary to test the hypothesis about the population parameter.
3. Identify the characteristics of the random sampling distribution for the sample statistic under the assumption that the hypothesis about the population parameter is true. This identifies the expected frequency of random sample outcomes if the initial hypothesis is true.

4. The hypothesis is either rejected or not rejected. Predetermined probability levels form the basis for deciding whether to reject the hypothesis.

12.1 THE ROLE OF STATISTICAL HYPOTHESES

Applying the hypothesis testing process within statistical inference and research involves first directly translating a research question or area of interest into a testable statistical hypothesis. This translation links the processes and allows us to generalize conclusions from tests of statistical hypotheses to the population research question of interest. Decisions about the validity of statistical hypotheses should relate directly to information about the research question under study.

For our present purposes, a statistical hypothesis is a statement about a population characteristic in terms of a specific value of a population parameter. For instance, if we wished to find the average IQ of a particular population, we would form a statement about its hypothesized value, μ_{IQ}. If the relationship between two variables were of interest, the statistical hypothesis would be about the parameter ρ_{xy}, the population correlation coefficient. In identifying the population parameter of interest, the statistical hypothesis determines the specifics of the hypothesis testing process. It determines the sample statistics to be calculated, the sampling distribution to be referenced, the statistical techniques to be applied, and the types of inferences to be made.

In hypothesis testing, the statistical hypothesis identifies an *assumed* value or relationship about a population. We will not know the exact population characteristic value, so we formulate a hypothesis about it. The goal of statistical inference is to be able to infer something about the truth of this hypothesis without collecting data from the entire population. Hypothesis testing allows us to make the inference, but we must start with some *hypothesized* value of a population characteristic. We *assume* that the hypothesized population value is correct until our sample data provide contradictory evidence. The hypothesis is a formal statement of what we assume initially to be true about the population. We test the likely validity of that hypothesis for the population through sample data collection and statistical analysis following the procedures of hypothesis testing.

12.2 FORMULATING HYPOTHESES

To formulate a *testable* statistical hypothesis, the research question must be translated into a null hypothesis statement about a population parameter. A null statistical hypothesis is a statement of equality, that is, no difference or no relationship exists among population parameters. A **null hypothesis** is symbolized as H_0. Null hypotheses are represented statistically using population parameter symbols. The following illustration translates a simple research question into verbal and symbolic null hypotheses.

- Research question: Is the level of this year's 11th-grade mathematics achievement better or worse than it was five years ago when the average test score was 40?
- Verbal H_0: There exists no difference between the mean level of mathematics achievement for today's 11th graders and that of 11th graders five years ago.
- Symbolic H_0: $\mu_{\substack{\text{current math}\\ \text{achievement}}} = 40$

Other examples of research questions, verbal statements of H_0, and symbolic representations of H_0 appear in Table 12.1.

TABLE 12.1
Examples translating research questions into null hypotheses.

Research Question	Verbal Null Hypothesis	Symbolic H_0
Did our college seniors score at a level equivalent to the national average of 500 on the GRE verbal test?	H_0: There is no difference between the mean GRE verbal score for our college seniors and the national average of 500.	H_0: $\mu_{\text{GRE}} = 500$
Does assertiveness training help make people more assertive?	H_0: There exists no difference in the mean level of assertiveness between persons who have received training and those who have not.	H_0: $\mu_{\text{training}} = \mu_{\text{no training}}$
Is there a relationship between school achievement and per-pupil expenditure?	H_0: There is no relationship between school achievement and per pupil expenditure.	H_0: $\rho_{xy} = .00$
Do student and peer evaluations order teachers the same on ratings of teaching effectiveness?	H_0: There exists no relationship between ratings by students (X) and ratings made by other teachers (Y) on teaching effectiveness for a group of teachers.	H_0: $\rho_{xy} = .00$
Does individual instruction result in more homogeneous achievement scores than group-paced instruction?	H_0: There is no difference in the variability of achievement scores between students receiving individualized instruction and those taught using a group-paced instructional approach.	H_0: $\sigma^2_{\text{individualized}} = \sigma^2_{\text{group-paced}}$

Translating research questions into corresponding null hypothesis statements may seem complex, but it is a reasonably straightforward process. The reason is that the majority of research questions can be categorized into questions about (a) frequencies, proportions, or percentages; (b) magnitude in terms of central tendency levels; (c) variability; or (d) relationships. The corresponding null hypotheses are statements about the equality of these familiar statistical terms, that is, hypotheses about the equality of proportions or means or variances or correlation coefficients. There are other null hypotheses, but they are often complex extensions of hypotheses involving these basic statistical parameters. Many advanced statistical methods do nothing more than focus on the various statistical techniques and procedures that test specific complex hypotheses. The remainder of this text will deal with introductory null hypotheses: hypotheses about single sample means (this chapter and the next), hypotheses about two sample means (Chapter 13), hypotheses about variances and correlation coefficients (Chapter 14), hypotheses about frequencies and proportions (Chapter 15), and hypotheses about multisample means (Chapter 16). The hypothesis testing process remains the same across different null hypotheses, but the statistical technique changes to fit the null hypothesis.

Verbally, H_0 can always begin "There exists no difference . . ." or "There is no relationship . . .". A statement of no difference or no relationship is a statement of equality. The null hypothesis about mean level of mathematics achievement was written as a statement of equality, H_0: $\mu = 40$. We could write the same H_0 symbolically as a statement of no difference between the population mean and the value of 40, H_0: $\mu - 40 = 0$. When a hypothesis is written in this form, as a null hypothesis of equality, we can test it using statistical procedures. In fact, a hypothesis must be written in null form to be testable. The reason is quite simple. The characteristics of a hypothetical sampling distribution can only be identified for hypothesized values of population parameters in null form. The sampling distribution associated with a particular H_0 tells us what we might expect about the long-run occurrences of specific sample results *if the null hypothesis is true.* We compare the expectations formed from the sampling distribution under an assumed-true H_0 to a probability distribution to determine how probable a specific sample result is. We then use the probability value attached to the sample result to support or refute the null hypothesis.

If the sample evidence leads us to reject a null hypothesis, we infer that some other value of the population parameter represents the true population characteristic. In this case, conclude that the null hypothesis statement is not correct, and therefore some other, alternative hypothesized statement is correct. An **alternative hypothesis** is implied and accompanies every null hypothesis statement. The alternative hypothesis is often called the research hypothesis and is symbolically represented by H_A. Whereas H_0 is a statement of equality, H_A is its complement, a statement of inequality. We form it using the symbols for "does not equal" ($\neq$), "less than" ($<$) or "greater than" ($>$). If the null hypothesis H_0: $\mu = 40$ is not true, then the alternatives are

$$H_A: \mu \neq 40; \quad H_A: \mu < 40; \quad \text{or} \quad H_A: \mu > 40$$

The alternative hypothesis does not state a specific alternative value for the population parameter. The process does not test a hypothesized alternative. It provides only for a test of H_0. If H_0 is not true, then it must be false: if H_0: $\mu = 40$ is not true, then the conclusion is that $\mu \neq 40$, that $\mu < 40$, or that $\mu > 40$. These are the only three alternative conclusions we can draw. Each implies that H_0 is false.

12.3 THE NATURE OF STATISTICAL DECISIONS

So far, the text has made numerous references to decision making in statistical inference and hypothesis testing and the role of probability as the basis for these decisions. This section attempts to synthesize that material. First, let's go through the specific steps of the hypothesis testing process with an example. Suppose we want to consider the research question about the current level of mathematics achievement among 11th-grade students relative to the performance level five years ago. Assume that we know the mathematics achievement for the population of 11th grade students five years ago. The mean test performance score was 40.0 ($\mu = 40$) and the standard deviation was 6.0 ($\sigma = 6$). If the current performance level is no different from that five years ago, the hypothesized mean performance level for the current population of 11th-grade students would be 40.0. By this reasoning, we translate the research question into the null hypothesis, H_0: $\mu = 40$. As part of conducting the research (the research process), we would draw a random sample of current 11th graders, administer the same test given five years ago, and calculate the relevant sample statistic ($\overline{X}$) to test H_0. Assume the size of the sample drawn at random from the current population was 64 students and the calculated sample mean performance level on the mathematics achievement test was 42.6. Direct comparison shows that the sample mean is greater than the hypothesized performance level of $\mu = 40$. Direct comparison is not an appropriate procedure, however. We must remember that we are looking at *sample data*, which we expect to deviate somewhat from the population characteristics. The relevant question is: "Does the sample mean of 42.6 provide sufficient evidence to conclude that the hypothesized population value of 40 is incorrect?" The answer to this question is the basic decision to be made in hypothesis testing. We need to decide what to conclude about H_0 based on the sample evidence. Obviously, we need both the mechanism for objectively providing information on which to base the decision and the criteria to be used in making the decision. As you might suspect, the sampling distribution for the sample statistic provides random sample result expectations that we use to make a decision about H_0. The criteria for making the decision are related to the probability of the sample event.

Chapter 11 demonstrated application of sampling distributions. For the current example, testing H_0: $\mu = 40$, the central limit theorem provides the necessary information about the sampling distribution of means. *If we assume H_0: $\mu = 40$ to be true,* then sample means from random samples of 64 from a population with $\mu = 40$ and $\sigma = 6$ are approximately normally distributed with a mean of 40 and a standard deviation of

$$\sigma_{\overline{X}} = \frac{\sigma_x}{\sqrt{n}} = \frac{6}{\sqrt{64}} = \frac{6}{8} = .75$$

How likely (probable) is a mean of 42.6 or greater for samples of 64 drawn at random from a population with a mean of 40 and standard deviation of 6? Our decision about the truth of H_0 is linked to the level of the probability attached to the occurrence of the sample event: $p(\overline{X} \geq 42.6,\ \mu = 40,\ \sigma = 6,\ n = 64)$. The logic is

- *If* the true population mean from which the random sample was drawn is as hypothesized (H_0 is true),
- *then* sample results can be expected to occur with a frequency determined by the sampling distribution based on a model of randomness.
- *If* a specific sample result occurs that is unlikely to happen based on our expectations assuming H_0 is true,
- *then* the evidence would support a decision that H_0 must not be true.

The lower the probability attached to the sample result, the less likely such a result is to have occurred by chance when sampling from a population defined by H_0. To obtain the probability attached to the occurrence of $\overline{X} \geq 42.6$, we use the unit-normal distribution. Changing $\overline{X} = 42.6$ to an observed z score,

$$z_{\text{obs}} = \frac{\overline{X} - \mu}{\sigma_{\overline{X}}} = \frac{42.6 - 40}{.75} = 3.47$$

The probability (from Table B.2) of getting a sample mean this large or larger as a result of random sampling from a population with $\mu = 40$ is .0002. A sample mean of 42.6 or larger occurring by chance as a result of random sampling is very unlikely. Based on the probability of .0002 attached to the sample outcome $\overline{X} \geq 42.6$, we might conclude that the population mean mathematics achievement level from which the sample was randomly drawn is not 40. The sample event of $\overline{X} = 42.6$ is just too unlikely to have happened if H_0: $\mu = 40$ really were true. The hypothesis-testing decision is to conclude that H_0: $\mu = 40$ is false. If H_0: $\mu = 40$ is false, then the alternative H_A: $\mu \neq 40$ must be true. The statistical inference from the sample data based on $\overline{X} = 42.6$ as a point estimate of the true population mean is that $\mu > 40$. The research generalization and conclusion based on the statistical inference is that the current 11th graders' level of mathematical achievement is higher than the level of achievement five years ago.

You should be able to see the logic relating hypothesis testing to statistical inference and generalized research conclusions demonstrated in the example. The decision about H_0: $\mu = 40$ was obvious. The probability of .0002 for the chance occurrence of $\overline{X} \geq 42.6$ was extremely low. You might wonder how one decides whether the probability of a sample event is low enough to reject H_0. In practice, this decision is reasonably straightforward but arbitrary. Common probability levels exist in behavioral sciences as acceptable conventions (standards) for rejecting H_0. You have probably heard or read the phrase "significant at the .05 level." The .05 refers to a probability level as a criterion for rejecting H_0.

Criteria for Rejecting H_0

We decide to reject H_0 depending on our expectations about sample outcomes defined by a sampling distribution *assuming that H_0 is true* and if a sample value is in line with our expectations. The extent to which the sample value coincides with our expectations is defined by the probability of the sample value occurring as a result of random sampling from the hypothesized population. We can set any probability level as a criterion for rejecting H_0. Any decision made based on probabilities is not going to be correct 100% of the time, however. It is always possible that an error has been made in the statistical inference resulting from the decision on whether to reject H_0. The probability level chosen as a criterion for rejecting H_0 directly affects the potential for making specific kinds of errors in the inferential process. To look at these potential errors, let's examine the consequences of our decisions resulting from hypothesis testing. We can make only two decisions about H_0: Reject it or do not reject it. Consequences of this decision relate directly to the true state of affairs concerning H_0—whether H_0 is true or whether it is false as a description of the population characteristic. Table 12.2 identifies the consequences (error or correct decision) of rejecting or not rejecting H_0 given the truth about H_0.

TABLE 12.2
Consequences of the hypothesis test.

	True State of Affairs: H_0 True	True State of Affairs: H_0 False
Decision: Reject H_0	Type I Decision Error (α)	Correct Decision ($1 - \beta$)
Do Not Reject H_0	Correct Decision ($1 - \alpha$)	Type II Decision Error (β)

The two kinds of errors that can result from our decisions are Type I and Type II errors. **A Type I error is a decision to reject H_0 when H_0 is true. A Type II error is a decision to not reject H_0 when H_0 is false.**

A Type I error can only occur when the null hypothesis is true. In the hypothesis-testing process, we always assume that H_0 is true initially. When a probability level is set as the criterion for rejecting H_0, we actually are identifying the *potential* probability of committing a Type I decision error. The process assumes that H_0 is true. We reject H_0 only if the sample result occurs with a probability below the criterion probability level. This is an extremely important point! The probability level that the researcher sets as the criterion for rejecting H_0 is the probability of committing a Type I decision error. The probability of making a Type I error is thus entirely under the researcher's control. In many instances, the Type I error is the more severe error. Therefore, we usually set the probability criterion level for rejecting H_0 by chance quite low, usually at .05 or lower. By setting the rejection probability criterion at .05 ($p = .05$), we would expect that 5% of our sample results would deviate by chance far enough from the hypothesized H_0 value to result in an incorrect decision to reject H_0.

The probability criterion level for rejecting H_0 is called the **level of significance.** The level of significance for testing H_0 defines the risk we are willing to take in possibly making a Type I error. The level of significance is referred to as the alpha (α) level in a research study. The Greek letter α refers to the level of significance but actually represents a probability value, the probability associated with making a Type I decision error in an experiment.

$$\text{level of significance} = \alpha = p(\text{committing a Type I error})$$

Selecting a value for α as the probability criterion level for rejecting H_0 objectively determines specific values in the probability distribution that we use to decide whether to reject H_0. These specific values in a probability distribution determined by the value of α are called **critical values.** For example, assume we selected an $\alpha = .05$ level of significance. This means that in testing the null hypothesis, we would reject H_0 when H_0 is true. We would thus make a Type I error if our sample mean were one of the 5% expected to be at extreme ends of the probability distribution. Figure 12.1 illustrates the H_0 rejection region of the normal probability distribution determined by $\alpha = .05$. Note that because the normal curve is symmetrical and because an equal number of random sample means are as likely to fall below μ in the sample distribution of means as above μ, half of the area determined by an α of .05 (area = .025) appears in the lower tail of the distribution and half in the upper tail. The area of the curve to the left of the lower critical value and to the right of the upper critical value is the **region of rejection.** Any sample value within these regions would lead us to reject H_0. The total proportion of sample means expected to fall beyond the critical values is still equal to the α value of .05; we would expect 2.5% to occur in the lower tail to the left of the lower

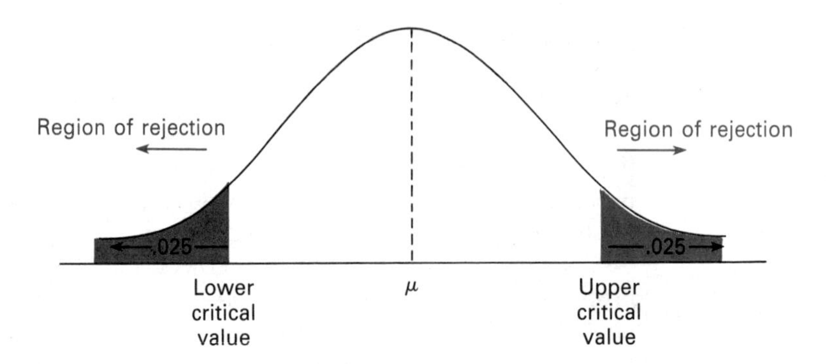

FIGURE 12.1
Rejection regions when $\alpha = .05$.

critical value and we would expect 2.5% to occur in the upper tail to the right of the upper critical value. What critical z score values from Table B.2 divide the normal distribution into the proportional areas illustrated in Figure 12.1?[1]

The logic behind the decision criterion for rejecting H_0 needs to be clear. By setting a low probability level for α, say .05, the probability is less than or equal to .05 that a random sample mean would lie beyond the critical values and within the rejection region. Ninety-five percent of the sample means would be expected to lie between the critical values. Because the probability is low (equal to α) that a sample mean would fall into the region of rejection *by chance,* any sample mean that does fall into the rejection region is considered evidence that H_0 must be wrong. If H_0 is true, the probabilities are in our favor that the sample mean would not fall in the region of rejection, so when it does, we risk concluding that H_0 is not true. We set the risk level in our conclusion (making a Type I error) by the α probability level.

The Decision Process

The decision process in hypothesis testing is straightforward. Table 12.3 outlines the general steps. An example of each step is provided testing the mathematics achievement level null hypothesis used previously.

Let's work through another hypothetical example to illustrate hypothesis testing. Suppose we want to test H_0: $\mu = 200$ against the alternative H_A: $\mu \neq 200$. Assume the population variance to be $\sigma^2 = 225$. A random sample of size 36 results in a mean of $\overline{X} = 194.6$. If we set our α level at .01, what are the critical z score values for rejecting H_0: $\mu = 200$?[2] Converting $\overline{X} = 194.6$

---

[1]Lower critical z score $= -1.96$
Upper critical z score $= +1.96$.

[2]The critical z score values separating the area of the normal curve into the lower .5% (.005), the middle 99% $(1 - \alpha)$, and the upper .5% (.005) are $z_{\text{crit}} = \pm 2.575$.

TABLE 12.3
Steps in decision making for hypotheses tests.

General Hypothesis Testing Process	Example
1. Identify H_0 and H_A.	1. H_0: $\mu = 40$ H_A: $\mu \neq 40$
2. Set a probability criterion level for rejecting H_0 by selecting a value for α.	2. Let $\alpha = .05$
3. Determine the region(s) of rejection by identifying the *critical standard score* values resulting from the α level set in the probability distribution appropriate for testing H_0.	3. $z_{crit} = -1.96$ $z_{crit} = 1.96$
4. Calculate the sample statistic value needed to test H_0.	4. $\overline{X} = 42.6$
5. Convert the sample statistic value to an *observed standard score* based on the characteristics of the sampling distribution for the statistic.	5. For $\mu = 40$, $\sigma = 6$ and $n = 64$; $z_{obs} = \frac{42.6 - 40}{\frac{6}{\sqrt{64}}} = 3.47$
6. Compare the *observed* sample standard score to the *critical* standard score value determined by the α level.	6. $(z_{obs} = 3.47) > (z_{crit} = +1.96)$
7. Reject H_0 if the observed standard score falls in the region of rejection beyond values of the critical standard score, for example, Reject H_0: If $z_{obs} > (+z_{crit})$ or $z_{obs} < (-z_{crit})$	7. $z_{obs} > (+z_{crit})$, so reject H_0: $\mu = 40$.
8. If H_0 is rejected, conclude that H_A is true knowing that your decision to reject H_0 *if it is true* has an α% probability.	8. Conclude that H_A: $\mu \neq 40$ is the true state of affairs. The probability is .05 that you would reach this decision when in fact H_0: $\mu = 40$ is true.

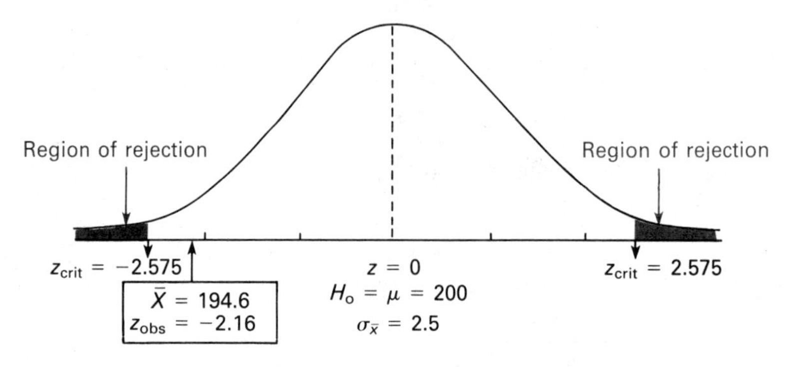

FIGURE 12.2
Deciding not to reject H_0 when a $-z_{crit} < z_{obs} < z_{crit}$.

to an observed standard score,

$$z_{obs} = \frac{194.6 - 200}{\frac{15}{\sqrt{36}}} = \frac{-5.4}{2.5} = -2.16$$

Comparing z_{obs} to z_{crit}, the z_{obs} value of -2.16 falls in the range between the z_{crit} values. Because z_{obs} does not fall in the rejection region, we decide *not to reject H_0* when $\alpha = .01$. The decision information is illustrated in Figure 12.2. Would our decision have been different if we had set an alpha level of .05 as our criterion?[3] What if we had drawn a sample of size 100 and obtained the same result, $\bar{X} = 194.6$? What would we decide about H_0: $\mu = 200$ using an α level of .01?[4]

Type II Errors and Power

The hypothesis testing paradigm leads to only one decision: whether to reject H_0 based on evidence from sample data. The statistical decision is objective once we set the probability criterion level by specifying a value for α and determine the resulting critical values. If we decide to reject H_0, one of two potential consequences results: an error (Type I) or a correct decision. Which consequence occurs depends entirely on whether H_0 is really true or really false. We never know whether our decision about H_0 is correct or incorrect.

[3]By changing the level of α, we change the critical values. For $\alpha = .05$, $z_{crit} = \pm 1.96$. The sample z_{obs} value of -2.16 is beyond the lower z_{crit} of -1.96. Using $\alpha = .05$ as a criterion, we would reject H_0: $\mu = 200$.

[4]Drawing a larger sample reduces the standard error of the sampling distribution, $\sigma_{\bar{X}} = 15/100^{1/2} = 1.5$. The sample means would not be expected to vary as much from $\mu = 200$ if H_0: $\mu = 200$ is true for $n = 100$ as for $n = 36$. The z_{obs} value for $n = 100$ is $z_{obs} = (194.6 - 200)/1.5 = -3.6$. The $z_{obs} = -3.6$ indicates that an $\bar{X} = 194.6$ for $n = 100$ is not likely to result from random sampling. The z_{obs} value of -3.6 is less than the critical value ($z_{crit} = -2.575$) for $\alpha = .01$, so the decision would be to reject H_0: $\mu = 200$.

Hypothesis testing, however, *does* indicate the probability level for a particular consequence. The probabilities of potential consequences indicate the confidence we should have in our decision.

Assume we decide to reject H_0. We know the probability of one consequence, that of making a Type I error. The probability equals α, the level of significance, set as the probability criterion level by the experimenter. The α probability is similar to a "comfort index" and serves as the basis (criterion) for decisions about H_0. It is "comforting" to know that *if H_0 is really true,* our random sample results will fall in the region of rejection only $\alpha\%$ of the time. The rest of the time, $(1 - \alpha)\%$, our sample results would fall between the critical values. We know that for any test of a hypothesis, the consequence of a Type I error will happen with a probability of α. The α level identifies the probability that an unlikely sample event will randomly occur when H_0 is true, leading to the incorrect decision to reject H_0.

The α probability is only relevant and applicable *if H_0 is really true.* If we decide to reject H_0 and H_0 is really false, we have made a correct decision. In fact, the desired outcome of hypothesis testing is making the correct decision. As the statistical hypothesis tested, the null hypothesis is a statement of equality. In many research situations, the inferences desired are those resulting from rejecting the null (equality) hypothesis. We can then infer that one group mean is higher or lower than another or that a relationship exists. In the mathematics achievement example, rejecting H_0: $\mu = 40$ led us to conclude based on the sample mean of 42.6 that current 11th-grade students demonstrated a higher achievement level. As this example shows, we usually infer positive implications in a study as a result of rejecting H_0. When we reject H_0, we want to be confident that in doing so, we are making a correct decision, that in fact *H_0 is false and should be rejected.* The probability of this outcome is called the **power** of a statistical test of the hypothesis H_0 and equals p(rejecting H_0 when H_0 is false). The power of a statistical test of the null hypothesis is the probability that we will correctly reject H_0 when it is truly false. Manipulating several design factors can increase the power when testing the null hypothesis so that *if H_0 is false,* the probability of rejecting it is high. Section 12.5 discusses the method for calculating power and the factors that affect it.

We have looked at the probabilities attached to the two consequences resulting from a decision to reject the null hypothesis. What if our sample result lies between our critical values and we therefore decide not to reject H_0? Again, two potential consequences result from our decision, one if H_0 is true and the other if H_0 is false. If we decide not to reject H_0 and H_0 is actually true, we have made a correct decision. When H_0 is true, the probability of making a correct decision—not rejecting H_0—equals $1 - \alpha$. This follows directly from the hypothesis testing model, which is built on the *assumption that H_0 is true.* The value of α determines the region of rejection. The remainder of the probability distribution between the critical values is the nonrejection region. When H_0 is true, we would expect $(1 - \alpha)\%$ of our sample means

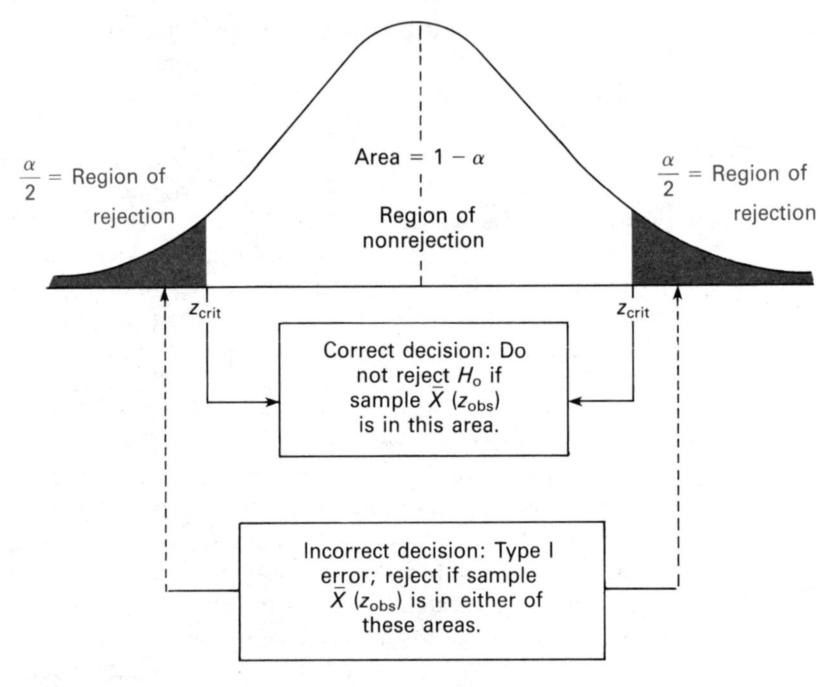

FIGURE 12.3
Decision probabilities when H_0 is true.

to fall someplace along the scale in the middle nonrejection region (see Figure 12.3).

Only one other decision consequence exists: deciding not to reject H_0 when, in fact, it is false and should be rejected. This decision results in a Type II error. Symbolically, the probability of making a Type II decision error is represented by the Greek letter beta, β.

$$\beta = p(\text{Type II error}) = p(\text{not rejecting } H_0 \text{ when } H_0 \text{ is false})$$

This decision error occurs when H_0 is actually false, but our sample results are insufficient to lead us to reject H_0. Even though H_0 is false, the sample result falls in the nonrejection region when testing H_0. The Type II error probability and the power probability are complementary. When H_0 is false, one of two outcomes still results from hypothesis testing—we either reject H_0 or do not reject H_0. If the probability of not rejecting H_0 based on the sample results equals β, then the probability of rejecting H_0 is $1 - \beta$ or the power of the statistical test of H_0.

Reexamining the Logic of Hypothesis Testing

So far, discussion of the decision process in hypothesis testing has focused on the consequences of the decision, that is, an error or a correct decision. We attached probabilities to each of the potential consequences. Actually, the probabilities are conditional probabilities dependent on the true state of affairs for H_0. They reflect the conditional probability that we will make a specific decision—to reject H_0 or not to reject H_0—*given* that H_0 is true *or given* that H_0 is false. The two conditional realities, H_0 true or H_0 false, are mutually exclusive events; their probabilities should be viewed and interpreted accordingly within hypothesis testing.

The logic of hypothesis testing starts with an assumed-true H_0 that we test using sample data. The sample results provide evidence we use to reject H_0 or not to reject H_0. The decision to reject H_0 or not to reject H_0 is the statistical decision of hypothesis testing. The rule for making this statistical decision is clear: Reject H_0 if the observed sample statistic lies in the region of rejection determined by the level of significance; otherwise, do not reject H_0. What are the probabilities that the researcher will reject H_0 or not reject H_0? This is where conditional probabilities enter into the decision process. If the true state of affairs is that H_0 is true, then α and $1 - \alpha$ are the controlling probabilities that determine the decision outcome. Figure 12.3 illustrated the operating conditions that are in effect and will determine the outcome of the test of H_0. These operating conditions are

- *When* the true state of affairs is *H_0 true,*
- *Then* the probability of deciding to reject H_0 is α, and to not reject H_0 is $1 - \alpha$

The operating conditions depicted in Figure 12.3 are the conditions assumed to be controlling the decision outcome during hypothesis testing. The researcher determines the level of α, which provides critical values for use in deciding whether to reject H_0 or not to reject H_0. By setting α at a low level (usually $\leq .05$), the researcher knows that when *H_0 is true,* H_0 will be correctly not rejected the vast majority of the time. In fact, H_0 will be correctly not rejected in the long run $(1 - \alpha)\%$ of the time.

What if H_0 really is false? The hypothesis testing model for decision making does not change. The model follows the procedures outlined for the operating conditions *when H_0 is assumed true.* The level of α is set, critical values are determined, the sample result is compared to the critical values, and a decision about H_0 is reached based on the comparison. Neither the hypothesis testing nor the decision process changes. The probabilities of making a decision to reject H_0 or not to reject H_0 do change, however. Using the hypothesis testing model for decision making, the probability of rejecting H_0 when H_0 is false is not the same as when H_0 is true. In fact, when H_0 is false, the model for testing H_0 makes it more probable that we will decide to reject H_0. This is the way a good model should work. When H_0 actually is false, the probability of rejecting H_0 (correct decision) should be greater than the probability of rejecting H_0 when H_0 is true (decision error). What is the conditional probability of deciding to reject H_0 when it really is false? Not surprisingly, it is the definition of power. When H_0 is false, the probability is $1 - \beta$ that the model will lead to the correct decision to reject H_0. Likewise, when H_0 is false, the probability that the model will lead to the decision error of not rejecting H_0 is β, the probability of making a Type II error. Section 12.5 will discuss these concepts in more detail and the factors that affect the probability levels of β and $1 - \beta$. An important point is that the levels of β and $1 - \beta$ *do not* affect the actual decision process in the hypothesis testing model. The levels set for α and $1 - \alpha$ *do* affect the process directly because the level of α determines the critical values used as criteria in deciding whether to reject H_0 or to not reject H_0. The importance of β and $1 - \beta$ is that they identify the true probabilities operating in hypothesis testing for rejecting H_0 $(1 - \beta)$ or not rejecting H_0 (β) *when H_0 is false.*

The logic of the hypothesis testing model for statistical decisions always produces conclusions about H_0 that have the higher probability of being correct. This is the model's advantage. Following the statistical decision process in testing H_0, what do we conclude if we decide to reject H_0 based on our sample results? We conclude that it must not be a correct description of the population. Why do we conclude that H_0 is false? Because the theory of the model used to test H_0 says that we are *more likely* to get a sample result leading us to reject H_0 when H_0 is actually false. Therefore, if the sample result leads us to reject H_0, we conclude that H_0 is false. We should have confidence that we have drawn a correct conclusion; the probabilities are in our favor.

12.4 DIRECTIONAL AND NONDIRECTIONAL HYPOTHESIS TESTING

In testing hypotheses about population parameters, we always test the hypothesis stated in null form. Based on sample results, we decide whether to reject the null hypothesis as a valid statement characteristic of the population. If we decide to reject H_0, we conclude that the null hypothesis statement of equality is probably false. If we conclude that H_0 is false, then we accept as true the alternative hypothesis as a complementary statement of inequality. As we have seen, alternative hypotheses as statements of inequality assume one of three relationships: $\neq$, $<$, or $>$. In testing H_0, the form of H_A has an important influence on the probability of making a correct decision to reject H_0 if H_0 really is false.

The three potential forms of H_A each imply an expectation about the true population state of affairs. An H_A using the sign $\neq$ (e.g., H_A: $\mu \neq 40$) indicates that we have no expectation or hypothesis about whether the true population mean is greater than or less than 40 if it differs from the H_0 value of 40. Formulating an alternative hypothesis using the $\neq$ results in a **nondirectional** alternative hypothesis. A nondirectional H_A does not specify whether we would expect the true parameter to be greater than or less than the null hypothesized value if H_0 is false. In contrast, using the inequality signs $<$ or $>$ in formulating H_A specifies an expected direction of the parameter below or above the H_0 value. Using one of the latter two inequality signs results in a **directional** alternative hypothesis.

The type of alternative hypothesis formed (directional or nondirectional) has implications in the hypothesis testing process. It identifies where we would anticipate sample results to occur if H_0 is not true, because if H_0 is not true, the random sample means will distribute themselves around the true population mean (μ_T). A nondirectional alternative hypothesis indicates that we cannot predict where the sample means are most likely to occur relative to the H_0 value of the population mean if H_0 is false. The implication is that when we test H_0, we need to provide for the potential possibility of the true population mean being either above or below the H_0 population mean (μ_0). If the actual mean is below the null hypothesis value, we are more likely to obtain sample means that would cause us to reject H_0 in the lower tail of the probability distribution. If it is above the null hypothesis value, more sample means will occur in the upper region of rejection. To provide for either of these eventualities, we conduct a **two-tailed** test of H_0 when H_A is *nondirectional.* A two-tailed test of H_0 will require us to provide for potential rejection of H_0 by establishing a critical value and a rejection region in both the lower and upper tails of the probability distribution. We halve the probability to determine the lower $\alpha/2$ critical value and the upper $\alpha/2$ critical value. Up to this point, all of our hypothesis testing examples have been two-tailed tests of H_0.

A directional H_A indicates that we anticipate the actual mean to be either above the null hypothesis mean (H_A: $\mu_T > a$) or below it ($H_A = \mu_T < a$). This anticipation implies that our sample mean will most likely cause us to reject H_0 in the upper tail for H_A: $\mu_T > a$ and in the lower tail for H_A: $\mu_T < a$. Because we can anticipate or predict the likely direction of the true mean relative to the hypothesized mean if H_0: $\mu_0 = a$ is false, we perform a **one-tailed** test of H_0 when H_A is *directional.* A one-tailed test of H_0 establishes a region of rejection and a critical value in only one tail of the probability distribution. The region of rejection and critical value for the one-tailed test of H_0 are determined by the area for the full value of α rather than by $\alpha/2$ as in a two-tailed test of H_0. Figure 12.4 illustrates the regions of rejection and critical values in the normal probability distribution for each of the three possible alternative hypotheses when $\alpha = .05$. Assuming a test of H_0: $\mu_0 = a$, what are the critical z score values for each of the following situations?[5]

1. H_A: $\mu \neq a$, $\alpha = .10$
2. H_A: $\mu < a$, $\alpha = .05$
3. H_A: $\mu > a$, $\alpha = .01$
4. H_A: $\mu \neq a$, $\alpha = .01$

---

[5]1. $z_{crit} = -1.645$ and 1.645
2. $z_{crit} = -1.645$
3. $z_{crit} = 2.33$
4. $z_{crit} = -2.575$ and 2.575

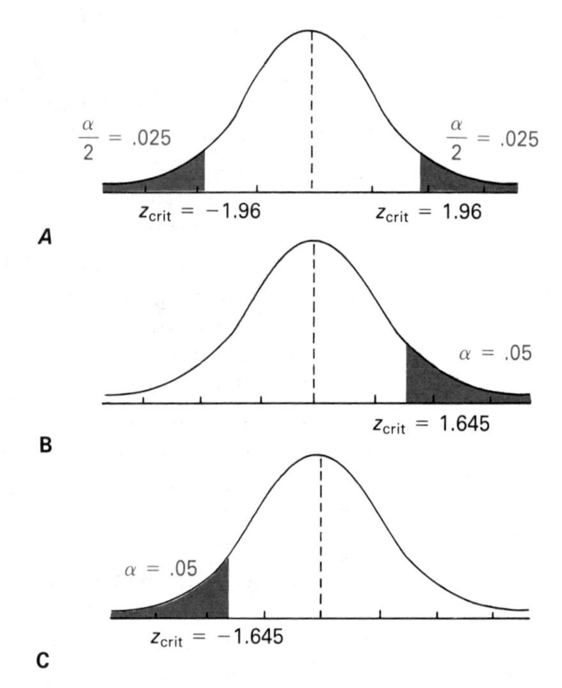

FIGURE 12.4
One- and two-tailed regions of rejection and critical values for $\alpha = .05$. **A**, H_0: $\mu_0 = a$, H_A: $\mu_T \neq a$; **B**, H_0: $\mu_0 = a$, H_A: $\mu_T > a$; **C**, H_0: $\mu_0 = a$, H_A: $\mu_T < a$.

As demonstrated in the examples, the effect of a directional or nondirectional H_A on the test of H_0 is in the resulting critical value used to reject H_0. A directional H_A results in a one-tailed test of H_0 and a lower critical value closer to μ_0 for testing H_0.

The mathematics achievement example provides a good example in which a directional H_A might be warranted. Based on the scores of 11th-grade students five years ago, a null hypothesis was formulated (H_0: $\mu = 40$). If the study is conducted because evidence exists that current students have achieved less, then we should use the directional alternative H_A: $\mu_T < 40$ to test H_0. If we just wanted to monitor the level of achievement in 11th-grade students and had no idea whether their achievement level would be up, down, or the same, the nondirectional alternative H_A: $\mu_T \neq 40$ would be used to test H_0. The nondirectional H_A: $\mu_T \neq 40$ would allow us to reject H_0 if the sample of students did worse or better than the hypothesized μ_0 value of 40. If the alternative hypothesis were H_A: $\mu_T < 40$, we could reject H_0 only if the current sample had a mean lower than 40.

12.5 FACTORS AFFECTING THE POWER OF A STATISTICAL TEST

When testing hypotheses, we naturally want to make statistical decisions about H_0 that lead to correct conclusions as often as possible. That is no problem when H_0 is true. As long as we employ proper research design procedures of sampling and measurement and do not violate any statistical assumptions when testing H_0, the hypothesis testing model indicates the probability of a correct decision. The probability of a correct decision when H_0 is true is $1 - \alpha$. In contrast, the probability of making a correct decision when H_0 is false is more elusive. This probability equals $1 - \beta$, the power of the statistical test of H_0. Calculating the power probability for a given test of H_0 poses a problem. To calculate $1 - \beta$, we must know the true population mean. For any one experiment, however, we never will know the exact value of the population mean, so we cannot calculate the exact value of $1 - \beta$. This really doesn't pose a serious problem in practice, however. We can do two things. First, we can calculate the power values for hypothetical μ_T values at fixed intervals along the scale. The graph of these power values as a function of how far μ_T deviates from μ_0 is called a power curve. Second, factors that affect power and that the experimenter has control over can be manipulated to increase the power of a given test of H_0, thereby increasing the probability of correctly rejecting H_0 if H_0 *is truly false.*

Power is the complement of the Type II error rate; a discussion of one automatically involves the other. Both are viable concepts only when H_0 is false. To understand power and Type II error probabilities, we need to visualize two different sampling distributions in the hypothesis testing process. We will call the first sampling distribution the **hypothesized sampling distribution** under an assumed-true H_0: $\mu_0 = a$. For a sampling distribution of means, the sample means will distribute approximately normally around the value μ_0. This is the sampling distribution we use to test H_0. When viewed as a probability distribution, the α level determines critical values as boundaries for the rejection region(s). The second sampling distribution is called the **true sampling distribution.** Sample means in this distribution distribute themselves exactly as the central limit theorem dictates, but they do so around the true population mean. The true sampling distribution thus depicts the true population mean; the hypothesized sampling distribution depicts the null hypothesized sampling characteristics. If the null hypothesis H_0: $\mu_0 = a$ is true, then the true sampling distribution and the hypothesized sampling distribution are the same. However, when the null hypothesis H_0: $\mu_0 = a$ is false, the two sampling distributions will be distinct distributions. The overlap between the two distributions is a function of how far μ_T is from μ_0. The shape and variability of each distribution is the same. Both are normally distributed with variance $\sigma_{\bar{x}}^2 = \sigma_x^2/n$. They differ only in location.

The overlap between the hypothesized and true sampling distributions can illustrate the probabilities associated with power $(1 - \beta)$ and Type II error (β). Recall that power is the probability of rejecting H_0 when H_0 is false. Power is the relative frequency (probability) with which sample means large or small enough occur (relative to μ_0) to reject H_0. Power is determined by the proportion of sample means in the true sampling distribution expected to lie beyond the hypothesized sampling distribution's critical values and in the rejection region. The shaded area of the true sampling distribution in Figure 12.5 illustrates power as a probability or area of the normal curve. The shaded area is the probability that a sample mean in the true sampling distribution will lie beyond the critical values established as the criteria for rejecting H_0, or $1 - \beta$. This is the power of the statistical test to reject H_0 when the population mean does not equal μ_0, but equals μ_T.

As the complement to power, the Type II error probability β is easy to illustrate. A Type II error occurs when H_0 is false, implying that $\mu_T \neq \mu_0$, but we decide not to reject H_0. We decide not to reject H_0 when the sample mean falls in the region of nonrejection between the critical values in the hypothesized sampling distribution. As with power, the frequency of this occurrence is defined by the proportion (area) of the true sampling distribution that overlaps into the region of nonrejection in the hypothesized sampling distribution. The middle, nonshaded area of the true sampling distribution normal curve in Figure 12.5 equals β, the probability of not rejecting H_0 when H_0 is false ($\mu_T \neq \mu_0$). Note that the areas depicting β and $1 - \beta$ are both in the true sampling distribution of means. The area in the regions of rejection is power $(1 - \beta)$ and the area in the region of nonrejection is β, the Type II error rate.

Computationally, we can find values for power and β probabilities for specific values of the true mean. Often power curves are generated by varying

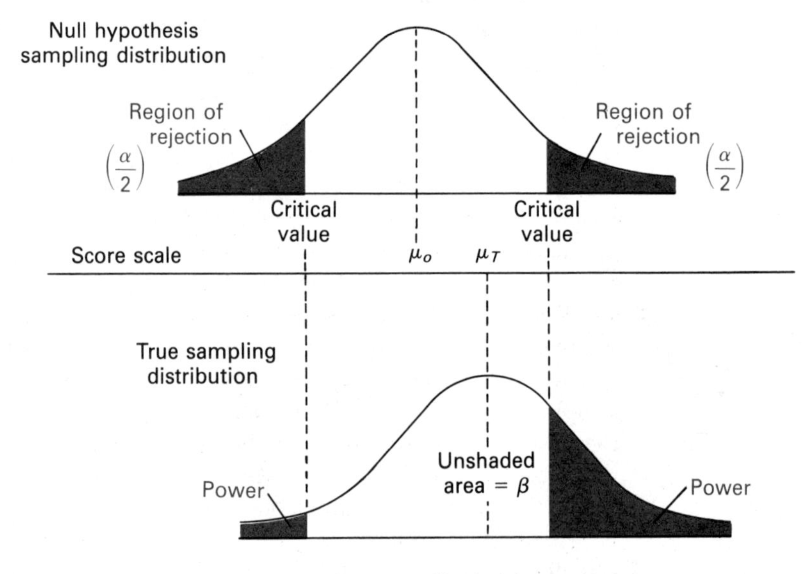

FIGURE 12.5
Illustration of power: Proportion of the true sampling distribution that lies beyond the rejection region.

the values of the mean across fixed intervals on a scale, calculating the power for each one, and plotting the values to form a curve. We will focus on the process of computing values of power and β for a fixed value of the true mean, not on the generation of power curves. Computing power and β should help understanding of these concepts in hypothesis testing. The easiest starting point is with the hypothesized sampling distribution critical values. Up to this point, the critical values for rejecting H_0 have been critical z score values determined by the level of α. To compute β and $1 - \beta$, we need to convert the critical z scores (z_{crit}) to critical mean values ($\overline{X}_{\text{crit}}$) in the hypothesized sampling distribution of means. The observed z score formula for testing H_0 can be converted algebraically to solve for $\overline{X}_{\text{crit}}$:

$$z = \frac{\overline{X} - \mu_0}{\sigma_{\overline{X}}} \longrightarrow z_{\text{crit}} = \frac{\overline{X}_{\text{crit}} - \mu_0}{\sigma_{\overline{X}}}$$

Solving for $\overline{X}_{\text{crit}}$,

$$\overline{X}_{\text{crit}} = \mu_0 + z_{\text{crit}}\sigma_{\overline{X}}$$

The value of $\overline{X}_{\text{crit}}$ gives the critical value equivalent of z_{crit} in terms of raw score sample means. The decision rule for testing H_0 is still the same: If the sample mean lies beyond $\overline{X}_{\text{crit}}$, reject H_0. Again we will use the mathematics achievement example to demonstrate the computations. The null hypothesis was H_0: $\mu_0 = 40$. The standard error of the mean was $\sigma_{\overline{X}} = .75$ for $n = 64$. If we assume an α level of .05 and a nondirectional H_A for testing H_0, the critical z scores are -1.96 and $+1.96$. Applying the formula for calculating $\overline{X}_{\text{crit}}$, the lower critical mean value for $z_{\text{crit}} = -1.96$ is

$$\text{lower } \overline{X}_{\text{crit}} = 40 + (-1.96)(.75) = 38.53$$

The upper critical mean value for $z_{\text{crit}} = 1.96$ is

$$\text{upper } \overline{X}_{\text{crit}} = 40 + (1.96)(.75) = 41.47$$

Using these critical values, if an observed sample mean is less than 38.53 or greater than 41.47, we would reject H_0: $\mu_0 = 40$. Now, assume that H_0: $\mu_0 = 40$ is false and that the true population mean μ_T is 42. The power of our test of H_0: $\mu_0 = 40$ given that the true mean equals 42 is the probability that either $\overline{X}_{\text{obs}} > 41.47$ or $\overline{X}_{\text{obs}} < 38.53$ *in the true sampling distribution* of means around $\mu_T = 42$ (see Figure 12.6). What is $p(\overline{X}_{\text{obs}} > 41.47)$ when $\mu_T = 42$ and $\sigma_{\overline{X}} = .75$? This is a normal curve probability problem.

$$z = \frac{41.47 - 42}{.75} = -.71$$

The area to the right of $z = -.71$ in the normal curve is .7611. The probability of getting a sample mean from the true population with $\mu_T = 42$ (which would exceed the critical value of 41.47 and result in rejecting H_0) is .7611. Because the test of H_0 was two-tailed, the potential for rejecting H_0 exists for

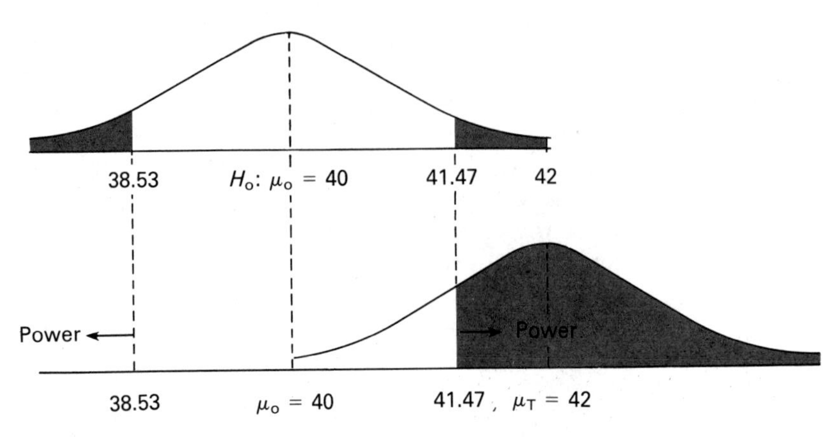

FIGURE 12.6
Power when testing H_0: $\mu_0 = 40$ given that $\mu_T = 42$.

sample means less than 38.53 also. What is the $p(\overline{X} \leq 38.53)$ given a true mean of 42 and a standard deviation of .75?[6] The probability of drawing a sample of size 64 from the population with $\mu_T = 42$ whose $\overline{X}_{obs}$ would be less than 38.53 is negligible. From the probabilities calculated for $\mu_T = 42$, the power of the statistical test of H_0: $\mu_0 = 40$ is .7611, indicating that if the true state of affairs were $\mu_T = 42$, we would decide to reject H_0: $\mu_0 = 40$ from approximately 76% of the potential random samples. The probability is that we would correctly reject H_0 76% of the time. The remaining 24% of the time, we would decide not to reject H_0 and make a Type II error ($\beta = .24$). What would the power of the test of H_0: $\mu_0 = 40$ be if $\mu_T = 39$?[7] The probability of rejecting H_0: $\mu_0 = 40$ when $\mu_T = 39$ is much less than when $\mu_T = 42$. The closer μ_T is to μ_0, the less the power. When $\mu_T = 39$, we will make a correct decision only about 26% of the time. The remaining 74% of the time we would expect to make a Type II error in our decision.

The last two examples demonstrate how the location of the true population mean relative to the hypothesized mean affects the values for power and β. This is one factor in testing H_0 that affects the probability of correctly rejecting it when it truly is false. When H_0 is false ($\mu_T \neq \mu_0$) we want to enhance the probabilities that we will detect the true population mean with

---

[6] $z = \dfrac{38.53 - 42}{.75} = -4.63$ The area in the normal curve to the left of $z = -4.63$ is very small (less than .0001).

[7] The critical mean values in the hypothesized sampling distribution are still 38.53 and 41.47. Changing each to a z score in the true sampling distribution with $\mu_T = 39$,

$$z = \frac{38.53 - 39}{.75} = -.63 \qquad z = \frac{41.47 - 39}{.75} = 3.29$$

The area to the left of $z = -.63$ is .2643. The area to the right of 3.29 is .0005. Power = .2643 + .0005 = .2648.

our sample estimate and reject H_0. The researcher does not have control over the value of the true mean, so it cannot be manipulated to increase power. The researcher does have control over other factors that affect the size of power (and β) and can be manipulated within the experiment to increase power and thereby decrease β. Increasing power within an experiment increases the probability that we will correctly reject H_0 rather than not rejecting H_0 (Type II error) when it is false.

The Location of the True Mean (μ_T) Relative to the Hypothesized Mean (μ_0)

It is important to understand how the relative distance of the true mean from the hypothesized mean affects power. As the scale distance between the two increases, the probability increases that the sample mean will fall in the region of rejection. Figure 12.7 demonstrates the increasing area of the true sampling distribution that lies beyond the critical mean values as the true mean increases in distance from the hypothesized mean. The magnitude of the difference between the two is directly related to the logic of rejecting H_0 if the sample mean exceeds the critical values for H_0. The larger the difference between them, the more likely a sample mean is substantially different from the hypothesized mean, so when a sample mean occurs beyond the critical values, we reject the hypothesized mean as the true population mean and conclude that the true mean must be some other value.

Increasing Precision: Increasing n or Reducing σ_x^2

Both the sample size (n) and the size of the population variance (σ_x^2) directly affect the size of $\sigma_{\bar{x}}$, the standard error of the mean. Reducing the standard error of the mean affects power in two ways, both as a function of increased precision. First, a smaller $\sigma_{\bar{x}}$ indicates that the sample means will vary less in both the hypothesized and true sampling distributions. The distributions will "shrink" toward the hypothesized and true means, respectively, creating less overlap. The sample means will be more precise estimates of the true population mean and will move closer to the true mean and away from the hypothesized one. Figure 12.8 illustrates the increase in power as $\sigma_{\bar{x}}$ is reduced in size. The area of the true sampling distribution beyond the critical values is greater when $\sigma_{\bar{x}}$ is smaller for the same value of μ_T. As we have seen previously, an increase in sample size will reduce $\sigma_{\bar{x}}$. Less variability in the population (smaller σ_x^2) also reduces $\sigma_{\bar{x}}$. The increase in precision has the effect of making the true mean further away from the hypothesized mean in standard error units. In raw score units the means are the same distance apart in both figures, but when $\sigma_{\bar{x}}$ is the smaller of the two examples, the true mean is a greater distance from the hypothesized one in standard score units, resulting in a higher power value.

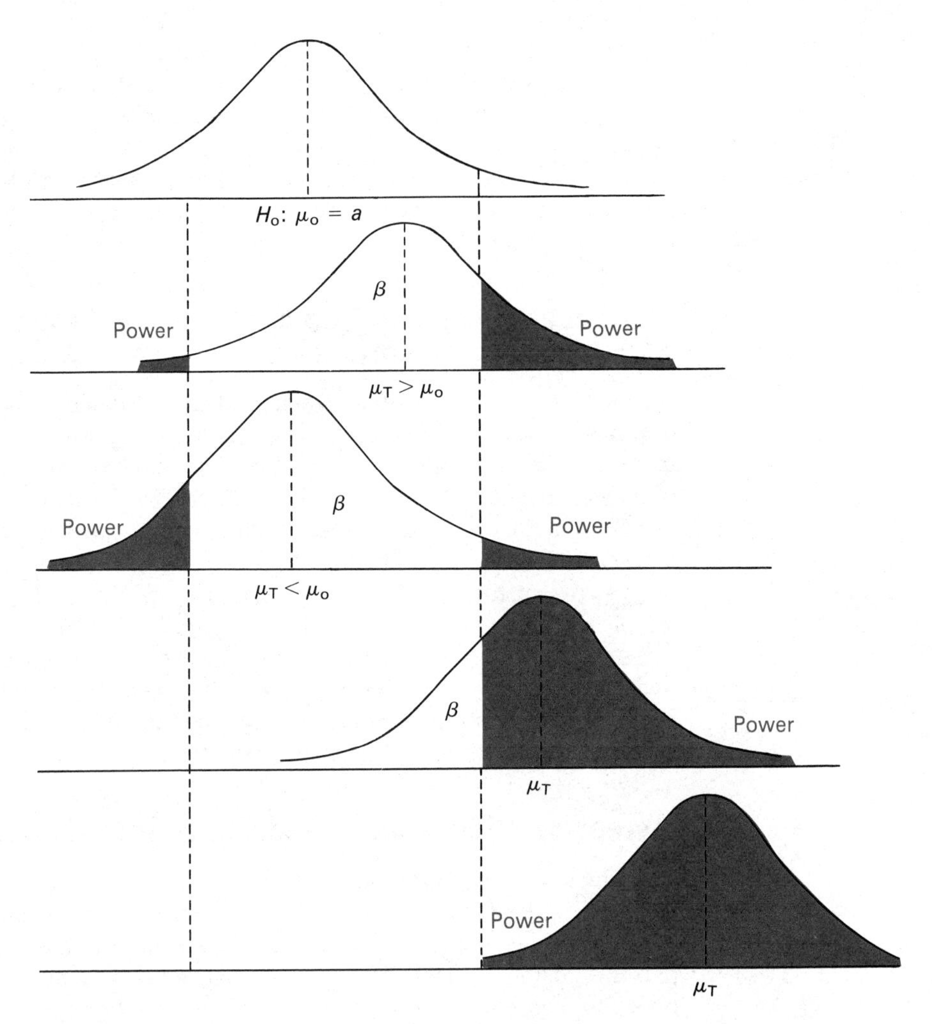

FIGURE 12.7
Effect of the relative distance of μ_T from μ_0 on power (α and n fixed).

In addition to bunching the hypothesized and true sampling distributions closer around the means, reducing the standard error of the mean also affects the critical value used to reject H_0. Note in Figure 12.8 that when $\sigma_{\overline{x}}$ is smaller, the value of $\overline{X}_{crit}$ is closer to the hypothesized mean and a greater distance from the true mean. This effect on $\overline{X}_{crit}$ is evident from the formula

$$\overline{X}_{crit} = \mu_0 \pm (z_{crit})\sigma_{\overline{x}}$$

If $\sigma_{\overline{x}}$ is reduced, the values for $\overline{X}_{crit}$ move closer to the hypothesized mean and

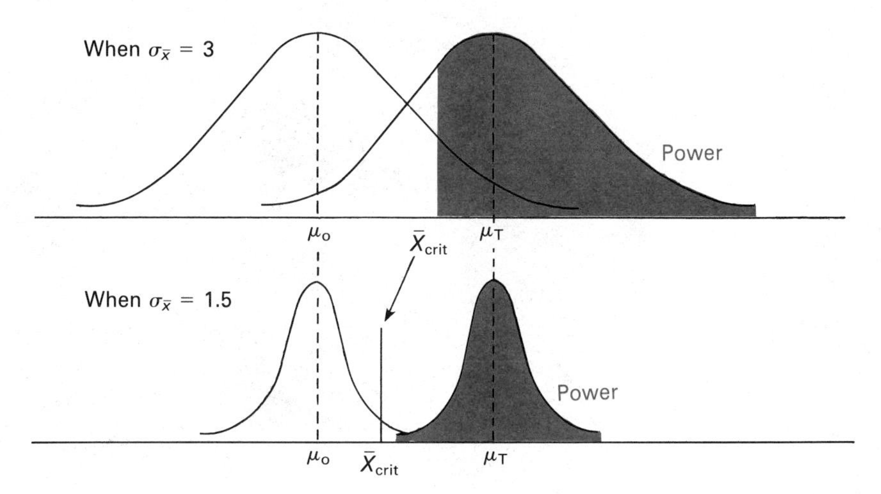

FIGURE 12.8
Increasing precision by reducing $\sigma_{\bar{X}}$ increases power.

away from the true mean. Because power is the area of the true sampling distribution that lies beyond $\bar{X}_{crit}$, it is increased as $\bar{X}_{crit}$ is farther away from the true mean.

To demonstrate how increased precision affects power, suppose we want to test the following hypothesis H_0: $\mu_0 = 50$ against the alternative H_A: $\mu \neq 50$ using an α level of .05. The population variance is $\sigma^2 = 100$. A random sample of 25 will be drawn to test H_0. What is the power of the test of H_0 if the true population mean is $\mu_T = 48$? Given the problem characteristics, the test of H_0 is a two-tailed test (H_A: $\mu \neq 50$), the critical z scores are ± 1.96 ($\alpha = .05$), and the standard error is $\sigma_{\bar{X}} = 10/25^{1/2} = 2.00$. The critical mean values are

$$\text{lower } \bar{X}_{crit} = 50 - (1.96)(2.00) = 46.08$$
$$\text{upper } \bar{X}_{crit} = 50 + (1.96)(2.00) = 53.92$$

Given these critical values, power is the area of the true sampling distribution around $\mu_T = 48$ that is less than 46.08 and greater than 53.92 (see Figure 12.9). To find the upper area, first change the upper $\bar{X}_{crit}$ to a z score.

FIGURE 12.9
Power as an area of the true sampling distribution ($\mu_T = 48$).

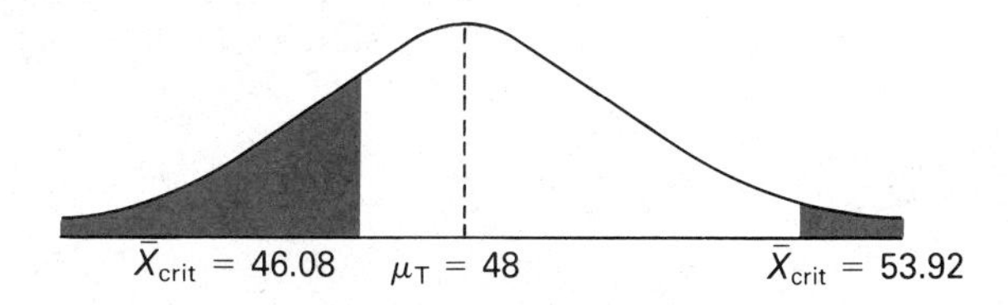

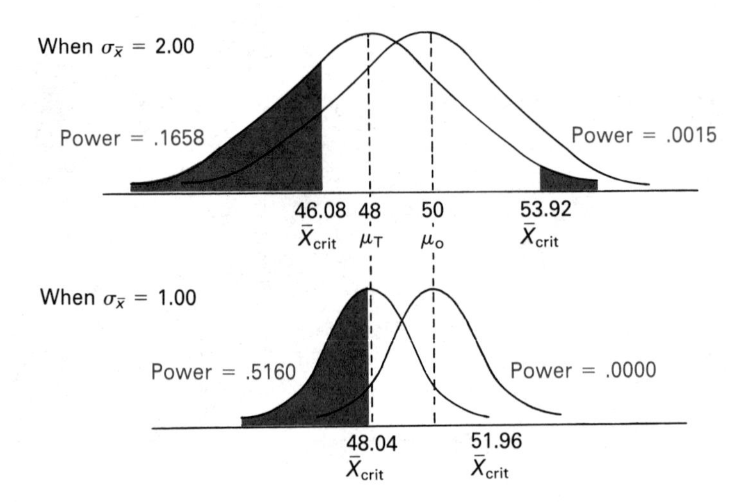

FIGURE 12.10
The effect of increased precision on the values of $\bar{X}_{crit}$ and power.

$$z = \frac{53.92 - 48}{2.00} = 2.96$$

The upper area equals $p(z \geq 2.96)$, which is .0015. The lower area is

$$z = \frac{46.08 - 48}{2.00} = -.96$$

and $p(z \leq -.96) = .1685$. The probabilities of finding a sample mean from a random sample of 25 with a value greater than 53.92 or less than 46.08 and given a true population mean of 48 is .0015 and .1685, respectively. The power for rejecting H_0: $\mu_0 = 50$ given $\mu_T = 48$ is the sum of the probabilities, $.0015 + .1685 = .1700$. The probability of getting a sample mean which would lead us to reject H_0: $\mu_0 = 50$ under the conditions outlined is .17, a 17% chance. The remaining 83% of the time, the sample mean would lead us to not reject H_0, resulting in a Type II error. What would the affect on power have been if we had drawn a random sample of 100, reducing the standard error ($\sigma_{\bar{X}} = 10/100^{1/2}$)?[8] The power is increased from .17 when $\sigma_{\bar{X}} = 2.00$ to .516 when $\sigma_{\bar{X}} = 1.00$. Increasing sample size increases precision, which in-

[8]Lower $\bar{X}_{crit} = 50 - 1.96(1.00) = 48.04$;
upper $\bar{X}_{crit} = 50 + 1.96(1.00) = 51.96$.

$$z = \frac{48.04 - 48}{1.00} = .04; \quad z = \frac{51.96 - 48}{1.00} = 3.96$$

$p(z \leq .04) = .5160$; $p(z \geq 3.96) = .0000$

power = .5160.

creases power. Figure 12.10 illustrates the effect on power of reducing the standard error by increasing n in the last example. The most obvious (and possibly best) way to increase power is to increase sample size.

Critical Values and Significance Level

We have seen that increasing precision through reducing $\sigma_{\bar{X}}$ results in critical mean values closer to the hypothesized mean and farther from the true mean in standard error units, thus increasing power. The same effect on the critical values will occur when α increases or when we make a one-tailed test of H_0 instead of a two-tailed test. The α level and the direction of the test of H_0 affect the critical value, thereby affecting the size of the power of the test of H_0. Any movement of $\bar{X}_{crit}$ closer to the hypothesized mean and farther from the true mean will increase power. For example, what happens to the critical value in testing any H_0 if we test H_0 at an α level of .10 rather than $\alpha = .05$? The critical values for a two-tailed test when $\alpha = .10$ are $z_{crit} = \pm 1.645$. When $\alpha = .05$, the critical values are $z_{crit} = \pm 1.96$. Power is increased by increasing the value of α. Testing H_0 at $\alpha = .10$ is more powerful than testing H_0 at $\alpha = .05$. Figure 12.11 illustrates the increase in power resulting from manipulating critical values by changing α levels in testing H_0. Manipulating α is not a good way to increase power, however. Remember that α is the probability of making a Type I error. To increase power, you must also increase the risk of making a decision error if H_0 is true. In most instances this is not a good trade-off.

One advantageous way to manipulate the critical value without changing the α level is to make a directional test of H_0 whenever legitimately possible. In

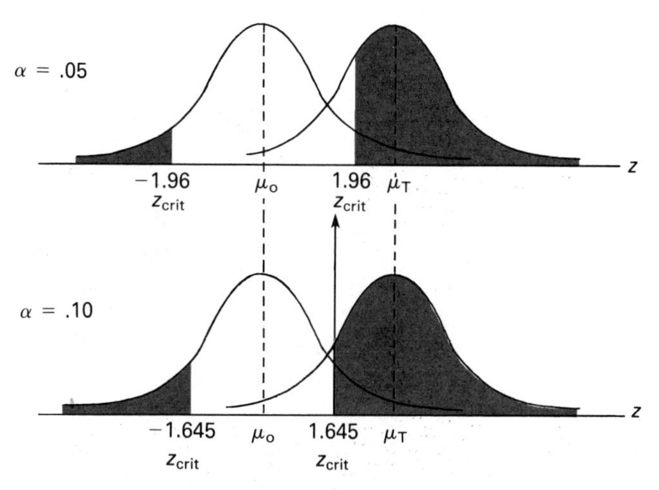

FIGURE 12.11
Effect of the significance level (α) on power.

a one-tailed test of H_0, we determine the critical value by putting the entire area defined by α in one tail of the probability distribution. A one-tailed test of H_0 at $\alpha = .05$ has about the same power as a two-tailed test at $\alpha = .10$. The critical value for rejection is $z_{crit} = 1.645$ in both instances. Making a directional test of H_0 when possible is a legitimate way to increase power without sacrificing the protection of a fixed α against making a Type I error. A directional test of H_0 will always be more powerful than a nondirectional test of H_0 for a fixed α level.

The Interaction of Factors Affecting Power

We have discussed five factors affecting the power of a statistical test of H_0. They are

1. The value of the true mean relative to the hypothesized mean: The greater the distance between μ_T and μ_0 either in raw score or standard error units, the greater the power.
2. Sample size: Increasing n reduces $\sigma_{\bar{x}}$, which increases precision, resulting in greater power.
3. Population variance: Sampling from a more homogeneous population (smaller σ_x^2) has the same effect on $\sigma_{\bar{x}}$, precision, and power as increasing n.
4. Level of significance: Larger α values result in greater power in testing H_0 when H_0 is false. When H_0 is true, however, the Type I error rate is larger as a result of the increased α.
5. Direction of the test of H_0: A one-tailed test of H_0 is more powerful than a two-tailed test of H_0 for the same α level.

We can see the interactive effects on power of these factors working in combination using an example. Table 12.4 describes three situations testing the same null hypothesis, H_0: $\mu = 32$, but with different values for H_A, μ_T, α, σ_x^2, and n.

TABLE 12.4
Three sets of values for the same null hypothesis.

	A	B	C
H_0	$\mu = 32$	$\mu = 32$	$\mu = 32$
H_A	$\mu \neq 32$	$\mu < 32$	$\mu > 32$
σ^2	36	100	36
n	16	64	64
α	.10	.05	.01
Assume $\mu_T =$	35	29.5	33.5

Two of the situations have approximately identical power in rejecting H_0 and both are more powerful than the third situation. Calculate the power for each.[9]

12.6 CONSIDERATIONS FOR DETERMINING SAMPLE SIZE

Probably novice researchers' most frequent question is "How large of a sample do I need?" The answer is not as simple as one might think. No one sample size is generally agreed to be sufficiently large or efficiently small to result in a valid test of H_0. We need to consider too many factors in estimating the sample size needed for a study. These factors preclude a simple single answer to the sample size question. Factors influencing hypothesis testing include the probability of a Type I error (α level), the power desired in rejecting H_0 if it is false ($1 - \beta$ or alternatively the probability of a Type II error, β), the size of the difference between μ_T and μ_0 that we want to be sure to detect, and the amount of variability in the population, σ_x^2. The size of the sample needed to test H_0 depends on values for these latter four factors, all of which can be set for a given study.

Determining sample size is a reversal of the process to determine power of β. The sample size, the α level, the relative distance of the true mean from

---

[9]For A,

$\sigma_{\bar{x}} = 1.5; \quad z_{\text{crit}} = \pm 1.645$

$\bar{X}_{\text{crit}} = 29.53 \quad \text{and} \quad 34.47$

$$p(\bar{X} \leq 29.53; \quad \mu_T = 35) = p(z \leq -3.65) = .0001$$

$$p(\bar{X} \geq 34.47; \quad \mu_T = 35) = p(z \geq -.35) = .6368$$

power = .6369.

For B,

$\sigma_{\bar{x}} = 1.25; \quad z_{\text{crit}} = -1.645$

$\bar{X}_{\text{crit}} = 29.94$

$$p(\bar{X} \leq 29.94; \quad \mu_T = 29.5) = p(z \leq +.35) = .6368$$

power = .6368.

For C,

$\sigma_{\bar{x}} = .75; \quad z_{\text{crit}} = 2.33$

$\bar{X}_{\text{crit}} = 33.75$

$$p(\bar{X} \geq 33.75; \quad \mu_T = 33.5) = p(z \geq .33) = .3707$$

power = .3707.

the sample mean, and $\sigma_{\bar{x}}^2$ were seen to affect the power of the test of H_0 if H_0 was false. We now reverse the process, fix the power level, and solve for n. Because sample size affects the precision of the test of H_0, we then ask "What size sample is needed to reject H_0: $\mu_0 = a$ using a specific α level with power equal to $1 - \beta$ when μ_T is a fixed distance from μ_0 and the population variance is $\sigma_{\bar{x}}^2$?" This a precision question. Precision is measured by the size of the standard error which equals $\sigma_x/n^{1/2}$. If $\sigma_{\bar{x}}^2$ is given, what size sample do we need to provide the precision necessary to reject H_0 with probability $1 - \beta$ given values of μ_T and α?

The problem is best solved by deriving a formula for n by algebraically manipulating the formulas for computing power values. Think through the process. For simplicity in developing the initial formula for n, we will assume a one-tailed test with the rejection region in the right tail. Power is determined by first calculating $\bar{X}_{\text{crit}}$ in the hypothesized sampling distribution for a given value of α.

$$\bar{X}_{\text{crit}} = \mu_0 + z_{1-\alpha}\sigma_{\bar{x}}$$

We then change $\bar{X}_{\text{crit}}$ to a z score in the true sampling distribution with a mean of μ_T. The area to the left of this latter z score is β and to the right is $1 - \beta$ or power. We will label the z score equivalent for $\bar{X}_{\text{crit}}$ in the true sampling distribution around the true mean z_β. Then,

$$z_\beta = \frac{\bar{X}_{\text{crit}} - \mu_T}{\sigma_{\bar{x}}}$$

Knowing z_β, we can solve for n using the following formula:

$$n = \frac{\sigma_{\bar{x}}^2(z_{1-\alpha} - z_\beta)^2}{(\mu_T - \mu_0)^2} \tag{12.3}$$

The quantities in formula 12.3 are illustrated in Figure 12.12 for the following situation:

$$H_0\text{: } \mu_0 = 40; \quad \alpha = .05; \quad \sigma^2 = 100$$

$$H_A\text{: } \mu_T > 40 \text{ (one-tailed test)}$$

Assume we want to determine the sample size necessary to detect a difference of three units ($\mu_T = 43$) between the true mean and the hypothesized one with power equal to .80 ($\beta = .20$). For this situation, $z_{1-\alpha} = 1.645$ and $z_\beta = -.84$.

$$n = \frac{100[1.645 - (-.84)]^2}{(43 - 40)^2}$$
$$= 68.61$$

We would need a sample size of at least 69 to ensure rejecting H_0: $\mu_0 = 40$ when $\mu_T > 43$ with 80% probability.

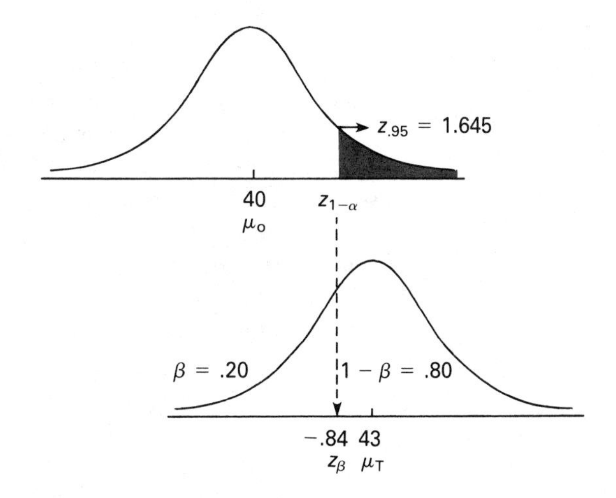

FIGURE 12.12
Quantities necessary to determine sample size.

In this case, z_β will always be negative when our desired power is greater than .50. A power value greater than .50 is not unreasonable because one would want β, the probability of a Type II error, to be less than .50. In practice, we usually set β at a value less than or equal to .20 which results in power $\geq .80$. If z_β is always negative, the formulas for n will result in always adding the two $z_{1-\alpha}$ and z_β values.

Formula 12.3 is specific to the case in which a one-tailed test is being made and the rejection region is in the right tail of the hypothesized distribution. If the region of rejection is in the left tail, $\overline{X}_{\text{crit}} = \mu_0 - z_{1-\alpha}\sigma_{\overline{x}}$ and the formula becomes

$$n = \frac{\sigma_x^2(-z_{1-\alpha} - z_\beta)^2}{(\mu_T - \mu_0)^2}$$

When our desired power is greater than .50, z_β will always be positive. Again, the values $z_{1-\alpha}$ and z_β will both be negative and will always be added. The sum of these two values will then be squared, resulting in only a positive quantity. Because of that, we can write a general formula, by summing the absolute values of $z_{1-\alpha}$ and z_β, $|z_{1-\alpha}| + |z_\beta|$, to handle both cases in which the rejection region is in the left or right tail of the distribution.

Another conventional change in the formula is to represent the difference between μ_T and μ_0 as d units: $d = \mu_T - \mu_0$. This generalizes the formula, since we need not actually know the values for the two means. Rather, the same size n can be determined for detecting a difference of d units with a given power. The generalized formula is

$$n = \frac{\sigma_x^2(|z_{1-\alpha}| + |z_\beta|)^2}{d^2} \qquad (12.4)$$

Applying formula 12.4, what sample size is needed to detect a difference of five units between the hypothesized mean and the true mean with power equal to

.90, an α level of .01, making a one-tailed test, and $\sigma^2 = 144$?[10] In contrast, if we had set an α level of .05, we would need a sample size of 50 ($n = 49.28$) to result in the same power. Check the calculation of n in the latter case.

If a two-tailed test is being made, the formula is the same, but $z_{1-\alpha}$ is changed to $z_{1-\alpha/2}$ to reflect the critical z score for a two-tailed test. This substitution assumes that the power value in one of the tails is negligible. We estimate the sample size assuming that all of the area determining power is relative to the critical value in one of the two tails. This assumption will be true for $\beta \leq .50$, which will exist in all practical situations. The formula for estimating the sample size needed in a two-tailed test of H_0 is

$$n = \frac{\sigma_x^2(|z_{1-\alpha/2}| + |z_\beta|)^2}{d^2} \quad (12.5)$$

What would the estimated sample size be for the last example if we had used a two-tailed test of H_0 ($d = 5$; $\alpha = .05$; $\sigma^2 = 144$; $\beta = .10$)?[11]

The size of the difference, d (between the true mean and the hypothesized mean), that we want to detect is the crucial determinant of sample size needed for a study. The size of this difference is usually based on the practical implications of rejecting H_0. Slight differences between the two means often are not practically meaningful. We can detect any difference between them, *no matter how small,* and reject H_0 if sample size is large enough. In the preceding example for a two-tailed test, a sample size of 24,187 would result in sufficient precision to detect a difference in μ_T and μ_0 as small as $d = .25$ 90% of the time. How practical is this and how meaningful is a difference as small as .25? In selecting sample size, we really want to ensure enough precision and thus sufficient power to detect a meaningful difference in the true and hypothesized means.

12.7 INTERPRETATIONS RESULTING FROM STATISTICAL DECISIONS

The critical values we use as criteria for statistical decisions about H_0 are determined by the level of significance (α), the probability of making a Type I error. When we reject H_0 using a specific value of α, we commonly say that the result is statistically significant at the α level. "Statistically significant at the .05

[10]The difference of $d = 5$ units could be ± 5 units. If it were a -5 units, μ_T would be less than μ_0 ($\mu_T - \mu_0 = d = -5$) and the region of rejection would be in the lower tail. In the latter case, the value of z_{crit} would be negative (-2.33) but $z_{1-\alpha} = 2.33$ would be positive in the formula. The same would be true for $z_\beta = -1.28$, but $|z_\beta| = 1.28$.

$$n = \frac{144(2.33 + 1.28)^2}{(-5)^2} = 75.06$$

[11]$$n = \frac{144(1.96 + 1.28)^2}{5^2} = 60.47$$

level" indicates that a sample mean ($\overline{X}$) occurred that resulted in rejection of H_0 when $\alpha = .05$ was the probability criterion level for determining the critical values. Research reported in the literature will often specify $p \leq .05$, indicating that the sample result had less than a 5% probability of occurring by chance (due to randomness) *if the statement H_0 identified the true characteristics of the population.* Rejecting H_0 at an α level leads us to infer or conclude that H_0 is false. When H_0 is rejected, $\overline{X}$ is considered a point estimate of the true population mean. The probability is higher that a sample mean falling in the rejection region in the hypothesized sampling distribution is an estimate of a population mean different from the hypothesized value. The conclusion based on the discrepant $\overline{X}$ is that H_0: $\mu_0 = a$ is false, that the true population mean differs significantly from the hypothesized mean at the α significance level.

The conclusion we draw when we decide not to reject H_0 is less clear. In a statistical probability sense, not rejecting H_0 does not mean we accept H_0 as true. From a probability standpoint, H_0 can never be *proven* true or false from sample data. The null hypothesis is often a hypothesis about a specific value of a population parameter. The variability expected as a result of random sampling makes it impossible to conclude that the population mean is some constant value. Rather than accepting H_0 as true if the decision is to not reject H_0, we usually conclude that the sample data were not sufficient to reject H_0 at the α level. While we can never prove that the null hypothesis is true, we can support it or that the true mean is very close to the hypothesized mean by increasing the value of α, say to .20 or .30. This effectively increases power and decreases β. By so doing, if H_0 is false, increased power will increase the probability of rejecting H_0. If you do not reject H_0 at an α level of .20 or .30, some support exists that H_0 may be true, or that the true mean is close to the hypothesized one.

The concept of randomness as a basis for expectations in sampling distributions and for the associated probabilities used as criteria in testing H_0 is crucial to the whole process. The hypothesis testing theory assumes randomness. The extent to which sampling and assignment of individuals to groups violates the assumptions of randomness reflects on the validity of decisions, inferences, and conclusions from the process. In practice, very few samples are actually selected at random from a defined population. If random sampling does not actually occur, the question in hypothesis testing is whether we can consider the sample data points used to test H_0 as a random sample from the population of concern. The sampling distribution of means only has the characteristics identified by the central limit theorem for assumed random samples. If representativeness through random sampling is not ensured in the sampling plan, then one has to guess to what population we can generalize the results of hypothesis testing. The assumptions of randomness in hypothesis testing suggest that this population is any one from which a sample could be considered randomly drawn.

12.8 SUMMARY

Statistical reasoning and procedures for making inferences about population parameters from sample data statistics were presented as the hypothesis testing process. Hypothesis testing applies statistical techniques and statistical reasoning to the inferential process.

The steps in testing hypotheses are constant and thus can apply to all hypothesis testing situations. The hypotheses will change and the statistical techniques applied will change depending on the hypothesis, but the process will not change. The general hypothesis testing process comprises the following steps:

- Step 1 Translate the research question into a testable null hypothesis (H_0) about a population parameter of interest.
- Step 2 Identify a directional or nondirectional alternative hypothesis (H_A).
- Step 3 Set a probability criterion level for rejecting H_0 by selecting a value for the level of significance α.
- Step 4 Calculate the sample statistic value needed to test H_0.
- Step 5 Convert the sample statistic value to an observed standard score based on the characteristics of the sampling distribution for the statistic. In this chapter, the observed standard score was

$$z_{obs} = \frac{\overline{X} - \mu}{\sigma_{\overline{x}}}$$

where

$$\sigma_{\overline{x}} = \frac{\sigma_x}{\sqrt{n}}$$

- Step 6 Determine the rejection region(s) by identifying the critical standard score value(s) z_{crit} resulting from the α level and H_A in the probability distribution appropriate for testing H_0.
- Step 7 Determine whether the sample value is likely when H_0 is true by comparing the observed sample standard score z_{obs} to the critical standard score value z_{crit}.
- Step 8 Decide to reject H_0 if z_{obs} falls in the rejection region of the probability distribution beyond values of z_{crit}; otherwise, do not reject H_0.
- Step 9 If H_0 is rejected, conclude that H_A is true knowing that rejecting H_0 if it is really true has an α probability of being a Type I decision error. If H_0 is not rejected, conclude that the sample data were not sufficient to reject H_0 at the α level.
- Step 10 Draw inferences and generalizations about the population relative to the initial research question based on your decision to reject or not to reject H_0.

PROBLEMS

Problem Set A

1. Which of the following would represent legitimate expressions of null hypotheses?
   a. $H: \bar{X} = 40$
   b. $H: \sigma^2 = 16.42$
   c. $H: \mu < 540$
   d. $H: \mu_x = 17.21$
   e. $H: \sigma_x^2 - \sigma_y^2 = 0$
   f. $H: \mu_x > \mu_y$
   g. $H: \rho_{xy} - .25 = 0$
   h. $H: \bar{X} = \bar{Y}$
   i. $H: s_x = 3.84$
   j. $H: \mu_x = \mu_y$

2. Which of the hypotheses in problem 1 could be legitimate expressions of alternative hypotheses? Write a compatible null hypothesis for each identified alternative hypothesis.

3. Several hypothesis testing situations are specified along with the true state of affairs concerning H_0. Test each hypothesis, then based upon the results decide whether an error or correct decision has resulted from the test of H_0. Classify the error by type when a decision error results.
   a. $H_0: \mu = 60$
      $H_A: \mu \neq 60$
      given: $\sigma = 8; n = 20$
      $\alpha = .10; \bar{X} = 56.9$
      $\mu_T = 60$
   b. $H_0: \mu = 60$
      $H_A: \mu < 60$
      given: $\sigma = 8; n = 20$
      $\alpha = .05; \bar{X} = 56.9$
      $\mu_T < 60$
   c. $H_0: \mu = 60$
      $H_A: \mu \neq 60$
      given: $\sigma = 8; n = 20$
      $\alpha = .05; \bar{X} = 56.9$
      $\mu_T = 60$
   d. $H_0: \mu = 60$
      $H_A: \mu < 60$
      given: $\sigma = 8; n = 20$
      $\alpha = .01; \bar{X} = 56.9$
      $\mu_T < 60$
   e. $H_0: \mu = 60$
      $H_A: \mu \neq 60$
      given: $\sigma = 15; n = 12$
      $\alpha = .05; \bar{X} = 67.8$
      $\mu_T = 60$

f. H_0: $\mu = 60$
H_A: $\mu > 60$
given: $\sigma = 15$; $n = 12$
$\alpha = .01$; $\overline{X} = 72$
$\mu_T > 60$

g. H_0: $\mu = 110$
H_A: $\mu \neq 110$
given: $\sigma = 12$; $n = 36$
$\alpha = .01$; $\overline{X} = 96$
$\mu_T = 110$

h. H_0: $\mu = 110$
H_A: $\mu > 110$
given: $\sigma = 12$; $n = 36$
$\alpha = .05$; $\overline{X} = 115$
$\mu_T > 110$

4. For each change specified, identify whether the change affects the probability of making a Type I error or the probability of making a Type II error and indicate whether the probability is increased or decreased. Note that neither error type may be affected.

a. H_0 is true, α is changed from .10 to .05.
b. H_0 is true, n is increased.
c. H_0 is false, n is decreased.
d. H_0 is true, σ^2 is increased.
e. H_0 is false, α is changed from .01 to .05.
f. H_0 is false, σ^2 is increased.
g. H_0 is true, α is changed from .01 to .05.
h. H_0 is true, H_A is directional rather than nondirectional.
i. H_0 is false, H_A is nondirectional rather than directional.
j. H_0 is false, α is changed from .05 to .10.

5. Which changes in problem 4 would affect the *power* of the statistical test? How would each change identified affect the power?

6. Determine which of the three hypothesis testing situations listed here would result in the most powerful test of H_0.

	Case A	Case B	Case C
H_0	$\mu = 30$	$\mu = 100$	$\mu = 70$
H_A	$\mu \neq 30$	$\mu > 100$	$\mu < 70$
μ_T	33	103	67
σ	5	12	8
n	20	40	30
α	.05	.01	.05

7. Calculate the power when testing H_0 for case B in problem 6 if only the following changes were made: (a) $\mu_T = 105$; (b) $\sigma = 10$; (c) $n = 20$; (d) $\alpha = .05$; (e) H_A: $\mu \neq 100$.

8. What factors affect the power of the test of H_0 in each part of problem 7? Summarize the effect each has on power based on the example.

9. What size sample would you need to reject H_0: $\mu = 70$ under case C conditions in problem 6 with power equal to .70? With power equal to .90?

10. What size sample would you need to reject H_0: $\mu = 30$ under case A conditions in problem 6 with power equal to .80? With power equal to .50?

Problem Set B

I. Use the summary data reported from the Streissguth (1983) study in Chapter 11, Problem Set B to respond to the following problems. This study examined the effects of alcohol use during pregnancy on the habituation and low-arousal behavior of newborn infants. Table 12.5 shows means and sample sizes per alcohol consumption (AA) category. Assume the population standard deviation for both behavioral variables was 4.0.

1. Assuming that the mean habituation score for the nonalcohol-consuming group is the population value if maternal alcohol consumption has no effect on infant habituation behavior, we can test the following hypothesis for each alcohol-consuming group:

- H_0: The mean habituation level of infants of mothers in the alcohol consumption Group ______ (AA interval that defines group) equals 17.0.

Symbolically, the four null hypotheses (one for each group) may be written (a) H_0: $\mu_1 = 17.0$, (b) H_0: $\mu_2 = 17.0$, (c) H_0: $\mu_3 = 17.0$, (d) H_0: $\mu_4 = 17.0$. Test each of the four hypotheses using the sample means and *n*s reported in the table. Assuming that H_A is $\mu < 17.0$ and $\alpha = .05$, what might you conclude from these tests?

2. If the true population mean habituation level for infants of alcohol-consuming mothers is 15.5 ($\mu_T = 15.5$), what type of decision resulted from each of your tests, a correct one or an error?

3. Calculate the power of the test of H_0 for group 2 and group 4 if $\mu_T = 15.5$.

4. What would have been a sample size sufficient to reject H_0: $\mu = 17$ if you had wanted to detect a difference of one unit between μ_0 and μ_T with a probability level of .50?

TABLE 12.5

Mean habituation and low-arousal scores of infants for different levels of maternal alcohol use (AA) during midpregnancy.

	Habituation		Low Arousal	
AA Interval	$\overline{X}$	N	$\overline{Y}$	N
0	17.0	88	9.3	74
(Group 1) .01–.10	16.0	103	8.8	102
(Group 2) .11–.99	15.9	119	9.2	107
(Group 3) 1.00–1.99	15.0	20	10.7	23
(Group 4) 2.00 or more	13.5	6	12.0	4
Total	16.1	336	9.2	310

5. Repeat the same analyses as in problem 1, but use the Low Arousal means reported. Also, let H_A be $\mu > 9.3$.

II. Use the data in Chapter 11, Problem Set B, Part II, for the following problems.

6. Test the hypothesis that the mastery students' mean achievement level was equal to 42.9. Use $\sigma^2 = 25$, $\alpha = .01$, and a nondirectional test to test H_0.

7. Test the same hypothesis, but assume that the scores in the population are less variable and that $\sigma = 3.0$.

8. Compute the power of each test of H_0: $\mu = 42.9$ in problems 6 and 7 if $\mu_T = 44.0$.

9. What sample size would we need to detect a 2-unit difference between the hypothesized mean and true mean with a probability of .75 under the conditions of problem 6? How would the sample size change to detect the same difference under the conditions of problem 7?

Problem Set C

1. The population mean on the "Sesame Street" Knowledge variable is 7.05 and the population variance is 9.00. A random sample of 8 individuals from the first 60 cases includes individuals 47, 24, 17, 38, 11, 57, 25, and 43. Record their "Sesame Street" Knowledge scores and use statistical inference procedures to test whether the sample mean is likely when random sampling from the population of scores. Use an $\alpha = .10$ and a nondirectional alternative hypothesis.

2. The mean posttest scores for individuals in viewing category 1 (VC1) in the "Sesame Street" data base may be viewed as the population parameters to be expected at the time of the posttest if no knowledge was gained from more frequent viewing of "Sesame Street" during the year. To formulate hypothesized values of population parameters, the posttest means for individuals in VQ1 on Body Parts (BP), Relational Terms (RT), and Classification Skills (CS) were calculated. For these individuals, $\mu_{BP} = 22.54$, $\mu_{RT} = 10.02$, and $\mu_{CS} = 12.90$. To test to see if more frequent viewing had any effect, a random sample of 10 was drawn from individuals in the third viewing category (individuals 003, 007, 029, 033, 034, 059, 067, 082, 105, and 145). Record the posttest scores for these individuals on BP, RT, and CS and use them to test the following hypotheses:

a. H_0: $\mu_{BP} = 22.54$
H_A: $\mu_{BP} > 22.54$
$\alpha = .05$
$\sigma = 5.4$

b. H_0: $\mu_{RT} = 10.02$
H_A: $\mu_{RT} \neq 10.02$
$\alpha = .05$
$\sigma = 2.73$

c. H_0: $\mu_{CS} = 12.90$
H_A: $\mu_{CS} > 12.90$
$\alpha = .01$
$\sigma = 5.06$

For each variable, what might you conclude based on the test of the hypothesis about the effect of more frequent viewing on posttest performance?

3. Draw a random sample of 12 individuals who have viewing category scores of 4 in the "Sesame Street" data base. Test the Body Parts null hypothesis in problem 2 using the BP scores for the individuals sampled.

4. If the true population BP mean (μ_T) for individuals in the fourth viewing category is 26.30, what is the probability that your sample mean in problem 3 should have resulted in rejecting H_0: $\mu = 22.54$?

13

Hypotheses about Population Means

Beginning with the discussion of probability, the previous three chapters have introduced the fundamental concepts of hypothesis testing and decision making using inferential statistics. Hypothesis testing is one of the primary roles of statistics within the total research process identified in Chapter 1. Given an identified need or area of investigation, we must define the variables of interest and purpose of the study. Testable hypotheses based on the purpose guide the study design in terms of sampling, data collection, and specific statistical procedures. Statistical techniques are chosen based on their appropriateness for testing the stated hypotheses. Hypotheses will differ depending on a particular study, (e.g., we may want to hypothesize differences in group means, variances, and proportions, or hypothesize relationships in terms of correlation coefficients). This and the following chapters describe specific statistical procedures for testing a variety of hypotheses.

13.1 DESIGNING RESEARCH FOR STATISTICAL INFERENCE

While choosing an appropriate statistical procedure is crucial to the research process, the other components cannot be overlooked. Statistical procedures are merely the tools we use to make decisions about the likely validity of hypotheses. The validity of the decisions and the generalizations of results (inferences) depend on the other facets in the research process as well. The definition and measurement of variables, sampling method, and how data are collected all interact to determine the population of variables, persons, and situations to which we can make inferences based on the results of statistical tests of hypotheses. As discussed in Chapter 12, the hypotheses tested employing appropriate statistical procedures are statements about population values (parameters). Research collects data from samples of individuals according to a research plan attempting to make inferences about the hypothesized population parameters from the observed sample indices. Because we are using results based on sample data, our inferences about the population values will not be exact. For any given sample, error or discrepancy will exist between the sample value and the population parameter it is estimating. As discussed in Chapters 11 and 12, the question becomes: "How large a difference can we tolerate before concluding that the sample was not drawn from a population with the hypothesized value?" This is where hypothesis testing procedures allow the researcher to decide whether to consider a specific discrepancy as a chance difference or whether it is real and not due to sampling error. We can determine the probability of a discrepancy occurring as large or larger than that between the sample index and the hypothesized population value if we know the characteristics of the sampling distribution of the sample index. When the sample value deviates so much from the hypothesized value that it is highly unlikely that this difference would have occurred by chance alone, the researcher concludes that the statement about the population value is not correct.

Translated into hypothesis-testing jargon, the first step is to identify the sampling (probability) distribution for the test statistic (ts) necessary to test the stated hypothesis. Second, set the level of significance that identifies the critical value of the test statistic (ts_{crit}), which determines the rejection region in the probability distribution. Next, calculate the necessary sample value. We then convert this sample value into standard score units for the probability distribution selected using the appropriate statistical formula. The index calculated using the statistical formula will give you an observed test statistic value (ts_{obs}). For testing many hypotheses, the statistical formula will take the following form:

$$\text{observed test statistic } (ts_{obs}) = \frac{\text{sample value} - \text{hypothesized population value}}{\text{standard error of the sample value}}$$

The standard error is the standard deviation of the sampling distribution of the specific statistic calculated to test the hypothesis. As can be seen in the formula, the numerator indicates how much in raw score units the sample value deviates from the hypothesized population value. We then change this deviation score or discrepancy to standard score units by dividing it by the standard deviation of the sampling distribution. The resulting standard score is the observed test statistic. It might be a z score, a t score, or some other standard score, depending on the particular probability distribution we use to test the hypothesis. So far we have only considered a z score as the test statistic ($ts = z$ score) for testing the null hypothesis H_0: $\mu_0 = a$.

Once we obtain ts_{obs}, we compare it to ts_{crit} in the probability distribution. If the absolute value of ts_{obs} is greater than the absolute value of ts_{crit} and is thus located in the rejection region, we conclude that the probability of the sample value occurring is too low for it to be a random sample from a population with the hypothesized value. The probability indicates that our sample is more likely a random sample from some other population; therefore, we reject the null hypothesis.

If you followed the steps in the last paragraphs, you realize that the general process for testing hypotheses does not change; instead, the appropriate test statistic and formula to use in calculating ts_{obs} change. The formula for testing many hypotheses also remains constant in its general form, but the values change depending on whether we are testing hypotheses about single sample means, two independent means, correlation coefficients, variances, or other indices. Specifically, the sampling distribution differs for each statistical index, meaning that the standard error (denominator in the general formula) changes when we test hypotheses about different values or situations. In addition, the appropriate test statistic (probability distribution) may change from one hypothesis to the next. With these alternatives in mind, we next consider specific hypothesis situations and the appropriate sampling distribution, test statistics, and formulas for testing a given hypothesis.

13.2 TESTING HYPOTHESES ABOUT SINGLE SAMPLE MEANS

The null hypothesis for this situation takes the familiar form of H_0: $\mu = a$, where μ represents the population mean, and a is a given hypothesized value. Hypotheses of this type are very rare in practice because most research considers more complex situations. Procedures for testing hypotheses about single sample means are given here, however, as a review and extension of material covered in Chapters 11 and 12. The research situation in which a test of H_0: $\mu = a$ would be desirable occurs when we have reason to believe that a population value equals our hypothesized value. If we draw a sample or have an available sample of individuals, we ask "Is our sample a representative or random sample from a population of individuals whose mean is the hypothesized value?" In deciding how to answer this question we use the hypothesis-testing model that asks "Will our sample mean deviate from the hypothesized population mean within a range we would expect by chance?" or "Will our sample mean deviate so far from the hypothesized population value that we conclude the probability is that it is not a chance deviation, but that our sample is more likely to be from some other population with a mean not equal to a (H_A: $\mu \neq a$)?"

To illustrate, two school-based examples identify situations in which the test of H_0: $\mu = a$ would be desirable. The first illustration involves a study in which the investigator wants to determine, using a district-wide standardized test of reading comprehension, if a sample of fourth graders from a school district is representative of the norm group. The norm group is assumed to be the population. The mean reading comprehension score for the norm group is 50; therefore, the null hypothesis is H_0: $\mu = 50$. The sample of fourth graders is then administered the reading comprehension test and the mean score is calculated. If the sample mean score deviates too much from 50 (too much is defined by the level of significance and critical values of the test statistic), then we would conclude that the sample is not representative of the hypothesized distribution, differing in its mean reading comprehension level from that of the norm group.

The second illustration is similar to the 11th-grade mathematics achievement example used in Chapter 12. Individuals' performance on a variable is compared with the performance of individuals from a prior period of time or another group. For example, we could compare the mean attitude toward education score for a sample of individuals against a known population mean from 20 years ago or for another population group. This latter mean score becomes the population value or baseline data point to formulate the null hypothesis. The current sample attitude toward education mean becomes the sample value. The question hypothesis testing asks is whether the current attitude toward education has changed from that of the population of 20 years ago or whether the current sample of scores can be considered part of the same population of attitude scores, implying no change. If we rejected the null hypothesis, the inference would be that the average attitude level has changed from 20 years ago. Rejecting H_0 would imply that the distribution of attitude scores has shifted.

Two test statistics exist for testing H_0: $\mu = a$. One is a z test, with which you are familiar. The z test as the test statistic uses the normal distribution as its probability distribution. The other appropriate test statistic is a t test, which uses the t distribution as the probability distribution for identifying ts_{crit} and the region of rejection. To determine when one test statistic is better suited than the other, we need to determine whether the population variance is known (z test is appropriate) or whether we must estimate the population variance from the sample data (t test situation). We will discuss each in turn.

The Single-Sample *z* Test

If we know the population variance, the test of H_0: $\mu_0 = a$ follows directly from the central limit theorem. The sampling distribution of means is normally distributed. The mean of the sampling distribution *if the null hypothesis is true* is the hypothesized value, $\mu_0 = a$, and the standard deviation is the standard error of the mean, $\sigma_{\overline{x}}$, which equals $\sigma_x/n^{1/2}$, the standard deviation of the population (σ_x) divided by the square root of the sample size. From the information about the sampling distribution, the observed test statistic formula is

$$ts_{\text{obs}} = z_{\text{obs}} = \frac{\text{sample value} - \text{hypothesized value}}{\text{standard error of the sample value}} = \frac{\overline{X} - \mu_0}{\sigma_{\overline{x}}}$$

The formula $(\overline{X} - \mu_0)/\sigma_{\overline{x}}$ transforms the sample mean, $\overline{X}$, to a standard score in the sampling distribution of means. The observed test statistic, ts_{obs}, is a standard score and for this hypothesis situation is normally distributed. Therefore, our observed test statistic is a z score and the probability distribution used for determining the critical values and region of rejection is the unit normal distribution whose values are given in Table B.2. In summary, to test a hypothesis about a single sample mean, H_0: $\mu_0 = a$, when the population variance is known, we use a z test formula:

$$z_{\text{obs}} = \frac{\overline{X} - \mu_0}{\dfrac{\sigma_x}{\sqrt{n}}}$$

The information necessary for testing the hypothesis is the sample mean, $\overline{X}$, the hypothesized value, $\mu_0 = a$, the population standard deviation or variance, σ_x or σ_x^2, and the sample size, n.

Let's return now to the example testing the null hypothesis that the population mean reading comprehension score from which the sample of fourth graders was drawn was equal to the norm group mean of 50 (H_0: $\mu = 50$). What other information about the population (norm group) would we need before we could use the z test formula as the observed test statistic? We need to know the population variance or standard deviation. Most standardized test manuals provide this information for the norm group, so it is likely that the population variance is available. Let's assume that $\sigma_x^2 = 100$ and therefore, $\sigma_x = 10$. What other information do we need before we can test the hypothesis? We need to know the sample size and the sample mean. Because we are collecting reading comprehension scores from the sample, we would know n and we would need to calculate $\overline{X}$. Let's assume that we have reading comprehension scores on 120 fourth graders and that the sample mean is 48.4. Do we need any other information before we can test H_0: $\mu = 50$? You should have identified two other pieces of necessary information, the level of significance α to be used and the alternative hypothesis H_A. Why are these necessary? The alternative hypothesis identifies whether the test is directional or nondirectional and when combined with the level of significance establishes the critical value and region of rejection. In this instance, we will specify $\alpha = .05$ and an alternative hypothesis of H_A: $\mu \neq 50$. You should now have enough information to test H_0: $\mu = 50$.

Steps to test H_0: $\mu = 50$ given H_A: $\mu \neq 50$

1. If H_0: $\mu = 50$ is true, the sampling distribution of sample means will have the following characteristics: The mean of the sampling distribution will be 50 and the standard deviation of the sampling distribution, $\sigma_{\overline{X}}$, will equal

$$\frac{\sigma_x}{\sqrt{n}} = \frac{10}{\sqrt{120}} = .913$$

2. Setting $\alpha = .05$ and identifying the alternative hypothesis as H_A: $\mu \neq 50$, the critical values are $z_{\text{crit}} = -1.96$ and $z_{\text{crit}} = +1.96$.
3. Using the formula to change $\overline{X} = 48.4$ to a standard score yields an observed z score of

$$z_{\text{obs}} = \frac{48.4 - 50}{.913} = -1.75$$

4. The normal curve in Figure 13.1A shows that z_{obs} does not lie beyond the critical values and thus is not in the region of rejection; therefore, we do not reject H_0: $\mu = 50$. The observed difference between the sample mean and μ_0 is considered to be a probable occurrence due to

FIGURE 13.1
Test of H_0 when σ^2 is known. **A**, H_0: $\mu = 50$ given $H_A \neq 50$; **B**, H_0: $\mu = 50$ given H_A: $\mu < 50$.

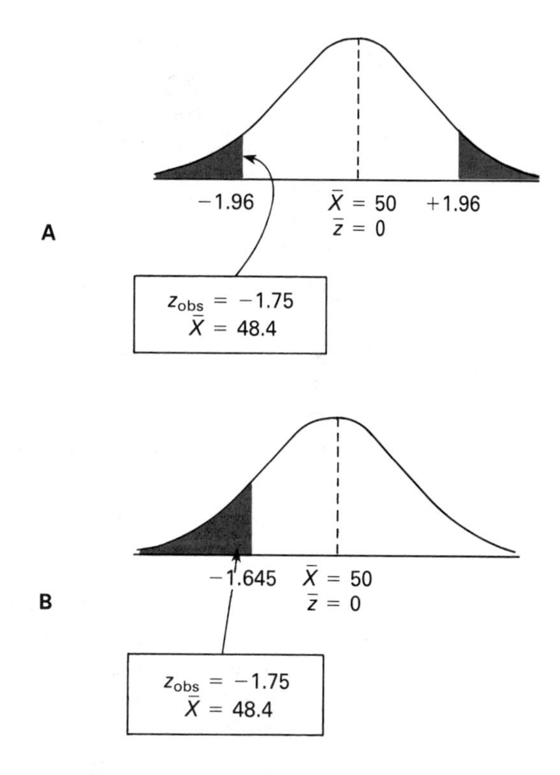

random sampling error. In not rejecting H_0, however, there is always the possibility that we have made a Type II error.

Would our decision to reject or not reject H_0 based on our sample data be different if we had had reason to suspect before our experiment that the fourth graders in the district were not up to par with the norm group. (H_A: $\mu < 50$)? The following list outlines the steps necessary to test the hypothesis. Figures 13.1A and B portray the decision-making process.

Steps to test H_0: $\mu = 50$ given H_A: $\mu < 50$ and $\alpha = .05$

1. The values from steps 1 and 3 for H_A: $\mu \neq 50$ remain the same in this situation. Make sure you understand why these values would not change.
2. Step 2 from H_A: $\mu \neq 50$ changes. The alternative hypothesis H_A: $\mu < 50$ is directional. The full 5% of the curve that constitutes the rejection region is now concentrated entirely in the left tail of the normal curve. The critical value becomes -1.645. This is the z_{crit} value that divides the curve into 5% and 95% proportions.
3. The normal curve in Figure 13.1B now shows that z_{obs} does lie beyond the critical value and in the rejection region. Therefore, we would reject H_0: $\mu = 50$ and conclude that H_A: $\mu < 50$ is more likely

the true state of affairs. The result of the test of the hypothesis is that the current fourth graders' mean score is significantly lower than the norm group's mean of 50 at the .05 level of significance. The inference is that the sample's fourth-grade reading comprehension scores are representative of some other population, a population in which scores tend to be lower on the average.

As a second illustration, suppose we examine the hypothesis testing process with a contrived example from the "Sesame Street" data. Assume that the null hypothesis to be tested is that the population mean Peabody mental age score (PMA) equals 48 (H_0: $\mu = 48$). Assume further that the directional alternative (H_A: $\mu < 48$) is specified for testing H_0 and that the level of significance is set at .05. Viewing all 240 children as a population, we know the true population mean and standard deviation are 46.78 and 15.99. Because we have contrived this unusual situation, we know in advance of any sampling or analysis that H_0 is false. To test H_0: $\mu = 48$, we have drawn a random sample of size 25 from Appendix A. The individuals sampled, their raw PMA scores, the raw-score sum, ΣX, and the sum of squared raw scores, ΣX^2, appear in Table 13.1. What would you expect the sample mean to be for the data in Table 13.1? For $n = 25$, what is the probability that $\overline{X}$ will be small enough to reject $\mu_0 = 48$ given that the true population mean (μ_T) is 46.78 (a question of power)? Asked in another form, what is the probability that $\overline{X}$ would fall in the region of nonrejection around $\mu_0 = 48$ and we would make a Type II decision error, not rejecting H_0 ($\mu_0 = 48$) when it is really false ($\mu_T = 46.78$)? The question on expectation is one of power and the Type II error probability. Figure 13.2 illustrates the probabilities for each in the situation described. Based on the probabilities, the likelihood is far greater that $\overline{X}$ for a random sample of size 25 will lead us not to reject H_0 even though it is false ($\beta = .8962$). There is not

TABLE 13.1
Random sample of Peabody mental age (PMA) scores ($n = 25$).

Student	PMA Score (X)	Student	PMA Score (X)	Student	PMA Score (X)
044	28	049	35	138	46
125	29	209	36	172	32
079	82	126	47	120	32
104	46	236	49	230	59
064	55	042	32	188	47
013	39	198	34		
021	55	058	37		
193	69	152	33	$\Sigma X = 1149$	
216	55	063	65	$\Sigma X^2 = 57{,}435$	
202	49	089	58		

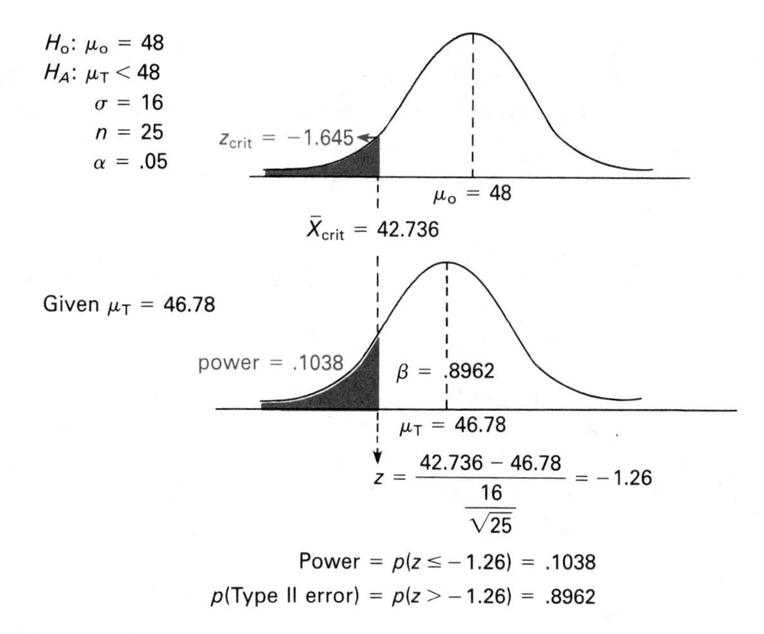

FIGURE 13.2
Type II error rate and power for test of PMA H_0: $\mu_0 = 48$ when $\mu_T = 46.78$.

sufficient power in the test of H_0 given $\mu_T = 46.78$ to expect that more than 10%–11% (power = .1038) of the sample $\bar{X}$s will lie in the region of rejection below $\bar{X}_{crit} = 42.736$ even though $\mu_0 = 48$ is false. Based on the value of $\bar{X}$ for the data in Table 13.1, would you correctly reject H_0: $\mu = 48$ or would the test of H_0 lead to a Type II error?[1]

The Single-Sample *t* Test

In the majority of instances when H_0: $\mu = a$ is the null hypothesis of interest, we are not likely to know the value of the population variance. We need some idea of its value because it is necessary in calculating the standard error of the mean. The central limit theorem states that the standard deviation of the sampling distribution of means equals $\sigma_x/n^{1/2}$. If σ_x^2 is not known, then we cannot calculate $\sigma_{\bar{x}} = \sigma_x/n^{1/2}$. If we do not know σ_x^2, our best bet is to use inferential statistics and estimate σ_x^2 using its sample point estimator, s_x^2. The sample variance is an unbiased estimator of σ_x^2, but it is only an estimate for a given sample and will vary from one sample to another due to sampling error. Over a large number of samples, the mean value of s_x^2 will equal σ_x^2 (it is an unbiased

[1] $\bar{X} = \frac{1149}{25} = 45.96$

A direct test of H_0 can be made by comparing $\bar{X}$ to $\bar{X}_{crit}$; 45.96 > 42.736. The sample mean lies in the region of nonrejection, therefore we would decide not to reject H_0 and make a Type II error.

estimator), but for any one sample, s_x^2 will deviate from σ_x^2 on a chance probability basis. The expected variability of sample variances is described by the sampling distribution of variances (see Table 11.4 and Figure 11.2).

If s_x^2 is used in place of σ_x^2 in the formula for the standard error of the mean, then $\sigma_{\bar{x}}^2$ is estimated by $s_{\bar{x}}^2$ and the standard error ($\sigma_{\bar{x}}$) by $s_{\bar{x}}$.

$$\sigma_{\bar{x}} = \frac{\sigma_x}{\sqrt{n}}$$

$$\text{Estimated } \sigma_{\bar{x}} = s_{\bar{x}} = \frac{s_x}{\sqrt{n}}$$

The estimated standard error of the mean will change depending on the variability of a specific sample. This variability in $s_{\bar{x}}$ as an estimate of $\sigma_{\bar{x}}$ from one sample to another affects the distribution of the observed test statistic, used in testing the hypothesis H_0: $\mu = a$. The observed standard score test statistic *when σ^2 is known* is a z score,

$$ts_{\text{obs}} = z_{\text{obs}} = \frac{\overline{X} - \mu_0}{\sigma_{\bar{x}}}$$

The z_{obs} scores will be normally distributed around a mean $\bar{z}_{\text{obs}}$ of zero when H_0 is true. The appropriate probability distribution to reference in setting critical values as criteria for testing H_0 when σ_x^2 is known is the unit normal distribution. When *σ_x^2 is not known* and $s_{\bar{x}}$ is used as an estimate of $\sigma_{\bar{x}}$, however, the observed test statistic calculated from the formula

$$ts_{\text{obs}} = \frac{\overline{X} - \mu_0}{s_{\bar{x}}}$$

where $s_{\bar{x}} = s_x/n^{1/2}$ is *not* normally distributed. The sampling distribution of means still is normally distributed, but when $s_{\bar{x}}$ is used in the formula to convert a sample mean to a standard score, the resulting distribution of standard score test statistics is defined by a t score distribution rather than a z score distribution. When σ_x^2 is unknown and s_x^2 from the sample in the study is used as an estimate of σ_x^2 in estimating the standard error, the resulting test statistic is a t score and the appropriate probability distribution to use when testing H_0: $\mu = a$ is a t distribution. The t score formula is the same as the z score formula, except $s_{\bar{x}}$ is substituted for $\sigma_{\bar{x}}$:

$$t_{\text{obs}} = \frac{\overline{X} - \mu_0}{s_{\bar{x}}} = \frac{\overline{X} - \mu_0}{\dfrac{s_x}{\sqrt{n}}} \tag{13.1}$$

The distribution of t scores is referred to as Student's t distribution after W.S. Gosset, who wrote under the pseudonym "Student." We can view the t distribution as a theoretical sampling distribution of t_{obs} scores or as a probability distribution similar in use as the normal. There is not just one t distribution, but rather a family of t distributions. The parameter that distinguishes one t distribution from another is a function of sample size and is called **degrees of freedom,** abbreviated ***df***. A different t distribution exists for every distinct number of degrees of freedom. We can readily determine the number of degrees of freedom for any null hypothesis for which a t distribution is appropriate. For example, when testing the single sample hypothesis H_0: $\mu = a$ and σ_x^2 is unknown, the t distribution to be used as the probability distribution has $n - 1$ degrees of freedom. The degrees of freedom for the single sample t test case is always the size of the sample selected minus 1 ($df = n - 1$).

We can conceptualize degrees of freedom in two ways. First, it represents the number of independent pieces of information on which a sample statistic is based as an estimator. For example, the sample mean is based on n independent raw score data points and has n degrees of freedom. Each of n data points provides an independent piece of information in the calculation of $\overline{X}$. The same is not true for the variance and standard deviation. The variance is defined by deviations of raw scores around the mean ($\overline{X}$), whose value is necessary to define and calculate s_x^2. If we know $\overline{X}$ for a sample set of data, all n scores are no longer free to vary because $\Sigma(X - \overline{X})$ must equal 0. Given the mean and $n - 1$ of the raw scores, the nth score is determined. It is the value necessary to make $\Sigma(X - \overline{X}) = 0$. In calculating s_x^2 as an estimate of σ_x^2, we lose one degree of freedom. Because the sample variance uses the sample mean in its calculation, only $n - 1$ data points provide independent pieces of information. The number of degrees of freedom associated with calculating s_x^2 is $df = n - 1$. The formula for the variance is the sum of squared deviations divided by its degrees of freedom. As a sample statistic, s_x^2 is an unbiased estimator of σ_x^2 when Σx^2 is divided by $df = n - 1$: $s_x^2 = \Sigma x^2/df$. Any time we use the mean or another statistic to estimate a population parameter in the calculation of a statistic (e.g., $\overline{X}$ in calculating the variance), we lose a degree of freedom. This is the second way to think of degrees of freedom. The number of df equals n minus the number of parameters estimated in calculating a sample statistic. As an additional example, the correlation coefficient r_{xy} uses both $\overline{X}$ and $\overline{Y}$ in its calculation. As an estimator of ρ_{xy}, it loses two degrees of freedom ($df = n - 2$). It should be easy to see why there are $n - 1$ degrees of freedom for the t test when testing H_0: $\mu = a$. In calculating the standard error as a sample statistic, we use the sample standard deviation, losing one degree of freedom.

Even though every distinct number of df defines a different t distribution, fortunately the shape, central tendency, and variability characteristics follow predictable patterns. The entire family of t distributions of concern here displays symmetrical, unimodal curves with a mean of 0, $\overline{t} = 0$. The variance of a t distribution with df degrees of freedom is $df/(df - 2)$, approaching 1 as df increases. All t distributions are slightly leptokurtic with the expected frequencies of t values more peaked around $\overline{t} = 0$ but with slightly more area in the tails of the distribution than the normal distribution. As *df* becomes larger and larger, the t distribution associated with *df* begins to approximate closer and closer the characteristics and areas of the normal distribution. The normal distribution is the theoretical limiting shape of the t distribution when *df* is infinitely large. The t distributions with $df = 1, 10$, and 25 are illustrated along with the normal distribution in Figure 13.3. Note particularly that the tails of the t distribution are higher than the tails of the normal distribution (i.e., the area identified in the tails of the t distribution is greater than in the normal distribution for identical values of t and z). One implication of this is that the critical values for t in hypothesis testing are larger in absolute value than critical values of z for a fixed α level.

When σ_x^2 is unknown and we use s_x to estimate the standard error of the sampling distribution of means, the t test formula (13.1) is the appropriate standard score formula for calculating the observed test statistic to test H_0: $\mu_0 = a$.

$$t_{obs} = \frac{\overline{X} - \mu_0}{s_{\overline{X}}}$$

The formula still transforms the sample mean ($\overline{X}$) to a standard score in the sampling distribution of means, but the standard score is no longer normally distributed. Because we are using $s_{\overline{X}}$ to estimate $\sigma_{\overline{X}}$, the standard score is distributed as a t distribution with $n - 1$ degrees of freedom.

The test statistic formula and the probability distribution change when we test H_0: $\mu = a$ with unknown σ_x^2, but the hypothesis testing process does not change. We still specify H_0 and H_A, set α as the criterion probability level for rejecting H_0, determine the critical values in the probability distribution associated with α for a directional or nondirectional test of H_0, calculate t_{obs}, and compare it to the critical values. If the t_{obs} value lies beyond the critical values (in the region of rejection) determined by the α level, then we reject H_0. Using the t test formula to calculate the observed sample test statistic implies that the t distribution is the appropriate probability distribution for establishing the critical values used in making decisions about H_0. Rather than finding z_{crit} values that partition the normal curve, we now need to find critical t values that partition the t probability distribution into α and $1 - \alpha$ areas. For a nondirectional test of H_0, t_{crit} values will partition the t distribution into areas

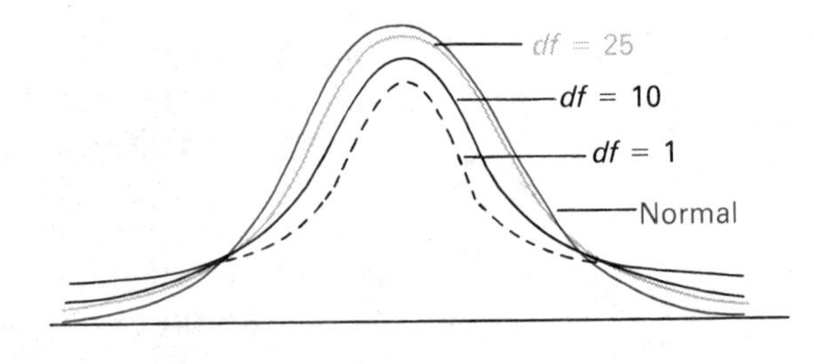

FIGURE 13.3
The t distribution with 1, 10, and 25 degrees of freedom and the unit normal distribution.

FIGURE 13.4
A t distribution—nondirectional test.

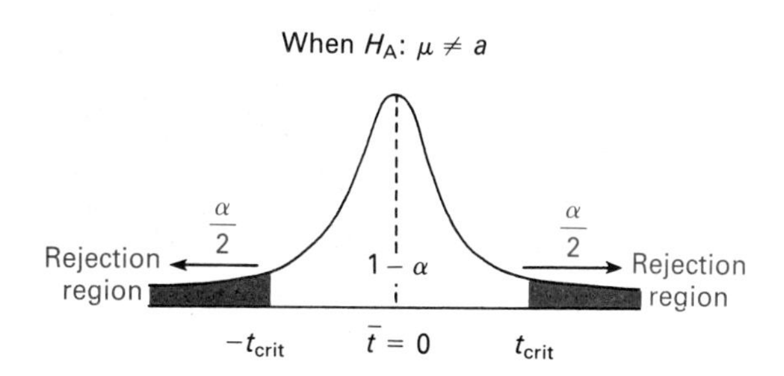

equal to $\alpha/2$ in the upper and lower tails of the t distribution and $1 - \alpha$ between the values of t_{crit}. Because the t distribution is symmetrical, the absolute values of the upper and lower critical values will be equal. As in the unit normal distribution in which the mean is 0, the lower critical value is negative and the upper critical value is positive (see Figure 13.4). For a directional test of H_0, t_{crit} is in one tail of the distribution that partitions the area of the t distribution into α and $1 - \alpha$ proportions (see Figure 13.5).

Finding a critical t value in the t distribution for a given α level is only slightly more complicated than finding z_{crit} in the normal distribution. The unit normal distribution is only one curve, so z scores and associated percentile areas can be tabled for z scores across the entire scale continuum. To do the same for the t distribution would result in a large number of tables, one for each distinct df number. To circumvent this problem, only the most used percentiles in the t distribution are tabled. These areas and equivalent t scores are located in the tail of the t distribution and correspond to selected values of α and $\alpha/2$ that are frequently used as criterion levels for testing H_0. The t values (t_{crit} values) for several levels of α and $\alpha/2$ appear in Table B.3, Appendix B for a full range of degrees of freedom, each defining a distinct t distribution. Only the upper percentile points (t values) appear in Table B.3. The symmetry of all t distributions makes it unnecessary to table the lower-tail critical values.

As an example, the t value representing the 95th percentile point in the

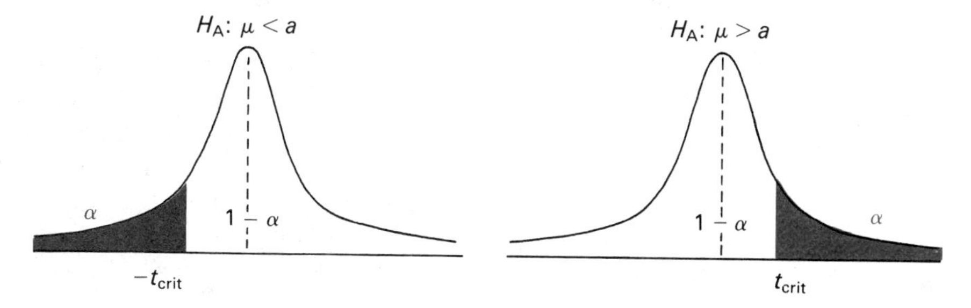

FIGURE 13.5
A t distribution—directional test.

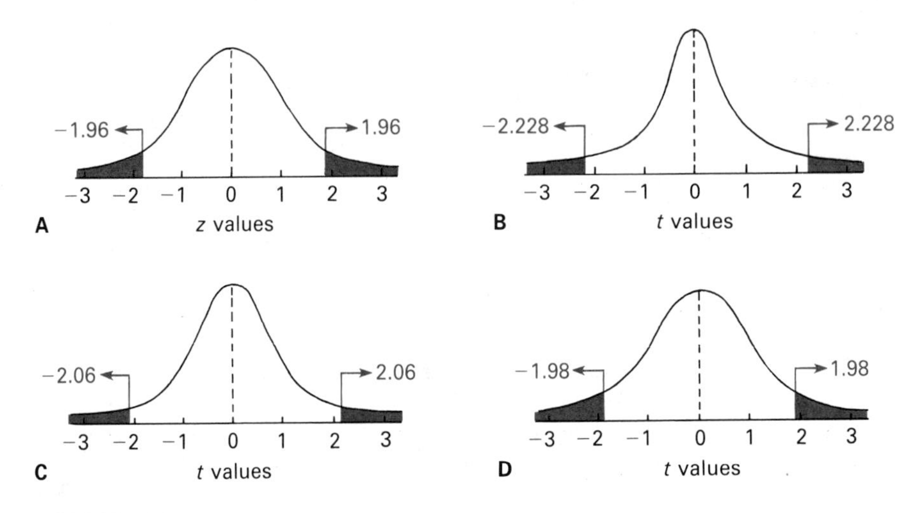

FIGURE 13.6
Critical values at $\alpha = .05$ for the unit-normal distribution and t distributions with $df = 10$, 25, and 100. **A,** unit-normal z distribution; **B,** t distribution ($df = 10$); **C,** t distribution ($df = 25$); **D,** t distribution ($df = 100$).

upper tail of a t distribution with 20 degrees of freedom is 1.725. Five percent of the area of the t distribution with $df = 20$ lies beyond a t value of 1.725. Similarly, the t value for the *5th* percentile point is -1.725. The t values of 1.725 and -1.725 would be the critical t values for making a two-tailed test of H_0: $\mu = a$ at the $\alpha = .10$ level of significance for a sample of size 21 ($df = 20$). What would the critical t values be if a sample of size 15 had been used to test H_0 under the same conditions?[2] Note that the values of t_{crit} do not change drastically for different sample sizes and degrees of freedom when α is fixed, but they do change from one t distribution to another to reflect the increasing areas in the tails as df decreases. The effect is that larger values of t_{crit} are needed to reject H_0 for smaller sample sizes because $|t_{crit}|$ is larger for smaller n. More samples are expected to have more deviant t values from smaller sized samples due to randomness when H_0 is true because of the need to estimate $\sigma_{\bar{x}}$ with $s_{\bar{x}}$. The greater areas in the tail of the t distribution reflect this expectation and the increased size of t_{crit} relative to z_{crit} is an adjustment in the t distribution to maintain a constant α level as the probability criterion level for testing H_0. In fact, the $|t_{crit}|$ will always be greater than $|z_{crit}|$ for the same α level. As n increases, t_{crit} will approach the value of z_{crit}. Note that when the degrees of freedom are infinite (∞), t_{crit} equals z_{crit} at all percentile points. Check this in Table B.3. Figure 13.6 illustrates the values of t_{crit} for a two-tailed test of H_0: $\mu = a$ at $\alpha = .05$ for sample sizes of 11, 26, and 101

[2] $df = n - 1 = 14$; The critical t values for a nondirectional test of H_0 with $\alpha = .10$ and 14 df are 1.761 and -1.761.

compared to the normal-curve z_{crit} values of ± 1.96 used to test the same hypothesis when σ_x^2 is known. What would the critical values be for each of the four distributions in Figure 13.6 if we made a directional test (H_A: $\mu < a$) of H_0 at the same α level?[3]

To illustrate use of the t test for testing H_0: $\mu = a$, consider the "Sesame Street" example for Peabody mental age scores testing H_0: $\mu = 48$ using the sample data in Table 13.1. Assume that the population variance of Peabody mental age scores is *not* known, that $\alpha = .05$, and that the alternative hypothesis is H_A: $\mu < 48$. Two pieces of additional information not used before with a z test are now required, the values of s_x and t_{crit}. From the summation values in Table 13.1,

$$s_x^2 = \frac{57435 - \dfrac{1149^2}{25}}{24} = 192.79$$

$$s_x = \sqrt{192.79} = 13.88$$

The s_x^2 value of 192.79 is merely a sample statistic value used to estimate σ_x^2. As expected due to randomness, it deviates from the true population variance value of $\sigma_x^2 = 255.68$. When used to compute $s_{\bar{x}}$, it yields an underestimate of the true standard error:

$$s_{\bar{x}} = \frac{13.88}{\sqrt{25}} = 2.78$$

$$\sigma_{\bar{x}} = \frac{15.99}{\sqrt{25}} = 3.20$$

Use of the t distribution corrects for the underestimation of $\sigma_{\bar{x}}$ expected in the majority of random samples (the sampling distribution of variances is positively skewed). The larger area in the tails of the t distribution (expected greater frequency of extreme t_{obs} values) results because of the more frequent underestimation of $\sigma_{\bar{x}}$. Using the value of $s_{\bar{x}}$ in formula 13.1,

$$t_{obs} = \frac{45.96 - 48}{2.78} = -.73$$

Is this result statistically significant at the .05 significance level?[4] Again we

---

[3]$z_{crit} = -1.645$; $t_{crit} = -1.812$ for $df = 10$; $t_{crit} = -1.708$ for $df = 25$; $t_{crit} = -1.66$ for $df = 100$.

[4]For a directional test of H_0 at $\alpha = .05$, $t_{crit} = -1.711$ with $df = 24$. The t_{obs} value of $-.73$ does not lie to the left of t_{crit}, so we do not reject H_0.

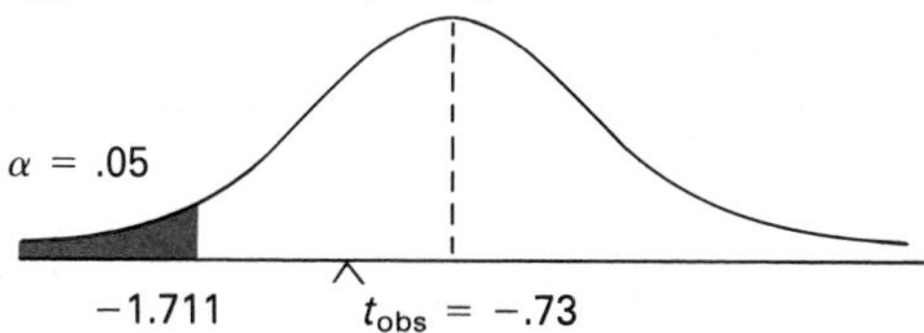

TABLE 13.2
Solution to t test of H_0: $\mu = 42$.

Individual	PMA Score	Calculations
012	58	$\Sigma X = 449 \quad \overline{X} = 44.9$
134	28	$\Sigma X^2 = 21997 \quad s_x = 14.29$
071	58	$s_{\overline{x}} = \frac{14.29}{\sqrt{10}} = 4.52$
114	43	
167	35	$t_{obs} = \frac{44.9 - 42}{4.52} = .641$
045	29	
088	59	For $\alpha = .01$, $t_{crit} = -3.25$ and $+3.25$
166	34	$\lvert t_{obs} \rvert < \lvert t_{crit} \rvert$; *do not reject* H_0
008	38	
109	67	

would decide not to reject H_0, but the test statistic, probability distribution, observed standard score, and critical value have changed from when we used a z test to test H_0, all due to the fact that we did not know the population variance and had to estimate it from our sample data.

For additional practice, try testing the null hypothesis, H_0: $\mu = 42$, using the data in Appendix A on PMA scores for the random sample of individuals 012, 134, 071, 114, 167, 045, 088, 166, 008, and 109. Use an alpha level of .01 and the alternative hypothesis, H_A: $\mu \neq 42$. Table 13.2 provides the solution to this problem. Notice that the standard error of the sampling distribution is larger for samples of size 10. The estimate of σ_x^2 is still reasonable for the random sample of $n = 10 (s_x^2 = 204.10, \sigma_x^2 = 255.68)$ but the smaller sample size increases the standard error, $s_{\overline{x}}$, indicating that we can expect more variability in sample means when $n = 10$ than when $n = 25$. Stated another way, only a sample mean more deviant from the hypothesized mean would be statistically significant for samples of size 10 given the same alpha level. In any event, the test of H_0: $\mu = 42$ is not powerful enough to reject H_0, even though the true population mean is quite discrepant from μ_0 ($\mu_T = 46.78$). Again, we end up making a Type II error.

13.3 TESTING HYPOTHESES ABOUT TWO INDEPENDENT SAMPLE MEANS

The hypothesis testing situation to be covered in this section occurs frequently in the literature. It applies to numerous research situations and questions, including

1. Does instructional method X produce greater achievement outcomes on the average than method Y?

2. Is reading comprehension equal on the average if students are taught to read using a sight versus a phonetic method? Rephrasing the question, does a group taught to read by a sight method differ in reading comprehension from a phonetic method–instructed group?
3. Is the mean aggression score for first-grade boys different in classes with a male teacher than for those with a female teacher?
4. Do children classified as impulsive differ in mean self-concept from reflective children?
5. Do adolescent girls exhibit greater conformity on the average than adolescent boys?

Notice that all these examples have one thing in common. Each identifies two different, independent groups on which data will be collected. The groups may be two representative samples who will receive different treatment conditions, as in the first two examples. Or the groups may not actually receive a treatment, but they were sampled and put into a group because they differ on some important characteristic. In the last two examples, individuals would not be sampled, assigned to groups, and then administered a treatment. The treatment is not induced, but is already a characteristic that differentiates the two groups. In the fourth example, children are not randomly assigned to groups and then made either impulsive or reflective; rather, they are sampled into a group based on the impulsive-reflective characteristics they already possess.

The first two and last two examples illustrate the difference between experimental studies that manipulate samples of individuals by subjecting them to a treatment and comparative studies that do not manipulate groups to make them different—rather, they differ on some characteristic and are assigned to groups based on that characteristic. Example 3 identifies a situation that could be experimental or comparative depending on how the researcher designed the study. If boys were randomly assigned to male and female teachers at the beginning of first grade, then the teacher's sex becomes the experimental treatment. If no random assignment took place at the beginning of the year, however, and students were assigned to groups at the end of the year based on whether they had a male or female teacher, then the study is comparative. Space does not permit a complete discussion of the advantages and disadvantages of each type of study. The reason for distinguishing between them is to create the awareness that both types exist. As long as two groups exist that are independent of each other, the test statistic for testing hypotheses about differences in their means is the same.

The Sampling Distribution for the Difference between Two Independent Sample Means

The null hypothesis formulated to compare two independent group means is a statement of equality or no difference. Symbolically, we can write it as H_0:

$\mu_x = \mu_y$ (equality) or H_0: $\mu_x - \mu_y = 0$ (no difference). To test the null hypothesis of no difference, we calculate the means for group X and group Y. We then subtract the two means $(\overline{X} - \overline{Y})$ to see how discrepant they are. In hypothesis testing we ask "How large a difference between two random sample means can we expect due just to chance random sampling error?" If the difference is too large (beyond what would be expected by chance) then we would reject the null hypothesis of equality and conclude that the groups differ—that, in fact, they are not representative random samples drawn from the same population. Rather, each must be a sample from populations with different means. We ask now "How do we decide whether a difference between $\overline{X}$ and $\overline{Y}$ is probably a chance difference or a real difference due to differential treatments or group characteristics?"

Again, the hypothesis testing process does not change. To decide whether a difference in $\overline{X}$ and $\overline{Y}$ is large enough to give us confidence in rejecting H_0: $\mu_x = \mu_y$, we need to know the characteristics of the sampling distribution for differences in independent group means, $\overline{X} - \overline{Y}$. To illustrate how the sampling distribution of $\overline{X} - \overline{Y}$ might be generated, think through the following process. Draw two random samples, X and Y, of size n_x and n_y, respectively, from an identified population of scores. The two samples need not be of the same size. Calculate the means $\overline{X}$ and $\overline{Y}$, on the criterion variable for each sample, then subtract the sample means to see how discrepant they are. For any given set of two random samples, calculate the mean difference score $(\overline{X} - \overline{Y})$. Continue drawing pairs of random samples of size n_x and n_y, calculating their mean difference scores until you have generated a large number of difference scores, one for each pair of random samples drawn. These difference scores between group means may now be graphed in the form of a frequency polygon. The resulting distribution described by the frequency polygon is the sampling distribution of differences between independent group means. Both sample X and sample Y in each pair were random samples from the same population, so the sampling distribution of $\overline{X} - \overline{Y}$ scores describes what to expect about the frequency of occurrence of $\overline{X} - \overline{Y}$ difference scores from two independent random samples. Note that all the mean difference score values in the frequency polygon would be chance occurrences due to random sampling error alone.

To help visualize this sampling distribution, we have empirically generated a partial sampling distribution for differences in mean Numbers posttest scores from the "Sesame Street" data in Appendix A. One hundred pairs of random samples X and Y, each having 12 scores, were drawn. Table 13.3 presents the means for each sample pair along with the difference score. Figure 13.7 displays the frequency distribution and the frequency polygon of the 100 difference scores.

To identify the characteristics of the sampling distribution, the mean and standard deviation of the difference scores appear at the bottom of Table 13.3. As indicated, the mean of the sampling distribution is approximately zero,

TABLE 13.3
Generated sampling distribution for the difference between two independent samples means.

Sample Pair	$\overline{X}$	$\overline{Y}$	$\overline{X} - \overline{Y}$	Sample Pair	$\overline{X}$	$\overline{Y}$	$\overline{X} - \overline{Y}$	Sample Pair	$\overline{X}$	$\overline{Y}$	$\overline{X} - \overline{Y}$
001	32.17	32.33	−.16	034	30.58	31.75	−1.17	067	30.33	29.08	1.25
002	26.00	27.50	−1.50	035	33.17	28.17	5.00	068	30.83	34.67	−3.84
003	33.42	34.00	−.58	036	32.25	31.83	.42	069	28.75	29.67	−.92
004	31.83	24.17	7.66	037	35.17	35.17	.00	070	32.33	34.50	−2.17
005	32.58	26.42	6.16	038	34.17	27.83	6.34	071	27.67	31.92	−4.25
006	26.25	31.09	−4.84	039	29.75	28.25	1.50	072	32.00	23.25	8.75
007	25.75	36.25	−10.50	040	34.83	23.33	11.50	073	29.92	29.58	.34
008	30.42	30.00	.42	041	31.17	32.25	−1.08	074	19.50	33.08	−13.58
009	24.75	31.42	−6.67	042	36.50	31.83	4.67	075	29.00	26.17	2.83
010	38.09	27.50	10.59	043	32.50	28.83	3.67	076	27.58	31.08	−3.50
011	29.00	29.67	−.67	044	31.83	37.42	−5.59	077	38.17	37.42	.75
012	30.09	33.25	−3.16	045	27.67	26.25	1.42	078	34.50	32.42	2.08
013	33.50	25.67	7.83	046	29.64	39.08	−9.44	079	30.83	29.75	1.08
014	30.09	28.17	1.92	047	28.08	34.67	−6.59	080	26.25	24.42	1.83
015	24.92	32.92	−8.00	048	32.17	30.00	2.17	081	28.25	32.92	−4.67
016	35.92	30.00	5.92	049	31.83	27.00	4.83	082	25.67	33.33	−7.66
017	27.84	33.84	−6.00	050	24.25	27.17	−2.92	083	32.67	31.33	1.34
018	25.50	27.17	−1.67	051	27.67	25.92	1.75	084	37.58	27.83	9.75
019	32.09	26.67	5.42	052	26.67	30.75	−4.08	085	28.08	31.17	−3.09
020	26.25	31.50	−5.25	053	25.92	25.42	.50	086	28.25	32.33	−4.08
021	35.92	36.42	−.50	054	36.42	29.50	6.92	087	31.17	33.58	−2.41
022	35.59	31.92	3.67	055	34.92	33.42	1.50	088	32.17	29.42	2.75
023	26.09	33.92	−7.83	056	31.92	30.00	1.92	089	30.17	28.00	2.17
024	35.75	29.84	5.91	057	34.17	22.92	11.25	090	32.33	24.00	8.33
025	28.00	30.42	−2.42	058	20.33	31.67	−11.34	091	28.42	28.83	−.41
026	30.00	29.64	.36	059	26.08	32.58	−6.50	092	30.75	29.25	1.50
027	32.17	28.42	3.75	060	35.92	33.33	2.59	093	36.00	29.17	6.83
028	26.50	25.50	1.00	061	28.83	29.17	−.34	094	27.75	31.25	−3.50
029	25.30	26.25	−.95	062	32.42	31.00	1.42	095	29.25	37.25	−8.00
030	27.09	28.09	−1.00	063	29.50	28.50	1.00	096	29.58	26.33	3.25
031	27.30	28.75	−1.45	064	33.75	27.92	5.83	097	27.50	31.58	−4.08
032	26.25	34.17	−7.92	065	27.83	28.67	−.84	098	26.42	23.75	2.67
033	24.75	26.08	−1.33	066	29.75	28.83	.92	099	30.00	33.50	−3.50
								100	26.92	25.75	1.17

Mean of $\overline{X}$ = 30.11 Mean of $\overline{Y}$ = 30.07 Mean of $\overline{X} - \overline{Y}$ = .04
$s_{\bar{x}}$ = 3.75 $s_{\bar{y}}$ = 3.47 $s_{\bar{x}-\bar{y}}$ = 5.04

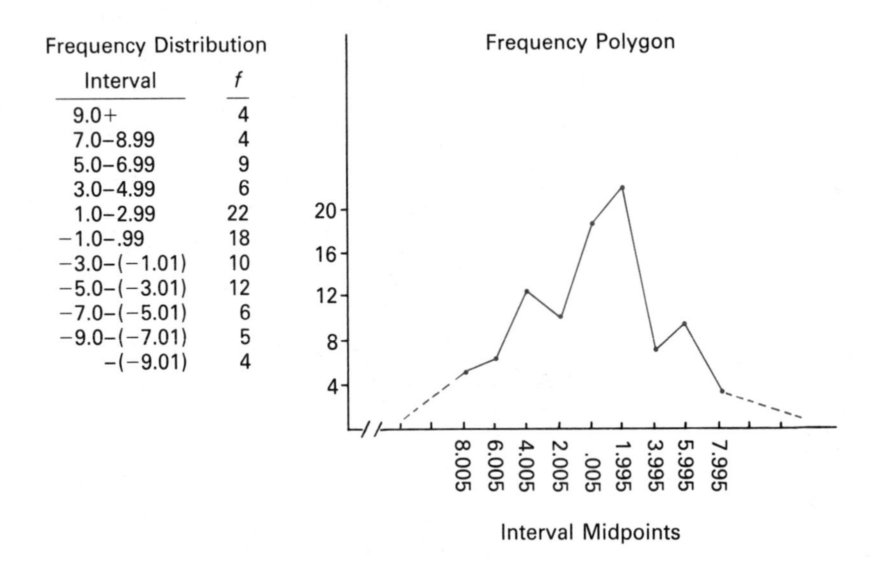
Frequency Distribution

Interval	f
9.0+	4
7.0–8.99	4
5.0–6.99	9
3.0–4.99	6
1.0–2.99	22
−1.0–.99	18
−3.0–(−1.01)	10
−5.0–(−3.01)	12
−7.0–(−5.01)	6
−9.0–(−7.01)	5
−(−9.01)	4

FIGURE 13.7
Frequency distribution and polygon for mean differences for 100 independent random sample pairs.

which we should expect. Randomly, $\overline{X}$ will be greater than $\overline{Y}$ for some sample pairs and less than $\overline{Y}$ for others as both samples were drawn at random from the same population. Because we always subtract $\overline{Y}$ from $\overline{X}$, we should expect the difference scores to fluctuate between negative and positive values at random, thus balancing each other out in the long run and leaving an average difference score of zero, the null-hypothesized value, (i.e., H_0: $\mu_x - \mu_y = 0$). This is only true when the null hypothesis is correct and the difference really is zero in the population. If each sample is representative of populations whose means really do differ, that is, $\mu_x \neq \mu_y$, then the differences in sample means will deviate at random around the true population difference.

As with single samples, the standard deviation of the sampling distribution is the standard error for the sample statistic and indicates how much we can expect difference scores to vary by chance alone. The sampling distribution and the standard deviation of the difference in mean scores ($s_{\overline{x}-\overline{y}}$) are appropriate only for samples of size 12 and the variability of the Numbers posttest scores that serve as our population data. If the samples drawn were larger, each sample mean would be a more precise estimate of the population mean. Therefore, the difference in sample means, $\overline{X} - \overline{Y}$, should be a more precise estimate of the population difference, thus making the standard deviation of the differences in means smaller. Likewise, the larger the criterion variable variability in the population, the larger the standard error. In both cases, the reverse is also true.

Fortunately, we need not empirically generate the sampling distribution each time we change the sample size or criterion variable. The mean and standard deviation of the sampling distribution for differences in two independent group means have been derived mathematically in the same manner as that described by the central limit theorem. The mean of the sampling distribution for samples of any size will be the actual value of $\mu_x - \mu_y$. If the null hypothesis H_0: $\mu_x - \mu_y = 0$ is true, then the mean of the sampling distribution is zero, and sample values of $\overline{X} - \overline{Y}$ will distribute themselves around zero at random in a known probability distribution. If H_0 is false, the sample values will distribute themselves around the true value of $\mu_x - \mu_y$.

To estimate the extent to which the difference between mean scores $\overline{X} - \overline{Y}$ will vary, we estimate $\sigma_{\overline{x}-\overline{y}}$, the standard deviation of the sampling distribution of $\overline{X} - \overline{Y}$. The estimated standard error of the mean differences for two independent group means is $s_{\overline{x}-\overline{y}}$. The size of $s_{\overline{x}-\overline{y}}$ indicates the amount of variability in $\overline{X} - \overline{Y}$ scores expected due to randomness in the sampling distribution. It is analogous to the standard error of the mean $(s_{\overline{x}})$ for single samples.

If σ_x^2 and σ_y^2 are known, then the standard error of the sampling distribution for differences between independent group means is

$$\sigma_{\overline{x}-\overline{y}} = \sqrt{\sigma_{\overline{x}}^2 + \sigma_{\overline{y}}^2} = \sqrt{\frac{\sigma_x^2}{n_x} + \frac{\sigma_y^2}{n_y}}$$

Because σ_x^2 and σ_y^2 are rarely known in practice, however, an unbiased estimate of $\sigma_{\overline{x}-\overline{y}}$ is usually obtained from sample information. The following formula for $s_{\overline{x}-\overline{y}}$ provides an unbiased estimate of $\sigma_{\overline{x}-\overline{y}}$.

$$s_{\overline{x}-\overline{y}} = \sqrt{s_{\overline{x}}^2 + s_{\overline{y}}^2} = \sqrt{\frac{s_x^2}{n_x} + \frac{s_y^2}{n_y}}$$

The true standard deviation of the sampling distribution of differences in independent group means, $\sigma_{\overline{x}-\overline{y}}$, is estimated by $s_{\overline{x}-\overline{y}}$. We can calculate the true standard error $\sigma_{\overline{x}-\overline{y}}$ directly if we know the population variances, σ_x^2 and σ_y^2, but we seldom know them. Therefore, we use the sample variances, s_x^2 and s_y^2 to calculate $s_{\overline{x}-\overline{y}}$, providing an estimate from sample data of the expected variability of $\overline{X} - \overline{Y}$ in the sampling distribution. While the formula for $s_{\overline{x}-\overline{y}}$ provides an unbiased estimate of $\sigma_{\overline{x}-\overline{y}}$, a more stable estimate can be made if we assume that the variances of the populations from which $\overline{X}$ and $\overline{Y}$ were sampled are equal. If we assume that $\sigma_x^2 = \sigma_y^2$, then we can pool (add) the sample sums of squared deviations, Σx^2 and Σy^2, and sample degrees of freedom to provide a more stable estimate of the equal population variance, σ^2. We assume homogeneity of population variances $(\sigma_x^2 = \sigma_y^2)$ in testing two independent sample means. The formula for $s_{\overline{x}-\overline{y}}$ in this case becomes

$$s_{\bar{x}-\bar{y}} = \sqrt{\frac{s_p^2}{n_x} + \frac{s_p^2}{n_y}} = \sqrt{s_p^2\left(\frac{1}{n_x} + \frac{1}{n_y}\right)} \tag{13.2}$$

where

$$s_p^2 = \frac{(n_x - 1)s_x^2 + (n_y - 1)s_y^2}{n_x + n_y - 2} = \frac{\Sigma x^2 + \Sigma y^2}{(n_x - 1) + (n_y - 1)}$$

$$= \frac{\left[\Sigma X^2 - \dfrac{(\Sigma X)^2}{n_x}\right] + \left[\Sigma Y^2 - \dfrac{(\Sigma Y)^2}{n_y}\right]}{n_x + n_y - 2}$$

Information from the population data in Appendix A, from the empirically generated sampling distribution in Table 13.3, and from a single sample pair in Table 13.3 gives us three different methods for obtaining the sampling distribution standard deviation (the standard error of $\bar{X} - \bar{Y}$). For samples of size 12, we can calculate the standard error of $\bar{X} - \bar{Y}$ using the known population variance for the Numbers posttest scores on all 240 children from Appendix A. The population variance of Numbers posttest scores is 158.71 and the standard error of $\bar{X} - \bar{Y}$ for $n_x = n_y = 12$ is

$$\sigma_{\bar{x}-\bar{y}} = \sqrt{\frac{158.71}{12} + \frac{158.71}{12}} = 5.14$$

The value 5.14 represents the standard deviation of the sampling distribution for differences between two independent group means if we sample an infinite number of sample pairs. For 100 random sample pairs generated in Table 13.3, the standard deviation of the difference scores $(\bar{X} - \bar{Y})$ is 5.04. This empirically calculated standard error is very close to the population value of 5.14.

Each of the preceding methods for obtaining the standard error of $\bar{X} - \bar{Y}$ is presented for conceptual understanding. In practice, we rarely know variances of the population from which the two independent samples were drawn, eliminating the first method as a possible alternative. The latter method of generating the sampling distribution empirically is not practical. Therefore, formula 13.2 becomes the most efficient, practical method for estimating the standard deviation of the sampling distribution of $\bar{X} - \bar{Y}$. The process is reasonably simple. Without the population variances or generating the sampling distribution, we calculate $s_{\bar{x}-\bar{y}}$ directly from the data for the two independent samples. To illustrate, consider sample pair 020 from Table 13.3. The standard deviation for the 12 scores in sample X was 13.17 and in sample Y, 12.71. The pooled variance is

$$s_p^2 = \frac{(12 - 1)(13.17)^2 + (12 - 1)(12.71)^2}{12 + 12 - 2}$$

$$= \frac{1907.94 + 1776.99}{22} = 167.50$$

Substituting in formula 13.2 yields the following value for the standard error of $\overline{X} - \overline{Y}$:

$$s_{\bar{x}-\bar{y}} = \sqrt{167.50\left(\frac{1}{12} + \frac{1}{12}\right)} = 5.28$$

As can be seen, the estimate of the exact value is reasonably close for the sample pair selected. With random sampling, extreme over- and underestimates will occur infrequently by chance. For larger samples the estimates would be more precise.

The *t* Test for Independent Sample Means

With independent sample means, the test statistic formula again follows the general format for testing all hypotheses. Calculate the sample statistic necessary to test H_0 (in this case $\overline{X} - \overline{Y}$). Determine how far the sample value deviates from the hypothesized population values, $(\overline{X} - \overline{Y}) - (\mu_x - \mu_y)$. Transform the difference between the sample and population values to an observed standard score (ts_{obs}) by dividing by the standard deviation of the sampling distribution, $s_{\bar{x}-\bar{y}}$. Decide to reject H_0 based on whether ts_{obs} lies beyond the critical value(s) of the standard score probability distribution determined by α and H_A.

For testing H_0: $\mu_x = \mu_y$ for independent groups, the appropriate test statistic (observed standard score) formula is a *t* test. The *t* distribution is the appropriate probability distribution for determining the critical value. The *t* test formula to be used for testing H_0: $\mu_x = \mu_y$ is

$$t_{\text{obs}} = \frac{(\overline{X} - \overline{Y}) - (\mu_x - \mu_y)}{s_{\bar{x}-\bar{y}}}$$

From the null hypothesis H_0: $\mu_x = \mu_y$, the value of $\mu_x - \mu_y$ in the formula is zero.

$$t_{\text{obs}} = \frac{(\overline{X} - \overline{Y}) - 0}{\sqrt{s_p^2\left(\frac{1}{n_x} + \frac{1}{n_y}\right)}}$$

$$= \frac{(\overline{X} - \overline{Y})}{\sqrt{\frac{\Sigma x^2 + \Sigma y^2}{n_x + n_y - 2}\left(\frac{1}{n_x} + \frac{1}{n_y}\right)}} \qquad (13.3)$$

The degrees of freedom associated with this *t* test are $n_x + n_y - 2$.

Formula 13.3 uses the pooled sums of squares or pooled variance s_p^2 in calculating $s_{\bar{x}-\bar{y}}$, thereby assuming that $\sigma_x = \sigma_y$. The consequences of violating the homogeneity of variance assumption on Type I error rates are discussed in the next section. A test of the null hypothesis H_0: $\sigma_x^2 = \sigma_y^2$ using sample data for groups X and Y is available so we can statistically test the homogeneity of variance assumption. We will postpone discussing this test until Chapter 14.

Two examples from the "Sesame Street" data set in Appendix A illustrate the t test for independent sample means. The first example tests the null hypothesis that $\mu_x = \mu_y$ for the sample pair 020 from Table 13.3. Because both X and Y were sampled from the same population with no treatment administered, we would expect approximately 95% of the sample pairs in Table 13.3 to result in nonsignificant tests while approximately 5% should be statistically significant by chance if $\alpha = .05$. The standard error for $\bar{X} - \bar{Y}$ based on the sample data was calculated previously as 5.28. The observed test statistic is

$$t_{\text{obs}} = \frac{26.25 - 31.50}{5.28} = -.99$$

Assuming H_A: $\mu_x \neq \mu_y$ and $\alpha = .05$, the degrees of freedom are 22 and $t_{\text{crit}} = \pm 2.074$. Comparing t_{obs} to t_{crit} results in a decision not to reject H_0: $\mu_x = \mu_y$. By chance, due to random sampling, $\bar{X}$ does differ from $\bar{Y}$ by -5.25 Numbers posttest score points, but a difference this large is probable based on the expectations of occurrences for the sampling distribution.

For the second example we will investigate whether children who are encouraged to view "Sesame Street" score higher on letters knowledge than children who are not encouraged to view the show. To gather relevant information, we drew two random samples from Appendix A. The first sample (X) consists of 15 children drawn at random from the population of children who were encouraged to watch "Sesame Street." The second sample (Y) of 12 children was drawn at random from the population not encouraged to watch. The samples are unequal to provide a variety in the example. The Letters posttest scores for the children sampled appear in Table 13.4. We can state the null hypothesis as H_0: There is no difference in mean Letters posttest scores for children who were or were not encouraged to view "Sesame Street." Symbolically, H_0 translates into H_0: $\mu_e = \mu_{ne}$. The alternative hypothesis can justifiably be directional, H_A: $\mu_e > \mu_{ne}$. The null hypothesis is a statement of no difference in mean knowledge about letters between the two groups. Before viewing of "Sesame Street" started, both groups of children belonged to the same population regarding knowledge of letters. The research design randomly assigned children to the encouraged or not encouraged treatment conditions. If viewing encouragement had no differential effect on learning, then H_0 is true. As random samples of the same population, then, the group mean Letter posttest scores should differ for the two groups only by chance with probability of occurrence described by the sampling distribution of $\bar{X} - \bar{Y}$ under a true H_0: $\mu_x - \mu_y = 0$. If encouragement to view "Sesame Street" increased children's knowledge about letters, then H_A: $\mu_x > \mu_y$ would be true, leading us to reject H_0 if the test of H_0 were powerful enough.

TABLE 13.4
Letters posttest data on random samples of children encouraged or not encouraged to view "Sesame Street."

Encouraged Group		Not Encouraged Group	
Child	Letters Posttest Score	Child	Letters Posttest Score
076	48	061	48
156	14	180	16
238	19	121	44
219	23	195	44
125	12	069	38
014	28	128	13
101	47	197	15
097	41	170	11
015	21	178	16
049	20	011	52
035	41	088	20
209	17	229	13
131	31		
024	27		
048	36		

Using an α level of .05, the test of the hypothesis is as follows:

Encouraged Group (X)

$n_x = 15$

$\Sigma X = 425$

$\Sigma X^2 = 14005$

$\overline{X} = 28.33$

$$\Sigma x^2 = 14005 - \frac{425^2}{15} = 1963.33$$

$s_x = 11.84$

Not Encouraged Group (Y)

$n_y = 12$

$\Sigma Y = 330$

$\Sigma Y^2 = 11920$

$\overline{Y} = 27.50$

$$\Sigma y^2 = 11920 - \frac{330^2}{12} = 2845$$

$s_y = 16.08$

The pooled variance is

$$s_p^2 = \frac{1963.33 + 2845}{15 + 12 - 2} = 192.33$$

The standard error of $\overline{X} - \overline{Y}$ is

$$s_{\overline{x}-\overline{y}} = \sqrt{192.33\left(\frac{1}{15} + \frac{1}{12}\right)} = 5.37$$

The observed t standard score value is calculated from formula 13.3.

$$t_{obs} = \frac{(\overline{X} - \overline{Y})}{s_{\overline{x}-\overline{y}}} = \frac{28.33 - 27.50}{5.37} = .15$$

Using sample data, the estimate of the standard error $\overline{X} - \overline{Y}$ is 5.37. This is the amount of variability in mean differences that we can expect due to randomness. Is the difference in our two sample means ($\overline{X} - \overline{Y} = .83$) statistically significant? The critical value, t_{crit}, at the .05 level for a directional test and $df = 25$ is 1.708. Comparing t_{obs} to t_{crit} leads us to not reject H_0: $\mu_x = \mu_y$. The inference is that our two samples, while representing two distinctly defined populations on the viewing encouragement treatment, do not differ statistically on average Letters posttest performance. For performance on the criterion variable, these samples can be assumed to be representative of the same population of Letters posttest scores, implying that the data were not sufficient to support the existence of a relationship between encouragement to view "Sesame Street" and subsequent knowledge about letters. Encouragement to view "Sesame Street" did not result in a significantly higher mean Letters posttest score than that attained under a "no encouragement" condition.

A last example is provided for you to work. The null hypothesis compares the performance of boys versus girls on classification skills knowledge at the time of the pretest from the "Sesame Street" evaluation data. It is H_0: There exists no difference in mean classification skills pretest scores between girls and boys (H_0: $\mu_{girls} = \mu_{boys}$). Assume a nondirectional alternative hypothesis H_A: $\mu_{girls} \neq \mu_{boys}$ and use an α level of .10 to test H_0. We drew random samples of 10 girls and 10 boys from Appendix A. The pretest Classification Skills raw score sums and the deviation sums of squares are given here for the two samples. Complete the necessary calculations and steps to test H_0: $\mu_{girls} = \mu_{boys}$. What conclusions would you draw based on the hypothesis-testing results?[5]

---

[5]Mean Classification Skills pretest score: for girls, $\overline{X} = 13.7$
for boys, $\overline{Y} = 10.5$

$$t_{obs} = \frac{13.7 - 10.5}{\sqrt{\frac{101.1 + 137.5}{10 + 10 - 2}\left(\frac{1}{10} + \frac{1}{10}\right)}} = \frac{3.2}{1.63} = 1.96$$

When $df = 18$ and $\alpha = .10$, $t_{crit} = \pm 1.734$.

The t_{obs} value of 1.96 is greater than t_{crit} of 1.734, so we would reject H_0. We conclude that at the time of the pretest, boys and girls cannot be assumed to be from the same population regarding classification skills. Rather, the inference based on the sample evidence is that girls' knowledge level is significantly higher than that of boys at the .10 level of statistical significance.

Girls (X)	**Boys (Y)**
$n_x = 10$	$n_y = 10$
$\Sigma X = 137$	$\Sigma Y = 105$
$\Sigma X^2 = 1978$	$\Sigma Y^2 = 1240$
$\Sigma x^2 = 1978 - \dfrac{(137)^2}{10}$	$\Sigma y^2 = 1240 - \dfrac{(105)^2}{10}$
$= 101.1$	$= 137.5$

Rejecting H_0: $\mu_{\text{girls}} = \mu_{\text{boys}}$ leads to an inference that the alternative hypothesis H_A: $\mu_{\text{girls}} \neq \mu_{\text{boys}}$ is true. Our sample statistics as point estimates of the true population means for girls and boys would support the conclusion that the true mean knowledge of classification skills for girls was higher than that for boys, $\mu_{T(\text{girls})} > \mu_{T(\text{boys})}$. The statistical inference is that the sampling distribution of mean differences between girls and boys for classification skills is not centered around a mean of 0. Rather, the sampling distribution is more likely to be centered around a mean difference greater than 0, $\mu_{T(\text{girls})} - \mu_{T(\text{boys})} > 0$. In drawing these inferences and conclusions, we must always keep in mind that the criterion we set ($\alpha = .10$) allows for rejection of H_0: $\mu_{\text{girls}} = \mu_{\text{boys}}$ from 10% of the mean differences between random samples when H_0 is actually true. The probability always exists that we have made a Type I decision error when we reject H_0.

Assumptions Underlying the *t* Test for Independent Groups

The appropriateness of the t distribution and accuracy of critical values used in testing H_0: $\mu_x = \mu_y$ for independent groups at a specified alpha level depends on certain assumptions. Assumptions are made due to mathematical considerations in developing the test statistic. The ratio of $(\overline{X} - \overline{Y})$ to $s_{\overline{x}-\overline{y}}$ is distributed as Student's t with $df = n_x + n_y - 2$ if the data meet the following assumptions:

1. *Score independence.* Scores within and across the two groups are assumed to be independent of each other. This means that one person's score is not determined or related in any way to another person's score and that all scores are free to vary across the total measurement continuum without restrictions other than the individual's performance or response level.
2. *Normality of population distributions.* The distribution of the criterion variable in each population is assumed to be normal. The normality assumption ensures that the mean and variance for each population for which the groups are representative are independent of each other. In this manner, the group mean may be raised or lowered by a

treatment condition, but the variability of scores should not be affected. Recall the discussions in Chapters 5 and 6 of the effect of adding constants to scores on the mean and variance. If the variability is affected by the treatment (as it might be in a skewed distribution where the mean and variance are related), it is no longer an accurate estimate of the population variance. In this case, $s_{\bar{x}-\bar{y}}$ will not accurately estimate the standard deviation of the sampling distribution of $\bar{X} - \bar{Y}$.

3. *Homogeneity of population variances.* The variances in the populations represented by the samples are assumed to be equal. This is the "homogeneity of variances" assumption and is necessary to use the pooled sum of squared deviations formula in calculating $s_{\bar{x}-\bar{y}}$ (formula 13.2). If the group variances are not approximately equal, both cannot be accurate estimates of the population variance, and $\sigma_{\bar{x}-\bar{y}}$ will be over- or underestimated by $s_{\bar{x}-\bar{y}}$.

The consequence of violating the assumptions is that the test statistic formula (formula 13.3) will produce an artificially inflated or deflated t value and will lead us to reject or not reject H_0: $\mu_x = \mu_y$ due to the assumption's violation rather than the presence or absence of a true difference in μ_x and μ_y. If we violate an assumption, we should not have as much confidence in the t test as a valid test of H_0: $\mu_x = \mu_y$ at the alpha level specified. Several empirical studies have investigated the consequences of violating the assumptions on Type I error rates. The consensus is that when the sample sizes for the two groups are equal ($n_x = n_y$), the t test is robust to violations with the exception of independence of scores. When $n_x \neq n_y$, we should pay specific attention to any violations, but particularly the variance homogeneity assumption.

Assumptions are made when using the t test for independent samples to test H_0: $\mu_x = \mu_y$. When performing a research study, the independence of scores should be examined, the distribution shape identified, and the variability of scores in each group should be calculated and compared. Visual inspection of the data and use of descriptive statistical indices do not take the place of the statistical tests available to determine whether assumptions have been violated. They do, however, provide descriptive information about the assumptions and should be used as such. Statistical tests for assumptions are available in Winer (1971). Glass, Peckham, and Sanders (1972) review the specific consequences on Type I error rates of violating the assumptions.

13.4 TESTING HYPOTHESES ABOUT TWO CORRELATED SAMPLE MEANS

Given the same group of individuals measured at time one and time two on the same variable, the question of interest often is "Did change occur?" "Is growth evident?" "Do the individuals differ on the variable of interest from time two to time one?" Each question identifies the same situation where a test of the hypothesis for *two correlated (dependent) sample means* would be appropriate. The null hypothesis to be tested is still the same as that for independent group means (i.e., no difference exists in the means for two sets of scores, H_0: $\mu_x = \mu_y$). The difference in the independent versus the dependent group case is that in the former, the sample means $\overline{X}$ and $\overline{Y}$ are obtained from measuring each of two separate, independent groups once. In the dependent group, the sample means are obtained by measuring a single group of individuals on the same criterion variable across two occasions.

Two research situations exist in which we might have data on the same variable for the same group of individuals measured at two occasions. The first is the typical pretest/posttest design in which a treatment has been administered between pre- and posttests or longitudinal growth or change is to be assessed. In the second case, the same group of individuals have been measured after receiving each of two different treatment conditions or measured in each of two different stimulus situations. The problem with this case is the carry-over effect of one treatment or stimulus situation to the other when the same individuals are used in both. To avoid this possible confounding, the two treatments or stimulus conditions should be given in a random order to counterbalance any order effects. With adequate experimental design control, this research design will allow for inferences about the equality or nonequality of the two treatment or stimulus conditions. In the former test situation, pre- and posttesting, we can make inferences about mean performance differences on the pretest and the posttest. We can make a probability statement about the degree of confidence that the observed difference in $\overline{X}$ and $\overline{Y}$ is a chance or nonchance difference. However, with the single-group pretest/posttest design, we cannot infer that the change, growth, or difference in pre-/posttest means is necessarily due to the intervening treatment effect. This simple design does not offer sufficient control of rival explanations of why or how the change might have occurred to allow us to make direct causal inferences.

Sampling Distribution for the Differences between Two Correlated Sample Means

The sampling distribution for the correlated sample means case plays the same role in hypothesis testing as in previous sampling distributions. If the null hypothesis is true, H_0: $\mu_x = \mu_y$, then the population mean difference between the two sets of scores should be zero, $\mu_x - \mu_y = 0$. Again, the sample mean difference, $\overline{X} - \overline{Y}$, will deviate around the value of zero due to sampling error if H_0 is true. Consistent with the previous sampling distributions we have considered, the sampling distribution of $\overline{X} - \overline{Y}$ for two correlated sample means is normally distributed with a true difference mean of $\mu_x - \mu_y$. The standard deviation of the sampling distribution indicates the amount of variability one can expect in $\overline{X} - \overline{Y}$ due to randomness. Because the scores for both sets of data are on the same individuals, the scores will be correlated and will affect the variability of $\overline{X} - \overline{Y}$. Taking this relationship into account, the standard error of $\overline{X} - \overline{Y}$, $\sigma_{\overline{x}-\overline{y}}$, for the two correlated sample means case is

$$\sigma_{\overline{x}-\overline{y}} = \sqrt{\sigma_{\overline{x}}^2 + \sigma_{\overline{y}}^2 - 2\rho_{xy}\sigma_{\overline{x}}\sigma_{\overline{y}}} \tag{13.4}$$

$$= \sqrt{\frac{\sigma_x^2}{n_x} + \frac{\sigma_y^2}{n_y} - 2\rho_{xy}\left(\frac{\sigma_x}{\sqrt{n_x}}\right)\left(\frac{\sigma_y}{\sqrt{n_y}}\right)}$$

Because we rarely know the population variances and correlation coefficient, the sample estimate of $\sigma_{\overline{x}-\overline{y}}$ becomes

$$s_{\overline{x}-\overline{y}} = \sqrt{s_{\overline{x}}^2 + s_{\overline{y}}^2 - 2r_{xy}s_{\overline{x}}s_{\overline{y}}} \tag{13.5}$$

$$= \sqrt{\frac{s_x^2}{n_x} + \frac{s_y^2}{n_y} - 2r_{xy}\left(\frac{s_x}{\sqrt{n_x}}\right)\left(\frac{s_y}{\sqrt{n_y}}\right)}$$

Fortunately, there is a simpler way to obtain $s_{\overline{x}-\overline{y}}$ that is mathematically equivalent to formula 13.5. Each individual in the sample will have two scores. We can obtain a raw difference score for each individual by subtracting one score from the other. Using the difference scores across individuals, we obtain a single set of scores that we can use to test the null hypothesis H_0: $\mu_D = 0$ using the same procedures as described in section 13.2 for a single sample. In the hypothesis H_0: $\mu_D = 0$, μ_D is the population mean for the set of difference scores, $\mu_D = \mu_x - \mu_y$. The sampling distribution of mean difference scores, $\overline{D} = \overline{X} - \overline{Y}$, is normally distributed around $\mu_D = 0$ if H_0 is true with the standard error of $\overline{D}$,

$$\sigma_{\overline{D}} = \frac{\sigma_D}{\sqrt{n}}$$

where σ_D is the population standard deviation of difference scores and n is the number of pairs in the sample. Estimating σ_D with sample values results in

$$s_{\overline{D}} = \frac{s_D}{\sqrt{n}} \tag{13.6}$$

Using sample values, we can summarize the standard deviation of the sampling distribution for differences between two correlated sample means by the following equivalent expressions:

$$s_{\overline{x}-\overline{y}} = \sqrt{s_{\overline{x}}^2 + s_{\overline{y}}^2 - 2r_{xy}s_{\overline{x}}s_{\overline{y}}}$$

$$= s_{\overline{D}} = \frac{s_D}{\sqrt{n}}$$

To illustrate the empirical generation of the sampling distribution for differences between correlated sample means, scores on Numbers pretest (Y) and Numbers posttest (X) were used for 100 random samples of 12 individuals from the "Sesame Street" data set in Appendix A. Table 13.5 presents the

TABLE 13.5
Sampling distribution for correlated sample Numbers posttest and pretest means.

Sample	Pretest $\overline{Y}$	Posttest $\overline{X}$	$\overline{X}-\overline{Y}$	Sample	Pretest $\overline{Y}$	Posttest $\overline{X}$	$\overline{X}-\overline{Y}$	Sample	Pretest $\overline{Y}$	Posttest $\overline{X}$	$\overline{X}-\overline{Y}$
001	19.92	33.33	13.41	034	16.75	28.83	12.08	067	20.25	29.17	8.92
002	19.67	29.50	9.83	035	20.42	31.00	10.58	068	24.58	32.42	7.84
003	18.25	28.50	10.25	036	21.50	33.45	11.95	069	17.92	27.92	10.00
004	19.75	29.75	10.00	037	17.58	28.67	11.09	070	19.17	27.83	8.66
005	18.58	28.83	10.25	038	26.83	35.82	8.99	071	21.00	30.33	9.33
006	20.83	34.67	13.84	039	22.00	30.83	8.83	072	21.58	29.08	7.50
007	23.33	28.75	5.42	040	22.83	29.67	6.84	073	25.33	32.33	7.00
008	19.08	23.25	4.17	041	21.00	31.92	10.92	074	22.08	34.50	12.42
009	20.92	32.00	11.08	042	22.00	29.92	7.92	075	17.50	27.67	10.17
010	21.58	29.00	7.42	043	21.83	33.08	11.25	076	21.42	29.58	8.16
011	23.33	26.17	2.84	044	18.17	27.58	9.41	077	14.50	19.50	5.00
012	21.58	34.50	12.92	045	19.42	37.42	18.00	078	22.17	31.08	8.91
013	25.25	32.42	7.17	046	19.92	30.83	10.91	079	22.00	38.17	16.17
014	18.50	24.42	5.92	047	20.50	26.25	5.75	080	23.00	29.75	6.75
015	21.00	28.25	7.25	048	22.92	32.92	10.00	081	19.17	25.67	6.50
016	22.50	31.33	8.83	049	21.50	32.67	11.17	082	22.25	33.33	11.08
017	22.92	27.83	4.91	050	24.33	37.58	13.25	083	18.92	28.08	9.16
018	26.92	32.33	5.41	051	14.83	28.25	13.42	084	16.25	31.17	14.92
019	23.08	29.42	6.34	052	23.08	31.17	8.09	085	20.50	33.58	13.08
020	19.25	30.17	10.92	053	19.33	28.00	8.67	086	23.58	32.33	8.75
021	19.67	28.83	9.16	054	22.42	28.42	6.00	087	16.17	24.00	7.83
022	20.00	30.75	10.75	055	20.67	29.25	8.58	088	24.42	36.00	11.58
023	21.67	31.25	9.58	056	17.58	27.75	10.17	089	18.50	29.17	10.67
024	19.58	29.25	9.67	057	28.33	37.25	8.92	090	21.25	29.58	8.33
025	18.33	31.58	13.25	058	18.50	27.50	9.00	091	20.00	26.33	6.33
026	18.25	26.42	8.17	059	16.42	23.75	7.33	092	17.83	30.00	12.17
027	19.25	25.75	6.50	060	18.08	26.92	8.84	093	24.33	33.50	9.17
028	17.33	30.75	13.42	061	22.75	34.92	12.17	094	19.58	29.67	10.09
029	18.25	26.50	8.25	062	25.17	34.67	9.50	095	20.58	26.33	5.75
030	17.17	27.00	9.83	063	23.00	31.92	8.92	096	15.50	26.17	10.67
031	21.67	30.08	8.41	064	20.83	27.17	6.34	097	19.42	27.67	8.25
032	21.42	30.08	8.66	065	26.08	38.58	12.50	098	20.58	32.25	11.67
033	21.83	34.42	12.59	066	25.58	32.17	6.59	099	26.83	33.92	7.09
								100	19.08	25.67	6.59

$\Sigma(\overline{X}-\overline{Y}) = 936.91$
$\Sigma(\overline{X}-\overline{Y})^2 = 9473.35$

$$\text{Mean of } (\overline{X}-\overline{Y}) = \frac{936.91}{100} = 9.37$$

$$s_{\overline{X}-\overline{Y}} = \sqrt{\frac{9473.35 - \frac{936.91^2}{100}}{99}} = 2.65$$

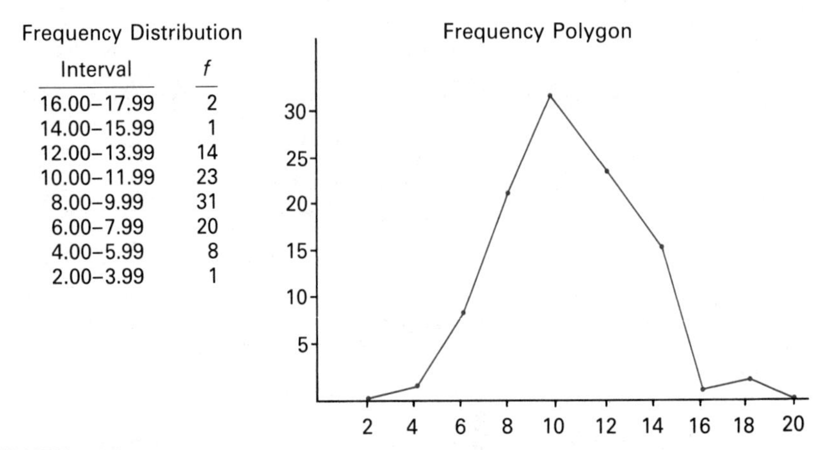
Frequency Distribution

Interval	f
16.00–17.99	2
14.00–15.99	1
12.00–13.99	14
10.00–11.99	23
8.00–9.99	31
6.00–7.99	20
4.00–5.99	8
2.00–3.99	1

FIGURE 13.8
Frequency distribution and polygon for the sampling distribution of $\overline{X} - \overline{Y}$ for correlated sample means.

means on each occasion for the 100 samples along with the difference in means ($\overline{X} - \overline{Y}$) for each sample. Figure 13.8 displays the frequency distribution and frequency polygon of the 100 mean difference scores. If the null hypothesis were true, H_0: $\mu_x = \mu_y$, the mean of the sampling distribution should be zero. Because we have the population means from Appendix A, however, we know that the true population mean difference in Numbers post- and pretest means is 30.43 − 20.90, or 9.53. Therefore, our partial sampling distribution of 100 difference scores in Table 13.5 should vary around the true population difference $\mu_x - \mu_y = 9.53$. As we can see from Table 13.5, the mean of the 100 difference scores is 9.37, slightly less than the true population value. The difference in our empirically generated sampling distribution mean difference of 9.37 and the true population mean difference of 9.53 would have been lower if more samples had been generated.

The empirical standard error of $\overline{X} - \overline{Y}$ is the standard deviation of the 100 difference scores in Table 13.5, or 2.65. This is an empirically generated standard error that we could have calculated from Formula 13.4, since we know the population variance values from Appendix A. For the Numbers posttest scores, $\sigma_x^2 = 158.71$, for Numbers pretest, $\sigma_y^2 = 114.18$ and the correlation coefficient (ρ_{xy}) between the two sets of scores for the 238 children with data is .669. Substituting in the formula,

$$\sigma_{\bar{x}-\bar{y}} = \sqrt{\frac{158.71}{12} + \frac{114.18}{12} - 2(.669)\left(\sqrt{\frac{158.71}{12}}\right)\left(\sqrt{\frac{114.18}{12}}\right)}$$

$$= \sqrt{7.731}$$

$$= 2.78$$

The standard deviation value of 2.65 for the sampling distribution of 100 samples in Table 13.5 is a reasonable approximation of $\sigma_{\overline{x}-\overline{y}} = 2.78$. As in the independent group case, we use neither of these methods to obtain the standard error. Rather, we use data from a single sample to estimate $\sigma_{\overline{x}-\overline{y}}$ using formula 13.6. This procedure is illustrated in the next section.

The *t* Test for Correlated Samples

For testing H_0: $\mu_x = \mu_y$ (or equivalently, H_0: $\mu_D = 0$) for correlated groups, the t distribution is the appropriate probability distribution when the standard error is estimated by $s_{\overline{x}-\overline{y}}$ or $s_{\overline{D}}$. Following the general formula format, the test statistic formula becomes

$$t_{\text{obs}} = \frac{(\overline{X} - \overline{Y}) - (\mu_x - \mu_y)}{s_{\overline{x}-\overline{y}}} = \frac{\overline{D} - \mu_D}{s_{\overline{D}}} = \frac{\overline{D}}{\dfrac{s_D}{\sqrt{n}}} \tag{13.7}$$

The observed test statistic, t_{obs}, is distributed as Student's t with $n - 1$ degrees of freedom, where n is the number of paired observations.

To illustrate the use of the t test for correlated groups, let's use data compatible with the samples and variables in Table 13.5. Scores on the Numbers posttest (X) and pretest (Y) are given for a random sample of 10 individuals in Table 13.6. The calculations are provided for testing the null hypothesis, H_0: $\mu_x = \mu_y$, against the alternative, H_A: $\mu_x > \mu_y$ with $\alpha = .05$. Note that the estimate of $\sigma_{\overline{x}-\overline{y}} = 2.78$ using data only from these 10 individuals is $s_{\overline{x}-\overline{y}} = s_D = 3.23$. Comparing t_{obs} to t_{crit} allows us to reject H_0 and conclude that mean posttest scores do differ from mean pretest scores. Thus, the increase in mean numbers knowledge is statistically significant at the .05 level.

The latter illustration also provides us with a unique opportunity to calculate the power of the statistical test. Power is an appropriate consideration when the null hypothesis is actually false. Because we have the population data, we know that H_0: $\mu_x = \mu_y$ is false because the true $\mu_x - \mu_y = 9.53$ for the 240 data cases in Appendix A. Figure 13.9 displays the power associated with the present test. The true standard error ($\sigma_{\overline{x}-\overline{y}} = 2.78$) is constant no matter what the value of $\mu_x - \mu_y$. Under the null hypothesis (H_0 is assumed true), $\overline{X} - \overline{Y}$ is distributed around zero with $\sigma_{\overline{x}-\overline{y}} = 2.78$. Since $t_{\text{crit}} = 1.83$, the critical mean difference needed for rejection is 1.83 standard score units from the mean of zero. The critical mean difference in raw-score units in the null hypothesis distribution is $(1.83)(2.78) = 5.087$. Now, to calculate power, we need to determine the probability of obtaining a mean difference as large

TABLE 13.6
Example: t test for correlated groups.

Individual	Numbers Posttest (X)	Numbers Pretest (Y)	$X - Y$ (D)	D^2
216	17	24	−7	49
222	49	27	22	484
137	38	17	21	441
048	20	8	12	144
169	8	9	−1	1
099	31	14	17	289
114	23	13	10	100
042	22	13	9	81
004	19	14	5	25
078	53	29	24	576
		TOTAL	D = 112	D^2 = 2190

$$\overline{X} - \overline{Y} = \overline{D} = \frac{112}{10} = 11.2; \quad s_D^2 = \frac{2190 - \dfrac{112^2}{10}}{9} = 103.96$$

$$s_D = \sqrt{103.96} = 10.20; \quad S_{\overline{D}} = \frac{10.20}{\sqrt{10}} = 3.23$$

$$t_{\text{obs}} = \frac{11.2}{3.23} = 3.47$$

or larger than 5.087 *given that the true mean difference is 9.53.* In fact, if $\mu_x - \mu_y = 9.53$, then the $\overline{X} - \overline{Y}$ random samples will distribute themselves around 9.53, as Figure 13.9 shows. The power is the proportion of the true sampling distribution (H_T) that exceeds the critical mean difference of 5.087. To determine this proportion, we change 5.087 to a standard t score in the H_T distribution:

$$t = \frac{5.087 - 9.53}{2.78} = -1.60$$

The probability of a t score exceeding a t value of −1.60 in a t distribution with $df = 9$ is approximately .93. The power or probability of rejecting H_0: $\mu_x - \mu_y = 0$ for samples of size 10 given that the *true* alternative H_T: $\mu_x - \mu_y = 9.53$ is .93. For random samples of size 10, we could expect to reject H_0 93% of the time. The other 7% of the tests would result in a Type II error. As an empirical check, 93 of the 100 mean differences in Table 13.5 are larger than 5.087 and would result in rejections. The approximate number of tests out of 100 that should result in rejecting H_0 for this example should

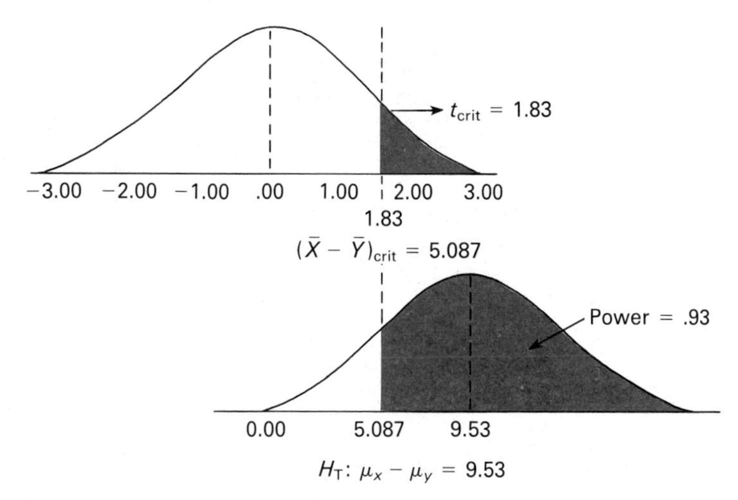

FIGURE 13.9
Power of a test for correlated group means.

approximate the power of the test, which it does. Based on the power, we would have expected approximately 93 of 100 samples to have mean differences between Numbers pretest and posttest greater than 5.087.

As another example from the "Sesame Street" data, the null hypothesis of no change in relational terms knowledge from pre- to posttest for children who seldom watched "Sesame Street" can be tested. These latter children are coded with a one (1) on the viewing category variable. Pretest data (variable 11) and posttest data (variable 19) on the test of relational terms are given in Table 13.7 for a random sample of 16 individuals from Appendix A who viewed "Sesame Street" an average of once a week or never. Given the summary data, test H_0: $\mu_{post} = \mu_{pre}$ against H_A: $\mu_{post} \neq \mu_{pre}$ at the .01 level of significance.[6] Based on the test of H_0 we conclude that children who seldom watched "Sesame Street" still gained statistically significant (at the .01 level) knowledge on relational terms from pre- to posttesting.

---

[6] $\bar{D} = \frac{43}{16} = 2.69$, $s_D^2 = \frac{223 - \frac{43^2}{16}}{15} = 7.16$, $s_D = \sqrt{7.16} = 2.68$

$t_{obs} = \frac{2.69}{\frac{2.68}{\sqrt{16}}} = \frac{2.69}{.67} = 4.01,$

$t_{crit} = 2.947$ ($df = 15$; $\alpha = .01$; nondirectional test)

Reject H_0: Conclude that posttest knowledge of relational terms is significantly higher than pretest knowledge even for children who seldom watched "Sesame Street."

TABLE 13.7
Relational Terms pretest and posttest scores for a random sample of children who seldom watched "Sesame Street."

Individual	Posttest (X)	Pretest (Y)	$D = X - Y$
128	9	9	0
051	10	8	2
207	13	3	10
119	12	9	3
195	15	14	1
038	9	5	4
220	12	11	1
211	13	13	0
228	11	12	−1
209	15	9	6
044	6	2	4
161	10	7	3
201	12	9	3
230	14	13	1
062	11	7	4
100	10	8	2
			$\Sigma D = 43$
			$\Sigma D^2 = 223$

13.5 A COMPARISON OF ANALYSES FOR INDEPENDENT AND CORRELATED SAMPLES

As part of the research process, the research question under investigation will usually dictate the sampling plan and method of statistical analysis to employ. For some research questions, sampling of two independent groups is the only possible way to conduct the study, for example, comparing mean scores on a criterion variable for males and females, or studying the effects of two different advanced organizers on comprehension of a given set of instructional materials. The need for two independent groups in the former case (males vs. females) is self-evident. In the latter case, once the same set of materials has been given to a group using one advanced organizer, it would be impossible to erase the learning that has already occurred and give the same set of materials to the same persons using the second advanced organizer as the treatment. If the same persons were to be used under both treatment conditions, we would need to use two different sets of materials *of equal difficulty, but unrelated content* to avoid carry-over between treatments. The confounding effect of the materials would be impossible to control using just a single experimental group.

Some research situations logically dictate collecting data from the same group of individuals on two occasions, such as pre- and posttest comparisons. For every correlated sample study, however, the study could have been conducted using independent groups as long as random samples were selected from the same population and data were collected on one sample at each of the two separate occasions. For example, in the pre-/posttest situation, two random samples of the available students could be selected with one sample tested at the time of the pretest and the other tested at posttest. Comparing pre- and posttest means would use the independent group t test in the latter case.

Advantages of using either independent or correlated samples in the research situation in which either is applicable relate to the precision of the statistical test and/or the control of possible carryover effects in the data collection from time one to time two. Precision in a statistical test is related to the variability we can expect in the difference between sample means $\overline{X}$ and $\overline{Y}$. The index of this variability is the standard error of the sampling distribution, $s_{\overline{x}-\overline{y}}$. The greater $s_{\overline{x}-\overline{y}}$, the less precise is our estimate of $\mu_x - \mu_y$ using $\overline{X} - \overline{Y}$; thus, we would need a larger difference in $\overline{X}$ and $\overline{Y}$ to result in a statistically significant difference. As you should have suspected, $s_{\overline{x}-\overline{y}}$ as an index of precision relates directly to the power of the statistical test. The smaller $s_{\overline{x}-\overline{y}}$ (greater precision), the more powerful is our test of H_0: $\mu_x = \mu_y$ given that H_0 is in fact false. Contrasting the two estimates of $s_{\overline{x}-\overline{y}}$ for the independent group and correlated group cases indicates that the correlated sample case will result in a smaller $s_{\overline{x}-\overline{y}}$ unless the scores on the two occasions are not correlated ($r_{xy} = .00$).

For the independent sample case,

$$s_{\overline{x}-\overline{y}} = \sqrt{s_{\overline{x}}^2 + s_{\overline{y}}^2} = \sqrt{\frac{s_x^2}{n_x} + \frac{s_y^2}{n_y}}$$

For the correlated sample case,

$$\begin{aligned} s_{\overline{x}-\overline{y}} &= \sqrt{s_{\overline{x}}^2 + s_{\overline{y}}^2 - 2r_{xy}s_{\overline{x}}s_{\overline{y}}} \\ &= \sqrt{\frac{s_x^2}{n_x} + \frac{s_y^2}{n_y} - 2r_{xy}s_{\overline{x}}s_{\overline{y}}} \end{aligned}$$

The first two terms in the correlated sample case are identical to $s_{\overline{x}-\overline{y}}$ for independent samples. Because scores are correlated in the correlated sample case, however, $2r_{xy}s_{\overline{x}}s_{\overline{y}}$ is subtracted as an adjustment, thus reducing $s_{\overline{x}-\overline{y}}$.[7]

When given the choice in designing a research study, correlated samples will result in more precision and therefore a more powerful test of H_0: $\mu_x = \mu_y$. The increase in precision, however, should not override the control of possible confounding effects in decisions to use independent or correlated samples.

[7]It is assumed that r_{xy} will always be positive in the correlated sample case. It is highly unlikely given that the same variable is being measured using the same or parallel instruments that scores would change such that r_{xy} would be negative.

13.6 SUMMARY

This chapter has stressed the continuity of concepts in the hypothesis testing process (introduced in Chapters 11 and 12) when testing hypotheses about one- and two-sample (independent and dependent) means. Components in the process include

1. stating the null and alternative hypotheses
2. determining the significance level for testing the hypothesis
3. identifying the test statistic formula appropriate for testing the null hypothesis (this also identifies the probability distribution to be used, e.g., z or t distributions)
4. calculating the observed test statistic
5. identifying the critical value in the probability distribution to be used as the criterion for rejection or nonrejection of the null hypothesis
6. deciding whether to reject the null hypothesis or not based on comparing the observed test statistic to the critical value

While the specifics of each component change from one research situation to another, the overall process never does change. This chapter discussed the statistical procedures appropriate for testing hypotheses about

1. a single sample mean when the population variance is known, H_0: $\mu = a$. The test statistic and formula are

$$z = \frac{\overline{X} - \mu}{\sigma_{\overline{x}}} = \frac{\overline{X} - \mu}{\frac{\sigma_x}{\sqrt{n}}}$$

2. a single sample mean when the population variance is *not* known, H_0: $\mu = a$. The test statistic and formula are

$$t = \frac{\overline{X} - \mu}{s_{\overline{x}}} = \frac{\overline{X} - \mu}{\frac{s_x}{\sqrt{n}}}$$

The degrees of freedom for the test are $n - 1$.

3. Two independent sample means assuming homogeneity of population variances, H_0: $\mu_x = \mu_y$. The test statistic and formula are

$$t = \frac{(\overline{X} - \overline{Y})}{s_{\overline{x}-\overline{y}}}$$

$$= \frac{\overline{X} - \overline{Y}}{\sqrt{\frac{\Sigma x^2 + \Sigma y^2}{n_x + n_y - 2}\left(\frac{1}{n_x} + \frac{1}{n_y}\right)}}$$

The degrees of freedom for this test are $n_x + n_y - 2$.

4. Two dependent sample means, H_0: $\mu_x = \mu_y$ or H_0: $\mu_D = 0$. The test statistic and formula are

$$t = \frac{\overline{X} - \overline{Y}}{s_{\overline{x}-\overline{y}}} = \frac{\overline{D}}{s_{\overline{D}}}$$

$$= \frac{\overline{D}}{\frac{s_D}{\sqrt{n}}}$$

The degrees of freedom for this test are $n - 1$, where n represents the number of paired observations.

Another important conceptual notion is the role of the sampling distributions and their associated standard errors in testing hypotheses. Knowing the characteristics of the sampling distribution for specific hypotheses provides the researcher with the luxury of *not* having to empirically generate the probability distribution to test each specific hypothesis. The researcher can then use the theoretical information about the shape, central tendency, and amount of variability for a sampling distribution to see if the sample data correspond to data that would be expected *if the null hypothesis were true.* This last phrase is essential to understanding the role of a sampling distribution in testing hypotheses. *If the null hypothesis is true,* the sampling distribution will have a mean equal to that hypothesized (e.g., H_0: $\mu = a$), and the sample value can be expected to deviate from this hypothesized value according to the size of the standard error. The standard error, $s_{\overline{x}}$, $s_{\overline{x}-\overline{y}}$, or $s_{\overline{D}}$, is the estimated standard deviation of the sampling distribution and is used in the denominator of the test statistic formula to determine how discrepant the sample value is from the hypothesized value in standard score units. If the sample value varies from the hypothesized value more than we would expect given the size of the standard error, then we conclude that the hypothesized value is incorrect and reject it. The criterion used to decide the range within which the sample value should fall, given the size of the standard error, is the critical value determined by the level of significance specified.

PROBLEMS

Problem Set A

For each of the following conditions in problems 1–10, identify the critical value(s) you would use to test the stated null hypothesis.

1. H_0: $\mu = a$
H_A: $\mu \neq a$
$\alpha = .05$
$n = 24$
σ^2 is unknown

2. H_0: $\mu = a$
H_A: $\mu \neq a$
$\alpha = .05$
$n = 24$
σ^2 is known

3. H_0: $\mu = a$
H_A: $\mu < a$
$\alpha = .01$
$n = 15$
σ^2 is known

4. H_0: $\mu = a$
H_A: $\mu > a$
$\alpha = .01$
$n = 15$
σ^2 is unknown

5. H_0: $\mu_x = \mu_y$
H_A: $\mu_x < \mu_y$
$\alpha = .01$
$n_x = 14$;
$n_y = 12$

6. H_0: $\mu_x = \mu_y$
H_A: $\mu_x \neq \mu_y$
$\alpha = .01$
$n_x = 14$;
$n_y = 12$

7. H_0: $\mu_x = \mu_y$
H_A: $\mu_x > \mu_y$
$\alpha = .01$
$n_x = 9$;
$n_y = 17$

8. H_0: $\mu_x = \mu_y$
H_A: $\mu_x \neq \mu_y$
$\alpha = .05$
$n = 18$
dependent groups

9. H_0: $\mu_x - \mu_y = 0$
H_A: $\mu_x - \mu_y < 0$
$\alpha = .05$
$n_x = 20$
$n_y = 31$

10. H_0: $\mu_x - \mu_y = 0$
H_A: $\mu_x - \mu_y \neq 0$
$\alpha = .10$
$n = 6$
dependent groups

Which of the following test situations in problems 11–15 would result in rejection of H_0?

11. H_0: $\mu = 30$
H_A: $\mu \neq 30$
$\alpha = .05$; $n = 16$
$\overline{X} = 26.4$; $s_x^2 = 6$

12. H_0: $\mu = 16$
H_A: $\mu < 16$
$\alpha = .01$; $n = 36$
$\overline{X} = 15.2$; $\sigma_x^2 = 2$

13. H_0: $\mu = 150$
H_A: $\mu > 150$
$\alpha = .05$; $n = 20$
$\overline{X} = 159$; $s_x = 20$

14. H_0: $\mu_x = \mu_y$
H_A: $\mu_x \neq \mu_y$
$\alpha = .05$
$n_x = 10$; $n_y = 14$
$\overline{X} = 64$; $\overline{Y} = 72$
$s_x = 12$; $s_y = 9$

15. H_0: $\mu_x = \mu_y$
H_A: $\mu_x < \mu_y$
$\alpha = .05$; $n = 14$
$\overline{D} = -2.8$; $s_D = 3$

16. A random sample of 10 independent observations from a normally distributed population yielded the following values: 51, 50, 49, 43, 56, 46, 45, 30, 55, 52. Estimate the standard deviation of the sampling distribution of means for samples like this one. Using $\alpha = .05$, test the hypothesis that the true mean is 50 against the alternative that the true mean is not 50. State the conclusion.

17. The mean for a random sample of 101 individuals was $\overline{X} = 38$. Is it reasonable to assume that this sample mean was obtained by random sampling from a population in which $\mu = 43$ if $\sigma^2 = 150$? (Use $\alpha = .01$)

TABLE 13.8
Mean judgment error for experimental and control groups.

Experimental	Control	Experimental	Control
41	22	34	11
95	17	43	05
82	11	08	49
44	19	57	23
64	32	06	22
76	97	05	45
12	17	09	28
41	79		

18. Given that the true population mean was 40 ($\mu = 40$), what is the power of the statistical test used in problem 17?

19. An experimenter wants to test the hypothesis that the presence of a stooge giving prearranged false reports would influence reports made aloud (experimental group, $n = 15$) more than reports written down privately (control group, $n = 15$). The experimenter presented each of the 30 participants with a set of stimuli to judge in terms of a physical dimension. The average amount of judgment error for each individual (the predicted influence of a stooge is to increase error scores) appears in Table 13.8.
a. State H_0 and H_A.
b. Analyze the data using an appropriate test statistic at $\alpha = .05$.

20. A study was conducted to compare auditory and visual reaction times. Twenty individuals were administered both auditory and visual stimuli in random order under controlled conditions. Reaction time data under each treatment condition for each individual appear in Table 13.9. Test the following hypothesis at the .01 level of significance: There exists no difference in mean reaction time when responding to visual or auditory stimuli.

Problem Set B

I. Weinraub and Wolf (1983) studied the social networks, coping abilities, life stresses, and amount of mother/child interaction for 28 mother/child pairs (14 single mothers and 14 matched married women). They focused on data collection and analysis to provide information on the following four questions:

1. Do single mothers who have been raising their children alone face more life changes and potential stresses and have fewer social supports and community ties than married mothers?

2. In comparison to married mothers, do single mothers experience more difficulty coping with the stresses and responsibilities in their lives?

3. Are life changes and available social supports related to mother/child interaction in single and two-parent families?

TABLE 13.9
Reaction time for auditory and visual stimuli.

Individual	Auditory	Visual
1	13	15
2	20	13
3	15	12
4	14	15
5	23	12
6	16	16
7	18	11
8	15	21
9	20	17
10	17	14
11	22	17
12	14	16
13	22	11
14	23	29
15	18	10
16	19	19
17	18	21
18	21	18
19	23	17
20	22	16

4. Does the relationship among stresses, supports, and mother/child interactions differ in single and two-parent families?

The sample from which data were collected consisted of 28 mother/child pairs of which 14 were single-parent mothers and their children and 14 were *matched* married women and their children. Participant selection followed a sampling plan that resulted in each of the 14 single-parent mothers being matched with a married mother on age, educational level, and family per capita income. The children for each matched parent pair had to be of similar sex, age, and birth order. Because of the matching to control for the effects of extraneous variables in the study, the two groups of mothers were viewed as correlated groups and each matched pair's data was treated as if it had come from the same individual. Dependent group t tests were performed testing hypotheses about group differences on each of the following variables:

1. the number of potentially stressful life events that had occurred in the preceding 12 months

2. the nature and extent of maternal social supports in five areas:
   a. social contacts
   b. emotional support
   c. parenting support
   d. practical help with child-care and household tasks
   e. satisfaction with the supports received

3. maternal perception of ability to cope in
   a. household responsibilities
   b. child-care responsibilities
   c. finances
   d. emotions
   e. overall ability to cope

4. quality and frequency of mother/child interactions in five areas

Data are simulated in Table 13.10 for matched individuals in each of the two groups on four of the variables investigated by Weinraub and Wolf:

V1: number of potentially stressful life events
V2: satisfaction level with maternal social support
V3: overall ratings of ability to cope
V4: frequency of mother/child communications

Assume that individual A and A′, B and B′, and so on represent the matched pairs. Use these data to answer the problems that follow.

1. Four questions were the focus of the current study. Which of the questions can you translate into hypotheses about mean differences?

2. For the four criterion variables on which simulated data were generated, state null hypotheses in both verbal and symbolic form to investigate potential differences in single-parent and married-women responses.

3. Assuming that the data are matched for each pair, perform dependent group *t* tests for the null hypothesis of no mean difference between the groups for each criterion

TABLE 13.10
Simulated data for two groups on four variables.

Single-Parent Group					Married Group				
Individual	V1	V2	V3	V4	Individual	V1	V2	V3	V4
A	14	2	3	10	A′	6	4	4	8
B	6	4	6	9	B′	6	3	4	11
C	9	1	1	15	C′	8	3	3	10
D	10	2	4	13	D′	8	3	2	10
E	16	3	5	12	E′	12	3	3	14
F	9	4	5	13	F′	5	3	4	7
G	12	5	7	10	G′	9	4	5	10
H	7	4	4	14	H′	5	5	6	9
I	8	4	2	8	I′	3	3	7	12
J	10	3	5	7	J′	7	5	6	13
K	15	2	1	11	K′	8	5	3	13
L	4	5	3	12	L′	5	5	5	12
M	7	3	5	10	M′	7	4	4	10
N	3	4	2	9	N′	7	4	2	12

variable. Use an $\alpha = .05$ for all tests. Use a directional alternative when analyzing data for V2 and V3 and a nondirectional test for V1 and V4.

4. Briefly identify the inferences/conclusions you would draw based on the results of your statistical tests.

5. The sampling plan for this study resulted in matched pairs of individuals across the two groups; thus, a dependent t test was used to test for mean differences. If individuals had not been matched during selection, but assumed selected at random from each of the two populations of parent groups, an independent group t test would have been appropriate. Assume the latter was the case for the data given and test each hypothesis using an independent group t test.

6. How does each of the following change when applying the independent as opposed to a dependent group t test?
   a. H_0
   b. H_A
   c. critical value(s)
   d. difference in group means
   e. standard error of the sampling distribution
   f. decision about H_0

II. An investigation by Bettencourt, Gillett, and Gall (1983) studied the effects of enthusiasm training for teachers on student academic performance. Seventeen first-year teachers were stratified by grade level taught and then randomly assigned within strata to either a training ($n = 8$) or a no-training condition ($n = 9$). After training, each teacher spent two weeks teaching the same curriculum unit to their own students, spending approximately 20 minutes per day in instructional activities. Data were collected on student achievement over the curriculum unit content for each teacher's class and on teacher enthusiasm level. Teacher enthusiasm was rated from videotapes of

TABLE 13.11
Simulated data for achievement class means and ratings of teacher enthusiasm.

	Enthusiasm Training Group			No Training Group	
Teacher	Enthusiasm Rating	Class Mean Achievement	Teacher	Enthusiasm Rating	Class Mean Achievement
1	2.12	24.56	1	1.58	28.52
2	4.16	30.71	2	2.06	28.67
3	3.78	32.44	3	2.35	30.22
4	3.18	29.10	4	2.60	33.58
5	2.28	28.02	5	1.74	24.21
6	2.32	26.51	6	1.92	25.37
7	2.46	26.92	7	1.86	26.85
8	2.71	27.39	8	2.18	27.52
			9	2.05	29.80

classroom instruction over a short lesson from the curriculum unit using a five-point rating scale. Eight teaching behaviors were rated for enthusiasm.

Data in the form of achievement class means and ratings of teacher enthusiasm are simulated in Table 13.11 for teachers in the training and no training groups. The teaching enthusiasm ratings are mean ratings over the eight teacher behaviors. Use these data to respond to the problems that follow.

1. We would expect the training group to demonstrate more enthusiastic behavior than the no training group if training was effective. Test the compatible null hypothesis to see if the data support this expectation. Use a directional alternative hypothesis and a .01 level of significance.

2. Test the effect of teacher enthusiasm training on student achievement outcomes. Use $\alpha = .05$. Would you be willing to infer that teacher-enthusiasm training for first-year teachers is likely to result in higher student achievement scores?

Problem Set C

The "Sesame Street" data base is to be used as the data source for the following problems.

1. The population mean on the Which Comes First Test is 5.95 and the population variance is 8.07. For a random sample of eight individuals from Site I (037, 020, 009, 056, 053, 023, 049, 028) use statistical inference procedures to test to see if the sample mean is likely to occur when random sampling from the population of scores. Use an $\alpha = .10$ and a nondirectional hypothesis.

2. Using the same eight individuals sampled for problem 1, test whether their mean Relational Terms pretest score is likely as a result of random sampling from the population set of scores. The only information given is that the population mean Relational Terms pretest score is 9.94. Use $\alpha = .05$ and a nondirectional test.

Two random samples of 10 individuals drawn from the data base are listed in Table 13.12. One sample was selected from those individuals in the first viewing category and the second sample from those individuals in the fourth viewing category. Record Numbers pretest *and* posttest scores for each individual selected. Then state and test null hypotheses (at .05 level) that will allow you to draw conclusions in problems 3–6.

TABLE 13.12
Random samples from two viewing categories.

Random Sample from Viewing Category 1		Random Sample from Viewing Category 4	
004	199	018	171
053	200	035	205
182	201	095	214
184	213	107	233
193	231	156	235

3. Did individuals who viewed "Sesame Street" less have the same or different mean Numbers pretest score than those who viewed "Sesame Street" more?

4. Did individuals who viewed "Sesame Street" less have the same or different mean Numbers posttest score than those who viewed "Sesame Street" more?

5. Did individuals who viewed "Sesame Street" less learn anything about numbers as measured by their pre- and posttest scores?

6. Did individuals who viewed "Sesame Street" more learn anything about numbers as measured by their pre- and posttest scores?

7. What conclusion would you draw based on the data analysis results for problems 3–6?

14

Hypotheses about Other Statistics

The general hypothesis-testing steps should be familiar by now. Consistency in applying the process across different hypotheses is important. The actual process and the statistical reasoning behind it are constant. So far we have used the hypothesis testing process to test null hypotheses for four distinct research situations:

1. a single sample mean where σ_x^2 is known, H_0: $\mu_0 = a$
2. a single sample mean where σ_x^2 is unknown, H_0: $\mu_0 = a$
3. two independent sample means where σ_x^2 and σ_y^2 are unknown, H_0: $\mu_x = \mu_y$
4. two dependent (correlated) sample means where σ_x^2 and σ_y^2 are unknown, H_0: $\mu_x = \mu_y$

This chapter expands that list to include the statistical methods for testing simple hypotheses about correlation coefficients, variances, and proportions. It emphasizes the appropriate test statistic to be used for testing a given H_0 and the statistical formula for calculating the observed test statistic (ts_{obs}). The statistical formula is designed to transform an observed sample statistic, appropriate for testing a given H_0, to a test statistic value having known probability distribution characteristics. Using critical values from a known probability distribution to compare with ts_{obs} allows us either to reject or not reject H_0. The correct alignment between the specific H_0, the observed test statistic formula, and the appropriate probability distribution for identifying critical values is essential for a valid test of any H_0. These elements require careful evaluation as we consider each different null hypothesis situation. What remains for testing any H_0 is to follow the steps discussed previously and reviewed in the next section.

14.1 REVIEW: THE HYPOTHESIS TESTING PROCESS

The statistical hypothesis testing process needs to be so ingrained that it becomes routine. For each distinct H_0 we need to know which test statistic formula and which probability distribution to use in identifying the critical values that are the criteria to test H_0. Let's review the sequence of hypothesis testing.

- Step 1 Translate the research question into a testable null hypothesis (H_0) about a population parameter of interest.
- Step 2 Identify a directional or nondirectional alternative hypothesis (H_A).

- Step 3 Set a probability criterion level for rejecting H_0 by specifying a level of significance, α.
- Step 4 Calculate the sample statistic value(s) needed to test H_0.
- Step 5 Convert the sample statistic value(s) to an observed test statistic based on the characteristics of the sampling distribution for the statistic. That is, compute ts_{obs} using the observed test statistic formula.
- Step 6 Determine the region(s) of rejection in the probability distribution appropriate for testing H_0. That is, given the α level and H_A, identify the critical test statistic value(s), ts_{crit}, to be used as the criteria against which we compare ts_{obs}.
- Step 7 Compare ts_{obs} to ts_{crit}. Decide to reject H_0 if ts_{obs} falls in the rejection region of the probability distribution; otherwise, do not reject H_0.
- Step 8 If you reject H_0, conclude that H_A is the true state of affairs knowing that your decision to reject H_0 if it is really true has an α% probability of being a Type I decision error. If you do not reject H_0, you cannot conclude that H_0 is true. Rather, conclude that there is insufficient evidence to reject H_0.
- Step 9 Form inferences and generalizations about the population relative to the initial research question based on your decision to reject or not to reject H_0.

14.2 REVIEW: THE IMPORTANCE OF SAMPLING DISTRIBUTIONS AS PROBABILITY DISTRIBUTIONS

To test statistically a null hypothesis, the sampling distribution of the appropriate sample statistic must be identifiable with a known probability distribution shape. The probability distribution forms a basis for our expectations about the relative frequency with which certain kinds of sample events should occur. Without knowing the characteristics of such distributions, we have no way to determine whether a specific sample event is probable or improbable when sampling from a population with parameters defined by the null hypothesis. The sampling distribution as a probability distribution provides the norm for evaluating the truth or falsity of H_0.

The population parameter(s) specified in the null hypothesis of Chapter 13 defined the sampling distributions of interest for testing H_0. When testing the single sample null hypothesis that the population mean is equal to some constant, H_0: $\mu = a$, the characteristics of the sampling distribution of single sample means (the $\bar{X}$s) were of interest. When testing the null hypothesis about the equality of two sample means H_0: $\mu_x = \mu_y$, the characteristics of the sampling distribution of mean differences $(\bar{X} - \bar{Y})$ were of interest. Recall that the form of this last hypothesis, H_0: $\mu_x = \mu_y$, is used to represent two separate and distinct sampling distributions, one in which $\overline{X}$ and $\overline{Y}$ as sample events are point estimates of μ_x and μ_y from two independent samples. The other null hypothesis sampling distribution of $\overline{X} - \overline{Y}$ occurs when two sets of data on the same variable have been collected from the same or paired individuals resulting in the dependent (correlated) sample case. The independent and dependent sample tests of H_0: $\mu_x = \mu_y$ both use knowledge about the sampling distributions of $(\overline{X} - \overline{Y})$s. The difference in the two sampling distributions is that the standard errors of $\overline{X} - \overline{Y}$, that is, the standard deviations of the sampling distributions of mean differences, are not the same. The variability of sample mean differences is greater for independent samples than it is for dependent samples ($\sigma_{(\overline{x}-\overline{y})\text{indep}} > \sigma_{(\overline{x}-\overline{y})\text{dep}}$).

While the sampling distributions of raw score sample statistical values are important in many instances, the companion probability distribution of test statistic values is most important. The theoretical probability sampling distribution of observed test statistic values (ts_{obs}) provides a standardized basis for determining if a sample outcome falls within the expected range of occurrences defined by a sampling distribution *when H_0 is assumed to be true.* Transforming sample data into an observed test statistic value using a statistical formula is a necessary step in hypothesis testing. That is why so much statistical content focuses on computation. We need to summarize raw sample data into a sample test statistic value, which we then compare to critical values in the appropriate probability sampling distribution. Based on comparison, we infer the truth or falsity of H_0.

The observed test statistic formulas are labeled according to their probability sampling distributions. For example, the statistical formula for testing H_0: $\mu = a$ when σ_x^2 is known results in an approximately normally distributed test statistic score. The test statistic in this case is a z score; the formula, a z score formula. The probability distribution used in the test of H_0 would naturally be the unit-normal distribution. The test of H_0: $\mu_x = \mu_y$ for independent groups where σ_x^2 and σ_y^2 are unknown is commonly referred to as a "t test". Using a t test implies that the statistical formula results in an observed test statistic that displays the characteristics of a t distribution. The t distribution with $df = n_x + n_y - 2$ is the probability distribution to use in this case in determining critical values for the t test of H_0: $\mu_x = \mu_y$. While we test many hypotheses using test statistics distributed as z (z tests) or t (t tests), all are not. Some hypotheses result in sampling distributions with standardized characteristics that do not conform to a z or t distribution. Rather, they often take the shape of one of two other common statistical probability distributions, the chi square (χ^2) distribution or the F distribution. The tests of some hypotheses result in "chi square tests" or "F tests" of H_0. The critical values used as criteria in testing H_0 will be found in a chi square–probability distribution or in an F probability distribution for these latter tests of H_0 in the same way as the critical values are tabled in the normal curve or t distribution for z tests or t tests, respectively.

14.3 PROBABILITY DISTRIBUTIONS: z, t, χ^2, AND F DISTRIBUTIONS

The z, t, χ^2, and F statistical probability distributions form the core of distributions needed to test many null hypotheses across all kinds of research situations. Their utility in hypothesis testing is in identifying *critical* test statistic values to be used as criteria when deciding whether to reject or not to reject H_0. Critical values of z (z_{crit}), t (t_{crit}), χ^2 (χ^2_{crit}), or F (F_{crit}) are the criteria against which we compare an observed test statistic in the decision process. Critical values in the χ^2 and F distributions partition the distribution into proportional regions of rejection and nonrejection equivalent to values of α and $1 - \alpha$, just as z_{crit} and t_{crit} do in their respective distributions.

The Unit-Normal z Distribution

The unit-normal distribution has been discussed at length. You are no doubt familiar with its characteristics: bell-shaped, $\bar{z} = 0$, and $\sigma_z = 1$. Its importance as a statistical distribution should be quite evident. First, it tends to describe accurately the standardized distribution of many variables in the behavioral sciences. Second, the assumption of a normally distributed set of scores is important in the mathematical development and interpretation of many statistical formulas and indices. Third, and most important for this chapter's focus, it has important applications as a theoretical probability distribution in testing certain statistical hypotheses. In addition, the normal curve because of its familiarity becomes the standard reference for interpreting other distribution shapes.

The t Distribution

Chapter 13 discussed the characteristics of the family of t distributions. The distribution of t values depends on the degrees of freedom (df). A different t distribution exists for every distinct df value. The shape of all t distributions is leptokurtic and symmetrical around a mean t score of 0. The variability of t scores is a function of df with $\sigma_t^2 = df/(df - 2)$. As df increases, the variance approaches a limit of 1.00 and the t distribution approaches the characteristics of the unit-normal distribution. For all practical situations, however, t scores will vary slightly more than z scores. The increased variability plus the leptokurtic shape of a t distribution result in an expected greater frequency of t scores in the tails of the t distribution than we would expect of z scores in the normal distribution. The implication for hypothesis testing is that the absolute value of t_{crit} will always be larger than the absolute value of z_{crit} for the same α level and H_A. To demonstrate, compare the critical values of t and z that we

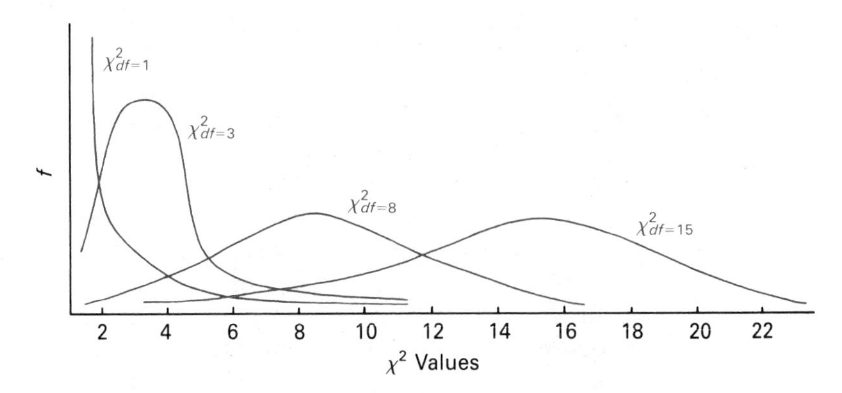

FIGURE 14.1
χ^2 distributions for 1, 3, 8, and 15 *df*.

would use in a two-tailed test of H_0: $\mu = a$ for an α level of .20. Assume a sample size of 18.[1]

The Chi Square (χ^2) Distribution

The χ^2 distribution is unlike the unit-normal z distribution and the t distribution because it is not symmetrically distributed around a mean of 0. A chi square distribution is a positively skewed distribution whose shape, mean, and variance depend on the degrees of freedom. It is like the t distribution, since a different chi square distribution exists for each distinct value of *df* rather than existing as a single constant distribution like the unit-normal distribution. The degree of positive skewness for a chi square distribution with *df* degrees of freedom is given by $(8/df)^{1/2}$ where a value of 0 indicates no skewness. This reveals that as *df* increases the χ^2 distribution becomes less and less skewed. The mean of a chi square distribution equals its number of degrees of freedom ($\overline{\chi^2} = df$), and its variance equals 2 *df*. As *df* increases, the distribution shape becomes less positively skewed and tends to approach symmetry, the center (mean) of the distribution moves to the right on the χ^2 scale, and the distribution becomes more variable. Figure 14.1 illustrates these characteristics for several different χ^2 distributions.

The best way to illustrate a specific χ^2 distribution accurately is using a distribution of squared z scores. Imagine a large population of normally distributed raw scores. Suppose we randomly select a single score (X) from the population and transform it to a z score:

$$z = \frac{X - \mu}{\sigma_x}$$

[1]$z_{crit} = \pm 1.28$; t_{crit} with 17 $df = \pm 1.333$. Note in the t table that only when the degrees of freedom associated with t reach infinity (∞) are t values and z values at specified αs identical. At all points less than infinity, t values are greater than z values are for fixed αs.

If we continue to do this over many repeated samples of single X scores, we know that the distribution of the z scores will be normal with a mean of 0 and standard deviation of 1. If we had squared each z score, however,

$$z^2 = \frac{(X - \mu)^2}{\sigma_x^2}$$

and graphed the frequency distribution of z^2, we would not have a normally distributed shape. Rather, the distribution of single z^2 scores over repeated random samplings would be extremely positively skewed and form a chi square distribution with one degree of freedom. Squared z scores for repeated random samples of size 1 from a normally distributed population of scores are distributed as chi square with 1 *df*: $\chi^2_{df=1} = z^2 = (X - \mu)^2/\sigma_x^2$. Chi square values are squared deviation values. The chi square distribution with one degree of freedom is one of the χ^2 distributions shown in Figure 14.1. The χ^2 distribution with 1 *df* has a mean of 1 and a variance of 2. Think about the distribution of squared z scores in terms of your own expectations. Approximately 68% of the z scores in a normal distribution lie between z values of -1.00 and 1.00. The square of these z scores will be less than or equal to 1, implying that 68% of the z^2 values (χ^2 values) will lie between 0 and 1 in the chi square distribution. Approximately what percentage of the z^2 scores or χ^2 values in the χ^2 distribution with 1 *df* would be less than or equal to 4.00?[2] Actually, the 95th percentile in the χ^2 distribution with 1 *df* equals $\chi^2 = z^2 = 1.96^2 = 3.84$. If we were using the χ^2 distribution with 1 *df* as our probability distribution for testing an H_0, the critical chi square value χ^2_{crit} for $\alpha = .05$ would equal 3.84.

Chi square values with two degrees of freedom can be illustrated by drawing two scores (X_1 and X_2) at random from a normally distributed population of scores, transforming the scores to z scores, squaring the z scores, and adding them:

$$\chi^2_{df=2} = \frac{(X_1 - \mu)^2}{\sigma^2} + \frac{(X_2 - \mu)^2}{\sigma^2}$$
$$= z_1^2 + z_2^2$$

For repeated random samplings of two scores, the χ^2 values with 2 *df* obtained would distribute themselves around a mean χ^2 value of 2 with a variance of 4. Similarly, a χ^2 distribution with *df* degrees of freedom may be thought of as the distribution of χ^2 values over repeated samplings formed by adding the z^2 quantities for a random sample of *df* scores.

$$\chi^2_{df} = z_1^2 + z_2^2 + \cdots + z_{df-1}^2 + z_{df}^2$$

[2]In the normal distribution, approximately 95% of the z scores lie in the interval between $z = -2.00$ and $z = 2.00$. The square of any z score in this range would be less than or equal to 4.00. Approximately 95% of the values with 1 *df* would be expected to be less than or equal to 4.00.

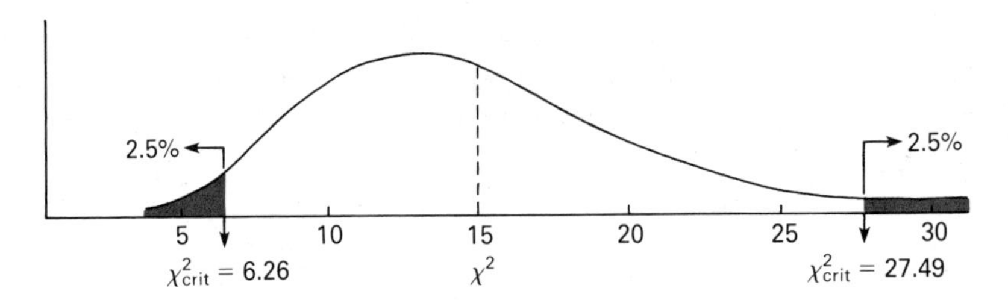

FIGURE 14.2
Hypothesis testing with a χ^2 distribution, $df = 15$.

The χ^2 values so formed would distribute themselves around a mean value of *df* with a variance equal to 2 *df*. Figure 14.1 illustrates the chi square distributions for 1, 3, 8, and 15 degrees of freedom.

Our interest in the χ^2 distribution centers on its use as a probability sampling distribution in testing statistical hypotheses. As such, the critical χ^2 values associated with specific levels of α become the essential information to obtain from a given χ^2 distribution. Because there is a family of chi square distributions, tabling χ^2 values at specific percentile points for distinct *df* values follows the pattern used to table percentile points in the *t* distribution: Only values of χ^2 associated with areas in the tails of a chi square distribution are tabled. These values of χ^2 for a variety of *df*s appear in Appendix B, Table B.4. Table B.4 lists χ^2 values at designated percentile points for a specific χ^2 distribution in each row. The specific χ^2 distribution is identified for the *df* value given at the beginning of each row in the first column. All other columns are headed by specific percentile values that identify the proportion of the chi square distribution above (to the right of) the χ^2 value given in the table. For example, the χ^2 value at the 90th percentile in the distribution with 10 *df* is found at the intersection of the row identified by 10 in the *df* column and the column headed by .100. The χ^2 value of 15.987 is the 90th percentile point in the chi square distribution with 10 degrees of freedom. Ninety percent of the χ^2 values with 10 *df* will be less than or equal to 15.987; 10% of the chi square values could be expected to be greater than 15.987 on a random basis. What would the chi square value be at the 10th percentile in the chi square distribution with $df = 10$?[3]

For use in testing hypotheses, we will be particularly interested in the χ^2 values at percentile points associated with fixed α levels. If α equaled .05 and we wanted equal rejection regions in each tail of the χ^2 distribution, the χ^2 values at the 2.5th and 97.5th percentiles become the critical values (χ^2_{crit}) used as criteria for comparing χ^2_{obs} when deciding whether to reject H_0. If the hypothesis situation results in, for instance, $df = 15$, the critical χ^2 values for the 2.5th and 97.5th percentiles are 6.26 and 27.49, as Figure 14.2 shows. In

[3]The chi square value at the 10th percentile with 10 *df* is 4.865.

practice, very few χ^2 tests of hypotheses require finding a critical χ^2 value in the lower tail of the distribution. The additive nature that defines most chi square formulas results in only large χ^2_{obs} values if H_0 is not true. For this reason, we treat the majority of χ^2 tests as though they were directional in nature with a single critical value such that the entire rejection region is in the upper tail of the χ^2 distribution. Find the value of χ^2_{crit} for $\alpha = .10$, $df = 2$, and the entire rejection region in the upper tail of the χ^2 distribution.[4]

The *F* Distribution

An *F* distribution is related to the prior three probability distributions. An *F* value can be defined as the ratio of two independent chi square values each divided by their respective degrees of freedom.

$$F = \frac{\frac{\chi^2_n}{df_n}}{\frac{\chi^2_d}{df_d}}$$

The df_n is the degrees of freedom associated with the numerator χ^2 value, and df_d is the degrees of freedom for the denominator χ^2 value. The distribution of *F* scores calculated over repeated random samples, resulting in the ratio of two independent chi square values for fixed values of df_n and df_d, defines an *F* distribution with df_n and df_d degrees of freedom. A distinct *F* distribution exists for the ratio of chi square values defined for all combinations of numerator and denominator degrees of freedom. The shape of an *F* distribution depends on the df_n and df_d values. An *F* distribution cannot be used without knowing values of both df_n and df_d. We will adopt the notation $F_{(df_n, df_d)}$ to identify a specific *F* distribution with df_n numerator degrees of freedom and df_d denominator degrees of freedom. Following this notation, $F_{(4,12)}$ refers to the *F* distribution formed by a ratio of chi square values with 4 degrees of freedom in the numerator and 12 degrees of freedom in the denominator.

All *F* ratios are positive. The distribution of *F* values for any pair of df_n and df_d is unimodal and positively skewed. The mean of an *F* distribution equals $df_d/(df_d - 2)$ for all distributions in which $df_d \geq 3$. You can see from the formula that the mean of all *F* distributions will be slightly greater than 1.00. How much greater depends on the number of degrees of freedom in the denominator of the *F* ratios forming the distribution. As either df_n or df_d increases in size, the *F* distribution is less skewed with the upper tail of the distribution being drawn closer to the center of the distribution around an approximate *F* ratio of 1. Figure 14.3 illustrates the *F* distributions for $F_{(2,5)}$, $F_{(5,5)}$, and $F_{(2,20)}$.

As with the χ^2 distribution, most tests of hypotheses using the *F* distribu-

[4] $\chi^2_{crit} = 4.605$.

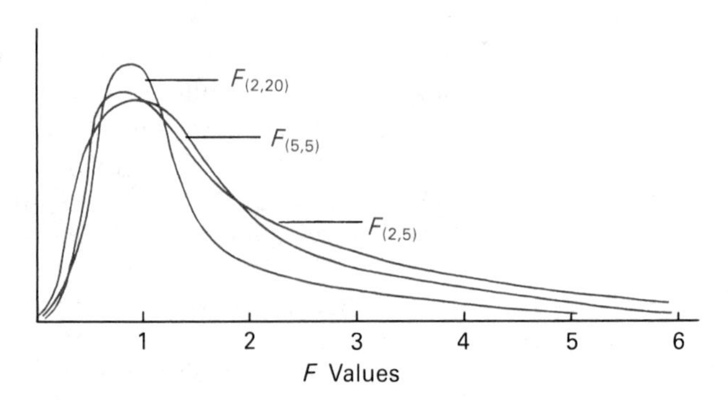

FIGURE 14.3
F distribution for $F_{(2,5)}$, $F_{(5,5)}$, and $F_{(2,20)}$.

tion as the probability distribution require determining a critical *F* value with the entire $\alpha\%$ rejection region in the upper tail of the *F* distribution. For this reason and because two degrees of freedom values need be specified for each *F* distribution, only the *F* values for the upper percentile points corresponding to standard α levels appear in Appendix B, Table B.5. The columns of Table B.5 correspond to the df_n values and the rows correspond to values of df_d. Assuming $\alpha = .05$, $df_n = 3$, and $df_d = 20$ and the entire 5% region of rejection in the upper tail of the distribution, the critical *F* ratio used as the criterion for testing H_0 would be $F_{crit} = 3.10$. For a hypothesis situation in which $F_{(3,20)}$ is the appropriate test probability distribution, we would expect random sampling to result in an observed sample *F* ratio (F_{obs}) greater than 3.10 approximately 5% of the time *when H_0 is true.* The remaining 95% of the time $(1 - \alpha)$, F_{obs} should be less than 3.10 if H_0 is true. Check the upper critical value at the .01 level of significance for $F_{(5,10)}$.[5]

While only the upper percentile points in the *F* distribution appear in Table B.5, the lower percentile points can be calculated from the following relationship. If ${}_pF_{(df_n,df_d)}$ is the *F* value at the upper *p*th percentile point, then the lower $(1 - p)$th percentile point is

$$_{1-p}F_{(df_n,df_d)} = \frac{1}{_pF_{(df_d,df_n)}} \text{(Note reversal of } df_n \text{ and } df_d\text{)}$$

For example, the *F* value at the 5th percentile for $F_{(2,10)}$ equals 1 divided by the tabled *F* value at the 95th percentile for $F_{(10,2)}$.

$$_{.05}F_{(2,10)} = \frac{1}{_{.95}F_{(10,2)}} = \frac{1}{19.40} = .0515$$

Relationships Among the *z*, *t*, χ^2, and *F* Distributions

Without mathematical derivation or proof, we now briefly identify some interesting interrelationships among the four families of distributions just pre-

[5] $F_{crit(5,10)} = 5.64$

sented. First, the t, χ^2, and F distributions are all based on the normal distribution. We have seen that the normal distribution is the limiting distribution for t as df becomes large. In fact, for $df > 40$ there is very little difference between the t distribution values and values of z in the unit-normal distribution. For the chi square distribution, we have seen that a chi square value is formed by summing squared z scores from a normal distribution. The F distribution is related to the normal distribution through the χ^2 values whose ratios form an F distribution.

In examining the t distribution more closely, we see that a t value with df degrees of freedom may be defined as the ratio of a unit normal variate z to the positive square root of a chi square variate with df degrees of freedom.

$$t_{df} = \frac{z}{\sqrt{\frac{\chi^2_{df}}{df}}}$$

Squaring this value relates t to an F distribution.

$$t^2_{df} = \frac{z^2}{\frac{\chi^2_{df}}{df}}$$

The numerator value z^2 is a chi square variable with 1 degree of freedom. The t^2 value is really the ratio of two χ^2 values that is distributed as F with one degree of freedom in the numerator and df degrees of freedom in the denominator.

$$_{(1-\alpha/2)}t^2_{df} = {}_{(1-\alpha)}F_{(1,df_d)}$$

We can confirm this relationship by checking the t values in Table B.3 and the corresponding F values for $df_n = 1$ in Table B.5. It turns out that the squared value of t_{df} at the $1 - \alpha/2$ percentile is equal to the value of $F_{(1,df)}$ at the $1 - \alpha$ percentile level. For example, the t value at the 97.5th percentile with $df = 20$ is 2.086. This value squared ($2.086^2 = 4.35$) is the value in Table B.5 for F with $df_n = 1$ and $df_d = 20$ at the 95th percentile. You should check a few more values to confirm this relationship between t^2 and F and your understanding of the relationship.

Each of the four distributions serves the same function as probability sampling distributions in testing null hypotheses. An observed test statistic calculated from a statistical formula corresponds to the appropriate distribution, and when compared to the probability distribution, it indicates the likelihood of the sample result occurring if H_0 is true. The test of an H_0 really asks "How probable is it that I would get an F ratio (or χ^2, t, or z value) this large merely as a result of random sampling if H_0 is true?" The appropriate probability distribution gives us the answer to the question and allows us to determine the truth or falsity of H_0.

The remainder of this chapter presents the statistical methods necessary to test a variety of different null hypotheses. The presentation of methods for testing each H_0 will systematically take the following form:

1. clarification of the null hypothesis
2. identification of the appropriate sampling distribution characteristics
3. specification of the test statistic formula for testing H_0
4. presentation of an example test of the specific H_0

It is assumed that testing a given H_0 would follow the steps outlined in Section 14.1. The test statistic formula will be presented as a z test, t test, χ^2 test, or F test, implying that the observed test statistic score resulting from application of the statistical formula is distributed as z, t, χ^2, or F. The z, t, χ^2, or F probability distribution should then be used to determine the appropriate critical value(s) to be used in testing H_0.

14.4 TESTING HYPOTHESES ABOUT CORRELATION COEFFICIENTS

The Single Sample Case, H_0: $\rho_{xy} = .00$

Null hypothesis. H_0: $\rho_{xy} = .00$ is the basic form of the null hypothesis of no relationship between two variables (X and Y) for a single sample of individuals where ρ_{xy} represents the population correlation index. Rejecting H_0 would imply that a nonzero relationship between variable X and variable Y probably exists. If the alternative hypothesis H_A: $\rho_{xy} < .00$ is specified and H_0 rejected, the evidence would support a negative relationship between X and Y in the population. The alternative H_A: $\rho_{xy} > .00$ would imply the existence of a positive relationship.

Sampling distribution. The sampling distribution of Pearson correlation coefficients when sampling at random from a population having $\rho_{xy} = .00$ is approximately normally distributed around a mean value of zero for sample sizes > 30. An approximate normal distribution only occurs when $\rho_{xy} = .00$, however. For all other values of ρ_{xy}, the sampling distribution will be negatively skewed for all $\rho_{xy} > .00$ and positively skewed for all $\rho_{xy} < .00$. The standard error of the sampling distribution of r_{xy} for random samples of size n from a bivariate normal population when $\rho_{xy} = .00$ is $\sigma_{r_{xy}} = 1/(n - 1)^{1/2}$. When $\rho_{xy} \neq .00$, the standard error of r_{xy} is $\sigma_{r_{xy}} = [(1 - \rho_{xy^2})/(n - 1)]^{1/2}$. In practice, the standard error of r_{xy} is estimated as

$$s_{r_{xy}} = \sqrt{\frac{1 - r_{xy}^2}{n - 2}}$$

Test statistic. The statistical formula for testing H_0: $\rho_{xy} = .00$ follows the general format of changing the amount the sample r_{xy} value deviates from the hypothesized value of .00 ($r_{xy} - .00$) to standard error units resulting in an observed standard score, $ts_{obs} = (r_{xy} - .00)/s_{r_{xy}}$. The division of $r_{xy} - .00$ by the standard error of r_{xy} is distributed as a t distribution with $n - 2$ degrees of freedom. The observed test statistic formula that exactly describes the sampling distribution of r_{xy} when $\rho_{xy} = .00$ is

$$t_{obs} = \frac{r_{xy}}{\sqrt{\dfrac{1 - r_{xy}^2}{n - 2}}} \tag{14.1}$$

When H_0: $\rho_{xy} = .00$ is true, t_{obs} from formula 14.1 is distributed as a t distribution with $df = n - 2$. The t distribution with $n - 2$ df is the appropriate probability distribution for determining the critical values at an α level for testing H_0: $\rho_{xy} = .00$.

Example. We can use data from the "Sesame Street" data base to investigate whether there is any relationship between parental educational aspiration level and children's learning of school-related content prior to the "Sesame Street" evaluation. Focusing on knowledge of letters as the school-related content, we translate the research question into the null hypothesis H_0: There exists no relationship between parental educational aspiration level (X) and knowledge about Letters pretest scores (Y) for children in the "Sesame Street" evaluation (H_0: $\rho_{xy} = .00$). The alternative hypothesis might be legitimately directional in predicting a positive relationship, such as H_A: $\rho_{xy} > .00$. Scores on parental education aspiration level (HEI variable 24) and Letters pretest (variable 8) for 15 children were randomly sampled from the Appendix A data base. The summary data for these scores were as follows:

Parental Education Aspiration Level (X)		Letters Pretest (Y)
$\Sigma X = 251$		$\Sigma Y = 203$
$\Sigma X^2 = 4303$		$\Sigma Y^2 = 3445$
	$\Sigma XY = 3574$	
$\overline{X} = 16.73$		$\overline{Y} = 13.53$
$s_x = 2.71$		$s_y = 7.06$
	$r_{xy} = .661$	

Substituting the value of $r_{xy} = .661$ in the test statistic formula 14.1,

$$t_{obs} = \frac{.661}{\sqrt{\dfrac{1 - .661^2}{15 - 2}}} = \frac{.661}{.208} = 3.18$$

Using an α level of .05, the critical value of t for a directional test of H_0 with 13 *df* from Table B.3 is

$$t_{crit} = 1.771$$

Because $t_{obs} = 3.18$ is greater than $t_{crit} = 1.771$, the decision is to reject H_0: $\rho_{xy} = .00$. Based on the sample data and test of H_0, the statistical inference is that a nonzero relationship between parental educational aspiration level and pretest knowledge of letters does exist and that the relationship is positive. The sample point estimate of $r_{xy} = .661$ indicates that the relationship is moderately high.

Separate data on the same two "Sesame Street" variables were randomly sampled from Appendix A for a sample of 10 individuals. The summation values for the data are given here. Test the same null hypothesis using a *nondirectional* test of H_0 at $\alpha = .05$.[6]

$$\Sigma X = 140 \qquad \Sigma Y = 141$$

$$\Sigma X^2 = 2172 \qquad \Sigma Y^2 = 2209$$

$$\Sigma XY = 2111$$

The Single Sample Case, H_0: $\rho_{xy} = a$

Null hypothesis. This null hypothesis differs from the previous hypothesis (H_0: $\rho_{xy} = .00$) in that the value of a may be nonzero. H_0: $\rho_{xy} = a$ where a is some constant specified correlation value between -1.00 and 1.00 hypothesizes that the population relationship as measured by the Pearson r_{xy} is equal to some constant value (a) such as H_0: $\rho_{xy} = .48$ or H_0: $\rho_{xy} = -.76$.

Sampling distribution. The sampling distribution of r_{xy} when $\rho_{xy} < .00$ is positively skewed. The degree of skewness increases as the true population ρ_{xy} gets closer to -1.00. When $\rho_{xy} > .00$, the sampling distribution exhibits an increasingly negative skew as the true value of ρ_{xy} approaches 1.00. The resulting skewness of the sampling distribution of r_{xy} should be intuitively apparent because of the ceiling placed on the absolute value of r_{xy} of 1.00. Figure 14.4 illustrates the sampling distributions of r_{xy} when $\rho_{xy} = -.92$ and $\rho_{xy} = -.35$. The sample r_{xy} values over repeated random samplings will distribute around the true population value ρ_{xy}.

Because of the skewed nature of the sampling distribution of r_{xy}, depend-

---

[6] $$r_{xy} = \frac{10(2111) - (140)(141)}{\sqrt{[10(2172) - 140^2][10(2209) - 141^2]}} = .633$$

$$t_{obs} = \frac{.633}{\sqrt{\dfrac{1 - .633^2}{10 - 2}}} = \frac{.633}{.274} = 2.31$$

t_{crit} with $df = 8$ at the .05 level is 2.306.

$t_{obs} \geq t_{crit}$; reject H_0.

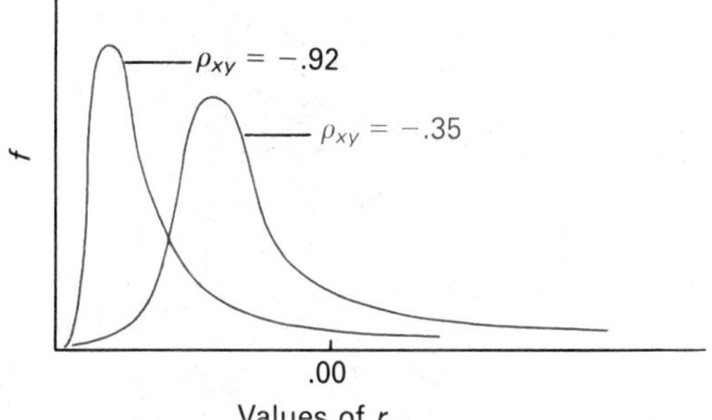

FIGURE 14.4
Sampling distribution of r_{xy} when $\rho_{xy} = -.92$ and $\rho_{xy} = -.35$.

ing on the value of ρ_{xy}, an appropriate direct test of H_0: $\rho_{xy} = a$ using the sampling distribution information is not available. Rather, we normalize the sampling distribution of r_{xy} using Fisher's Z transformation. The sample values of r_{xy} after Fisher's Z transformation into Z_r scores, form a sampling distribution of Z_r having an approximate normal distribution, a mean equal to ρ_{z_r} (the Fisher's Z transformation of the true population correlation coefficient), and a standard deviation (standard error of Z_r) of $1/(n - 3)^{1/2}$. The equivalent Fisher's Z transformation scores (Z_r scores) for values of r_{xy} in intervals of .01 appear in Appendix B, Table B.6. Figure 14.5 illustrates the equivalent sampling distribution of Z_r after Fisher's Z transformation when $\rho_{xy} = -.92$.

Test statistic. To test H_0, we first calculate r_{xy} from the sample data and then transform both the sample value of r_{xy} and the hypothesized value $\rho_{xy} = a$ to Fisher's Z scores, Z_r and Z_a, respectively. We divide the discrepancy between Z_r and Z_a by the standard error of Z_r to change it to an observed test statistic score. The observed test statistic formula is

$$z_{\text{obs}} = \frac{Z_r - Z_a}{\dfrac{1}{\sqrt{n - 3}}} \tag{14.2}$$

where Z_r is the Fisher's Z-transformed value of the sample r_{xy} value, Z_a is the Z-transformed value for the null hypothesized value $\rho_{xy} = a$, and n is the

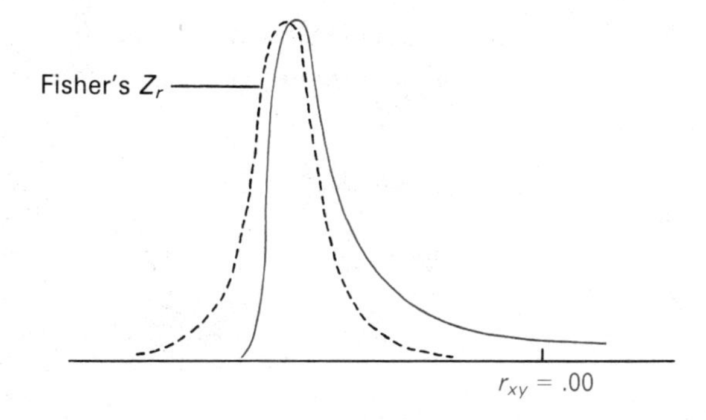

FIGURE 14.5
Fisher's Z_r transformation of r_{xy} values when $\rho_{xy} = -.92$.

sample size. Values of z_{obs} are distributed as a unit-normal distribution that becomes the appropriate probability distribution for determining the critical value(s) (z_{crit}) needed to test H_0: $\rho_{xy} = a$ at the α level of significance.

Example. The literature indicates that ability accounts for approximately 50% of the variability in achievement scores. If this is true for a population of preschool children, the population correlation between scores on an ability measure and an achievement measure would be hypothesized to be about .70, i.e., H_0: $\rho_{xy} = .70$. Let's test this hypothesis using "Sesame Street" data on Peabody mental age (X) and Classification Skills pretest (Y) scores. Summary information for a random sample of 19 individuals is provided. The hypothesis testing conditions are H_0: $\rho_{xy} = .70$, H_A: $\rho_{xy} \neq .70$, and $\alpha = .10$. Summary sample information ($n = 19$) is

$$\Sigma X = 718 \qquad \Sigma Y = 205$$
$$\Sigma X^2 = 28{,}304 \qquad \Sigma Y^2 = 2437$$
$$\Sigma XY = 8023$$
$$\overline{X} = 37.79 \qquad \overline{Y} = 10.79$$
$$s_x = 8.07 \qquad s_y = 3.54$$
$$r_{xy} = .54$$

To test H_0: $\rho_{xy} = .70$, we transformed the sample r_{xy} value of .54 and the null-hypothesized value of .70 to Fisher's Z scores using Table B.6, $Z_{.54} = .604$, and $Z_{.70} = .867$. The test statistic value is

$$z_{\text{obs}} = \frac{.604 - .867}{\dfrac{1}{\sqrt{19 - 3}}} = \frac{-.263}{.25} = -1.05$$

The critical values of z for rejecting H_0 at the .10 level of α and a nondirectional test are $z_{\text{crit}} = \pm 1.645$. The z_{obs} value of -1.05 lies in the nonrejection region. We should decide not to reject H_0: $\rho_{xy} = .70$. While our sample correlation $r_{xy} = .54$ differs descriptively from the null-hypothesized value $\rho_{xy} = .70$, the difference is not sufficient to allow us to reject H_0 and conclude that $\rho_{xy} \neq .70$. For random samples of size 19 drawn from an assumed population with $\rho_{xy} = .70$, a sample value of .54 is not improbable enough using our criterion of $\alpha = .10$ and H_A: $\rho_{xy} \neq .70$. Would we have arrived at the same conclusion if we obtained the same r_{xy} value of .54 from a sample of size 52?[7]

[7]When $n = 52$, the standard error of r_{xy} becomes smaller, increasing the precision with which H_0 is tested.

$$z_{\text{obs}} = \frac{.604 - .867}{\dfrac{1}{\sqrt{52 - 3}}} = \frac{-.263}{.143} = -1.84$$

The z_{obs} value of -1.84 lies beyond $z_{\text{crit}} = -1.645$. We would decide to reject H_0: $\rho_{xy} = .70$. For samples of size 52, an r_{xy} of .54 is improbable if H_0: $\rho_{xy} = .70$ is really true given $\alpha = .10$ and a two-tailed test.

What if the sample size were the same ($n = 19$), but $r_{xy} = .86$?[8] A sample r_{xy} value of .86 is the same absolute discrepancy of .16 units from $\rho_{xy} = .70$ as $r_{xy} = .54$ ($.54 - .70 = -.16$), but when transformed to Fisher's Z_r scores, it represents a greater discrepancy from the mean of the sampling distribution of Z_r scores. In a negatively skewed sampling distribution of r_{xy} for $\rho_{xy} = .70$, a random sample with an r_{xy} value of .86 or greater (.16 units above $\rho_{xy} = .70$) is less likely to occur than r_{xy} values of equal distance from ρ_{xy} but below the value of .70 in the sampling distribution (e.g., $r_{xy} = .54$). As a standard generalization, sample correlations beyond ρ_{xy} have disproportionately larger Z_r scores as they get closer to ± 1.00 and are less likely to occur as a result of random sampling relative to r_{xy} values of equal distance below ρ_{xy}.

The Two Independent-Sample Case, H_0: $\rho_{xy_1} = \rho_{xy_2}$

Null hypothesis. The null hypothesis, H_0: $\rho_{xy_1} = \rho_{xy_2}$, compares two correlation coefficients each identifying the relationship between the same two variables but for different populations. It is a statement that the relationship between variables X and Y for population 1 is the same as the magnitude and direction of the relationship between X and Y for population 2. H_0: $\rho_{xy_1} - \rho_{xy_2} = 0$ is the same hypothesis, written as a statement of no difference rather than equality. The potential alternative hypothesis are the nondirectional H_A: $\rho_{xy_1} \neq \rho_{xy_2}$ or the directional H_As: $\rho_{xy_1} < \rho_{xy_2}$ or $\rho_{xy_1} > \rho_{xy_2}$.

Sampling distribution. Rather than using the sampling distribution of differences between two independent sample correlation coefficients ($r_{xy_1} - r_{xy_2}$) to test H_0: $\rho_{xy_1} = \rho_{xy_2}$, Fisher's Z transformation is again used. The sampling distribution of $Z_{r_1} - Z_{r_2}$ is of interest. The sampling distribution of $Z_{r_1} - Z_{r_2}$ (differences in Z_r transformations for r_{xy_1} and r_{xy_2}) is approximately normally distributed around a mean difference of $Z_{\rho_1} - Z_{\rho_2}$ with a standard deviation (error) of $[1/(n_1 - 3) + 1/(n_2 - 3)]^{1/2}$. Here, Z_{ρ_1} is the Z transformation for ρ_{xy_1}, Z_{ρ_2} is the Z transformation for ρ_{xy_2}, and n_1 and n_2 are the sample sizes for groups 1 and 2, respectively.

Test statistic. To test H_0: $\rho_{xy_1} = \rho_{xy_2}$, the sample values of r_{xy} are calculated separately for each of the independent samples 1 and 2. The two sample correlation coefficients r_{xy_1} and r_{xy_2} are then transformed to Fisher's Z_r scores, Z_{r_1} and Z_{r_2}, as tabled in Appendix B.6. The test statistic formula for testing H_0: $\rho_{xy_1} = \rho_{xy_2}$ is

$$z_{obs} = \frac{Z_{r_1} - Z_{r_2}}{\sqrt{\dfrac{1}{n_1 - 3} + \dfrac{1}{n_2 - 3}}} \qquad (14.3)$$

---

[8]If $r_{xy} = .86$; $Z_{.86} = 1.293$

$$z_{obs} = \frac{1.293 - .867}{\dfrac{1}{\sqrt{19 - 3}}} = \frac{.426}{.25} = 1.70$$

$z_{obs} > z_{crit} = 1.645$, so reject H_0: $\rho_{xy} = .70$.

The unit-normal distribution is the probability distribution to use in determining critical values at the α level of significance for testing H_0: $\rho_{xy_1} = \rho_{xy_2}$.

Example. We might hypothesize that mental ability is not as highly related to achievement for children from high-educational-aspiration homes as it is for children from homes in which educational aspirations are lower. Again, the "Sesame Street" data can be used to test the null hypothesis, H_0: There exists no difference in the relationship between Peabody mental age (X) scores and Classification Skills pretest (Y) scores for children who are above the median on the home education index and children who are below the median, or H_0: $\rho_{xy_{\text{Hi}}} = \rho_{xy_{\text{Lo}}}$. Here, $\rho_{xy_{\text{Hi}}}$ is the population correlation coefficient between Peabody mental age scores (X) and pretest Classification Skills scores (Y) for children from homes high on an educational aspiration index (above the median on variable 24), and $\rho_{xy_{\text{Lo}}}$ is the population correlation between X and Y for children from homes low on the educational aspiration index (below the median on variable 24). The alternative hypothesis based on the original statement is directional (H_A: $\rho_{xy_{\text{Hi}}} < \rho_{xy_{\text{Lo}}}$). Data to test H_0 are given in Table 14.1 for two groups, a random sample of 10 children from those who had home education index (HEI) scores of 16 or above (the median home education index score is 15.167) and a random sample of 12 children from those who had HEI scores of 15 or less. These two groups constitute samples from the high educational aspiration group and the low educational aspiration group. The summary values also are given in Table 14.1. The calculated value of $r_{xy_{\text{Hi}}}$ is .54 and of $r_{xy_{\text{Lo}}}$ is .63. To test H_0: $\rho_{xy_{\text{Hi}}} = \rho_{xy_{\text{Lo}}}$, each sample correlation coefficient is transformed to a Fisher's Z score using Table B.6, $Z_{.54} = .604$ and $Z_{.63} = .741$. We then calculate the observed test statistic using formula 14.3:

$$z_{\text{obs}} = \frac{.604 - .741}{\sqrt{\dfrac{1}{10 - 3} + \dfrac{1}{12 - 3}}}$$

$$= \frac{-.137}{.504} = -.272$$

Assuming a significance level of $\alpha = .05$, $z_{\text{crit}} = -1.645$ for a directional test of H_0. The observed z value is in the hypothesized alternative direction ($r_{xy_{\text{Hi}}}$ is less than $r_{xy_{\text{Lo}}}$), but it is not sufficient to reject H_0 at the .05 level of significance. Test the same hypothesis for values of $r_{xy_1} = .42$ and $r_{xy_2} = .68$ and sample sizes of 18 and 23, respectively, for n_1 and n_2.[9]

---

[9] $Z_{.42} = .448$ and $Z_{.68} = .829$

$$z_{\text{obs}} = \frac{.448 - .829}{\sqrt{\dfrac{1}{18 - 3} + \dfrac{1}{23 - 3}}} = \frac{-.381}{.342} = -1.11$$

$z_{\text{obs}} = -1.11$ lies in the region of nonrejection. Even a difference in sample r_{xy} values as large as .42 versus .68 is not statistically significant for these sample sizes at the .05 level.

TABLE 14.1
Sample data to test H_0: $\rho_{xy_1} = \rho_{xy_2}$.

Group 1: High Home Educational Aspiration (HEI ≥ 16)				Group 2: Low Home Educational Aspiration (HEI ≤ 15)			
ID	HEI	Peabody Mental Age Score (X)	Classification Skills Pretest (Y)	ID	HEI	Peabody Mental Age Score (X)	Classification Skills Pretest (Y)
232	19	31	11	184	10	27	4
064	20	55	13	007	14	28	12
087	20	85	15	138	15	46	14
167	17	35	7	131	13	42	16
137	18	29	8	033	10	62	24
072	17	42	15	176	15	47	8
130	16	38	13	236	15	49	10
225	18	59	10	100	13	53	7
159	16	69	15	156	11	42	17
071	19	58	21	170	9	29	5
				146	8	48	14
				125	11	29	6
Group 1: ($n_1 = 10$)		$\Sigma X = 501$ $\Sigma X^2 = 28{,}091$ $\Sigma XY = 6775$ $\bar{X} = 50.1$ $s_x = 18.23$ $r_{xy} = .54$	$\Sigma Y = 128$ $\Sigma Y^2 = 1788$ $\bar{Y} = 12.8$ $s_y = 4.08$	Group 2: ($n_2 = 12$)		$\Sigma X = 502$ $\Sigma X^2 = 22{,}406$ $\Sigma XY = 6190$ $\bar{X} = 41.83$ $s_x = 11.30$ $r_{xy} = .63$	$\Sigma Y = 137$ $\Sigma Y^2 = 1947$ $\bar{Y} = 11.42$ $s_y = 5.90$

When testing H_0: $\rho_{xy_1} = \rho_{xy_2}$, the greater the magnitude of the relationship between X and Y, the less the discrepancy in sample values of r_{xy_1} and r_{xy_2} needed to reject H_0. For example, if $r_{xy_1} = .90$ and $r_{xy_2} = .80$ in one case and $r_{xy_1} = .30$ and $r_{xy_2} = .20$ in another case, they would appear to be equally discrepant $(.90 - .80 = .30 - .20)$. In light of the skewness of the sampling distribution of $r_{xy_1} - r_{xy_2}$, however, the transformation to Z_r scores gives more weight to higher values of r_{xy}. When transformed to Z_{r_1} and Z_{r_2} scores for substitution in the test statistic formula 14.3, $Z_{.90} - Z_{.80} = 1.472 - 1.099 = .373$ and $Z_{.30} - Z_{.20} = .310 - .203 = .107$. In calculating z_{obs} by formula 14.3, the effective difference in $r_{xy_1} = .90$ and $r_{xy_2} = .80$ is almost four times that of $r_{xy_1} = .30$ and $r_{xy_2} = .20$.

The Case of Testing Two Dependent Correlations, H_0: $\rho_{13} = \rho_{23}$

Null hypothesis. Testing hypotheses about the equality of two correlation coefficients when both correlations are from the same population is a very complex problem. For example, a complementary hypothesis to the last case using *independent* groups 1 and 2 would be H_0: $\rho_{xy_1} = \rho_{xy_2}$ where ρ_{xy_1} and ρ_{xy_2} are the correlations between X and Y at times 1 and 2 for the same sample. Unfortunately, the sampling distribution for $r_{xy_1} - r_{xy_2}$ is not known when r_{xy_1} and r_{xy_2} are correlated (based on the same sample). The only dependent sample null hypothesis about the equality of two correlation coefficients is H_0: $\rho_{13} = \rho_{23}$, which compares the correlation between variables 1 and 3 (ρ_{13}) to the correlation between variables 2 and 3 (ρ_{23}) for the same sample. Note in the hypothesis that three variables are involved. Each variable is represented by a number. The basic hypothesis tests whether one variable (e.g., variable 3) correlates to the same extent with each of the other two variables. One variable (3) may be thought of as a criterion variable, while the other two serve as predictors. The hypothesis tests whether variable 1 is a better predictor of variable 3 (ρ_{13}) than variable 2 is (ρ_{23}), H_0: $\rho_{13} = \rho_{23}$.

Sampling distribution. The sampling distribution of interest is that of $r_{13} - r_{23}$ when r_{13} and r_{23} are based on data from the same sample. When H_0: $\rho_{13} = \rho_{23}$ is true, the values of $r_{13} - r_{23}$ are distributed approximately as a normal distribution around a mean difference of zero. The standard error of $r_{13} - r_{23}$ is estimated by a complex formula given in the observed test statistic formula (14.4).

Test statistic. One observed test statistic formula for testing H_0: $\rho_{13} = \rho_{23}$ is a t test after Hotelling.

$$t_{\text{obs}} = \frac{(r_{13} - r_{23})\sqrt{(n-3)(1 + r_{12})}}{\sqrt{2(1 - r_{13}^2 - r_{23}^2 - r_{12}^2 + 2r_{13}r_{23}r_{12})}} \tag{14.4}$$

Formula 14.4 appears complicated because of the complexity of the standard error estimate. The formula results in an observed test statistic distributed as an appropriate t distribution with $n - 3$ degrees of freedom ($df = n - 3$).

Example. An example from the "Sesame Street" data can be used to illustrate a test of H_0: $\rho_{13} = \rho_{23}$. Suppose we want to determine whether Peabody mental age scores (X_1) are a better predictor of posttest knowledge of letters (X_3) than is the Home Education Index (X_2), using an α level of .05. The null and alternative hypotheses are H_0: There exists no difference in the relationship between Peabody mental age scores (X_1) and posttest Letters scores (X_3) and that between scores on the home education index (X_2) and posttest Letters scores for children in the "Sesame Street" evaluation (H_0: $\rho_{13} = \rho_{23}$ and H_A: $\rho_{13} \neq \rho_{23}$. A random sample of 16 individuals from the "Sesame Street" data base was selected to provide data for the analysis. Scores on each of the three variables were used to obtain values of r_{12}, r_{13}, and r_{23}. The three pairwise sample correlation coefficients are needed as input into formula 14.4. The values for each of the three coefficients were $r_{13} = .797$, $r_{23} = .336$, $r_{12} = .239$. Substituting these coefficients in formula 14.4,

$$t_{\text{obs}} = \frac{(.797 - .336)\sqrt{(16 - 3)(1 + .239)}}{\sqrt{2[1 - .797^2 - .336^2 - .239^2 + 2(.797)(.336)(.239)]}} = 2.303$$

The critical values for a two-tailed test of H_0 at the .05 level of significance are $t_{crit} = \pm 2.16$. The t_{obs} value of 2.303 is greater than $z_{crit} = 2.16$, therefore we would reject H_0: $\rho_{13} = \rho_{23}$. PMA scores are significantly better predictors of posttest knowledge of letters than are scores on the HEI at the .05 level of statistical significance.

14.5 TESTING HYPOTHESES ABOUT VARIANCES

While the sampling distribution for a raw score sample statistic is the basis for the test of any H_0, the transformation of the sample statistic(s) to a specific probability distribution (i.e., z, t, χ^2 or F) becomes the focus in any test of an H_0. We will change our presentation format for the hypotheses in this section to emphasize the probability sampling distribution for each H_0.

The Single Sample Case, H_0: $\sigma_x^2 = a$

Null hypothesis. The null hypothesis to be tested is that a population has a variance (σ_x^2) equal to some constant value $a(H_0$: $\sigma_x^2 = a)$. The form of this null hypothesis is identical to that for a single sample mean (H_0: $\mu = a$) or correlation coefficient (H_0: $\rho_{xy} = a$).

Test statistic. The test statistic for testing H_0: $\sigma_x^2 = a$ is a χ^2 statistic with $df = n - 1$ defined by the formula

$$\chi_{obs}^2 = \frac{(n - 1)s_x^2}{a} \tag{14.5}$$

where s_x^2 is the sample variance and a is the null-hypothesized population variance.

Sampling distribution. When H_0: $\sigma_x^2 = a$ is true, the χ_{obs}^2 statistic resulting from formula 14.5 is distributed as χ^2 with $n - 1$ degrees of freedom. When H_0: $\sigma_x^2 = a$ is not true and the population variance is different from a, the true sampling distribution of χ^2 will shift to the left if $\sigma_x^2 < a$ and to the right if $\sigma_x^2 > a$. Because of the potential for σ_x^2 to be less than the hypothesized value of a, this is one of the few χ^2 tests in which a critical value in the lower tail of the distribution may need to be determined depending on H_A. The implication of an alternative H_A: $\sigma_x^2 \neq a$ or the alternative H_A: $\sigma_x^2 < a$ is that critical χ^2 values in the lower tail of the χ^2 distribution will need to be determined. Most other H_0s tested by the χ^2 test statistic identify the region of rejection in the upper tail of the χ^2 distribution.

Example. The variance of mental ability in the population for Peabody mental age (PMA) scores is $\sigma_x^2 = 225$ ($\sigma_x = 15$). Using a random sample of 16 individuals from Appendix A, we can test the hypothesis H_0: $\sigma_x^2 = 225$ to see if our sample evidence would indicate that the "Sesame Street" children display more or less variability in PMA scores. Here, H_0: $\sigma_x^2 = 225$, H_A: $\sigma_x^2 \neq 225$, and

$\alpha = .05$ The sample variance calculated from the random sample of 16 PMA scores was 379.08($s_x = 19.47$). Then,

$$\chi^2_{\text{obs}} = \frac{(16 - 1)(379.08)}{225} = 25.27$$

With regions of rejection equal to 2.5% of the distribution in each tail, the critical values of χ^2 in the χ^2 distribution with $n - 1 = 15$ *df* are lower $\chi^2_{\text{crit}} = 6.26$ and upper $\chi^2_{\text{crit}} = 27.49$. The χ^2_{obs} value of 25.27 is less than $\chi^2_{\text{crit}} = 27.49$ and falls in the nonrejection region. The sample s^2_x value of 379.08 was not sufficiently different from the hypothesized value of $\sigma^2_x = 225$ to reject H_0.

Consider the alternative hypothesis that children from homes with low educational aspiration would exhibit less variability in PMA scores than those in the norm population. Test the null hypothesis H_0: $\sigma^2_x = 225$ against the alternative H_A: $\sigma^2_x < 225$ at the .05 level of significance using the sample PMA scores (variable X) for the Group 2 random sample of 12 children in Table 14.1.[10]

The Two Independent-Sample Case, H_0: $\sigma^2_1 = \sigma^2_2$

Null hypothesis. The null hypothesis H_0: $\sigma^2_1 = \sigma^2_2$ specifies the equality of two independent population variances. The alternative H_A would specify that the variability of scores on a variable in population 1 is not equal to (or is less than or greater than) the variability of scores *on the same variable* for population 2. Populations 1 and 2 are assumed to be normally distributed with mean μ_1 and μ_2 and variances σ^2_1 and σ^2_2, respectively.

Test statistic. To test H_0: $\sigma^2_1 = \sigma^2_2$, draw a random sample of size n_1 from population 1 and a random sample of size n_2 from population 2. The sample variances s^2_1 and s^2_2 are calculated and compared to test H_0. The test statistic for testing H_0 is an F value formed by taking the ratio of s^2_1 to s^2_2.

$$F_{\text{obs}} = \frac{s^2_1}{s^2_2} \tag{14.6}$$

Sampling distribution. The value of F_{obs} obtained from taking the ratio of s^2_1/s^2_2 is distributed as an F distribution with $n_1 - 1$ *df* in the numerator ($df_n = n_1 - 1$) and $n_2 - 1$ *df* in the denominator ($df_d = n_2 - 1$) when H_0: $\sigma^2_1 = \sigma^2_2$ is true. When H_0: $\sigma^2_1 = \sigma^2_2$ is not true, the distribution of F_{obs} will shift

---

[10]For the 12 PMA scores, $s_x = 11.30$. The sample variance is $s^2_x = 11.30^2 = 127.69$

$$\chi^2_{\text{obs}} = \frac{(12 - 1)(127.69)}{225} = 6.24$$

$\chi^2_{\text{crit}} = 4.57$ with 11 *df* and $\alpha = .05$, given a region of rejection only in the lower tail of the distribution. The observed χ^2 value of 6.24 does *not* lie beyond $\chi^2_{\text{crit}} = 4.57$; therefore, do *not* reject H_0.

to the left when $\sigma_1^2 < \sigma_2^2$ and to the right when $\sigma_1^2 > \sigma_2^2$. Designating one sample as sample 1 and the other as sample 2 is arbitrary, but it needs to be done prior to examining any of the data for a nondirectional test of H_0. That is, for this test deciding which variance to place in the numerator and which into the denominator should be done randomly. If we review the data first and always designate the sample with the largest variance as sample 1, then s_1^2 in the numerator of $F_{obs} = s_1^2/s_2^2$ would always be greater than s_2^2. Following this convention would effectively double the probability of a Type I error when a nondirectional test was performed. For directional alternative hypotheses, the expected larger variance is always divided by the expected smaller variance when calculating the F ratio.

Like the previous χ^2 test of H_0: $\sigma_x^2 = a$, the test of H_0: $\sigma_1^2 = \sigma_2^2$ is one of the few instances when we might need to determine critical values of F in the lower tail of the F distribution. These lower percentile values are not shown in Table B.5, but they may be determined from the relationship between F values in the lower and upper tails of F distributions defined by the reversal of numerator and denominator degrees of freedom.

$$\text{lower } F_{\text{crit}(df_n, df_d)} = \frac{1}{\text{upper } F_{\text{crit}(df_d, df_n)}}$$

Example. The Peabody mental age data for the two random samples in Table 14.1 can illustrate the test of H_0: $\sigma_1^2 = \sigma_2^2$. The null hypothesis may be stated H_0: There exists no difference in the variability of PMA scores between children from homes with higher educational aspirations (Group 1) and those from homes with lower educational aspirations (Group 2), or H_0: $\sigma_{\text{Hi}}^2 = \sigma_{\text{Lo}}^2$, H_A: $\sigma_{\text{Hi}}^2 \neq \sigma_{\text{Lo}}^2$, and $\alpha = .05$. From Table 14.1, the variance of PMA scores for the High HEI Group 1 is $s_{\text{Hi}}^2 = 18.23^2 = 332.33$ and for the Low HEI Group 2, $s_{\text{Lo}}^2 = 11.30^2 = 127.69$. Taking the ratio,

$$F_{\text{obs}} = \frac{332.33}{127.69} = 2.60$$

For a two-tailed test of H_0 and $\alpha = .05$, the upper critical F value is found at the 97.5th percentile in the F distribution with $df_n = n_1 - 1 = 9$ and $df_d = n_2 - 1 = 11$. The upper $F_{\text{crit}} = 3.59$. The lower critical F value is found by first determining the F value at the 97.5th percentile in the upper tail with the degrees of freedom reversed, $F_{(11,9)} = 3.91$. Note that this value is not tabled, but rather estimated between the values for $F_{(10,9)}$ and $F_{(12,9)}$. The lower critical value is then

$$\text{lower } F_{\text{crit}} = \frac{1}{3.91} = .256$$

The observed F value of 2.60 lies in the nonrejection region between the F_{crit} values of 3.59 and .256. Based on the sample evidence, the values of

$s_1^2 = 332.33$ and $s_2^2 = 127.69$ do not differ sufficiently to reject $H_0: \sigma_1^2 = \sigma_2^2$ at the .05 level of significance.

Testing Two Correlated Variances, H_0: $\sigma_1^2 = \sigma_2^2$

Null hypothesis. The null hypothesis ($H_0: \sigma_1^2 = \sigma_2^2$) appears to be the same as in the independent-groups case. It is the same hypothesis of the equality of two population variances on the same variable, but in the present correlated (dependent) case, the variances σ_1^2 and σ_2^2 are assumed to be indices of variability for two sets of scores on the same variable for the same population or matched populations, but collected on two distinct occasions. For example, a learning strategy is hypothesized to decrease the variability of achievement scores for persons instructed under the model. The null hypothesis would suggest equality of achievement score variances prior to and after instruction. The alternative hypothesis would be directional with achievement variability prior to instruction being greater than after instruction on the same or equivalent measures of achievement. Thus, $H_0: \sigma_{\text{pre}}^2 = \sigma_{\text{post}}^2$ and $H_A: \sigma_{\text{pre}}^2 > \sigma_{\text{post}}^2$. To test this hypothesis, we would collect achievement data for the same individuals on two occasions, pre- and postinstruction.

Test statistic. The test statistic used to test $H_0: \sigma_1^2 = \sigma_2^2$ in the correlated sample case is a t test with $df = n - 2$

$$t_{\text{obs}} = \frac{s_1^2 - s_2^2}{\sqrt{\dfrac{4(s_1^2)(s_2^2)}{n-2}(1 - r_{12}^2)}} \tag{14.7}$$

where s_1^2 is the sample variance for the first set of scores, s_2^2 is the sample variance for the second set of scores, n is the number of paired scores, and r_{12} is the Pearson correlation coefficient between the paired scores in sets 1 and 2. While our preceding example used pre- and postachievement scores for the same group of individuals, it is possible to have dependent sets of scores because of other types of pairing or matching (i.e., brother-sister, husband-wife, worker-supervisor, etc.)

Sampling distribution. The test statistic (t_{obs}) calculated from formula 14.7 is distributed as a t distribution with $df = n - 2$.

Example. We might hypothesize that children who often watch "Sesame Street" during the week should become more homogenous in their knowledge and skill level over certain content after an extended viewing period. We can test the null hypothesis by examining a sample of pre- and posttest achievement data for children in the "Sesame Street" data base who watched the show frequently. For our purposes we have randomly sampled pre- and posttest Body Parts scores for 10 children who reported viewing "Sesame Street" four or more times per week. Table 14.2 reports the data along with summary statistics. The null and alternative hypotheses are $H_0: \sigma_{\text{pre}}^2 = \sigma_{\text{post}}^2$ and H_0: $\sigma_{\text{pre}}^2 > \sigma_{\text{post}}^2$ ($\alpha = .05$). From the data in Table 14.2, $s_{\text{pre}}^2 = 7.99^2 = 63.84$ and

TABLE 14.2
Body Parts pre- and post-test scores for a random sample of ten children viewing "Sesame Street" frequently.

ID	Body Parts Pre (X_1)	Body Parts Post (X_2)
118	14	18
019	25	32
046	14	29
131	31	31
219	8	31
203	19	22
153	14	18
005	32	32
239	23	29
150	25	26
	$\Sigma X_1 = 205$	$\Sigma X_2 = 268$
	$\Sigma X_1^2 = 4777$	$\Sigma X_2^2 = 7460$
	$\Sigma X_1 X_2 = 5678$	
	$\bar{X}_1 = 20.5$	$\bar{X}_2 = 26.8$
	$s_1 = 7.99$	$s_2 = 5.55$
	$r_{12} = .461$	

$s_{\text{post}}^2 = 5.55^2 = 30.80$. The correlation coefficient needed in formula 14.7 is $r_{12} = .461$.

$$t_{\text{obs}} = \frac{63.84 - 30.80}{\sqrt{\dfrac{4(63.84)(30.80)}{8}(1 - .461^2)}}$$

$$= \frac{33.04}{27.82} = 1.19$$

For the directional test, t_{crit} with $df = 8$ and $\alpha = .05$ is 1.86. The sample variance of Body Parts posttest scores is less than the sample variance of pretest scores, but the difference is not sufficient to reject $H_0\colon \sigma_1^2 = \sigma_2^2$ at $\alpha = .05$.

14.6 TESTING HYPOTHESES ABOUT PROPORTIONS

As we have seen in past chapters, proportions are relative frequencies, the number of "units" in a population possessing an identified characteristic relative to all elements in the population. The symbol used as the population parameter is P. The sample statistic used as an estimator of P is the lowercase p. The hypotheses about proportions to be covered in this section are relatively simple hypotheses involving one and two samples. More complex hypotheses about frequencies and proportions are covered in the next chapter.

The Single Sample Case, H_0: $P = a$

Null hypothesis. The null hypothesis H_0: $P = a$ specifies that the population proportion of elements P with a given characteristic equals some constant value a. An example might be a test of the effect of a new reading curriculum purporting to increase student interest in reading. One index of effectiveness might be an increase in the number of children who check out 10 or more books from the public library during the summer vacation months compared to the proportion over the past five-year period. If 20% of the population has a five-year history of checking out 10 or more books during the summer months, then H_0: $P = .20$ would be the null hypothesis to test against the alternative H_A: $P > .20$.

Sampling distribution. When H_0: $P = a$ is true, the sampling distribution of sample proportions p is normally distributed for large sample sizes with a mean proportion equal to the population parameter, $\bar{p} = a$, and a standard deviation equal to $\sigma_p = [P(1 - P)/n]^{1/2}$. If $P = a$, then $\sigma_p = [a(1 - a)/n]^{1/2}$. The one restriction on the sampling distribution of p is that the sample size must be large. Large in this context is defined the same as in the normal approximation to the binomial distribution: nP and $n(1 - P)$ must both be greater than 5.

Test statistic. The test statistic to test H_0: $P = a$ is a z test,

$$z_{\text{obs}} = \frac{p - a}{\sqrt{a(1 - a)/n}} \tag{14.8}$$

where p is the sample proportion of units possessing the characteristic, a is the null hypothesized population proportion value, and n is the sample size drawn to test H_0.

Example. Suppose as an example we test H_0: $P = .20$—the proportion of children who check out 10 or more books from the public library is .20. The alternative is H_A: $P > .20$ for our data to support the reading program effectiveness. We draw a random sample of 100 children who were exposed to the new reading curriculum. We then check the public library's records to see how many of the 100 children checked out 10 books or more over the summer. If the number equals 28, the sample proportion is $p = 28/100 = .28$.

$$z_{\text{obs}} = \frac{.28 - .20}{\sqrt{\dfrac{.20(1 - .20)}{100}}}$$

$$= \frac{.08}{.04} = 2.00$$

using $\alpha = .05$, $z_{\text{crit}} = 1.645$ for a one-tailed test. We reject H_0: $P = .20$ and assume H_A: $P > .20$ is true. The statistically significant finding based on the sample data would support the program's effectiveness using a statistical hypothesis-testing model as the criterion for effectiveness.

The Two Independent-Sample Case, H_0: $P_1 = P_2$

Null hypothesis. The null hypothesis $H_0: P_1 = P_2$ states that the proportion of individuals with a particular characteristic is equal for two independent populations. For example, we might test the hypothesis that the proportion of females with high mathematics aptitude who take four years of high school math courses is the same as the proportion of their male counterparts (H_0: $P_{\text{female}} = P_{\text{male}}$).

Sampling distribution. For reasonably large sample sizes n_1 and n_2 in each of the two independent groups, the sampling distribution of the difference between two independent sample proportions $(p_1 - p_2)$ is an approximate normal distribution with a mean of $P_1 - P_2 = 0$ when H_0: $P_1 = P_2$ is true. Otherwise, the distribution centers around the true value of $P_1 - P_2$. Much like the sampling distribution of the differences between sample means $(\overline{X} - \overline{Y})$, the sample estimate of the standard error of $p_1 - p_2$ is a pooled estimate based on information from both samples. If we let f_1 be the frequency or number of individuals in group 1 possessing the characteristic, then the proportion of individuals in group 1 possessing the characteristic is $p_1 = f_1/n_1$. Similarly, $p_2 = f_2/n_2$, the proportion of individuals in group 2 possessing the characteristic of interest. If $H_0: P_1 = P_2$ is true, then P_1 and P_2 both equal a common value of P. We can estimate this common proportion by combining frequency information from both samples. Using the pooled formula, $p_p = (f_1 + f_2)/(n_1 + n_2)$. We then estimate the population variance using $p_p(1 - p_p)$, the standard formula for the variance of a proportion. The estimate of the standard error of $p_1 - p_2$ is

$$s_{p_1-p_2} = \sqrt{p_p(1 - p_p)\left(\frac{1}{n_1} + \frac{1}{n_2}\right)}$$

Test statistic. The test statistic for testing $H_0: P_1 = P_2$ is a z test in which the difference in sample proportions $(p_1 - p_2)$ is divided by the standard error of the sampling distribution $(s_{p_1-p_2})$.

$$z_{\text{obs}} = \frac{p_1 - p_2}{\sqrt{p_p(1 - p_p)\left(\frac{1}{n_1} + \frac{1}{n_2}\right)}} \tag{14.9}$$

where

$$p_p = \frac{f_1 + f_2}{n_1 + n_2}$$

The values of z_{obs} follow a unit-normal distribution when n_1 and n_2 are reasonably large. Reasonably large is defined by the values of n_1p_p, $n_1(1 - p_p)$, n_2p_p, and $n_2(1 - p_p)$, which must all be greater than 5 for z_{obs} to have an approximate normal distribution. The latter assumes that p_p is an accurate estimate of P_1 and P_2 when they are equal under a true H_0.

Example. We can use some hypothetical data to test the hypothesis about high math aptitude female and male enrollment in advanced mathematics courses. The null hypothesis is one of equal proportions of enrollment in the female and male populations, $H_0: P_{\text{female}} = P_{\text{male}}$. The alternative might be directional based on past survey data indicating that mathematics is a male-dominated field, $H_A: P_{\text{female}} < P_{\text{male}}$. We will use $\alpha = .05$ to test H_0 against H_A. Assume that from a random sample of 200 high mathematics–aptitude female seniors in high school, 128 were reported to have been enrolled in an advanced math sequence. For a random sample of 150 high mathematics–aptitude male seniors, 120 were reported to have been enrolled in the advanced math sequence. Using this information, $f_1 = 128$, $n_1 = 200$, $f_2 = 120$, and $n_2 = 150$.

$$p_1 = \frac{128}{200} = .64 \qquad \text{and} \qquad p_2 = \frac{120}{150} = .80$$

The pooled estimate of P_1 and P_2 is

$$p_p = \frac{128 + 120}{200 + 150} = \frac{248}{350} = .709$$

The observed test statistic is

$$z_{\text{obs}} = \frac{.64 - .80}{\sqrt{(.709)(1 - .709)\left(\dfrac{1}{200} + \dfrac{1}{150}\right)}} = \frac{-.16}{.049} = -3.27$$

The observed z value of -3.27 based on the difference in sample proportions is beyond the critical value of z ($z_{\text{crit}} = -1.645$) in the probability distribution. The statistical decision is to reject H_0 and infer from the results that a disproportionately lower number of high-school females equally capable in mathematics actually enroll in the advanced math sequence than do their male counterparts.

14.7 REVIEW: INTERPRETATIONS BASED ON TESTS OF HYPOTHESES

The logic of the hypothesis testing decision process, the role and meaning of the level of significance (α), and the concepts of β and power hold true for the test of any null hypothesis. If, as a result of the test of an H_0, the decision is *not* to reject H_0, we conclude that the sample data provided insufficient evidence to reject the H_0 statement of equality, no difference or no relationship involving the population parameters of interest. Nonrejection of H_0 does not prove that the exact population relationship(s) or value(s) specified by H_0 are in fact true.

If the decision is to reject H_0 based on the sample evidence, we conclude that the value of the observed sample statistic was too discrepant from the H_0 value to have been expected with any reasonable probability if H_0 were true; therefore, H_0 is most probably false and H_A is most probably true. For example, rejecting H_0: $\rho_{xy_1} = \rho_{xy_2}$ implies that the relationship between variables X and Y is not the same in the two populations represented by group 1 and group 2.

Deciding whether to reject or not to reject H_0 is based on statistical procedures and criteria. We have stressed before, however, that the validity of the inferences and conclusions resulting from the decision made about H_0 depends on sampling and measurement considerations in the research process. The model underlying the test of each H_0 is based on certain assumptions, most notably randomness and normality of the population distribution. Our discussion of sampling distributions for all the statistics used to test the null hypotheses always specified the distribution of the sample statistic ($\overline{X}, \overline{X} - \overline{Y}$, $r_{xy}, Z_r, Z_{r_1} - Z_{r_2}, r_{13} - r_{23}, s_x, s_1^2/s_2^2, s_1^2 - s_2^2, p$ or $p_1 - p_2$) *for repeated random samplings* of samples of size n. The probability sampling distribution used as the criterion comparison distribution in the hypothesis-testing process describes the sample results expected if randomness is operating when sampling from a population(s) described by H_0. The extent to which sample results are biased because of the sampling or measurement process affects the error rate in the test of H_0. Violating assumptions underlying the statistical test of an H_0 can lead us to reject H_0 with greater or less frequency than that set by the α level when H_0 is actually true. We should be aware of the consequences of violating the assumptions underlying tests of specific H_0s. The consequences are usually defined by the effect of an assumption violation on the levels of α or β, the Type I or Type II error rates operating when H_0 is tested.

14.8 SUMMARY

This chapter introduced two additional probability distributions: the chi square and the F distribution. Like the t distribution, chi square and F distributions represent a family of distributions; each unique distribution is defined by the degrees of freedom associated with it. These distributions are employed with various statistical tests to define critical values associated with the significance level being applied when evaluating the null hypothesis. The chi square distribution is a probability sampling distribution that relates an observed sample statistic as a squared discrepancy from an expected or hypothesized parameter. The F distribution, also a probability sampling distribution, is derived as a ratio of two independent chi square values. The importance in applied statistics of sampling distributions remains unaltered regardless of the probability sampling distribution being considered: z, t, χ^2, or F. Although they have been shown to be related, each plays a unique role in testing specific null hypotheses by detailing the norm of expectation for a single event. They offered a basis for determining if an outcome was probable or improbable in relation to the null hypothesis being tested. Bear in mind, however, that the decision point for rejecting or failing to reject the null hypothesis resides with the researcher and is not controlled by the sampling distribution being used. The specific probability sampling distribution only provides the referent for the decision.

The test statistics considered in the current chapter treated hypotheses for correlation coefficients, variances, and proportions. Applications for single-sample hypotheses, two independent samples, and correlated or dependent samples were presented. Appropriate use of these statistics requires use of the correct probability sampling distribution to evaluate results. The following chart details the probability sampling distribution to be used for each statistical test presented.

	Case		
Statistic	Single Sample	Two Independent Samples	Two Dependent Samples
Correlation Coefficient	H_0: $\rho_{xy} = 0$ t - - - - - - - - H_0: $\rho_{xy} = a$ z	H_0: $\rho_{xy_1} = \rho_{xy_2}$ z	H_0: $\rho_{23} = \rho_{14}$ t
Variance	H_0: $\sigma^2 = a$ χ^2	H_0: $\sigma_1^2 = \sigma_2^2$ F	H_0: $\sigma_1^2 = \sigma_2^2$ t
Proportion	H_0: $P = a$ z	H_0: $P_1 = P_2$ z	

PROBLEMS

Problem Set A

1. For each of the following alternative hypothesis situations, identify the appropriate probability sampling distribution and the critical value(s) at the .05 level of significance necessary to test the corresponding null hypothesis (a) that the correlation between income and years of schooling is not equal to zero in a sample of 30 adults, (b) that the proportion of smokers in a sample of 90 persons in 1983 was less than the proportion of smokers in a sample of 130 drawn in 1973, (c) that the variance of

pretest scores for 25 persons did not equal their posttest scores following treatment, (d) that the proportion of 100 children traveling at least one mile to school in a rural community exceeds the state average of .43, (e) that the variance of SAT scores for 101 first-year students at a private college does not equal the variance in the national norm, and (f) that the variance of test scores for 31 males does not equal the variance for a sample of 16 females.

2. Under the conditions specified for each of the following, test the null hypothesis using critical values at α = 10, .05 and .01. State your conclusions for each criterion level.
   a. H_0: $\rho_{xy} = 0$, H_A: $\rho_{xy} \neq 0$, $r = -.56$, $n = 32$
   b. H_0: $\sigma^2 = 75$, H_A: $\sigma^2 \neq 75$, $s^2 = 86.5$, $n = 66$
   c. H_0: $P = .25$, $H_A = P > .25$, $p = .33$, $n = 90$
   d. H_0: $\sigma_1^2 = \sigma_2^2$, H_A: $\sigma_1^2 \neq \sigma_2^2$, $s_1 = 4.13$, $n_1 = 25$, $s_2 = 5.81$, $n_2 = 11$.

3. Professor Jones finds that the correlation for his 35 students between grade point average and their final grade in his class is .45. Professor Smith finds that for these same students the correlation between grades and GPA is .68. Upon learning of this discrepancy, grades in the two classes were correlated and found to be .72. Test the hypothesis at the .05 level that Professor Smith's grades are better predictors of GPA than those assigned by Professor Jones. State the statistical null and alternative hypotheses. Remember to state your conclusions.

4. In one study conducted on 12 individuals the correlation between two variables was found to be .49. In a second study carried out at a different location but using 38 persons the correlation was .57. Test the hypothesis at the .05 level of significance that the observed correlations in the two sites are equivalent. Use a nondirectional test.

5. A researcher studying student course evaluations found for a group of 21 students taught by a graduate assistant that $\overline{X} = 18.15$ and $s^2 = 3.54$. In a similar course, instructed by a professor with a class size of 31, $\overline{X} = 18.94$ and $s^2 = 2.81$. Test the hypothesis that the variance of the ratings obtained were equal for the two classes (use $\alpha = .05$ and a nondirectional test).

6. A mail survey was sent out with hopes of a 90% return rate. Of the 200 surveys, 154 were returned. Test the hypothesis that the goal had been achieved using an alpha level of .01.

7. On a particular intelligence test the standard deviation in the norm population is 16. Using this device, a researcher finds that in her sample of 82 persons, $s = 12.47$. Test the hypothesis that the researcher's sample scores have the same variability as that in the population (use $\alpha = .05$ and a nondirectional test).

Problem Set B

1. Callenbach (1973) performed a study of the effects of instruction and practice on test-taking techniques. In his investigation, 24 test-naive second-grade students received eight 30-minute periods of instruction and practice in standardized test-taking techniques over a four-week period. Immediate (first posttest) and delayed (second posttest) treatment effects were assessed through an analysis of improvement in

TABLE 14.4
Treatment affects on test-taking techniques.

Group		Immediate Posttest	Delayed Posttest
Experimental	n	24	22
	$\bar{X}$	132.58	91.73
	s	19.16	27.01
Control	n	24	22
	$\bar{Y}$	127.08	84.45
	s	23.32	29.16

standardized test reading scores as measured by differences between experimental ($n = 24$) and control ($n = 24$) groups. Table 14.4 reports summary statistics from the study. Note that delayed posttest data existed for 22 of the original 24 participants. Based on the information presented respond to the following questions.

a. For the dependent variable, the standardized reading test, the test manual reports that the standard deviation for the national norm sample is 32.5. Test the hypothesis that at the time of the delayed posttest, the variance for the control group equalled that in the population. Test this hypothesis at $\alpha = .01$. State your conclusions.
b. Using information from above, test a hypothesis that the experimental and control groups had equal variabilities on the immediate posttest. Test this hypothesis at $\alpha = .05$ and state your conclusions.
c. Statistically evaluate a hypothesis that the variability of test scores remained the same in the experimental group for the immediate and delayed posttests. Assume a correlation between immediate and delayed posttest scores of .60. Test this hypothesis at $\alpha = .05$ and state your conclusions.

2. A research investigation by Gable, Roberts, and Owen (1977) studied affective and cognitive correlates of measured classroom achievement. Final social studies grades served as the criterion measure in the investigation. For the 431 11th-grade students from two predominantly white high schools, a subset of the assessments made included standardized test measures of vocabulary (VOC) and comprehension (COMP) and attitudinal measures of interest in (INT) and perceived usefulness of (USE) social studies. The following correlation matrix reports some of their findings.

	Correlations			
	VOC	COMP	INT	USE
COMP	+.37			
INT	.03	.11		
USE	−.11	.13	.31	
Class Grade	.40	.37	−.08	−.21

Use the data presented to answer the following questions.

a. Test the hypotheses that interest in and perceived usefulness of social studies are related to grades obtained. Test each coefficient at $\alpha = .01$ and state your conclusions.
b. The manual for the standardized test of comprehension states that the correlation between test scores and class grades in non-English subjects averages $+.42$ in the population. Test the hypothesis that for the sample in the Gable study the correlation between comprehension and social studies grades equals the population value. Test this hypothesis at $\alpha = .05$ and state your conclusion.
c. Is a statistical difference in the correlations found between vocabulary and class grades and comprehension and class grade? Report your findings for $\alpha = .05$ and state your conclusions.

Problem Set C

1. Draw a random sample of 20 children from the "Sesame Street" data base. For each selected, record their Body Parts and Letters posttest scores. Compute the Pearson correlation coefficient and test the hypothesis that there is no relationship between performance on these measures. Test the hypothesis at $\alpha = .05$ using a nondirectional test.

2. It is hypothesized that pre- and posttest scores on the Classifications Skills test have correlation of .42 in the population. Draw a random sample of 15 children from the data base and test this hypothesis at $\alpha = .01$ using a nondirectional test.

3. Select a random sample of 10 children from those who viewed "Sesame Street" at home and a sample of 12 children from those who viewed at school. For each sample, compute the Pearson correlation between Numbers pretest and posttest scores. Test the null hypothesis at $\alpha = .05$ that the correlations from the two samples are equal. Use a nondirectional test.

4. From the data analyzed in problem 3, test the hypothesis that the variances of the Numbers Pretest scores are equal for the two samples. Use $\alpha = .05$ and a nondirectional test.

5. The standard deviation for the Classification Skills posttest scores for the population of children in the data base is 5.06. Using data selected in problem 2, test the hypothesis that the variance of the sample equals the variance in the population on the Classification posttest. Use $\alpha = .10$ and a nondirectional test.

6. In the original "Sesame Street" evaluation, 19% of the population participating in the evaluation were from Site III. Test the hypothesis that the proportion of cases in the data base from Site III equals their level of inclusion in the total evaluation sample. Use $\alpha = .05$ and a nondirectional test to test this hypothesis.

7. From the data base select a sample of 15 children who were coded as 1, 2, or 3 on the Reading frequency variable (13). Select a second sample of 20 children coded as 4 or 5 on this variable. Test the hypothesis that the proportion of males is equally represented in both samples. Use $\alpha = .05$ and an alternative hypothesis that the proportion of boys in the lower reading-frequency categories is less than in the higher categories.

15

The χ^2 Distribution—Analysis of Frequencies

The last chapter discussed the sampling distribution of chi square (χ^2) as a standard score probability distribution along with the z, t, and F distributions. The χ^2 probability distribution was used to test only one hypothesis, that for single-sample variances (H_0: $\sigma^2 = a$). While the chi square distribution is the appropriate probability distribution when testing this latter null hypothesis, its primary use in introductory statistics is as the probability distribution for testing hypotheses involving frequencies or proportions.

In Chapter 14, the tests of null hypotheses about a single sample proportion H_0: $P = a$ and two independent sample proportions H_0: $P_1 = P_2$ were given as z tests. The z test formulas given are specific to these two hypotheses about proportions. The χ^2 distribution is in fact the more general probability distribution for testing expanded hypotheses about proportions (frequencies) in the single-sample and multisample cases.

15.1 REVIEW: MEASUREMENT LEVEL AND DATA TYPES

The data for which we perform statistical analyses result from quantifying information through the measurement process. We have learned that measurement can result in data classified at one of four levels: nominal, ordinal, interval, or ratio. The majority of the statistical hypotheses and techniques covered so far have assumed measurement resulting in data at the interval (or quasi-interval) or ratio level. Computing the mean, variance, or Pearson correlation coefficient results in an interpretative index only for data at the continuous levels of measurement. With the exception of dichotomous variables (two categories), continuous, rank ordered, assumed equal-interval data are necessary for the results to have interpretative meaning when scores on a variable are added, subtracted, multiplied, or divided (as when $\bar{X}$, s_x^2, or r_{xy} are calculated). Hypotheses and statistics about the mean site location or the variance of school-attendance scores from the "Sesame Street" data are not appropriate. The variables "site location" and "school attendance" are nominal-level variables. The coded scores of 1, 2, 3, and so on only identify group membership. We could mathematically compute the mean site score for a sample of individuals by adding the scores and dividing by n, but how would we interpret a site $\bar{X}$ value of 3.21? We can't. An $\bar{X}$ site location value of 3.21 has no interpretive meaning.

When nominal-level variables with more than two categories are to be statistically analyzed, hypotheses about means, variances, or Pearson correlation coefficients are not appropriate. With nominal-level data, we can merely tally frequencies within mutually exclusive categories. We can then change the frequencies to relative proportions or percentages, but the frequencies themselves are the primary piece of statistical information from nominal-level variables. Tests of hypotheses about frequencies (or proportions) within and across categories become the emphasis of statistical applications when nominal-level variables are of interest. We may want to investigate the career interests shown in the selection of curriculum area major by undergraduate college students. The variable of interest is the nominal-level variable "curricular area major." The data collected on each individual in a sample would be the coded score for their major curricular area. The only statistical manipulation of these data would be to tabulate (count) the frequency (number) of individuals with the same coded score and possibly transform the raw frequency in each curricular major category to a relative proportion based on the total number in the sample. Any null hypothesis to be tested over the selection of curricular area majors by undergraduate college students would involve comparing frequencies or proportions to some expected values defined by H_0. The chi square distribution plays a central role in testing hypotheses about frequencies resulting from variable measurement at the nominal level.

15.2 REVIEW: THE CHI SQUARE DISTRIBUTION

Chapter 14 defined the family of chi square distributions as a summation of squared standard scores resulting in positively skewed distributions of χ^2 scores. The defining characteristics of a χ^2 distribution depended on the degrees of freedom associated with a particular problem. For example, in testing the hypothesis H_0: $\sigma^2 = a$, the observed χ^2 value calculated by the test statistic formula

$$\chi^2_{\text{obs}} = \frac{(n - 1)s_x^2}{a}$$

was distributed as a χ^2 distribution with $n - 1$ *df* when H_0 was true. The $df = n - 1$ defined the specific χ^2 distribution to be used in testing H_0. Critical χ^2 values at different α levels are given in Appendix B, Table B.4 for various χ^2 distributions defined by specific *df* values.

Similar to the test of any H_0 about means, variances, or Pearson correlation coefficients, the test of a null hypothesis about frequencies or proportions requires examining the discrepancy between an observed sample statistic (frequency or proportion) and the null hypothesized population parameter that is the *expected* value if H_0 is true. When testing hypotheses about frequencies, the observed sample frequency in a category may be represented by f_{oi}, the observed sample frequency for category *i*. The expected frequency for category *i* based on the hypothesized value or relationship specified in H_0 is represented by f_{ei}. The discrepancies between f_{oi} and f_{ei} across categories of the nominal level variable(s) become the primary evidence used to decide whether to reject H_0. If H_0 is a true statement about the frequencies or proportions in the population, then f_{oi} for a random sample of individuals should be an accurate estimate of f_{ei} and the quantities $f_{oi} - f_{ei}$ across categories of the nominal level variable should fluctuate within ranges expected based on sampling distribution probabilities. For tests of hypotheses about frequencies, the quantities $f_{oi} - f_{ei}$ do not directly define the sampling distribution needed to test an H_0. Rather, the squared discrepancy quantity $(f_{oi} - f_{ei})^2$ divided by f_{ei} and summed over the k categories is the test statistic of interest. The quantity $\Sigma_{i=1}^{k} [(f_{oi} - f_{ei})^2/f_{ei}]$ is distributed as a χ^2 distribution and we use it as the observed test statistic formula when testing hypotheses about frequencies or proportions. The χ^2_{obs} value calculated over repeated random samples from the general formula

$$\chi^2_{\text{obs}} = \sum \frac{(f_{oi} - f_{ei})^2}{f_{ei}} \tag{15.1}$$

would have the characteristics of a chi square distribution. The degrees of freedom defining the exact χ^2 distribution for χ^2_{obs} values from formula 15.1 depends on the specific hypothesis situation.

15.3 THE CHI SQUARE GOODNESS OF FIT TEST

The chi square test statistic as defined by formula 15.1 is a measure of the accumulated squared discrepancies between observed and expected frequencies across categories of a nominal-level variable. Although the size of the observed χ^2 test statistic depends on the difference between observed and expected frequencies, we can view the null hypothesis best as a test about proportions across categories or about a hypothesized distribution shape for the nominal-level variable. Consider again the example of career interest patterns as demonstrated by the declared curricular major area of college undergraduates. Assume that five curricular area categories define the nominal-level variable. The first task in defining the null hypothesis is to identify the hypothesized population proportions of students *expected* to select each curricular area as a major area. You may want to use data from the past years' previous college students to establish these proportional expectations. If you use data from the past decade, you are really testing the null hypothesis of no difference in career interest patterns for current college enrollees and those enrolled over the past decade as demonstrated by declared major curricular area. Another approach to forming hypothesized proportions rather than using past data or some other data as a norm to establish the hypothesized expectations is to hypothesize that all categories are equally likely; that is, that the proportion of students expected to select a particular major area is the same as for all other areas.

The preceding null hypotheses certainly are different in intent, but their statement forms are the same. For hypotheses about category proportions for a single nominal-level variable, the null statement actually specifies a set of subhypotheses that we test simultaneously under one comprehensive null hypothesis. The statement of H_0 for the single nominal-variable case is

$$H_0\colon P_1 = a_1;\ P_2 = a_2;\ \cdots\ P_{k-1} = a_{k-1};\ P_k = a_k$$

Translated, H_0 specifies that the population proportion (P_1) of individuals in category 1 of the nominal variable equals the hypothesized constant proportion a_1, that the population proportion in category 2 (P_2) equals the hypothesized constant proportion a_2, and so forth for each of the k categories of the nominal-level variable. The hypothesized proportion a_i need not be equal, but must be exhaustive, that is, $\Sigma a_i = 1.00$, so that all individuals can be classified into one of the k categories. We can view this null hypothesis as a series of independent single-category hypotheses (of the form $H_0\colon P_i = a_i$) that we test all at once using a single χ^2 test. We can also view it as a single test of a hypothesized distribution shape for the nominal-level variable, a shape defined by the hypothesized proportions (frequencies for a fixed n) across consecutive categories of the nominal variable.

The test of the null hypothesis about categorical proportions for a single nominal-level variable is called the **chi square goodness of fit** test. First, the χ^2 test statistic defined by formula 15.1 is used to test H_0. Second, the test of H_0 examines the "goodness of fit" (lack of discrepancy) between the sample categorical frequencies f_{oi} and the expected frequencies f_{ei} based on the hypothesized proportions. We determine the goodness of the fit (or lack thereof) by comparing the magnitude of the χ^2_{obs} value, calculated by formula 15.1, to the critical chi square value (χ^2_{crit}) determined by the level of α in a χ^2 distribution with $df = k - 1$. If the χ^2_{obs} value is greater than χ^2_{crit} $(\chi^2_{\text{obs}} > \chi^2_{\text{crit}})$, the discrepancies between the f_{oi}s and f_{ei}s are judged to be too large to have resulted as a function of random sampling given our α criterion level if H_0 were true. In this situation, we reject H_0. When χ^2_{obs} is less than χ^2_{crit} $(\chi^2_{\text{obs}} < \chi^2_{\text{crit}})$, the sample frequencies are judged to conform to the hypothesized H_0 distributional shape within expected limits set as criteria. In this situation the discrepancies between f_{oi} and f_{ei} are not sufficiently large to reject H_0.

To illustrate, let's work through an example set of data for the hypothetical study of college career interest patterns as measured by selection of majors. Assume that all careers can be subsumed under five mutually exclusive, exhaustive majors labeled A through E. We will begin with a null hypothesis that all five majors are equally likely to be selected by college undergraduates. Because *five* categories define the nominal-level variable (major) and each area is hypothesized to contain the same proportion of students, the null hypothesis is that each population category proportion P_i should equal .20, that is, H_0: $P_1 = .20$; $P_2 = .20$; $P_3 = .20$; $P_4 = .20$, and $P_5 = .20$. The distributional shape hypothesized by H_0 is rectangular.

To test H_0 at a given α level ($\alpha = .05$), we select a random sample of n undergraduates. We would classify each person sampled into one of the five curricular major area categories and tabulate the observed sample frequencies f_{oi} for each category. We obtain the expected f_{ei} frequencies from the sample size n and the hypothesized a_i proportions in H_0. We can compute the expected frequency in category i by multiplying the hypothesized category proportion a_i times n,

$$f_{ei} = a_i n \tag{15.2}$$

Assuming we draw a sample of size 500 at random to test H_0, the expected number of students selecting each curricular area as a major area would be

$$f_{e1} = f_{e2} = f_{e3} = f_{e4} = f_{e5} = (.20)500 = 100$$

If H_0 is true, we would expect the 500 students sampled to be distributed equally across the five categories. The f_{ei}s define our expectations when H_0 is true. Due to randomness in our sampling, we would not expect the observed f_{oi} frequencies to all equal 100 exactly if H_0 were true, but we would expect the f_{oi}s to deviate minimally from the f_{ei}s such that if H_0 were true and $\alpha = .05$, 95% of the χ^2_{obs} values calculated from formula 15.1 would be expected to be less than χ^2_{crit} at the α level in the appropriate chi square distribution. For the chi square goodness of fit tests being considered in this chapter, the appropriate chi square distribution for testing H_0 has $k - 1$ degrees of freedom, where k represents the number of categories in the classification variable of interest.

To illustrate the computation of χ^2_{obs} and the test of H_0, assume we obtained the frequencies in Table 15.1 from the sample of $n = 500$. Table 15.1 also provides expected frequencies. Substituting the values of f_{oi} and f_{ei} in formula 15.1,

$$\begin{aligned}
\chi^2_{\text{obs}} &= \sum \frac{(f_{oi} - f_{ei})^2}{f_{ei}} \\
&= \frac{(80 - 100)^2}{100} + \frac{(105 - 100)^2}{100} + \frac{(150 - 100)^2}{100} + \frac{(75 - 100)^2}{100} \\
&\quad + \frac{(90 - 100)^2}{100} \\
&= 4.00 + .25 + 25.00 + 6.25 + 1.00 \\
&= 36.50
\end{aligned}$$

Several points should be noted about the calculation of χ^2_{obs}. First, each discrepancy $(f_{oi} - f_{ei})$ is squared in the formula and always results in a positive

TABLE 15.1
Expected and observed frequencies in five majors.

Major Area					
A	B	C	D	E	
$f_{oA} = 80$	$f_{oB} = 105$	$f_{oC} = 150$	$f_{oD} = 75$	$f_{oE} = 90$	$\Sigma f_{oi} = 500$
$f_{eA} = 100$	$f_{eB} = 100$	$f_{eC} = 100$	$f_{eD} = 100$	$f_{eE} = 100$	$\Sigma f_{ei} = 500$

quantity. The larger an $f_{oi} - f_{ei}$ discrepancy, the greater the resulting quantity that is added to the sum determining the size of χ^2_{obs}. Any discrepancy, whether negative or positive, always results in a positive quantity, increasing the size of χ^2_{obs}. Small values of χ^2_{obs} indicate a better "fit" between f_{oi} and f_{ei} and support a true H_0. When χ^2_{obs} is large, the sample evidence would support rejection of H_0. For this reason, we use a single critical value in the upper tail of the χ^2 distribution to test H_0. We determine χ^2_{crit} such that $(1 - \alpha)\%$ of the area of a χ^2 distribution lies below χ^2_{crit} and $\alpha\%$ of the area is above or to the right of it.

The second point to note in the computation of χ^2_{obs} is that the impact of a discrepancy between f_o and f_e on the size of χ^2_{obs} is relative to the size of f_e. The larger the value of f_e for a category, the less a constant discrepancy contributes to the size of χ^2_{obs}. A discrepancy of 10 frequencies, for example, $(f_o - f_e = 10)$ contributes 4.00 units to χ^2_{obs} if $f_e = 25$, but only 1.00 unit if $f_e = 100$. We expect larger discrepancies in $f_o - f_e$ on a random basis for larger values of f_e in a category if H_0 is true. Dividing $(f_o - f_e)^2$ by f_e "corrects" for this expectation, resulting in standardized χ^2 sampling distributions for the same *df*, but different sample sizes across research situations.

The last point of interest involves the relationships among the $f_{oi} - f_{ei}$ discrepancy values across the *k* categories. The categories of the nominal-level variable must be independent and exhaustive. Because the *k* categories are exhaustive and the single sample drawn at random is of fixed size, the Σf_{oi} and Σf_{ei} must equal each other and both equal the sample size. The dependencies among the f_{oi}s, the f_{ei}s, and *n* result in the loss of one *df* among the *k* discrepancy $f_{oi} - f_{ei}$ scores used in the χ^2_{obs} formula. One of the $(f_o - f_e)$ discrepancy scores is not free to vary. The χ^2_{obs} values resulting from formula 15.1 over repeated samplings when H_0 is true are therefore distributed as a χ^2 distribution with $df = k - 1$.

To continue with the test of H_0, we find the critical value χ^2_{crit} in the upper tail of the chi square distribution with $df = k - 1 = 4$ for the test of H_0 at the .05 level of significance. In the chi square distribution with 4 *df*, the value of χ^2_{crit} at the 95th percentile is 9.49. When H_0 is true, we would expect 95% of the χ^2_{obs} values calculated by formula 15.1 to be less than or equal to 9.49; we would expect 5% of the χ^2_{obs} values to be greater than 9.49 as a function of randomness. The χ^2_{obs} value of 36.50 is greater than $\chi^2_{\text{crit}} = 9.49$, so we reject H_0. The observed distribution of individuals selecting one of the five majors was too discrepant from the hypothesized distribution based on the probability criterion level selected. The sample frequencies (proportions) over categories are statistically different from the hypothesized frequencies at the .05 level of statistical significance.

Another approach to this example demonstrates the flexibility of chi square goodness of fit statistics. Assume we obtained the same observed frequencies for the sample of 500 students across the five categories, but now the hypothesized proportions are based on information assembled on declared majors of undergraduates for the past 10 years. Suppose the null hypothesis is $H_0: P_A = .15; P_B = .25; P_C = .25; P_D = .15; P_E = .20$. That is, over a 10-year period, data have shown that 15% of the students declare Category A for their major, 25% elect to major in Category B, another 25% in Category C, 15% in Category D, and 20% in Category E. Note that the sum of the a_is must equal 1.00. Given a sample size of 500, what are the expected frequencies for individuals with declared majors in each curricular area A through E?[1] The observed χ^2 is calculated as follows.

$$\chi^2_{\text{obs}} = \frac{(80 - 75)^2}{75} + \frac{(105 - 125)^2}{125} + \frac{(150 - 125)^2}{125} + \frac{(75 - 75)^2}{75} + \frac{(90 - 100)^2}{100}$$

$$= .33 + 3.20 + 5.00 + 0 + 1.00$$

$$= 9.53$$

The number of categories has not changed, so $\chi^2_{\text{crit}} = 9.49$ for $df = 4$ and $\alpha = .05$. The χ^2_{obs} value of 9.53 is still greater than χ^2_{crit} of 9.49, leading us to reject H_0. The discrepancies between the observed frequencies and the expected frequencies under the latter H_0 are still statistically significant at $\alpha = .05$.

As in the test of other null hypotheses, the χ^2 goodness of fit test assumes randomness and independence of observations. In addition, the test assumes that the observed frequencies are normally distributed around the expected frequencies over repeated samplings. While this latter assumption is often violated, it only affects the sampling distribution of χ^2_{obs} when the expected frequencies in a category are small. A small f_e is likely to occur when n is relatively small, or when the hypothesized proportion a_i is small, since the value of f_{ei} depends on each (i.e., $f_{ei} = a_i n$). There is some disagreement in the literature over how small any f_e can be before the sampling distribution of χ^2_{obs} is affected drastically, but we will take a conservative approach and recommend that all f_{ei}s be greater than 5.

[1] $f_{eA} = (.15)(500) = 75; f_{eB} = (.25)(500) = 125; f_{eC} = 125; f_{eD} = 75; f_{eE} = 100.$

15.4 THE CHI SQUARE TEST OF INDEPENDENCE

The chi square goodness of fit test focused on the category proportions (frequencies) for a single nominal-level variable. It expanded the single sample z test of $H_0: P = a$ to a simultaneous χ^2 test of $H_0: P_i = a_i$ across k categories of a single variable. Application of the χ^2 test statistic can be expanded further to a test of hypotheses about proportions in the two-variable case where two nominal-level variables are involved. Consider this situation. A school district was considering closing one of its attendance centers because of low enrollment. Prior to making its decision, the district surveyed residents in the neighborhood to get their reaction to the proposed closing. Residents were asked, "Do you favor or are you opposed to closing the school?" Their positive or negative responses constitute a dichotomous variable. Residents were also asked how often they had moved during the preceding five years. Persons responded to this question on a four-point scale in which 1 = zero, 2 = once or twice, 3 = three to six times, and 4 = more than six times. From the data available they wanted to determine if a relationship existed between these two variables. Was one's position on closing the school related to the frequency of residential moves? If we measured the number of moves using a continuous scale such as the actual number of times the family moved, we could address this question by computing a point biserial correlation for a sample. As a Pearson coefficient, we could then test the point biserial value using the form of the null hypothesis, $H_0: \rho_{xy} = 0$. Number of moves was *not* measured using a continuous interval-level scale, however. At best, we might consider it ordinal, but it certainly has the properties of a nominal-level variable. A test of H_0 examining the equality of proportions in each number of moves category for the two groups in favor of or opposed to the school closing is more appropriate to determine whether the position on the school closing is related to the frequency of residence moves.

The null hypothesis is one of no relationship between the two nominal-level variables. It states that there is no difference in the proportion of individuals classified in each of the four move frequency categories whether they favored or opposed the school closing. This is a null hypothesis of **independence,** that is, no relationship between the two variables.

A table of the bivariate frequency distribution provides the information needed to test H_0. The table showing the bivariate frequency distribution for two categorical variables is called a **contingency table.** Table 15.2 reports results from the survey in a 2×4 (read two by four) contingency table implying two categories for the variable defining the rows and four categories for the variable defining the columns. In all, 188 area residents sampled randomly participated in the survey.

Using the observed frequency information presented in the table, 78 residents favored closing the school. Fourteen had not moved in the preceding

TABLE 15.2
Contingency table for position on school closing by number of residence moves.

	Number of Moves				
	Category 1 zero	*Category 2* once or twice	*Category 3* 3 to 6	*Category 4* more than 6	Row Total
Position on Closing					
Favorable (Cat. 1)	$f_o = 14$	$f_o = 18$	$f_o = 22$	$f_o = 24$	78
Opposed (Cat. 2)	$f_o = 43$	$f_o = 26$	$f_o = 21$	$f_o = 20$	110
Column Total	57	44	43	44	188

five years, 18 had moved once or twice in that period, 22 had moved three to six times, and 24 had moved more than six times. The proportions of respondents in each category are $p_1 = 14/78 = .18$, $p_2 = 18/78 = .23$, $p_3 = 22/78 = .28$, and $p_4 = 24/78 = .31$. For the 110 residents who opposed the proposed school closing, the observed frequencies are 43, 26, 21, and 20, respectively, in the four number of moves categories. The respective proportions are $p_1 = 43/110 = .39$, $p_2 = 26/110 = .24$, $p_3 = 21/110 = .19$, and $p_4 = 20/110 = .18$. The test of the null hypothesis by the χ^2 test of independence between the two categorical variables simultaneously compares the distributional proportions for the two groups across the four categories and for the four categories across the two groups. Stated in terms of the sample data, does the set of proportions for the group favoring the school closing ($p_1 = .18$, $p_2 = .23$, $p_3 = .28$, and $p_4 = .31$) represent the same distributional shape as the proportions for the group opposed to the closing ($p_1 = .39$, $p_2 = .24$, $p_3 = .19$, and $p_4 = .18$)? Or are these sample distributional proportions too discrepant from each other? The null hypothesis of independence or no relationship is a statement of no difference in the two distribution sets of proportions, viewed either over rows or over columns:

$$H_0: \begin{bmatrix} P_{1f} = P_{1o} \\ P_{2f} = P_{2o} \\ P_{3f} = P_{3o} \\ P_{4f} = P_{4o} \end{bmatrix} \quad \text{or} \quad H_0: \begin{bmatrix} P_1 = P_2 = P_3 = P_4 \\ P_1 = P_2 = P_3 = P_4 \end{bmatrix}$$

Appropriate application of the chi square test of independence depends on recognizing that it is a single-sample statistical test of no relationship between two categorical variables. Its use is keyed to nominal-level variables on one sample in which the distribution into categories on either variable is unknown prior to forming the contingency table. That is, the marginal frequencies of each variable can vary. Suppose in the last example we had sampled 78 residents known to favor closing the school and 110 persons who were known to be opposed. Under these conditions, the school closing marginals would have been fixed by our sampling procedure. Two samples of 78 and 110 would have been drawn and we would then be asking whether the number of residence moves differs between the two samples. This approach yields a two independent–sample test of equivalence or homogeneity on the dependent variable, number of moves. A later section will discuss this particular design approach. For the chi square test of independence, neither marginal in the contingency table is restricted as a consequence of sampling. We select a sample from a population randomly and construct a table to analyze the variables under investigation.

15.5 CALCULATING CHI SQUARE FROM A BIVARIATE CONTINGENCY TABLE

Testing the null hypothesis of independence between two categorical variables uses the magnitude of the discrepancy between the observed cell frequencies in a contingency table and the expected cell frequencies determined under an assumed model of independence. The general χ^2_{obs} test statistic given by formula 15.1 provides the standardized index of the magnitude of the discrepancies in $f_{oi} - f_{ei}$ for all cells in a contingency table. Once we determine the observed cell frequencies from the sample data and arrange them in a contingency table (as in Table 15.2), we can calculate the expected frequencies. Unlike the χ^2 goodness of fit test for a single variable, H_0 does not specify the expected or hypothesized proportion values. The null hypothesis of independence is a statement of the equality of proportions among categories of one of the variables within each category of the other variable. This assumed independence depicted by equal within-category proportions under a true H_0 provides the model for determining the expected frequencies from the observed frequencies.

To begin the process of determining the expected frequencies, let's reexamine Table 15.2. If we assume H_0 is true (that there is no relationship between one's position on the school closing and number of residential moves), then we can best estimate the true population proportion by combining the observed frequency information contained in both groups (as we did in Chapter 14 when testing H_0: $P_1 = P_2$). The pooled estimate of the proportion of total individuals who did not move during the preceding five years (number of residence moves, category 1) is

$$P_{\text{pooled}_1} = \frac{f_1 + f_2}{n_1 + n_2}$$

$$= \frac{14 + 43}{78 + 110}$$

$$= \frac{57}{188} = .30$$

For categories 2, 3, and 4, respectively, the pooled proportions are

$$P_{\text{pooled}_2} = \frac{18 + 26}{78 + 110} = .23$$

$$P_{\text{pooled}_3} = \frac{22 + 21}{188} = .23$$

$$P_{\text{pooled}_4} = \frac{24 + 20}{188} = .23$$

Each of these pooled proportions is calculated by dividing the marginal column observed frequencies by the total number of frequencies in the table:

$$P_{\text{pooled}_i} = \frac{f_{\text{col}_i}}{N}$$

These marginal column observed frequencies and proportions depict the null hypothesized H_0 distribution of proportions under an assumed model of independence between the two variables. Based on the sample data in Table 15.2, our best estimate is that, *assuming no difference between the favor/oppose groups,* individuals in the population would distribute themselves across the Number of Moves categories, in proportion to the values .30, .23, .23, and .23. Given the column proportions, we can calculate the individual cell expected frequencies. If we expect .30 or 30% of all residents to be in Number of Moves category 1, how many of the 110 opposed respondents would we expect in category 1? We would expect $f_e = (.30)(110) = 33.00$. How many of the 78 residents favoring closing would we expect in Number of Moves category 1 if .30 is the proportion expected?[2] The remaining expected cell frequencies are calculated similarly, multiplying the column pooled proportion by the observed row frequency corresponding to a particular cell. The expected frequencies in Table 15.3 indicate how individuals should be distributed across the contingency table cells if H_0 is true, that number of moves and position on the school closing are indeed independent of each other.

[2] $f_{ei} = (.30)(78) = 23.40$.

TABLE 15.3
Position on school closing and number of moves.

		Number of Moves				
		1	2	3	4	$fo_{row(i)}$
Position on closing	1	$f_o = 14$	$f_o = 18$	$f_o = 22$	$f_o = 24$	$fo_{row(1)} = 78$
		$f_e = 23.65$	$f_e = 18.26$	$f_e = 17.84$	$f_e = 18.26$	
	2	$f_o = 43$	$f_o = 26$	$f_o = 21$	$f_o = 20$	$fo_{row(2)} = 110$
		$f_e = 33.35$	$f_e = 25.74$	$f_e = 25.16$	$f_e = 25.74$	
	$fo_{col(j)}$	$fo_{col(1)} = 57$	$fo_{col(2)} = 44$	$fo_{col(3)} = 43$	$fo_{col(4)} = 44$	188

We have demonstrated the concept of independence and calculated f_e for each cell by focusing on the marginal column observed frequencies and proportions to define the H_0 hypothesized distribution under assumed independence. The marginal row frequencies and proportions are just as appropriate a focus. Recall that both row and column marginals are free to vary. Nevertheless, the same values for the cell f_es would have been obtained using the row frequency totals. The pooled row proportions are $fo_{row(i)}/N$ or 78/188 = .415 for row 1, the group favoring closing, and 110/188 = .585 for row 2, the group opposed to closing. Under a null hypothesis for independence, we would *expect* the frequencies in each of the number of moves categories to be distributed across the two position on school closing groups according to these relative proportions: 41.5% of the frequencies in the favor closing cell, and 58.5% in the oppose closing cell. Fifty-seven individuals are classified in number of moves category 1. Of these 57, we would expect 41.5% or an f_e of (.415)(57) = 23.65 to be from the favoring group and 58.5% or an f_e of (.585)(57) = 33.35 from the opposed group. The expected frequencies in each cell are the same within rounding error.

We have presented these computational procedures for determining the expected cell frequencies to facilitate understanding of the chi square test of independence. In practice, we calculate expected frequencies by the same format, but more mechanically.

A general formula for f_e can be developed based on the probability rules covered in Chapter 10. You may recall that we can define the probability of an event occurring through random sampling as the frequency with which the event occurs relative to the total number of possible events. In a contingency table, we can interpret each of the marginal column proportions and each of the marginal row proportions as probabilities of classification into one of the column or row categories defining each of the two separate variables for an individual sampled at random. For example, the probability is .415 that an individual selected at random from the 188 individuals in Table 15.2 would belong to the group favoring closing. The probability is .585 that an individual would belong to the group opposed to closing. Similarly, for the column values the probability is .30 that a randomly sampled individual would be classified in category 1, .23 in category 2, .23 in category 3, and .23 in category 4.

The column and row marginal proportions provide the probabilities of classification for the categories of each variable separately, but the probability of interest is the cell or joint probability of categories combined across the two variables occurring—for instance, that a randomly sampled individual would be classified as both favoring closing (row 1) and having moved more than six times (column 4). The latter probability is that of being classified in the cell defined by the category labels for the intersection of row 1 and column 4. Assuming independence of the categories for variables X and Y under H_o, the probability of the joint event occurring is the product of the two independent events. You may recall the multiplication rule for the probability of two independent events (presented in Chapter 10):

$$P\begin{pmatrix}\text{Being classified in} \\ \text{row 1 and column 4}\end{pmatrix} = P\begin{pmatrix}\text{Being classified} \\ \text{in row 1}\end{pmatrix} \times P\begin{pmatrix}\text{Being classified} \\ \text{in column 4}\end{pmatrix}$$

Stated in terms of frequencies,

$$\frac{\text{An expected cell frequency, } fe_{cell(i,j)}}{N} = \frac{\text{The observed row frequency, } fo_{row(i)}}{N} \times \frac{\text{The observed column frequency, } fo_{col(j)}}{N}$$

$$\frac{fe_{cell(i,j)}}{N} = \frac{fo_{row(i)}}{N} \times \frac{fo_{col(j)}}{N}$$

Under the assumption of independence stated by the null hypothesis, the general formula for calculating an expected cell frequency in row i, column j is

$$fe_{cell(i,j)} = \frac{(fo_{col(j)})(fo_{row(i)})}{N} \qquad (15.3)$$

Formula 15.3 merely states symbolically the process we used previously to find an expected cell frequency.

$$fe_{cell(i,j)} = \frac{(fo_{col(j)})}{N} \times (fo_{row(i)}) = \begin{pmatrix}\text{column } j \\ \text{proportion}\end{pmatrix}\begin{pmatrix}\text{row } i \\ \text{frequency}\end{pmatrix}$$

$$fe_{cell(i,j)} = \frac{(fo_{row(i)})}{N} \times (fo_{col(j)}) = \begin{pmatrix}\text{row } i \\ \text{proportion}\end{pmatrix}\begin{pmatrix}\text{column } j \\ \text{frequency}\end{pmatrix}$$

It is usually easier to multiply the row and column frequencies first and then divide by N, as in formula 15.3. Using formula 15.3, we would obtain the same expected cell frequencies as given in Table 15.3. For example, the expected cell frequency in row 1, column 4 would be

$$f_{e_{cell(1,4)}} = \frac{(44)(78)}{188}$$

$$= \frac{3432}{188} = 18.26$$

We can check the accuracy of the expected f_e frequencies. The sum of the expected frequencies for any row or any column $\Sigma f_{e_{cell(i,j)}}$ for fixed values of i or j always equals the observed marginal frequency for that row or column (see Table 15.3):

$$\sum_j f_{e_{cell(1,j)}} = f_{o_{row(1)}} \quad \text{and} \quad \sum_i f_{e_{cell(i,1)}} = f_{o_{col(1)}}$$

To illustrate from Table 15.3, the sum of the expected frequencies in row 1 across the j columns is

$$\sum_j f_{e_{cell(1,j)}} = 23.65 + 18.26 + 17.84 + 18.26$$

$$= f_{o_{row(1)}} = 78.01 \text{ or } 78$$

Once we determine the expected cell frequencies for every cell, we can use the observed χ^2 test statistic defined by the general χ^2 formula 15.1 to test H_0. Modifying the symbols to reflect the bivariate χ^2 test of independence,

$$\chi^2_{obs} = \sum \frac{(f_{o_{cell(i,j)}} - f_{e_{cell(i,j)}})^2}{f_{e_{cell(i,j)}}} \qquad (15.4)$$

The squared discrepancies between observed and expected cell frequencies divided by the expected cell frequency are summed across all r times c cells to get the value of χ^2_{obs}. When H_0 is true, the categorical variables X and Y are independent of each other. The χ^2_{obs} values over repeated random samples are distributed as a chi square distribution with degrees of freedom equal to the number of rows − 1 times the number of columns − 1 in the contingency table: $df = (r - 1)(c - 1)$. The critical value for testing H_0 using formula 15.4 is again determined in the upper tail of the χ^2 distribution.

Using the data on observed and expected frequencies in Table 15.3, the χ^2 test of independence would proceed as follows:

$$\chi^2_{obs} = \frac{(14 - 23.65)^2}{23.65} + \frac{(18 - 18.26)^2}{18.26} + \frac{(22 - 17.84)^2}{17.84}$$
$$+ \frac{(24 - 18.26)^2}{18.26} + \frac{(43 - 33.35)^2}{33.35} + \frac{(26 - 25.74)^2}{25.74}$$
$$+ \frac{(21 - 25.16)^2}{25.16} + \frac{(20 - 25.74)^2}{25.74}$$
$$= 3.94 + .00 + .97 + 1.80 + 2.79 + .00 + .69 + 1.28$$
$$= 11.47$$

Assuming an α level of .05, χ^2_{crit} with $df = (2 - 1)(4 - 1) = 3$ equals 7.81 from Table B.5 of chi square values in Appendix B. The observed chi square value $\chi^2_{obs} = 11.47$ is greater than $\chi^2_{crit} = 7.81$, so we would reject the null hypothesis of independence and conclude that position on school closing and number of moves are related. The sample data thus support a conclusion that one's position on school closing is related to the number of residential moves. The general χ^2 test cannot determine the exact nature—the direction—of the relationship. Section 15.7 discusses appropriate interpretations and limitations of this general χ^2 test of independence.

As a second illustration of the χ^2 test of independence from the "Sesame Street" data, we will investigate the relationship between the frequency with which the mother reads to the child (variable 13) and scores on the Home Education Index (variable 24). The reading frequency variable is coded with six categories in Appendix A. To illustrate, we will collapse categories 1 (never), 2 (less than once a week), and 3 (once a week) into a single category of low reading frequency (an average of once or fewer times a week), keep category 4 as defined (several times a week), keep category 5 as defined (once a day), and eliminate category 6 (don't know) from the analysis. To simplify the labels of the categories, we will operationally define the frequency of reading categories as

- low frequency = any child with a coded score of 1, 2, or 3
- medium frequency = any child with a coded score of 4
- high frequency = any child with a coded score of 5

To be appropriate for the χ^2 test, we must reduce the Home Education Index (HEI) variable to a categorical variable. We will operationally define three categories as follows:

- low on HEI = any child with an HEI score less than or equal to 13
- medium on HEI = any child with a score greater than or equal to 14 and less than or equal to 17
- high on HEI = any child with a score greater than or equal to 18

We drew a random sample of 120 children from Appendix A. Based on the coded reading frequency score and HEI score, we classified each child into one of the cells in the 3 × 3 contingency table illustrated in Table 15.4. The observed bivariate cell frequencies are given based on the classification data for the 120 children sampled. The computed expected frequencies for the three cells in the low reading frequency row are, respectively,

$$f_{e_{cell(1,1)}} = \frac{(42)(30)}{120} = 10.5$$

$$f_{e_{cell(1,2)}} = \frac{(40)(30)}{120} = 10.0$$

$$f_{e_{cell(1,3)}} = \frac{(38)(30)}{120} = 9.5$$

Table 15.4 also displays the expected frequencies for the remaining cells in the contingency table. Check the computations of $f_{e_{cell(2,2)}}$ and $f_{e_{cell(3,1)}}$.[3]

Computing χ^2_{obs} from the observed and expected cell frequencies in Table 15.4 involves calculating and summing nine quantities (one for each of the 3 × 3 cells),

$$\begin{aligned}\chi^2_{\text{obs}} &= \frac{(15-10.5)^2}{10.5} + \frac{(9-10)^2}{10} + \frac{(6-9.5)^2}{9.5} + \frac{(17-19.25)^2}{19.25} \\ &\quad + \frac{(21-18.33)^2}{18.33} + \frac{(17-17.42)^2}{17.42} + \frac{(10-12.25)^2}{12.25} \\ &\quad + \frac{(10-11.67)^2}{11.67} + \frac{(15-11.08)^2}{11.08} \\ &= 1.93 + .10 + 1.29 + .26 + .39 + .01 + .41 + .24 + 1.39 \\ &= 6.02\end{aligned}$$

The critical χ^2 with 4 df [$df = (3-1)(3-1)$] assuming $\alpha = .05$ is 9.49. The χ^2_{obs} value of 6.02 is not sufficiently large to reject H_0. The sample evidence indicates that the observed frequencies do not deviate sufficiently from the bivariate distribution of frequencies expected under the null hypothesis of independence. The test of H_0 permits us to conclude that no statistically significant relationship exists between the categories of HEI scores and the frequency with which the mother reports reading to the child as coded for this

---

[3] $$f_{e_{cell(2,2)}} = \frac{(40)(55)}{120} = 18.33$$

$$f_{e_{cell(3,1)}} = \frac{(42)(35)}{120} = 12.25.$$

TABLE 15.4
Contingency table for reading frequency and HEI scores categories.

		Home Education Index Score Categories			
		Low	Medium	High	$f_{o_{row(i)}}$
Frequency of Reading	Low	$f_{o_{cell(1,1)}} = 15$ $f_e = 10.5$	$f_o = 9$ $f_e = 10.0$	$f_o = 6$ $f_e = 9.5$	$f_{o_{row(1)}} = 30$
	Medium	$f_o = 17$ $f_e = 19.25$	$f_o = 21$ $f_e = 18.33$	$f_o = 17$ $f_e = 17.42$	$f_{o_{row(2)}} = 55$
	High	$f_o = 10$ $f_e = 12.25$	$f_o = 10$ $f_e = 11.67$	$f_o = 15$ $f_e = 11.08$	$f_{o_{row(3)}} = 35$
	$f_{o_{col(j)}}$	$f_{o_{col(i)}} = 42$	$f_{o_{col(2)}} = 40$	$f_{o_{col(3)}} = 38$	$N = 120$

analysis. Check the totals in Table 15.4 for columns 1, 2, and 3 when summing the expected cell frequencies across the rows.[4]

One additional comment should be made about the size of the expected cell frequencies necessary for χ^2_{obs} to have a sampling distribution approximating that of a chi square distribution with $df = (r - 1)(c - 1)$ when H_0 is true. A very conservative recommendation would be that all expected cell frequencies be greater than 5. No empirically determined rule has indicated how many cells can have f_e values < 5 without affecting the distribution of χ^2_{obs}. The larger the df in the χ^2 test (the greater the number of cells in the contingency table), the more tolerant the χ^2 test is to the number of f_e values < 5. As a general rule, one or two cells with f_e values < 5 will not affect the distribution of χ^2_{obs} too seriously under a true H_0 *except when* $df = 1$. Chi square tests with 1 df are special cases and will be addressed in the next section.

15.6 THE 2 × 2 CONTINGENCY TABLE: YATES' CORRECTION

A special case of the χ^2 test of independence occurs when both categorical variables are dichotomous variables forming a 2 × 2 contingency table. In this case, we perform the test of H_0 with 1 df. A special computational formula using only the observed cell frequencies facilitates the computation of χ^2_{obs} in

[4] $\Sigma_i f_{e_{cell(i,1)}} = 10.5 + 19.25 + 12.25 = f_{o_{col(1)}} = 42$
$\Sigma_i f_{e_{cell(i,2)}} = 10.0 + 18.33 + 11.67 = f_{o_{col(2)}} = 40$
$\Sigma f_{e_{cell(i,3)}} = 9.5 + 17.42 + 11.08 = f_{o_{col(3)}} = 38.$

this special case. If the observed cell frequencies are denoted by the *a*, *b*, *c*, and *d* in the following 2 × 2 contingency table,

		Variable Y		
		Cat. 1	Cat. 2	$f_{o_{row(i)}}$
Variable X	Cat. 1	a	b	a + b
	Cat. 2	c	d	c + d
	$f_{o_{col(j)}}$	a + c	b + d	N = a + b + c + d

then we can calculate the χ^2_{obs} value by

$$\chi^2_{\text{obs}} = \frac{N(ad - bc)^2}{(a + b)(c + d)(a + c)(b + d)} \tag{15.5}$$

The χ^2_{obs} value from formula 15.5 is distributed as a chi square distribution with $df = 1$ over repeated random samplings *if none of the expected frequencies are less than 5.* If any of the cell f_es are less than 5, we apply a correction developed by Yates (surprisingly called Yates' Correction) to the observed χ^2 value during calculation.

The χ^2_{obs} value must be corrected because calculated χ^2_{obs} values from formulas 15.4 or 15.5 do not form continuous distributions. Discrepancies between f_o and f_e will always result in discrete data points. The discrete sampling distribution of χ^2_{obs} approximates the continuous theoretical tabled values of χ^2 to which χ^2_{obs} is compared when *df* are greater than 1 and the values of f_e are large (≥ 5). When $df = 1$ and $f_e < 5$, Yates' correction is applied to χ^2_{obs} to compensate for the error involved in comparing discrete χ^2_{obs} values to the continuous tabled values of χ^2_{crit}. Yates' correction is called the correction for discontinuity: Essentially, it reduces each cell discrepancy by .5 units before squaring. The χ^2_{obs} test statistic formula with Yates' correction for testing H_0 in a 2 × 2 contingency table is

$$\chi^2_{\text{obs}} = \frac{N\left(|ad - bc| - \frac{N}{2}\right)^2}{(a + b)(c + d)(a + c)(b + d)} \tag{15.6}$$

where $|ad - bc|$ represents the absolute value (positive value) of the difference in the cell frequencies *ad* and *bc*. We subtract $N/2$ from the absolute value before squaring. It may be wise to apply Yates' correction through formula 15.6 in all 2 × 2 situations. The correction will improve the accuracy of the sampling distribution of χ^2_{obs} when H_0 is true, and if always applied, we need not calculate the expected cell frequencies to see if any are less than 5.

To illustrate the χ^2 test of independence in a 2×2 contingency table, we will look at the relationship between the sex of a child and preschool attendance or nonattendance for children in the "Sesame Street" data base. The child's sex is obviously a dichotomous variable. The school-attendance variable (14) has six categories, but categories 2–6 all specify attendance at some type of preschool care center or experience. Category 1 is the nonattendance category. The following cell frequencies were observed for a random sample of 40 children from Appendix A.

		Sex of Child		
		Male	Female	
Preschool Attendance	Yes	$f_o = 10$	$f_o = 11$	21
	No	$f_o = 8$	$f_o = 11$	19
		18	22	$N = 40$

Applying Yates' correction formula,

$$\chi^2_{obs} = \frac{40\left(|(10)(11) - (11)(8)| - \frac{40}{2}\right)^2}{(21)(19)(18)(22)}$$

$$= \frac{40(22 - 20)^2}{158004}$$

$$= \frac{160}{158004} = .001$$

The χ^2_{obs} value is extremely small and does not exceed the χ^2_{crit} value of 3.84 with $df = 1$ and $\alpha = .05$. The data do *not* support a statistically significant relationship between sex and whether a child attended a preschool of some type.

15.7 INTERPRETATIONS

The basic χ^2 test for "goodness of fit" or "independence" simultaneously tests subhypotheses about proportions across categories of a nominal level variable. The goodness of fit hypothesis takes the general form $H_0\colon P_i = a_i$ for each of k categories of the variable under investigation. It compares the distribution of observed frequencies across categories to the distribution of frequencies expected based on the hypothesized category proportions stated in H_0. Rejecting H_0 leads to a general conclusion that the distribution of observed frequencies *as a set* deviates significantly from the null hypothesized distribution. Rejecting H_0 does *not* imply that each or any of the individual category subhypotheses (e.g., $P_1 = a_1$ or $P_3 = a_3$) is statistically significant. The χ^2 goodness of fit test is a general test that does not allow for specific inferences about the statistical significance of individual category hypotheses.

The same conditions operate for interpretations resulting from the χ^2 test of independence. The null hypothesis is one of no relationship or independence of the frequency of classification across levels of two nominal-level variables. While all cell differences between f_o and f_e contribute to the magnitude of χ^2_{obs}, the test is *not* a test of each individual cell difference. Rather, the χ^2 test is a test of the accumulated differences across all cells and as such, it acts as an overall test of the equality of distributions. Rejecting H_0 leads to the general conclusion that the two nominal-level variables are systematically related, that the distributions of category frequencies for one variable are not the same across categories of the other variable. The shape of the distribution for one variable *depends* on which category individuals are in for the other variable. The row and column distributions are not independent of each other.

15.8 MEASURES OF ASSOCIATION

When we reject the H_0 for a chi square test of independence, that is, when $\chi^2_{obs} > \chi^2_{crit}$ at the specified α level, we conclude that a relationship exists between the levels of the nominal variables examined. Whatever the magnitude of the relationship, it is considered statistically significant at the α level. Two common coefficients describe the strength of the relationship in specific contingency tables: the contingency coefficient (C) and Cramer's (φ') coefficient.

Contingency Coefficient (C)

The contingency coefficient, C, describes the degree of association in a contingency table. It is

$$C = \sqrt{\frac{\chi^2_{obs}}{N + \chi^2_{obs}}} \tag{15.7}$$

where χ^2_{obs} is the resulting observed value of χ^2 and N is the number of observations tallied in the contingency table. If we apply this formula to the analysis presented in Table 15.3, we have

$$\begin{aligned} C &= \sqrt{\frac{11.47}{188 + 11.47}} \\ &= \sqrt{.06} \\ &= .24 \end{aligned}$$

This coefficient describes the extent of the association between position on school closing and number of moves. Although the relationship is statistically significant—$\chi^2_{obs} = 11.47$ is greater than $\chi^2_{crit} = 7.81$ at $\alpha = .05$ with $df = 3$—the degree of association is moderate at best.

Use of C is quite common. One of its characteristics is that it can only be a positive quantity. Since χ^2 is always positive, C will always be positive and can only indicate the degree of association, not the direction. Because the variables being examined are nominal, information on the relationship's direction is not possible. The exception to this situation is when the variables can be ordered and the contingency table is 2×2. Then, by inspecting observed frequencies, we can recognize the direction of the relationship.

Another characteristic of the contingency coefficient is that its minimum value is 0, signifying the absence of any relationship between the variables. The maximum value of C is $+1$, but this value may occur only with an infinite number of row and column categories. In practice, the maximum value of C is restricted as a function of the number of rows and columns in the table. Thus, contingency coefficients across different analyses are comparable only for tables containing the same number of rows and columns.

Cramer's Coefficient (φ')

Cramer's coefficient (φ') provides a measure of association between 0 (indicating no relationship) and $+1.00$ (complete dependence). To find it,

$$\varphi' = \sqrt{\frac{\chi^2_{obs}}{N(L - 1)}} \tag{15.8}$$

where χ^2_{obs} is the calculated observed chi square value, N is the number of observations tallied in the contingency table, and L is the lesser of R, the number of rows, or C, the number of columns. The advantage of φ' over C is that it can attain a fixed upper bound ($+1.00$) and this allows comparison across tables of differing dimensions.

For the example data presented in Table 15.4, let's compute φ' even though the relationship was not statistically significant. Because the analysis was formed with three categories for frequency of reading (rows) and three categories for HEI scores (columns), we set $L = 3$. The calculated value of χ^2_{obs} was 6.02 and N, 120. We can substitute these values into formula 15.8 to obtain φ' for the contingency table.

$$\begin{aligned}\varphi' &= \sqrt{\frac{6.02}{120(3 - 1)}} \\ &= \sqrt{.0251} \\ &= .158\end{aligned}$$

Note that for these data χ^2_{obs} was not statistically significant using $\alpha = .05$ as the criterion. The result supports the absence of a relationship between the two variables. Given the prior χ^2 test of independence, we would conclude that the relationship between frequency of reading and HEI is not sufficiently high to differ significantly from zero.

15.9 CHI SQUARE TEST OF HOMOGENEITY

The χ^2 test of homogeneity is a k sample test of the equality of proportionate distributions for the levels of a categorical variable. It is one of the family of chi square tests and is computed in similar fashion to the tests of independence and goodness of fit. This test differs from other tests treated thus far because it can evaluate more than two independent samples simultaneously. We apply the χ^2 test of homogeneity when two or more (k samples) independent samples have been chosen at random and for each element sampled the same categorical variable has been observed. To illustrate a situation suitable for the test of homogeneity, consider the following. An investigator sets out to determine if the members of three populations—those raised in an urban community, those raised in suburban communities, and those raised in a rural environment—evidence similar educational levels. To address this question a sample of members from each population is surveyed and asked to indicate highest level of education attained. Each person surveyed is classified into one of nine categories as follows: completed eighth grade or less, some high school training, high school diploma, some college, college degree, postbachelor's training, master's degree, postmaster's training, and doctoral degree. The dependent variable in this case, level of educational attainment, is a categorical variable and as such computing means or variances for groups of individuals is without justification. Using the χ^2 test of homogeneity in this case will permit the investigator to test the null hypothesis that educational attainment is distributed equally across the three independent samples.

For this test to be employed, classification on the categorical (dependent) variable must be mutually exclusive and independent and sampling must be carried out at random. Each sample does not need to be of equivalent size. Once data have been gathered, they are assembled into an $R \times C$ classification table. Each cell of the table contains the number (frequency) of elements for a particular sample at a specific level of the categorical variable. Table 15.5 displays the data. We can see, for example, that 18 of the 62 persons sampled from an urban area reported completing high school, 6 individuals of the 95 sampled from a rural area had obtained master's degrees, and over all three samples, 55 of the 306 individuals surveyed indicated they had completed college. Note that for the χ^2 test of homogeneity, groups of a specific size are sampled, then elements within each sample are categorized. In contrast to the test of independence, the frequencies in one of the margins are fixed as a result of sampling and the frequencies in the other margin may vary in the test of homogeneity. In the test of independence, a single sample is selected and both marginal frequencies may vary.

TABLE 15.5
Sample data for χ^2 test of homogeneity.

Educational Attainment	Urban	Location Suburban	Rural	Totals
≤ eighth grade	6	10	7	23
Some high school	8	21	7	36
High school graduation	18	30	25	73
Some college	10	16	12	38
College graduation	7	30	18	55
Postbachelor's work	6	17	10	33
Master's degree	4	12	6	22
Postmaster's work	2	8	7	17
Doctoral degree	1	5	3	9
TOTAL	62	149	95	306

The null hypothesis that is tested using the χ^2 test of homogeneity is that the proportional distributions for the *k*-samples are equivalent. The statistical form of this hypothesis is

$$H_0: \underset{\text{Group 1}}{\begin{bmatrix} P_{cat.1} \\ P_{cat.2} \\ \vdots \\ P_{cat.i} \\ \vdots \\ P_{cat.R} \end{bmatrix}} = \underset{\text{Group 2}}{\begin{bmatrix} P_{cat.1} \\ P_{cat.2} \\ \vdots \\ P_{cat.i} \\ \vdots \\ P_{cat.R} \end{bmatrix}} = \underset{\text{Group } k}{\begin{bmatrix} P_{cat.1} \\ P_{cat.2} \\ \vdots \\ P_{cat.i} \\ \vdots \\ P_{cat.R} \end{bmatrix}}$$

The elements of the null hypothesis in statistical form represent conditional probabilities. The statistical hypothesis specifies that category proportional membership is equivalent over samples.

Once we have tabulated the data, we compute the χ^2 test of homogeneity as we did for the test of independence. We find expected frequencies for each cell by taking the product of the row and column marginals and dividing the product by *N*. Using the example data in Table 15.5, compute the expected frequencies for the category "some college" for each sample.[5] Requirements

[5] $$f_{e_{\text{urban}}} = \frac{62 \times 38}{306} = \frac{2356}{306} = 7.70$$

$$f_{e_{\text{suburban}}} = \frac{149 \times 38}{306} = \frac{5662}{306} = 18.50$$

$$f_{e_{\text{rural}}} = \frac{95 \times 38}{306} = \frac{3610}{306} = 11.80$$

for expected frequencies are identical to the conditions discussed for the χ^2 test of independence. Once all the expected frequencies have been completed, we apply formula 15.4. (It is repeated here for convenience.)

$$\chi^2_{obs} = \sum \frac{(f_{o_{cell(i,j)}} - f_{e_{cell(i,j)}})^2}{f_{e_{cell(i,j)}}} \quad (15.4)$$

We then compare the resulting χ^2_{obs} to χ^2_{crit} having df equal to $(R - 1) \times (C - 1)$ at the specified α level of significance. When χ^2_{obs} is greater than χ^2_{crit}, we reject H_0, concluding that the sample proportional distributions are not statistically equivalent. If χ^2_{obs} is less than χ^2_{crit}, we do not reject H_0 and we conclude that the distributions on the categorical variable do not differ significantly over samples.

For the data in Table 15.5, $\chi^2_{obs} = 9.75$. The contingency table has $df = 16$. Evaluating the null hypothesis at $\alpha = .05$, $\chi^2_{crit} = 26.30$. Based on this information, we decide not to reject H_0, since χ^2_{obs} is less than χ^2_{crit} at the test α level. The observed value of χ^2 is in the nonrejection region. From analyzing these data, we conclude that the distributions of educational attainment across urban, suburban, and rural locations do not differ sufficiently to reject H_0. Verify these findings by solving for χ^2_{obs} for the data in Table 15.5.

As another demonstration of the χ^2 test of homogeneity, assume we want to determine if the distributions across the five evaluation-site locations are the same for selected groups of females and males. Our null hypothesis is "Are females distributed in the same proportion as males across the evaluation sites?" We will test this hypothesis at the $\alpha = .01$ level. In testing this hypothesis, we selected at random 40 females and 30 males from the "Sesame Street" data base. Table 15.6 lists the numbers (f) from each group classified into each

TABLE 15.6
Classification table for location and sex of children.

Location	Sex				N
	Female		Male		
	f_o	f_e	f_o	f_e	
Site I	7	8.57	8	6.43	15
Site II	10	8.57	5	6.43	15
Site III	14	13.71	10	10.29	24
Site IV	6	5.14	3	3.86	9
Site V	3	4.00	4	3.00	7
TOTAL	40		30		70

location category. The table reports both f_o and f_e. To obtain χ^2_{obs}, we substitute values into 15.4 and solve:

$$\chi^2_{\text{obs}} = \sum \frac{(f_{oi} - f_{ei})^2}{f_{ei}} = \frac{(7 - 8.57)^2}{8.57} + \frac{(8 - 6.43)^2}{6.43} + \frac{(10 - 8.57)^2}{8.57}$$

$$+ \frac{(5 - 6.43)^2}{6.43} + \frac{(14 - 13.71)^2}{13.71} + \frac{(10 - 10.29)^2}{10.29} + \frac{(6 - 5.14)^2}{5.14}$$

$$+ \frac{(3 - 3.86)^2}{3.86} + \frac{(3 - 4.00)^2}{4.00} + \frac{(4 - 3.00)^2}{3.00}$$

$$= 2.16$$

$$df = (R - 1)(C - 1)$$
$$(5 - 1)(2 - 1)$$
$$= 4$$

For this analysis with 4 df and $\alpha = .01$, $\chi^2_{\text{crit}} = 13.28$. Since $\chi^2_{\text{obs}} = 2.16$, we do not reject the null hypothesis ($\chi^2_{\text{obs}} < \chi^2_{\text{crit}}$ at $\alpha = .01$) and conclude that the distribution of males and females does not differ significantly across the five sites.

15.10 SUMMARY

This chapter introduced three test statistics following the chi square probability distribution: the χ^2 goodness of fit test, the χ^2 test of independence, and the χ^2 test of homogeneity. Each tests specific hypotheses and each is commonly used in behavioral science research to analyze frequencies for categorical or nominal variables. All the chi square test statistics were derived to evaluate the magnitude of the discrepancy between observed and expected frequencies. To the extent that discrepancies are large, this leads us to reject the null hypothesis.

The χ^2 goodness of fit test is uniquely suited as a single-sample test of distribution shape. The test is capable of testing H_0 for any hypothesized distribution shape. The χ^2 test of independence is also a single-sample test, but it tests a null hypothesis of no relationship between two nominal-level variables. Rejecting the null hypothesis suggests the presence of a statistical association between variables. The 2×2 contingency table with Yates' correction was presented as a special case.

Two association statistics, the contingency coefficient and Cramer's coefficient, were provided to estimate the strength of the association. The χ^2 test of homogeneity is a k independent sample test. This statistic tests the H_0 of equivalence of the sample distributions on a categorical variable.

PROBLEMS

Problem Set A

1. For each of the following situations, identify the degrees of freedom and the critical values of χ^2 at the α level specified.
   a. a contingency table having two rows and two columns, $\alpha = .05$
   b. a table having one row and six columns, $\alpha = .01$
   c. a table having four rows and five columns, $\alpha = .01$
   d. a table having three rows and six columns, $\alpha = .10$
   e. a table having five rows and two columns, $\alpha = .05$

2. A department head scheduled four sections of a statistics class at the same time but on different days. In all, a total of 135 students enrolled for the class: section 1—26, section 2—32, section 3—38, section 4—39, total—135. Test the hypothesis that the number of students enrolled is equal across the four sections. Use $\alpha = .01$ and state your conclusion.

3. In a particular community, voter registration records show that 45% of the registered voters are Democrats, 30% are Republican, 15% are independents, and the remaining 10% are unaffiliated. A researcher drew a random sample of 250 voters from these records and obtained the following distribution: Democrat—126, Republican—63, Independents—40, Unaffiliated—21. Test the hypothesis that the researcher's sample matches the political composition of the population. Test the hypothesis at $\alpha = .01$. State your conclusion.

4. A survey asked respondents whether either of their parents had graduated from college. Of the males, 29 said yes and 16, no; of the females, 20 responded yes and 12, no. Test the hypothesis that the respondent's sex is independent of parents' graduating from college at the .05 level of significance.

5. One hundred twenty students took a multiple-choice test. Question ten had four choices. They picked choice 1 20 times; choice 2, 30 times; choice 3, 27 times; and choice 4, 43 times.
   a. Test the hypothesis at $\alpha = .05$ that all students were guessing on this question.
   b. Choice 4 was the correct answer. Test the hypothesis at $\alpha = .01$ that the rate of correct and incorrect responses is what we would expect by chance.

TABLE 15.7
Time students reported watching television.

	School		
	A	B	C
Not at all	7	2	5
Less than one hour	8	5	11
One to two hours	9	7	16
Two to three hours	6	7	10
Three to four hours	9	6	8
More than five hours	6	6	6

TABLE 15.8
College graduates intending to enroll in graduate courses by major.

Major	Yes	No	Not Sure
Physical Science	25	8	3
Engineering	10	12	9
Business	14	11	10
Humanities	15	14	16
Education	20	7	12
Fine Arts	11	16	5
Agriculture	14	9	11

6. In schools A, B, and C samples of 45, 33, and 56 students were drawn. Students in each sample were asked how much time they watched television each day. Table 15.7 reports the results. Test the hypothesis that the rate of television viewing is the same across the schools. Use $\alpha = .05$ and state your conclusion.

7. A sample of college graduates was asked if they intend to continue their studies by taking graduate coursework. Table 15.8 provides their responses by academic area. Test the hypothesis that no relationship exists between plans for graduate studies and academic area. Test the hypothesis at $\alpha = .05$. What is the degree of association between these two variables? Discuss.

TABLE 15.9
Staff frequencies in three school districts.

	A	B	C
Teachers	155	297	73
Administrators	16	42	10
Support staff	28	34	9

8. Table 15.9 lists the number of teachers, administrators, and support staff in three school districts. Test the hypothesis that the distribution of school personnel is the same across the districts. Use $\alpha = .05$.

Problem Set B

I. A study reported by Walden (1967) considered changes in school-board membership and superintendent turnover. He hypothesized that there would be more turnover in the office of superintendent following school board elections in which incumbents were defeated than following elections in which no incumbents were unseated. His investigation gathered data over a 10-year period in 177 school districts. The table following reports some of his findings based on information collected.

Election Results	Superintendent Turnover	
	Voluntary	Involuntary
Incumbents reelected	31	15
Incumbents defeated	15	27

For this initial classification, Walden further investigated to verify through interviews whether the superintendent had voluntarily resigned, that is, without evidence of conflicts with the board, or had involuntarily resigned, that is, left due to conflict with the school board. Given the subjective nature of the superintendent turnover classification, Walden reports the following data classifying only the most clear-cut cases of voluntary and involuntary resignation.

Election Results	Superintendent Turnover	
	Voluntary	Involuntary
Incumbents reelected	16	10
Incumbents defeated	5	23

For both contingency tables, test Walden's hypothesis that there is no relationship between superintendent turnover and school-board elections. Test each table using Yates' correction at $\alpha = .01$. How do the results compare?

II. An article by Gallup (1977) contains many insights into the public opinions regarding public education. Table 15.10 reports the grades those sampled over four years gave to the education children are receiving. Assume that the values given are observed frequencies. Test the hypothesis that the distribution of grades assigned remained constant over the four years of the survey. Test the hypothesis at $\alpha = .01$.

TABLE 15.10
Grades assigned by respondents to the quality of public education.

Grade Assigned	Year			
	1977	1976	1975	1974
A	11	13	13	18
B	26	29	30	30
C	28	28	28	21
D	11	10	9	6
F	5	6	7	5
Don't Know	19	14	13	20

Problem Set C

Select a sample of 60 children from the "Sesame Street" data base. Record their scores for age, sex, and frequency of reading. Use these data for problems 1 and 2.

1. Classify your sample on the basis of age. Use as ranges 42 months or below, 43–48 months, 49–54 months, and 55 months or above. Test the hypothesis that the proportion of children represented within age ranges is the same. Use an $\alpha = .05$.

2. For the sample of 60 selected in problem 1, prepare a contingency table based on scores of 3 or less, 4, and 5 for Frequency of Reading (13) and sex. Test the hypothesis that frequency of reading defined by these three categories is independent of the child's sex. Use $\alpha = .10$ and discuss your findings.

3. Using only children from Sites I and II, classify everyone into a contingency table on the basis of the School Attendance (14) and Viewing Setting variables. Calculate the expected frequencies for each cell. Eliminate school attendance categories in which $f_e < 5$ for both cells. Then test the hypothesis ($\alpha = .05$) that the setting in which "Sesame Street" was viewed (home or school) is independent of school attendance. (Note: f_e will need to be calculated from new table.)

4. Draw a random sample of 80 children from the data base. Classify each child selected in terms of setting (home or school) in which "Sesame Street" was viewed and viewing category. Test the hypothesis that setting is independent of viewing category. Use $\alpha = .05$.

The One-Way Analysis of Variance

Testing a null hypothesis of equal independent group means is common in research. Thus far, we have examined the simplest hypothesis about group mean differences, that for the *two*-group situation. Obviously, the test of H_0: $\mu_x = \mu_y$ for two independent groups is limited. It allows us to compare only two means using the t test.

To expand the research situation, we frequently need to test the equality of more than two means across categories (groups) of an independent variable with more than two levels. For example, we might want to study the effects of different-colored word print on the reading-comprehension scores of first-grade children. To do so, we would instruct independent groups of children using materials printed with various ink colors. If we studied the effects of four colors, we would want to compare the mean reading comprehension scores for four groups, each instructed using a specific color. Assuming that the four word colors were black, green, blue, and red, the null hypothesis would be H_0: $\mu_{\text{black}} = \mu_{\text{green}} = \mu_{\text{blue}} = \mu_{\text{red}}$; that is, no difference in mean reading-comprehension scores occurred among the four groups instructed using different material. At first glance, it might seem that we could test this hypothesis using a series of independent group t tests between all possible pairs of group means:

$$H_0\colon \mu_{\text{black}} = \mu_{\text{green}};\quad H_0\colon \mu_{\text{black}} = \mu_{\text{blue}};\quad H_0\colon \mu_{\text{black}} = \mu_{\text{red}};$$
$$H_0\colon \mu_{\text{green}} = \mu_{\text{blue}};\quad H_0\colon \mu_{\text{green}} = \mu_{\text{red}};\quad \text{and } H_0\colon \mu_{\text{blue}} = \mu_{\text{red}}$$

It seems logical that if we reject any of these six hypotheses, we should also reject the null hypothesis about the equality of all four group means. Using a series of multiple two-group t tests to test a hypothesis involving more than two independent group means is fraught with statistical problems, however, particularly the resulting probability level of making a Type I error. Performing multiple two-group t tests is *not* the way to proceed.

In testing the equality of more than two independent group means, the statistical procedure should test the equality of all group means simultaneously. Later in the chapter an F test will be developed to fill our needs. Using the F distribution as the probability distribution for testing the equality of multiple-group means is a logical extension of using the t test in the two-group case. Chapter 14 showed that the F values in an F distribution with one degree of freedom in the numerator ($df_n = 1$) and df_d degrees of freedom in the denominator were equivalent to squared values of t in the t distribution with df_d degrees of freedom:

$$_{(1-\alpha/2)}t^2_{df_d} = {}_{(1-\alpha)}F(1, df_d)$$

The F distribution can test a null hypothesis about the equality of group means, even in the two-group case! The t test for the two-group case is really just a specific subcase of the more general *F* test. Applying the *F* distribution to test the equality of group means for more than two groups will simply involve changing the specific *F* distribution we use as the probability distribution by changing the numerator degrees of freedom (df_n) to reflect the number of groups.

In fact, extension of the two-group case to the multiple-group test of mean differences forms a basis for expanding the test of mean differences to far more sophisticated and complex research situations. The statistical procedures appropriate for testing both simple and complex hypotheses about mean differences fall into a category of statistics called **analysis of variance (ANOVA)** techniques. Comprehensive study of analysis of variance techniques includes both research design and statistical considerations and can fill an entire semester of study. This chapter introduces the simplest ANOVA design, the one-way analysis of variance or single-factor design with multiple independent groups. Understanding the concept and statistical procedures of the one-way ANOVA design is prerequisite to study of all other ANOVA designs.

16.1 THE RESEARCH DESIGN FOR THE MULTISAMPLE PROBLEM

We have looked at the statistical procedures for comparing two independent group means (*t* test), but we have not discussed in any detail the research design that might dictate testing such a hypothesis. The basis for the two-group mean comparison design is the same as that for the multigroup mean comparison design. The primary purpose of any group-mean difference design is to describe the effects of certain experimental treatments upon some characteristic of a particular population or to test some hypothesis about this treatment effect. From a design perspective, in the one-way ANOVA case, we must first identify a single treatment variable. The treatment variable is the independent variable in the design whose effect on some dependent variable is of interest. As in the chi square designs of the last chapter, we must define the treatment variable as a categorical variable in any ANOVA design. The categories or levels of the treatment variable identify the treatment conditions or group characteristics whose effects on a criterion variable we are investigating.

Consider as an example the study of the effects of the print color of instructional material on the reading comprehension of first-grade students. **Word color** would be the treatment or independent variable. The specific word colors chosen for investigation would define the color treatment conditions as categories or levels of the treatment variable. The one-way ANOVA design investigates the effects of a single factor, a single dimension, a single treatment variable as the independent variable. Several treatment conditions, levels, or categories may define the treatment variable, but a single unifying characteristic links all levels as categories in the measurement of a single variable. We can determine the effects of the print color treatment conditions by comparing group mean performances on the criterion variable, reading-comprehension scores. Differences in mean reading comprehension scores across independent groups receiving different print color treatment conditions would imply that different word colors have different effects. The goal of this chapter is to develop the statistical procedures necessary for determining whether observed sample mean differences of a particular magnitude are expected or are highly improbable based upon sampling distribution expectations. Again, we employ the general format for the hypothesis testing process.

Before examining the statistical basis for determining our expectations in testing the single-factor ANOVA null hypothesis (H_0: $\mu_1 = \mu_2 = \cdots = \mu_k$), the concept of a **treatment** should be further clarified within the context of ANOVA designs. For our purposes, a *treatment* can be defined (following from Lindquist, 1953) as any *induced* or *selected* variation in the research procedures or conditions whose effect is to be observed or evaluated. **Induced** variation results from directly manipulating some group of individuals in the research study by the experimenter, that is, directly administering the treatment to individuals involved in the research study. Examples of treatment variables in which we could induce variation include teaching methods, levels of anxiety, level of reinforcement, drug dosage, perceived amount of fear in a confederate companion, amount of practice, and color of printed material. Treatments differing in amount or type would define the levels of each of the treatment variables with each level or type of the treatment variable administered to a given independent group. Studies in which manipulation of the treatment variable results in induced variation are "true" experiments (Campbell & Stanley, 1963). Cause and effect conclusions are possible only from such true experiment designs in which the experimenter has direct control over manipulation of the treatment conditions that define the categories (groups) in the one-way ANOVA design.

In contrast to induced variation, **selected** variation occurs through sampling and assignment to a level of the treatment variable based on some common characteristic that defines a specific category of the independent variable. Amount of viewing from the "Sesame Street" data is a prime example of a treatment variable resulting in selected variation. The amount of time spent viewing "Sesame Street" was not manipulated directly. Children were not assigned at random to viewing-time categories prior to the experiment and monitored so that they watched only the assigned amount of time. Such procedures would have defined a treatment resulting in induced variation. Rather, viewing time was assessed on a post hoc basis and children were assigned to a treatment condition (group) based on the common characteristic of amount of time spent viewing "Sesame Street." In the induced variation case, the experimenter directly controls assignment to treatment groups and administration of the treatment conditions. In the selected variation case, the experimenter forfeits experimental control. Examples of establishing levels of the treatment variable based on selected variation include class size, marital status, educational level, number of children in the home, type of handicapping condition, and socioeconomic level.

Whether induced or selected variation defines the treatment dimension in the one-way ANOVA design, the same hypothesis is tested and the same statistical procedures are used to test H_0. The need to distinguish between the two types of variation as they define the treatment variable is reflected in the subsequent interpretations and valid inferences possible. It is important that one be aware of these differences from a research design perspective. Statistically, there is no need to differentiate between induced or selected variation in the design. As long as there are multiple independent group defining levels of a single treatment variable, the null hypothesis of no difference among group means on the dependent criterion variable can be tested using one-way ANOVA statistical procedures.

16.2 SOURCES OF VARIABILITY IN THE SINGLE-FACTOR EXPERIMENT

In the test of the null hypothesis in the two-group case, $H_0: \mu_x - \mu_y = 0$, any difference in group sample means is reflected in the simple subtraction of the two means, $\overline{X} - \overline{Y}$. When more than two groups exist, as when we test the one-way ANOVA null hypothesis,

$$H_0: \mu_1 = \mu_2 = \mu_3 = \cdots = \mu_k$$

the differences among the k group means cannot be represented by a single subtraction statement. A familiar index does exist that will capture the magnitude of any differences in group means, however. That familiar index is the variance for a set of scores, except in this case, the scores are the means for k independent groups. The larger the variance index for the set of group means, the greater the difference (spread) among them. If they are all equal, the variance would be zero, indicating no difference. Analysis of variance designs use this concept of variability among group sample means to test the null hypothesis of equality among the means. In doing so, the conceptual foundation remains the same as when testing other hypotheses. How much difference (variability) among sample means might be expected when H_0 is true as a function of random sampling? Note that the sample distribution of the F statistic will form the basis for this expectation. Is the observed difference (variability) among sample means a probable or improbable occurrence given our expectations under a true H_0? A critical F value will allow us to make this decision at a prescribed α level.

To develop the statistical procedures to be applied in the one-way ANOVA design, let's begin by examining the logic underlying the test of

H_0. Under H_0, we hypothesize that the population means represented by treatment groups are equal. If these population means are all equal ($\mu_1 = \mu_2 = \mu_3 = \cdots = \mu_k$), then each of the k treatment groups is representative of the same population set of scores. When testing any H_0, the null hypothesis statement is assumed to be true. The sampling distribution expectations are formed under the assumed-true H_0. In the one-way ANOVA case, each treatment group is assumed to be a random sample from the same parent population *prior to treatment.* If each sample group is drawn from the same population of criterion scores at random prior to treatment, H_0 will be true prior to treatment. However, both the individual scores *within the groups* will vary because of individual differences in the population and the *sample means will differ* due to sampling error (created by the score variability resulting from individual differences). The individual differences reflected in the criterion variable population variance become an important source of score variability in the one-way ANOVA design. This source is termed **experimental error variance** and is equated with variability resulting from sampling error. Error variance merely reflects sampling variability in a design due to individual differences on the criterion variable. As such, we can equate error variance with the population criterion score variance.

Error variance is an important concept in testing the null hypothesis in the one-way ANOVA design. When H_0 is true, the only source of score variability in the design is sampling error or variability due to individual differences inherent in the population. As the only source of variability in the design when H_0 is true, sampling error results in variability among two independent sets of scores. Individual scores within groups will vary and the group means will vary, both as a function of random sampling. We use the variability of these two sets of scores as independent estimates of the same error variance (equivalent to the population score variance, σ_x^2). Because each group is assumed to be a random sample from the same population prior to treatment, the variance of scores within each group (s_j^2) is considered an unbiased estimate of the population variance, σ_x^2. Similarly, the variability among group means is given by the variance error of the mean (square of the standard error), $\sigma_{\bar{x}}^2 = \sigma_x^2/n$. Using the sample variability among group $\bar{X}$s in place of $\sigma_{\bar{x}}^2$, we can estimate the population variance σ_x^2 by $n(s_{\bar{x}}^2)$, the variance among group means multiplied by the group sample size, assuming equal ns at each treatment level. We use these two estimates of σ_x^2 to test H_0. The estimate based on the variability among group means is a measure of how much the group means differ as a function of random sampling error. The estimate based on within-group score variability is a direct estimate of the error variance and serves as the criterion against which we measure the among-group mean estimate. When H_0 is true, both are unbiased estimates of the same population variance and should be approximately equal, differing only by chance. To determine how similar or different the two estimates are, a ratio is formed by placing the among-group means estimate over the within-group estimate. The sampling distribution of this ratio of two variance estimates is distributed as an F distribution that provides the probability distribution to test H_0.

$$F = \frac{\text{variance estimate (based on variability } \textit{among group means}\text{)}}{\text{variance estimate (based on variability } \textit{among scores within groups}\text{)}}$$

Table 16.1 illustrates both the within-group variability of scores and the variability among group means resulting strictly from random sampling error. Assume we are studying the effects of four print colors (black, green, blue, and red) on the reading comprehension of first-grade students. Four separate treatment groups of size 10 were randomly sampled from a population with a mean reading comprehension score of 50 and a variance of 30 to form the data in Table 16.1. As indicated by the group descriptive statistics, the group means range from 48.5 to 50.8 and the group variances from 17.34 to 45.17 as a result of random sampling error.

The data in Table 16.1 reflect variability in reading comprehension scores due solely to sampling error. In any ANOVA design, however, the potential always exists for another source of variability—that due to differential treatment effects across the treatment groups. Note the phrase "differential treatment effects." Treatment effects could exist in a study, but be the same for all treatment conditions. If this were the case, no variability would be added among scores in the design. To illustrate, assume that we applied the separate

TABLE 16.1
One-way ANOVA random data with no treatment effects.

		Print-Color Treatment Group				
		Black	Green	Blue	Red	
		45	47	46	43	
		45	58	49	51	
		41	55	50	45	
		49	46	59	49	
		45	44	49	52	
		56	52	57	54	
		44	58	44	52	
		59	43	42	57	
		49	41	59	52	
		55	41	53	48	
(Raw Score Sums)	$\Sigma X_{.j}$*	488	485	508	503	$\Sigma X_{..} = 1984$
(Squared Score Sums)	$\Sigma X_{.j}^2$	24136	23929	26138	25457	$\Sigma X_{..}^2 = 99660$
(Group Mean)	$\overline{X}_{.j}$	48.8	48.5	50.8	50.3	$\overline{G} = 49.6$
(Group Variance)	s_j^2	35.73	45.17	36.84	17.34	

*Here, we have expanded the summation symbols. This notation will be discussed in section 16.3.

print color treatment conditions to each of the four groups in Table 16.1 and that each treatment added three score points to everyone's reading-comprehension score. Table 16.2 lists the resulting group descriptive statistics. Adding a constant to each score in a group does not change the group variance but does change the mean by an amount equal to the constant. In this case, each group mean reading comprehension score went up three units, but because the group treatment effects were constant, the variability among the group means did not change.

What would happen to the data in Table 16.1 if differential treatment effects were added? Assume that the black print condition had the effect of adding four points to everyone's score in Group 1, that the green print condition decreased each score in group 2 by three points (treatment effect $= -3$), that the treatment effect for the blue print group was to add one point, and the treatment effect for group 4, the red print, was to subtract one reading-comprehension point. Table 16.3 reflects the changes in the data. Note the changes in the group summary statistics. *Only* the group means changed. Again, the constant treatment effect assumed for each individual within a specific treatment group did not affect the within-group score variability. The differential treatment effects did, however, increase the variability among the treatment group means substantially.

The intent of this example is to illustrate that differential treatment effects affect only the variability among treatment group means. The treatment group means become more variable. They not only vary (differ) because of sampling error, but also because of the differences in the treatments. When testing H_0, the increased variability among group means due to the differential treatment effects results in an F ratio with a variance estimate in the numerator larger than expected under a true H_0. The larger-numerator variance estimate leads to a larger observed F ratio, resulting in rejection of H_0. A larger observed F ratio implies that more variability exists among treatment group means than that expected based solely on sampling error. We conclude that the larger variability exhibited among sample means is due to real differences in some or all of the treatment effects. This evidence indicates that the groups after treatment are no longer all representative of the same population. Rather, some or all are representative of separate treatment populations with different means; therefore, we reject H_0. This is the logic behind the test of H_0 in the one-way ANOVA.

TABLE 16.2
Effects of a constant treatment on group means and variances.

	Print Color Treatment Group			
	Black	Green	Blue	Red
$\overline{X}$	51.8	51.5	53.8	53.3
s_X^2	35.73	45.17	36.84	17.34

TABLE 16.3
One-way ANOVA data with differential treatment effects.

	Print Color Treatment Group				
	Black (Effect = 4)	Green (Effect = −3)	Blue (Effect = 1)	Red (Effect = −1)	
	49	44	47	42	
	49	55	50	50	
	45	52	51	44	
	53	43	60	48	
	49	41	50	51	
	60	49	58	53	
	48	55	45	51	
	63	40	43	56	
	53	38	60	51	
	59	38	54	47	
$\Sigma X_{.j}$	528	455	518	493	$\Sigma X_{..} = 1994$
$\Sigma X^2_{.j}$	28,200	21,109	27,164	24,461	$\Sigma X^2_{..} = 100{,}934$
$\overline{X}_{.j}$	52.8	45.5	51.8	49.3	$\overline{G} = 49.85$
$s^2_{.j}$	35.73	45.17	36.84	17.34	

16.3 PARTITIONING THE TOTAL SUMS OF SQUARES AND DEGREES OF FREEDOM

To test the null hypothesis in the one-way ANOVA design, we must calculate variance estimates from among-group means and among-individual scores within groups. In fact, all variability among scores in the total design is either among group–means variability or within group–score variability. The statistical procedures that analyze data in the one-way ANOVA design partition the total score variability in the design into separate parts, that due to differences among group means and that due to individual score differences within groups. Rather than partition total score variance, however, we partition the numerator and denominator of the variance index separately. From Chapter 6, we know the variance for a set of scores is the sum of squared deviations of individual scores from the mean divided by $n - 1$. We later learned that $n - 1$ really represents the degrees of freedom for the set of scores. Symbolically,

$$\text{variance} = s_x^2 = \frac{\Sigma(X - \overline{X})^2}{df}$$

The numerator of the variance formula represents the sum of squared deviations and has been represented by Σx^2. In analysis of variance notation, the

TABLE 16.4
Layout for the one-way ANOVA design.

	Treatment Variable Levels					
	Group 1	Group 2	Group 3	⋯	Group k	
	X_{11}	X_{12}	X_{13}		X_{1k}	
	X_{21}	X_{22}	X_{23}		X_{2k}	
	X_{31}	X_{32}	X_{33}		X_{3k}	
	⋮	⋮	⋮		⋮	
	$X_{n_{.1}1}$	$X_{n_{.2}2}$	$X_{n_{.3}3}$		$X_{n_{.k}k}$	
						TOTAL
$\Sigma X_{.j} \rightarrow$	$\Sigma X_{.1}$	$\Sigma X_{.2}$	$\Sigma X_{.3}$		$\Sigma X_{.k}$	$\Sigma X_{..}$
$\Sigma X^2_{.j} \rightarrow$	$\Sigma X^2_{.1}$	$\Sigma X^2_{.2}$	$\Sigma X^2_{.3}$		$\Sigma X^2_{.k}$	$\Sigma X^2_{..}$
$\overline{X}_{.j} \rightarrow$	$\overline{X}_{.1}$	$\overline{X}_{.2}$	$\overline{X}_{.3}$		$\overline{X}_{.k}$	$\overline{G}$

sum of squared deviations of any set of scores around a mean is *SS*. In this section, we want to partition the sum of squared deviations for all the scores in the design, SS_{total} (SS_T), and their degrees of freedom, df_{total}, into separate components representing different sources of variability–among groups and within groups. Before we analyze SS_T and df_{total}, however, we need to describe the notation system.

Table 16.4 presents the layout for the one-way ANOVA design with k treatment groups. The symbols to be used are given in the table. Each individual score is represented by X_{ij}, where j references the treatment group and i represents the ith individual score in the jth group. For example, X_{32} represents the score for the third individual in the second treatment group. We will use "dot" notation to indicate that the scores have been summed over the subscript values where the dot appears. For example, the notation $\Sigma X_{.3}$ identifies the sum of all raw scores X_{i3} in the third treatment group. The sum of the squared scores would be $\Sigma X^2_{.3}$. Why might we want to compute $\Sigma X_{.3}$ and $\Sigma X^2_{.3}$?[1]

We find the mean for a group, $\overline{X}_{.j}$, by

$$\overline{X}_{.j} = \frac{\Sigma X_{.j}}{n_{.j}}$$

where $n_{.j}$ is the sample size for group j. The only deviation from the dot notation is the symbol for the mean of all scores in the design and the total number of scores. This mean is called the "grand mean" and is represented by

[1]The quantities $\Sigma X_{.3}$ and $\Sigma X^2_{.3}$ are needed to compute the mean, sum of squares, variance, and standard deviation for group 3.

$\overline{G}$. The total number of scores across all groups in the design will be represented by N.

$$\overline{G} = \overline{X}_{..} = \frac{\Sigma X_{..}}{N} = \frac{\sum_i \sum_j X_{ij}}{\Sigma n_{.j}}$$

Using the new symbol system, the total sum of squared deviations ignores group membership and can be represented as squared deviations of individual scores (X_{ij}) from the grand mean ($\overline{G}$) summed over all individuals:

$$SS_T = \sum_i \sum_j (X_{ij} - \overline{G})^2$$

Algebraically manipulating the deviation SS_T formula results in the following.

$$\begin{aligned} SS_T &= \sum_i \sum_j (X_{ij} - \overline{G})^2 \\ &= \sum_i \sum_j [(X_{ij} - \overline{X}_{.j}) + (\overline{X}_{.j} - \overline{G})]^2 \\ &= \sum_i \sum_j [(X_{ij} - \overline{X}_{.j})^2 + 2(X_{ij} - \overline{X}_{.j})(\overline{X}_{.j} - \overline{G}) + (\overline{X}_{.j} - \overline{G})^2] \\ &= \sum_i \sum_j (X_{ij} - \overline{X}_{.j})^2 + 2\sum_i \sum_j (X_{ij} - \overline{X}_{.j})(\overline{X}_{.j} - \overline{G}) + \sum_j n_{.j}(\overline{X}_{.j} - \overline{G})^2 \end{aligned}$$

Since simple summed deviations of scores around a group mean will equal zero, the middle term in the last expression equals zero and we are left with

$$\begin{aligned} SS_T &= \sum_i \sum_j (X_{ij} - \overline{G})^2 \\ &= \sum_i \sum_j (X_{ij} - \overline{X}_{.j})^2 + \sum_j n_{.j}(\overline{X}_{.j} - \overline{G})^2 \end{aligned}$$

This last expression partitions the ANOVA design total sums of squares into two independent component parts. The first term, $\Sigma_i \Sigma_j (X_{ij} - \overline{X}_{.j})^2$, represents the sum of squared deviations (variability) of individual group scores (X_{ij}) from their respective group means ($\overline{X}_{.j}$). The sums of squared deviation quantities are then summed (pooled) over all groups. This quantity represents the individual *within-group* sums of squares pooled over all groups (SS_W).

$$SS_{\text{within groups}} = SS_W = \sum_i \sum_j (X_{ij} - \overline{X}_{.j})^2$$

The second term in the expression $\Sigma_j n_{.j}(\overline{X}_{.j} - \overline{G})^2$ represents the sum of squared deviations (variability) of group means $\overline{X}_{.j}$ from the grand mean $\overline{G}$ weighted by the individual group sample sizes $n_{.j}$ prior to summing. This quantity is commonly referred to as the between-group means sum of squares (SS_B) and reflects variability among group means.

$$SS_{\text{between groups}} = SS_B = \sum_j n_{.j}(\overline{X}_{.j} - \overline{G})^2$$

In summary, the total variability in the one-way ANOVA design as represented by SS_{total} can be partitioned into two independent components representing variability of different sets of scores.

$$SS_T = SS_B + SS_W$$

Similar to the partitioning is SS_T, we can partition the total degrees of freedom in the design into the degrees of freedom associated with SS_B and SS_W. One need only look at the set of scores contributing to each sum of squares to calculate the appropriate degrees of freedom. In general, the degrees of freedom equal the number of scores contributing to the variability minus one. All X_{ij} scores are used to calculate SS_T. The $df_{\text{total}} = N - 1$, the total number of scores in the design minus one. For SS_B, the k treatment group means are the scores for which the squared deviations are calculated; therefore, $df_B = k - 1$, the number of treatment groups minus one. The last quantity (SS_W) is a pooled source of variability. The sum of squared deviations is calculated for scores within each treatment group and then summed over groups.

$$SS_W = SS_{\text{within Group 1}} + SS_{\text{within Group 2}} + \cdots + SS_{\text{within Group } k}$$

The degrees of freedom within also are calculated within each group first and then summed over the k groups.

$$\begin{aligned} df_W &= df_{\text{within Group 1}} + df_{\text{within Group 2}} + \cdots + df_{\text{within Group } k} \\ &= (n_{.1} - 1) + (n_{.2} - 1) + \cdots + (n_{.k} - 1) \\ &= \sum_j n_{.j} - k \\ &= N - k \end{aligned}$$

The summary formula for calculating df_W is $df_W = N - k$, the total number of scores minus the number of groups. Conceptually, however, it is best to remember that we find df_W by summing the degrees of freedom within each group across all groups. You should not be surprised that the total degrees of freedom in a design equals the sum of df_B and df_W.

$$
\begin{array}{ccccc}
df_{\text{total}} & = & df_B & + & df_W \\
\downarrow & & \downarrow & & \downarrow \\
N - 1 & = & (k - 1) & + & (N - k)
\end{array}
$$

What are the values of df_{total}, df_B, and df_W for the data in Table 16.1?[2]

16.4 RAW SCORE COMPUTATIONAL FORMULAS

The formulas given for SS_T, SS_B, and SS_W in the last section are in deviation summation form. As we saw when the variance was introduced in Chapter 6, the deviation representation is appropriate for defining a variability quantity conceptually, but computationally, finding deviation scores from a mean is inefficient. Using equivalent raw-score formulas saves time and labor. The raw-score formulas for each of the desired sum of squares appear here without their derivation.

$$SS_B = \sum_j n_j(\overline{X}_{.j} - \overline{G})^2$$

$$= \sum_j \frac{(\Sigma X_{.j})^2}{n_{.j}} - \frac{(\Sigma X..)^2}{N} \tag{16.1}$$

This formula looks complicated, but really is straightforward.

$\Sigma X_{.j}$ = sum of the scores for group j

$n_{.j}$ = sample size for group j

$\Sigma X..$ = total sum of all scores in the design

N = total number of scores in the design

$$SS_B = \overbrace{\frac{(\text{Sum of Gp. 1 Scores})^2}{\text{No. of Scores in Gp. 2}} + \frac{(\text{Sum of Gp. 2 Scores})^2}{\text{No. of Scores in Gp. 1}} + \cdots + \frac{(\text{Sum of Gp. } k \text{ Scores})^2}{\text{No. of Scores in Gp. } k}}^{\sum_j \frac{(\Sigma X_{.j})^2}{n_j}} - \overbrace{\frac{(\text{Sum of All Scores})^2}{\text{Total No. of Scores}}}^{\frac{(\Sigma X..)^2}{N}}$$

---

[2] $df_{\text{total}} = 40 - 1 = 39$; $df_{\text{between}} = 4 - 1 = 3$;

$df_{\text{within}} = 40 - 4 = 36$

or $= (10 - 1) + (10 - 1) + (10 - 1) + (10 - 1)$

$= 36$ (the sum of the nine degrees of freedom within each of the four groups).

Using the data from Table 16.1, we would calculate SS_B as follows:

$$SS_B = \underbrace{\frac{488^2}{10} + \frac{485^2}{10} + \frac{508^2}{10} + \frac{503^2}{10}}_{\sum_j \frac{(\Sigma X_{.j})^2}{n_{.j}}} - \underbrace{\frac{1984^2}{40}}_{\frac{(\Sigma X_{..})^2}{N}}$$

$$= 98444.2 - 98406.4$$

$$= 37.8$$

The raw-score computational formula for SS_W is derived by pooling the standard sum of squares formula within each group across groups. The within-group formulas are

$$SS_{\text{within Gp. 1}} = \Sigma X_{.1}^2 - \frac{(\Sigma X_{.1})^2}{n_{.1}}$$

$$SS_{\text{within Gp. 2}} = \Sigma X_{.2}^2 - \frac{(\Sigma X_{.2})^2}{n_{.2}}$$

$$\vdots$$

$$SS_{\text{within Gp } k} = \Sigma X_{.k}^2 - \frac{(\Sigma X_{.k})^2}{n_{.k}}$$

Summing across the k groups,

$$SS_W = \sum_i \sum_j (X_{ij} - \overline{X}_{.j})^2$$

$$= \Sigma X_{..}^2 - \sum_j \frac{(\Sigma X_{.j})^2}{n_{.j}} \qquad \textbf{(16.2)}$$

where $\Sigma X_{..}^2$ is the sum of all squared scores in the design (square the scores first, then sum) and the last term $\Sigma_j[(\Sigma X_{.j})^2/n_{.j}]$ is identical to the first term in formula 16.1 for calculating SS_B.

$$SS_B = \sum_j \frac{(\Sigma X_{.j})^2}{n_{.j}} - \frac{(\Sigma X_{..})^2}{N}$$

same

$$SS_W = \Sigma X_{..}^2 - \sum_j \frac{(\Sigma X_{.j})^2}{n_{.j}}$$

To calculate SS_W from the data in Table 16.1,

$$SS_W = 45^2 + 45^2 + \cdots + 48^2 - \left(\frac{488^2}{10} + \frac{485^2}{10} + \frac{508^2}{10} + \frac{503^2}{10}\right)$$

$$= 99660 - 98444.2$$

$$= 1215.8$$

To demonstrate the equality, we could have calculated SS_W by finding each $SS_{\text{within Gp } j}$ and summing across the j groups.

$$SS_{\text{within Gp 1}} = 24136 - \frac{488^2}{10} = 321.6$$

$$SS_{\text{within Gp 2}} = 23929 - \frac{485^2}{10} = 406.5$$

$$SS_{\text{within Gp 3}} = 26138 - \frac{508^2}{10} = 331.6$$

$$SS_{\text{within Gp 4}} = 25457 - \frac{503^2}{10} = 156.1$$

The sum of $SS_{\text{within Gp}j} = 321.6 + 406.5 + 331.6 + 156.1 = 1215.8$, the same value as computing SS_W by formula 16.2. It should be noted that the calculated values for $SS_{\text{within Gp}j}$ are the sum of squares values needed to compute the group variances and standard deviations:

$$\text{variance for Gp. 1} = s_{.1}^2 = \frac{321.6}{9} = 35.73$$

We can obtain the raw-score computational formula for SS_T by adding the SS_B and SS_W formulas. Combining the two formulas cancels out the similar term,

$$\begin{aligned} SS_T &= SS_B + SS_W \\ &= \sum_i \sum_j (X_{ij} - \overline{G})^2 \\ &= \Sigma X_{..}^2 - \frac{(\Sigma X_{..})^2}{N} \end{aligned} \tag{16.3}$$

Formula 16.3 treats the data as one group and computes the sum of squared deviations for all scores from the grand mean. For the data in Table 16.1,

$$SS_T = 99{,}660 - 98{,}406.4 = 1253.6$$

Not surprisingly,

$$\begin{aligned} SS_T &= SS_B + SS_W \\ &= 37.8 + 1215.8 \\ &= 1253.6 \end{aligned}$$

For the data in Table 16.3, compute all three sums of squares.[3] Are you

---

[3] $SS_T = 100934 - \frac{1994^2}{40} = 1533.1$

$$SS_B = \frac{528^2}{10} + \frac{455^2}{10} + \frac{518^2}{10} + \frac{493^2}{10} - \frac{1994^2}{40} = 317.3$$

$$SS_W = 100934 - \left(\frac{528^2}{10} + \frac{455^2}{10} + \frac{518^2}{10} + \frac{493^2}{10}\right) = 1215.8.$$

surprised by the value for SS_W? You shouldn't be. Adding the differential treatment effects to the data did not affect within-group variability. The between-group variability, however, was affected as demonstrated by the values of SS_B for the data in Tables 16.1 and 16.3. For data in Table 16.1, $SS_B = 37.8$ and for data in Table 16.3, $SS_B = 317.3$. This clearly demonstrates which sum of squares is affected by differential treatment effects in ANOVA data.

16.5 MEAN SQUARES AS VARIANCE ESTIMATES AND THE TEST OF H_0

In the analysis of variance framework, SS_B and SS_W are converted to variance estimates through division by their respective degrees of freedom. These variance estimates are called **mean squares** (MS).

$$MS_B = \frac{SS_B}{df_B} \tag{16.4}$$

$$MS_W = \frac{SS_W}{df_W} \tag{16.5}$$

The magnitude of MS_B is based on the amount of variability among the group means. As we saw in section 16.2, treatment group means may differ (vary) as a result of two sources of variability, sampling error and differential group treatment effects. When H_0 is true (the treatment population means are indeed equal), MS_B is affected solely by sampling error variance in the design and is an estimate of the population variance due to individual differences. The population individual difference score variability that creates variability in ANOVA designs as a result of sampling is termed *error* variability.

- *When H_0 is true,* MS_B is an unbiased estimate of *error* variance.

If the treatments that differentiate the groups in a one-way ANOVA design are differentially effective, they will increase variability among group means over that due to error. The size of MS_B will reflect this additional variability among group means due to differential treatment effects. Because it is not possible to separate the MS_B due to error from the portion due to differential treatment effects, MS_B must be viewed as potentially composed of both types.

- *If H_0 is not true,* MS_B is an estimate of both *error* variance *and differential treatment effect* variability.

How can we judge whether MS_B represents both sources of variability or whether it reflects only error variance? Fortunately, MS_W offers a comparative criterion. The magnitude of MS_W is based entirely on variability among the individual scores within each group and pooled over groups. We have demonstrated (in Tables 16.1 and 16.3) that it is unaffected by differential treatment effects across groups. Whether H_0 is true or not, MS_W is a direct estimate of population individual differences on the criterion variable, better identified as error variance.

- MS_W always is an unbiased estimate of *error* variance.

As a pure estimate of error variance, we can use MS_W as the criterion against which to compare MS_B. We can decide whether MS_B reflects variability due to both error and differential treatment effects or due just to error based on the size of MS_B relative to MS_W. Comparing MS_B to MS_W serves as the test of H_0. ***If H_0 is true,*** both are estimates of the same error variance and their ratio has a known sampling distribution. As the ratio of two variance estimates, MS_B/MS_W is distributed as an F distribution with df_B and df_W degrees of freedom when H_0 is true. The mean of any F distribution is $df_d/(df_d - 2)$, slightly greater than 1.00. As estimates of the same error variance when H_0 is true, the ratio of MS_B to MS_W will fluctuate randomly around the mean F value. The test statistic formula we use to test H_0: $\mu_1 = \mu_2 = \cdots \mu_k$ is an F test,

$$F_{\text{obs}} = \frac{MS_B}{MS_W} \tag{16.6}$$

A critical F value for judging the size of F_{obs} in testing H_0 is found in the F probability distribution with $df = (k - 1, N - k)$. The values for selected levels of α are found in Appendix B, Table B.5. If F_{obs} is large ($>F_{\text{crit}}$), the implication is that MS_B contains not just error variance, but also variability due to differential treatment effects. The statistical inference if $F_{\text{obs}} > F_{\text{crit}}$ is that H_0 must be false. The treatment groups do not all represent samples from the same population, but rather are representative of populations with different means, reflecting the differential effects of the treatments. Note that the logic of the test of H_0 results in a directional test. Only if MS_B is substantially larger than MS_W would we reject H_0.

The components necessary to test the null hypothesis of equality among independent group means have been introduced and discussed. The test of H_0: $\mu_1 = \mu_2 = \cdots = \mu_k$ proceeds as follows:

1. Compute SS_B, SS_W, and SS_T from the sample data.
2. Compute $df_B = k - 1$ and $df_W = N - k$.
3. Convert SS_B and SS_W to mean squares:

$$MS_B = \frac{SS_B}{df_B}$$

$$MS_W = \frac{SS_W}{df_W}$$

4. Form the ratio of MS_B to MS_W to obtain an observed F value:

$$F_{\text{obs}} = \frac{MS_B}{MS_W}$$

TABLE 16.5
General format for an ANOVA summary table.

Source of Variability	*df*	*SS*	*MS*	*F*
Treatment variable effect (between groups)	$k - 1$	SS_B	MS_B	$\frac{MS_B}{MS_W}$
Within groups (error)	$N - k$	SS_W	MS_W	
TOTAL	$N - 1$	SS_T		

5. Identify the critical F value (F_{crit}) at a specified α level in the F probability distribution with $df = k - 1$ and $N - k$, assuming a directional test of H_0.
6. Compare F_{obs} to F_{crit}. If $F_{obs} > F_{crit}$, reject H_0. If $F_{obs} < F_{crit}$, do not reject H_0.
7. If H_0 is rejected, conclude that the sample means are statistically different at a significance probability of α. The inference is that some or all of the treatments defining each of the separate groups as a composite were differentially effective (statistically significant).

In practice, a standard format is used to report the results from these steps in the form of an **ANOVA summary table.** Table 16.5 displays the general form of an ANOVA summary table. The table includes the computed values for *df*, *SS*, *MS*, and *F* for the appropriate sources of variability. The *F* value in the table is the computed F_{obs} value. Table 16.6 illustrates the ANOVA summary table for the data from Table 16.1. The SS_B and SS_W were computed in section 16.4. The other values needed are as follows.

$$MS_B = \frac{37.8}{3} = 12.6$$

$$MS_W = \frac{1215.8}{36} = 33.77$$

$$F_{obs} = \frac{12.6}{33.77} = .373$$

TABLE 16.6
ANOVA summary table using data from Table 16.1.

Source of Variability	*df*	*SS*	*MS*	*F*
Print color (between groups)	3	37.8	12.6	.373
Within groups	36	1215.8	33.77	
TOTAL	39	1253.6		

TABLE 16.7
Numbers pretest scores for random samples of children classified by reading frequency.

	Reading of Frequency Group				
	Never	Less than Once a Week	Once a Week	Several Times a Week	Once a Day
	8	38	14	23	39
	4	23	13	20	18
	16	13	11	47	16
	17	14	25	23	18
	19	19	11	16	15
	14	14	5	21	27
			23	17	17
			11	32	14
				36	12
				17	16
$\Sigma X_{.j}$	78	121	113	252	192
$\Sigma X_{.j}^2$	1182	2895	1907	7262	4264
$\bar{X}_{.j}$	13.00	20.17	14.13	25.20	19.20
$s_{.j}$	5.80	9.54	6.66	10.06	8.01
$n_{.j}$	6	6	8	10	10

If $\alpha = .05$, F_{crit} with 3 and 36 degrees of freedom is 2.88. $F_{obs} < F_{crit}$, therefore we do not reject H_0. Given that the data in Table 16.1 represent only random sampling error, what percentage of the time would you expect F_{obs} to be less than or equal to F_{crit} using similarly generated examples?[4] Complete the ANOVA summary table for the data from Table 16.3. The SS_B and SS_W were computed previously as 317.3 and 1215.8, respectively.[5] Not surprisingly, given that differential treatment effects were built into the data in Table 16.3, we would reject H_0: $\mu_{black} = \mu_{green} = \mu_{blue} = \mu_{red}$ based on the size of F_{obs}. The data suggest that the variability among the group means is not likely (at the .05 level of significance) to have happened just as a result of random

[4]Given that $\alpha = .05$, and H_0 is true, our expectation is that F_{obs} should be $\leq F_{crit}$ 95% of the time.

[5]Source of Variability	*df*	*SS*	*MS*	*F*
Print Color Effect	3	317.3	105.77	3.13*
Within Groups	36	1215.8	33.77	
TOTAL	39	1533.1		

*Significant at the .05 level or ($p < .05$).

TABLE 16.8
ANOVA computations and summary table.

$$\Sigma X_{..} = 78 + 121 + 113 + 252 + 192 = 756$$

$$\Sigma X^2_{..} = 1182 + \cdots + 4264 = 17{,}510$$

$$SS_T = 17{,}510 - \frac{756^2}{40} = 3221.60$$

$$SS_B = \frac{78^2}{6} + \frac{121^2}{6} + \frac{113^2}{8} + \frac{252^2}{10} + \frac{192^2}{10} - \frac{756^2}{40}$$

$$= 15{,}087.09 - 14{,}288.40 = 798.69$$

$$SS_W = 17{,}510 - 15{,}087.09 = 2422.91$$

Source of Variability	*df*	*SS*	*MS*	*F*
Reading frequency group	4	798.69	199.67	2.88*
Within groups	35	2422.91	69.23	
TOTAL	39	3221.60		

*Significant at the .05 level [($F_{crit(4,35)} = 2.65$)].

sampling. The inference is that at least some of the color treatments must have been differentially effective, thereby increasing the variability among the group means.

As one additional illustration, Numbers pretest scores are given in Table 16.7 for random samples of students from the "Sesame Street" data base classified by the frequency with which the mother reported reading to the child. Test the null hypothesis at the .05 level of significance using these data—H_0: There exists no difference in mean Numbers pretest scores for children who were classified in different frequency of reading categories. (or H_0: $\mu_{never} = \mu_{<once} = \mu_{once} = \mu_{several} = \mu_{daily}$). Type of school attended is the treatment variable in this example and is operationally defined by the five categories listed. The criterion variable is Numbers pretest scores. The calculations and the ANOVA summary table appear in Table 16.8. The F_{obs} value of 2.88 exceeds the F_{crit} value of 2.65 resulting in the decision to reject H_0.

16.6 ASSUMPTIONS UNDERLYING THE *F* TEST OF H_0

Using the *F* distribution to test hypotheses about two or more means makes the same assumptions as those for the *t* test for two independent sample means. When H_0 is true, the ratio of MS_B to MS_W is distributed as an *F* distribution if the following assumptions are met in the data and the population(s) from which the data are gathered.

1. *Score independence.* Scores within and across the k groups are assumed to be independent of each other. This means that one person's score is not determined or related in any way to another person's score and that all scores are free to vary across the total measurement continuum without restrictions other than the individual's performance or response level.
2. *Normality of population distributions.* The distribution of the criterion variable in each population is assumed to be normal. The normality assumption ensures that the mean and the variance for each population that the groups represent are independent of each other. This way the group mean may be raised or lowered by a treatment condition, but the variability of scores within the group should not be affected. If the variability is affected by the treatment, as it might be in a skewed distribution in which the mean and variance are related, it is no longer an accurate estimate for the population. In this case, MS_W will not accurately estimate the error variance due to individual differences in the population.
3. *Homogeneity of population variance.* The variances in the populations that are represented by the samples are assumed to be equal. This assumption is labeled as the "homogeneity of variances" assumption and is necessary for the pooled sums of squared deviations formula in calculating SS_W. If the group variances are not approximately equal, all cannot be accurate estimates of the population variance and the error variance will be over or underestimated by MS_W.

The consequences of violating any or a combination of the assumptions is that the test statistic formula (formula 16.6) will result in an artificially inflated or deflated F_{obs} value and will lead us to reject or not reject H_0 due to violation of the assumption rather than the presence or absence of a true treatment effect. If we violate an assumption, we do not have as much confidence in the F test as an accurate test of H_0 at the alpha level specified. Several empirical studies have investigated how violating assumptions affects Type I error rates. Generally, when the sample sizes for the groups are equal, the F test is robust to violations of the normality and homogeneity of variance assumptions. When sample sizes are not equal, we should pay specific attention to any violations, particularly of the variance homogeneity assumption. A review by Glass, Peckham, and Sanders (1972) offers a comprehensive discussion of the consequences of violating the assumptions when sample sizes are unequal. While the F test is robust in testing the ANOVA H_0, we should still consider extreme violations in our data. Visually inspecting the data and using descriptive statistical indices do not take the place of the statistical tests available to see whether assumptions have been violated. Descriptive statistics, however, *do* provide some information regarding the assumptions and should be used as a first step in detecting extreme violations. Winer (1971) contains statistical tests of the assumptions.

16.7 ESTIMATING THE MAGNITUDE OF AN EFFECT

The overall F test for the one-way ANOVA indicates whether the observed differences in treatment group means are due to sampling error entirely (a nonsignificant F_{obs}) or support differential treatment effects (a statistically significant F_{obs}). In the latter case, many researchers *erroneously* interpret the size of the F_{obs} value or its associated significance level as an indication of the "importance" or "strength" of an effect. They inappropriately infer that an F_{obs} value large enough to be significant at say the .001 level indicates that the effect is larger or stronger than one with an F_{obs} value that would be significant at "only" the .05 level. This inference is not accurate. The problem with equating the size of F_{obs} with importance or strength is that the size of F_{obs} is affected by other factors in the design besides size of the treatment effects. These other factors largely involve design precision. Increased precision in a design (particularly through larger sample sizes) will increase the size of F_{obs} even if the treatment effects remain constant. This caution applies in t test evaluations as well. Cohen (1977) provides a comprehensive discussion of this point.

Several measures of the treatment effect magnitude are not influenced by sample size. One appropriate measure is an index called **omega squared** (ω^2). When applied to the one-way ANOVA design, ω^2 is the ratio of the variance estimate due to differential treatment effects relative to the total amount of variability in the design. As a squared index, the interpretation of ω^2 is similar to the interpretation of r_{xy}^2. It reflects the relative proportional amount of total variability in the design *accounted for* by the experimental treatment effects. For example, in the print color experiment, ω^2 would indicate the proportion of reading comprehension score variability accounted for or explained by the levels of the print color treatments.

Computationally, ω^2 is estimated by using quantities already included in the ANOVA summary table. The formula for computing ω^2 when sample sizes are equal is

$$\omega^2 = \frac{SS_B - (k - 1)MS_W}{SS_T + MS_W} \tag{16.7}$$

Halderson and Glasnapp (1972) provide the basis for formula 16.7 as the ratio of an estimate of differential treatment effect variance to an estimate of total variation. When treatment effects are absent, $\omega^2 = 0$ and when they are present, ω^2 may range between 0 and 1.00. When F_{obs} is less than 1.00, formula 16.7 will produce an ω^2 value less than zero. In this case, the evidence supports no differential treatment effect and by convention we set ω^2 equal to zero. Calculating ω^2 from the ANOVA summary table (in Table 16.6) values for data in Table 16.1 illustrates this last point.

$$\omega^2 = \frac{37.8 - 3(33.77)}{1253.6 + 33.77} = -.049$$

Calculate ω^2 value for the ANOVA design when treatment effects were added to the data (Table in footnote 5).[6] The print color treatment effects account for 14% of the reading comprehension variability in this latter set of data.

16.8 TESTING SUBHYPOTHESES WITHIN THE ANOVA DESIGN

The test of the null hypothesis in the one-way ANOVA is an overall, simultaneous test of the equality of the k treatment population means. Much like the conclusions from the χ^2 tests for goodness of fit, independence, and homogeneity, a statistically significant observed F test statistic for the one-way ANOVA null hypothesis indicates that the group means *as a set* differ from each other. Information on which group mean or combination of means differs significantly from another group mean or combination of means is *not* a result of the overall F test.

In designing a research study, particularly when specifying the null hypothesis, we must consider whether rejecting H_0 will lead to the desired conclusions. Too often, researchers do not think carefully enough about this aspect of their research design. Unfortunately, if several treatment groups fall across levels of a single independent variable, we can automatically apply the one-way ANOVA F test to analyze the data. While this is an appropriate procedure for many studies, automatic application of the overall F test is restrictive. We should consider whether subhypotheses among subsets of the treatment group means are more important than the test of overall equality in terms of the information provided. If we can specify subhypotheses of interest during the design of the study, we can test these subhypotheses without performing an overall one-way ANOVA F test. Advance specification of subhypotheses to be tested follows much the same reasoning and has the same statistical consequences as making a directional test of H_0 in the hypothesis-testing process. If theory, experience, or the literature support research hypotheses about specific differences, it is always best to specify these differences in advance through corresponding null subhypotheses. In the ANOVA design framework, specific subhypotheses identified in advance that compare means for some subset or combination of treatment group means are called **planned comparisons.**

In contrast, a researcher often engages in a study in which advanced specification of hypothesized mean differences is not possible. Rather, the experimenter is interested in first determining if general overall differences in treatment group means exist. If F_{obs} is statistically significant given the speci-

---

[6] $$\omega^2 = \frac{317.3 - (3)(33.77)}{1533.1 + 33.77} = .14.$$

fied α level, then the researcher will perform subanalyses or tests of subhypotheses specified in null form to explore where possible statistically significant differences exist among subsets or a combination of treatment group means. These subhypotheses tested after a statistically significant overall F_{obs} is obtained are called **post hoc comparisons.**

We must distinguish between *planned* and *post hoc* comparisons made as subhypotheses within an ANOVA design. The statistical decision criteria applied in testing the two types of subhypotheses differ: The test of planned subhypotheses is more powerful than the test of post hoc subhypotheses. For this reason, we should make planned comparisons whenever possible in planning the design of research within the ANOVA context. Situations requiring post hoc tests seem to occur more frequently in the research literature, however. The remainder of this section presents a foundation for forming and testing subhypotheses that compare or contrast a subset of the group means in the ANOVA design framework, emphasizing post hoc comparisons. Hays (1981), Keppel (1982), or Winer (1971) contain more detailed discussions of statistical procedures for testing both preplanned and post hoc comparisons.

Statistical Comparisons or Contrasts

Subhypotheses that may be tested during subanalyses in a one-way ANOVA are identified through comparisons or contrasts among group means. A statistical comparison (C) is defined as a weighted composite of treatment group means:

$$C = w_1\overline{X}_1 + w_2\overline{X}_2 + w_3\overline{X}_3 + \cdots + w_k\overline{X}_k$$

The value and sign of the weights (w_j) identify which treatment group means are involved in a comparison and which means are being contrasted against each other. For a comparison to be legitimate—that is, to test statistically the mean differences specified by a comparison—the sum of the weights defining the comparison must equal zero. This latter definition of a legitimate comparison assumes equal sample sizes across groups. When the n_js are equal, a weighted composite of group means forms a testable comparison if $\Sigma w_j = 0$. When sample sizes are unequal, for a comparison to be statistically testable, $\Sigma n_j w_j = 0$.

We should always keep in mind that a comparison is nothing more than a restatement of a null hypothesis using sample values. We use the values in the comparison statement to test the companion null hypothesis. The easiest illustration of a comparison is to look at the null hypothesis that compares only the means for two of the k groups. Assume we want to compare the mean for group 1 with the mean for group 3. The null hypothesis statement may be written as a statement of equality (H_0: $\mu_1 = \mu_3$) or no difference

(H_0: $\mu_1 - \mu_3 = 0$). The comparison statement will always follow the null hypothesis of no difference.

$$C = (1)\overline{X}_1 + (0)\overline{X}_2 + (-1)\overline{X}_3 + (0)\overline{X}_4 + \cdots + (0)\overline{X}_k$$
$$= \overline{X}_1 - \overline{X}_3$$

Note that any mean not included in the null hypothesis statement has a weight of zero ($w_j = 0$) in the comparison. Is the comparison $C = \overline{X}_1 - \overline{X}_3$ testable? Yes, the weights sum to zero. (We assumed equal sample sizes.)

$$\Sigma w_j = 1 + 0 + (-1) + 0 + \cdots + 0 = 0$$

Consider the print color treatment example again. Suppose we want to test a subhypothesis comparing the standard black print treatment condition to the average effect of the other three color conditions. The null hypothesis would be

$$H_0\text{: } \mu_{\text{black}} = \frac{\mu_{\text{green}} + \mu_{\text{blue}} + \mu_{\text{red}}}{3}$$

$$\text{or} \quad H_0\text{: } \mu_{\text{black}} - \frac{\mu_{\text{green}} + \mu_{\text{blue}} + \mu_{\text{red}}}{3} = 0$$

Stating the comparison using sample means:

$$C = (1)\overline{X}_{\text{black}} + \left(-\frac{1}{3}\right)\overline{X}_{\text{green}} + \left(-\frac{1}{3}\right)\overline{X}_{\text{blue}} + \left(-\frac{1}{3}\right)\overline{X}_{\text{red}}$$

Do these weights permit a legitimate comparison?[7] We can eliminate fractions in the comparison by multiplying the comparison by 3. Thus, $C = 3\overline{X}_{\text{black}} - \overline{X}_{\text{green}} - \overline{X}_{\text{blue}} - \overline{X}_{\text{red}}$ identifies the same comparison and satisfies the condition that $\Sigma w_j = 0$. How would you specify the weighted composites for the comparisons corresponding to the following two null hypotheses?

$$(1)\ H_0\text{: } \mu_{\text{blue}} = \frac{\mu_{\text{green}} + \mu_{\text{red}}}{2} \qquad (2)\ H_0\text{: } \mu_{\text{green}} = \mu_{\text{red}}$$

Rewriting H_0 (1) as a statement of no difference,

$$H_0\text{: } \mu_{\text{blue}} - \frac{\mu_{\text{green}} + \mu_{\text{red}}}{2} = 0$$

The contrast corresponding to this latter statement is

$$C = (0)\overline{X}_{\text{black}} + \left(-\frac{1}{2}\right)\overline{X}_{\text{green}} + (1)\overline{X}_{\text{blue}} + \left(-\frac{1}{2}\right)\overline{X}_{\text{red}}$$

$$\text{or } C = -\overline{X}_{\text{green}} + 2\overline{X}_{\text{blue}} - \overline{X}_{\text{red}}$$

---

[7]Yes, the sample sizes are equal, $n_j = 10$ and $\Sigma w_j = 1 - \frac{1}{3} - \frac{1}{3} - \frac{1}{3} = 0$.

For H_0 (2), the H_0 statement of no difference is

$$H_0\text{: } \mu_{\text{green}} - \mu_{\text{red}} = 0 \quad \text{and}$$
$$C = (0)\overline{X}_{\text{black}} + (1)\overline{X}_{\text{green}} + 0\overline{X}_{\text{blue}} + (-1)\overline{X}_{\text{red}} \quad \text{or} \quad C = \overline{X}_{\text{green}} - \overline{X}_{\text{red}}$$

The Observed Test Statistic for an ANOVA Subhypothesis

We use the values of the weights and sample means in a comparison to compute a value for C, which is a sample estimate of the difference in population means specified by the corresponding null hypothesis. The test statistic for a comparison is an observed F ratio.

$$F_{\text{obs}} = \frac{C^2}{MS_w\left(\dfrac{w_1^2}{n_{.1}} + \dfrac{w_2^2}{n_{.2}} + \cdots \dfrac{w_k^2}{n_{.k}}\right)} \tag{16.8}$$

where C is the value of the comparison when group mean values are substituted for the $\overline{X}_{.j}$s, MS_w is the mean square within groups from the ANOVA table, w_j^2 is the squared weight for the jth mean in the weighted composite, and $n_{.j}$ is the sample size for the jth group.

As an illustration, consider the print color treatment example in Table 16.3. The calculated observed F ratio for the test of the null hypothesis H_0: $\mu_{\text{black}} = (\mu_{\text{green}} + \mu_{\text{blue}} + \mu_{\text{red}})/3$ is as follows:

$$\begin{aligned} C &= 3\overline{X}_{\text{black}} - \overline{X}_{\text{green}} - \overline{X}_{\text{blue}} - \overline{X}_{\text{red}} \\ &= 3(52.8) - (45.5) - (51.8) - (49.3) \\ &= 11.8 \end{aligned}$$

The value of MS_w from the ANOVA summary table in footnote 5 is 33.77.

$$\begin{aligned} F_{\text{obs}} &= \frac{(11.8)^2}{33.77\left[\dfrac{3^2}{10} + \dfrac{(-1)^2}{10} + \dfrac{(-1)^2}{10} + \dfrac{(-1)^2}{10}\right]} \\ &= \frac{139.24}{40.524} \\ &= 3.44 \end{aligned}$$

Does an $F_{\text{obs}} = 3.44$ represent a statistically significant difference among the means in the comparison? Ordinarily, we would use the F probability distribution as the sampling distribution to find an F_{crit} value against which to compare F_{obs}. In testing hypotheses through post hoc comparisons in the one-way ANOVA, however, slight modifications in the critical values are recommended if more than one post hoc comparison is to be made in any single experiment. We will discuss these modifications next, but first you may want to practice the calculation of F_{obs} for different subhypotheses. Before moving ahead, calculate F_{obs} for the following three subhypotheses from the data in Table 16.3.[8]

$$(1)\ H_0: \mu_{blue} = \frac{\mu_{green} + \mu_{red}}{2}; \quad (2)\ H_0: \mu_{green} = \mu_{red}; \quad (3)\ H_0: \mu_{black} = \mu_{green}$$

The Scheffe Test

Testing post hoc comparisons requires an adjustment of the critical value. This adjustment is necessary because the Type I error rate for an entire set of post hoc comparisons is much higher than that specified by the level of α set per comparison. Usually in the post hoc situation, we test more than one subhypothesis. Because subsets of the data from the entire experiment are being analyzed the probability increases that *at least one of the many tests conducted would be statistically significant by chance* using standard critical values. This probability that *at least one* post hoc test would be significant by chance is called the **experimentwise** or familywise Type I error rate. Without an adjustment in the testing procedures, the probabilities of rejecting at least one subhypothesis *by chance* and committing at least one Type I error at $\alpha = .05$ when the H_0s are true are approximately .09 for two post hoc comparisons, .14 for three, .19 for four, and .23 when testing five comparisons!

Several procedures protect the researcher by reducing the experimentwise error rate. A widely accepted and versatile procedure is that developed by and named after Scheffe. The **Scheffe test** permits comparisons between pairs of means or any other combination of group means while controlling the experimentwise Type I error rate. It adjusts the critical value used as the criterion for comparing F_{obs} for any individual comparison. The Scheffe critical value is denoted by F'_{crit} and is calculated as follows:

$$F'_{crit} = (k - 1)[{}_{(1-\alpha)}F_{(df_B,\, df_W)}] \tag{16.9}$$

where k is the number of groups in the total design and ${}_{(1-\alpha)}F_{(df_B,\, df_W)}$ is the critical F value used in the ANOVA summary table as the overall F_{crit}. The

---

[8](1) $C = (-1)(45.5) + 2(51.8) + (-1)(49.3) = 8.8$

$$F_{obs} = \frac{(8.8)^2}{33.77\left(\frac{(-1)^2}{10} + \frac{2^2}{10} + \frac{(-1)^2}{10}\right)} = \frac{77.44}{20.262} = 3.82$$

(2) $C = 45.5 - 49.3 = -3.8$

$$F_{obs} = \frac{(-3.8)^2}{33.77\left(\frac{1^2}{10} + \frac{(-1)^2}{10}\right)} = \frac{14.44}{6.754} = 2.14$$

(3) $C = 52.8 - 45.5 = 7.3$

$$F_{obs} = \frac{(7.3)^2}{33.77\left(\frac{1^2}{10} + \frac{(-1)^2}{10}\right)} = \frac{53.29}{6.754} = 7.89$$

maximum experimentwise error-rate level for F'_{crit} is the same as the level set for the overall ANOVA test of the general H_0.

The F_{obs} value for the comparison that tested

$$H_0: \mu_{black} = \frac{\mu_{green} + \mu_{blue} + \mu_{red}}{3}$$

was 3.44. The Scheffe critical value used to test the post hoc comparison at an α level of .05 would be $F'_{crit} = (4 - 1)(2.88) = 8.64$. F_{obs} (3.44) is less than F'_{crit} (8.64), so we would *not* reject H_0. Note that F'_{crit} is a constant critical value for any post hoc comparison within the same experiment. Are any of F_{obs} values for the three comparisons in Footnote 8 statistically significant? The comparison $C = \overline{X}_{black} - \overline{X}_{green}$ comes close ($F_{obs} = 7.89$), but still does not exceed F'_{crit}.

While it may appear that no subhypotheses would be rejected in the preceding example, the Scheffe procedure guarantees that if the overall ANOVA F test is significant, at least one comparison among some combination of means will also be statistically significant. For these data, the comparison $(\mu_{black} + \mu_{blue})/2 - \mu_{green}$ results in a statistically significant Scheffe contrast ($F_{obs} = 9.13$). Check this result to confirm your computational understanding. Keppel (1982) and Winer (1971) discuss the Scheffe test along with other post hoc comparison procedures.

Orthogonal Contrasts

In the prior presentation, we were not concerned with the number of contrasts or the interdependence of contrasts being tested. As long as each null subhypothesis for a contrast weights the group means in such a way that a legitimate comparison was defined, the Scheffe test ensures that the experimentwise probability of making one or more Type I errors is at most equal to α. Any set of subhypotheses could be legitimately tested regardless of whether they defined contrasts involving overlapping variability among means. Dependency among the subhypotheses in terms of the variability among group means being examined was not an issue. If we wanted to test a set of subhypotheses independent of each other, however, we would be limited in the number and type of potential subhypotheses we could test. Independence of a set (two or more) of subhypotheses implies that the test of each subhypothesis would involve variability among group means different from the variability examined in testing other subhypotheses. Tests of independent subhypotheses have some interpretive advantages because as a set, they are not confounded.

The criterion that identifies whether two subhypotheses are independent of each other involves the weights assigned to the accompanying statements of each subhypothesis as a contrast. If the criteria for the contrasts' weights are satisfied for two subhypotheses, we say they are independent or

orthogonal to each other. Assuming equal sample sizes, two contrasts (C_i and $C_{i'}$) are orthogonal if $\Sigma_j w_{ij} w_{i'j} = 0$. Verbally stated, two contrasts are orthogonal if the crossproducts of the respective weights assigned each group mean sum to zero. Assume we have the following two contrasts from a one-way ANOVA design with four groups.

$$C_i = w_{i1}\overline{X}_1 + w_{i2}\overline{X}_2 + w_{i3}\overline{X}_3 + w_{i4}\overline{X}_4$$

$$\updownarrow w_{i1}w_{i'1} \quad \updownarrow w_{i2}w_{i'2} \quad \updownarrow w_{i3}w_{i'3} \quad \updownarrow w_{i4}w_{i'4} \longrightarrow$$

$$C_{i'} = w_{i'1}\overline{X}_1 + w_{i'2}\overline{X}_2 + w_{i'3}\overline{X}_3 + w_{i'4}\overline{X}_4$$

To be orthogonal, these crossproducts must sum to zero: $\Sigma w_{ij} w_{i'j} = 0$.

The weights involved in the crossproducts are the two for each group mean from the respective contrasts (C_i and $C_{i'}$). For example, consider the following two contrasts.

$$C_i = 3\overline{X}_1 - \overline{X}_2 - \overline{X}_3 - \overline{X}_4$$

$$C_{i'} = 0\overline{X}_1 - 2\overline{X}_2 + \overline{X}_3 + \overline{X}_4$$

First, are these two contrasts legitimate?[9] Are they orthogonal? The weights for each contrast need to be paired as follows:

C_i: 3 −1 −1 −1

$C_{i'}$: 0 −2 1 1

$$\Sigma w_{ij} w_{i'j} = 0(3) + (-2)(-1) + 1(-1) + 1(-1) = 0$$

The weights for these contrasts satisfy the criterion for orthogonality. Note that this criterion is only appropriate for designs in which sample sizes are equal.

Two subhypotheses are orthogonal if the weights that identify legitimate contrasts for testing the subhypotheses satisfy the preceding criterion. A set of subhypotheses is **mutually orthogonal** if *every* pair of subhypotheses within the set satisfies the condition for orthogonality. The upper bound to the number of mutually orthogonal subhypotheses within any one set for a given one-way ANOVA design is the degrees of freedom between groups, or $k - 1$ ($df_B = k - 1$). We cannot find any more mutually orthogonal subhypotheses within any given set than the number of groups minus one. We can identify different sets of mutually orthogonal subhypotheses, but the number of potential subhypotheses within each set is limited to $k - 1$. Identifying a set of mutually orthogonal subhypotheses is not an easy task if we use a trial and error method. The following systematic procedure is suggested and illustrated using a four-group ANOVA design:

1. Start with a subhypothesis that involves a contrast in which all group means are assigned nonzero weights.

---

[9] $\Sigma w_{ij} = 3 - 1 - 1 - 1 = 0$
$\Sigma w_{i'j} = 0 - 2 + 1 + 1 = 0$
Yes, both satisfy the criterion for defining legitimate contrasts.

$$H_{o1}: \frac{\mu_1 + \mu_2}{2} = \frac{\mu_3 + \mu_4}{2}$$

$$C_1 = \frac{1}{2}\overline{X}_1 + \frac{1}{2}\overline{X}_2 - \frac{1}{2}\overline{X}_3 - \frac{1}{2}\overline{X}_4$$

$$\text{or } C_1 = \overline{X}_1 + \overline{X}_2 - \overline{X}_3 - \overline{X}_4$$

2. Form contrasts involving means that are on the same side of the equal sign in H_{0i}.

$$H_{o2}: \mu_1 = \mu_2 \quad \text{and}$$
$$H_{o3}: \mu_3 = \mu_4$$
$$C_2 = \overline{X}_1 - \overline{X}_2 + (0)\overline{X}_3 + (0)\overline{X}_4$$
$$C_3 = (0)\overline{X}_1 + (0)\overline{X}_2 + \overline{X}_3 - \overline{X}_4$$

3. If more than one mean still exists on one side of an equal sign in H_{oi}, continue specifying subhypotheses until only two means are contrasted.
4. Check to see if the number of subhypotheses specified equals $k - 1$. In this case, of four groups, $df_B = 3$. Three H_{oi}s were specified.
5. Check that each H_{oi} identifies a legitimate contrast and that they are mutually orthogonal pairwise.

$$C_1: \Sigma w_{1j} = 1 + 1 - 1 - 1 = 0$$
$$C_2: \Sigma w_{2j} = 1 - 1 + 0 + 0 = 0$$
$$C_3: \Sigma w_{3j} = 0 + 0 + 1 - 1 = 0$$

$$C_1 \text{ vs. } C_2: \Sigma w_{1j}w_{2j} = 1 - 1 + 0 + 0 = 0$$
$$C_1 \text{ vs. } C_3: \Sigma w_{1j}w_{3j} = 0 + 0 - 1 + 1 = 0$$
$$C_2 \text{ vs. } C_3: \Sigma w_{2j}w_{3j} = 0 + 0 + 0 + 0 = 0$$

A different set of mutually orthogonal subhypotheses could have been identified. For the following two sets, one is mutually orthogonal and one is not. Make sure you understand why.[10]

Set A	**Set B**
$H_{o1}: \mu_3 = \frac{\mu_1 + \mu_2 + \mu_4}{3}$	$H_{o1}: \frac{\mu_1 + \mu_2 + \mu_3}{3} = \mu_4$
$H_{o2}: \frac{\mu_1 + \mu_4}{2} = \mu_2$	$H_{o2}: \mu_1 = \mu_3$
$H_{o3}: \mu_1 = \mu_4$	$H_{o3}: \mu_1 = \mu_2$

[10]Set B is not mutually orthogonal. The weights are

C_1:	1	1	1	−3
C_2:	1	0	−1	0
C_3:	1	−1	0	0

Each forms a legitimate contrast, C_1 is orthogonal to C_2 and C_3, but C_2 and C_3 are not orthogonal, $\Sigma w_{2j}w_{3j} = 1$.

When testing a mutually orthogonal set of subhypotheses, the sum of squares associated with a specific contrast (SS_{ci}) is independent of the sum of squares for any other contrast ($SS_{ci'}$) in the set. In fact, if a complete set ($k - 1$) of mutually orthogonal subhypotheses is specified, the sum of the individual contrast sum of squares equals the sum of squares between groups from the overall ANOVA design.

$$\Sigma SS_{Ci} = SS_B$$

The mutually orthogonal set of subhypotheses partitions the overall SS_B into independent component parts. The sum of squares for a contrast, SS_{Ci}, can be identified from the formula for F_{obs} (formula 16.8).

$$F_{\text{obs}} = \frac{C^2}{MS_w\left(\frac{w_1^2}{n_{.1}} + \frac{w_2^2}{n_{.2}} + \cdots + \frac{w_k^2}{n_{.k}}\right)}$$

Actually, the sum of squares for a contrast is the part of the formula exclusive of MS_w.

$$SS_{Ci} = \frac{C^2}{\frac{w_1^2}{n_{.1}} + \frac{w_2^2}{n_{.2}} + \cdots + \frac{w_k^2}{n_{.k}}}$$

Assuming equal samples of size n, the formula can be written

$$SS_{Ci} = \frac{C^2}{\frac{1}{n}\left(\sum_j w_{ij}^2\right)}$$

To illustrate that $\Sigma SS_{Ci} = SS_B$, we calculate the SS_{Ci} for the following set of mutually orthogonal subhypotheses using the print color treatment data from Table 16.3. The values for C were calculated in material presented earlier.

Subhypotheses	**Values of C_i**
$H_{o1}: \mu_{\text{black}} = \frac{\mu_{\text{green}} + \mu_{\text{blue}} + \mu_{\text{red}}}{3}$	$C_1 = 11.8$
$H_{o2}: \mu_{\text{blue}} = \frac{\mu_{\text{green}} + \mu_{\text{red}}}{2}$	$C_2 = 8.8$
$H_{o3}: \mu_{\text{green}} = \mu_{\text{red}}$	$C_3 = -3.8$

Computing each SS_{Ci},

$$H_{o1}: SS_{C_1} = \frac{(11.8)^2}{\frac{1}{10}[3^2 + (-1)^2 + (-1)^2 + (-1)^2]} = 116.03$$

$$H_{o2}: SS_{C_2} = \frac{(8.8)^2}{\frac{1}{10}[0^2 + 2^2 + (-1)^2 + (-1)^2]} = 129.07$$

$$H_{o3}: SS_{C_3} = \frac{(-3.8)^2}{\frac{1}{10}[0^2 + 0^2 + 1^2 + (-1)^2]} = 72.20$$

Note that $SS_{C_1} + SS_{C_2} + SS_{C_3} = SS_B$ from the ANOVA summary table in footnote 5 ($116.03 + 129.07 + 72.20 = 317.3$).

The procedures for testing orthogonal contrasts follow those in this section. The critical value used to test a contrast depends on whether they are planned or post hoc contrasts. If the contrasts are planned, the critical F value has numerator degrees of freedom equal to one and denominator df equal to degrees of freedom within groups (df_W), that is, $F_{\text{crit}} = {}_{(1-\alpha)}F_{(1,df_W)}$. If a post hoc orthogonal set of contrasts is tested, the Scheffe critical values should be used.

16.9 SUMMARY

The one-way ANOVA design is the basic analysis of variance design. It enables us to test a single hypothesis about the equality of means on a single criterion variable for a single independent treatment variable. Within the design, the groups are assumed to contain independent samples of individuals. The generalized null hypothesis for k independent groups is stated as no difference among the k population means, or H_0: $\mu_1 = \mu_2 = \mu_3 = \cdots = \mu_k$. To test this hypothesis, we collect data from independent random samples from populations with different characteristics (selected variation) or from random samples that have been treated differentially (induced variation). The total variability in the scores for all individuals in the design is partitioned into sources of variability *between* group means and among scores *within* groups, pooled over groups. Following is the step-by-step process for testing H_0.

- Step 1 Compute SS_B, SS_W, and SS_T from the sample data using the following formulas.

$$SS_B = \sum \frac{(\Sigma X_{.j})^2}{n_{.j}} - \frac{(\Sigma X_{..})^2}{N} \tag{16.1}$$

$$SS_W = \Sigma X_{..}^2 - \sum \frac{(\Sigma X_{.j})^2}{n_{.j}} \tag{16.2}$$

$$SS_T = \Sigma X_{..}^2 - \frac{(\Sigma X_{..})^2}{N} \tag{16.3}$$

- Step 2 Compute the degrees of freedom for each source of variability: $df_B = k - 1$; $df_W = N - k$; $df_T = N - 1$

- Step 3 Convert SS_B and SS_W to mean squares:

$$MS_B = \frac{SS_B}{df_B}; \; MS_W = \frac{SS_W}{df_W}$$

- Step 4 Form the ratio of MS_B to MS_W to obtain an observed F value:

$$F_{obs} = \frac{MS_B}{MS_W}$$

- Step 5 Identify the critical F value (F_{crit}) at a specified α level in the F probability distribution with $df = k - 1$ and $N - k$, assuming a directional test of H_0.
- Step 6 Compare F_{obs} to F_{crit}. If $F_{obs} > F_{crit}$, reject H_0. If $F_{obs} < F_{crit}$, do not reject H_0.
- Step 7 If H_0 is rejected, conclude that the sample means are statistically different at a significance probability of α. The inference is that some or all of the treatments defining each of the separate groups as a composite were differentially effective (statistically significant).
- Step 8 Summarize results in the form of an ANOVA summary table.

Rejecting H_0 merely indicates that the group means overall vary in excess of what one would expect due just to random sampling. How much the group means or which specific group means or combinations of means differ are not indicated by a test of H_0. Rather, we need to perform additional analyses. A measure of the extent to which the group means differ (magnitude of the effect) was presented in the form of an index called omega squared (ω^2). Analyses to identify which group means might differ were presented in the context of post hoc analyses using the Scheffe test. The Scheffe procedure tests subhypotheses as specific contrasts or comparisons among the group means.

As the most basic design, the analysis of variance terms and concepts introduced with the one-way ANOVA design form the foundation for expansion to a series of increasingly more complex ANOVA designs. Hypotheses about mean differences are tested in all ANOVA designs, but more advanced designs allow for multiple tests of hypotheses for several treatment variables and their interactions within the context of a single design. Independent groups, correlated groups, and mixed designs are available. The analysis of variance provides a tool necessary to a researcher's repertoire. The knowledge gained from study of this text will apply to studying analysis of variance procedures and other advanced statistical procedures.

PROBLEMS

Problem Set A

1–3. The number of independent groups and sample sizes per group are given for three ANOVA situations. Use the information to determine the degrees of freedom between groups and within groups and the total for testing the null hypotheses of no mean differences among groups.

1. for treatment group A, $n_{.A} = 6$; B, $n_{.B} = 6$; C, $n_{.C} = 6$; D, $n_{.D} = 6$; E, $n_{.E} = 6$; F, $n_{.F} = 6$.
2. for 13, treatment conditions, each with a sample size of 11.
3. for group type 1, $n_{.1} = 6$; 2, $n_{.2} = 11$; 3, $n_{.3} = 11$; 4, $n_{.4} = 9$; 5, $n_{.5} = 15$; 6, $n_{.6} = 8$; 7, $n_{.7} = 8$.

4. Demonstrate for the design in problem 3 how to calculate the df_W by pooling the within-group degrees of freedom across the seven groups.

5. Identify the critical F values you would use in problems 1–3 to test H_0 at both the .05 and .01 levels.

6. The following raw scores between 0 and 9 were randomly selected and distributed to each of three treatment groups. Calculate the necessary quantities to test H_0: $\mu_1 = \mu_2 = \mu_3$. Would you reject H_0 at the .05 level of significance? Treatment group 1–4, 4, 8, 2, 4; group 2–7, 4, 4, 0, 0; group 3–0, 5, 6, 4, 5.

7. Add the following treatment effects to each of the scores in the three problem 6 groups respectively: (+2) for group 1, (+1) for group 2, and (+4) for group 3. For the new set of data, test H_0: $\mu_1 = \mu_2 = \mu_3$ at the .05 level of significance.

8. Compute the magnitude of the effect for the data in problem 7.

9–12. Test each of the following subhypotheses for the data in problem 7 using the Scheffe procedure.

9. H_0: $\mu_1 = \mu_2$

10. H_0: $\mu_1 = \mu_3$

11. H_0: $\mu_2 = \mu_3$

12. $H_0 = \frac{\mu_1 + \mu_2}{2} = \mu_3$

13. Determine the subhypotheses in problems 9–12 that would define a mutually orthogonal set of contrasts. Demonstrate that the sum of SS_{Ci} for these contrasts equals SS_B from problem 7.

Problem Set B

Deaton, Glasnapp, and Poggio (1980) conducted a test of a one-way ANOVA hypothesis as one part of a study examining the effects of indeterminate frequency modifiers on responses to forced-choice scale items. An indeterminate frequency modifier is a word (usually an adverb) used in writing a statement for a self-report instrument that is purposefully vague and void of precise meaning (e.g., the words *sometime* or *often*). The researchers hypothesized that when modifiers of different intensities were used in writing inventory items, the modifiers themselves, and not just the content of the item, resulted in differential responses among individuals.

To test this hypothesis, nine indeterminate frequency modifiers based on the degree of "oftenness" were scaled on intensity level using a paired-comparison technique. The modifiers *rarely, occasionally,* and *sometimes* were categorized into the *low* level of intensity condition. The *moderate* level of intensity condition included the modifiers *frequently, generally,* and *often,* and the *extreme* level of intensity was characterized by the modifiers *usually, very often,* and *always.*

Twelve external items from Rotter's Internal-External Control Scale that contained modifiers in original form were selected as the stimulus instrument to be manipulated. Five forms of the instrument were constructed. Form 1 contained the item statements from the original Rotter's scale. In Form 2, the modifiers were deleted in the 12 treatment items, producing a treatment condition with absolutely worded items. Forms 3, 4, and 5 contained the low, moderate, and extreme levels of modifiers in the 12 experimental items, respectively. These five forms that defined the levels of the treatment condition were randomly distributed to college students in several sections of an undergraduate class. Students responded to the form of the instrument randomly assigned to them. Scale scores across the 12 items were computed for each individual. We have simulated scale score values for eight individuals under each of the five treatment conditions. Use Table 16.9 to respond to the following problems.

1. Compute group means and standard deviations.

2. Test the hypothesis of no differential treatment effect as demonstrated by the differences in group means (use $\alpha = .05$). Report the results in an ANOVA summary table. Also compute ω^2.

3–6. Test each of the following null subhypotheses using Scheffe contrasting procedures.

3. $H_0\colon \mu_{\text{orig}} = \dfrac{\mu_{\text{abs}} + \mu_{\text{lo}} + \mu_{\text{mod}} + \mu_{\text{ext}}}{4}$

4. $H_0\colon \dfrac{\mu_{\text{lo}} + \mu_{\text{mod}} + \mu_{\text{ext}}}{3} = \mu_{\text{abs}}$

5. $H_0\colon \mu_{\text{lo}} = \dfrac{\mu_{\text{mod}} + \mu_{\text{ext}}}{2}$

TABLE 16.9
Simulated scores values for eight individuals under five treatment conditions.

Modifier Intensity Treatment Conditions				
Original (ORIG)	Absolute (ABS)	Low (LO)	Moderate (MOD)	Extreme (EXT)
33	34	36	33	28
51	28	42	30	39
36	23	46	40	40
43	37	39	36	38
48	29	50	43	34
41	32	42	36	31
44	26	40	32	27
39	35	47	42	33

6. All pairwise contrasts.

7. Identify a set of four subhypotheses from all those tested in problems 3–6 that form a mutually orthogonal set. Add the sum of squares for these mutually orthogonal contrasts to demonstrate that the $\Sigma SS_{ci} = SS_{\text{between groups}}$.

8. What conclusions might you draw from the analysis?

Problem Set C

1. Using ANOVA procedures, test the following null hypotheses separately at the .05 level of significance.
   - H_0: There exists no difference in mean Forms pretest scores among individuals who viewed "Sesame Street" for different amounts of time per week.
   - H_0: There exists no difference in mean Forms posttest scores among individuals who viewed "Sesame Street" for different amounts of time per week.

   Draw random samples of six to test these hypotheses. Use pretest and posttest scores for the same samples to test the separate hypotheses.

2. Using the Scheffe procedure, test to see if the mean Forms posttest score for Viewing Categories 1 and 2 differs significantly from the mean for categories 3 and 4.

3. Identify and test two contrasts that would be orthogonal to the one specified in Problem 2. Demonstrate using the SS_{c_i} for these contrasts that the SS_B from the overall ANOVA is equal to ΣSS_{c_i} for contrasts that are orthogonal.

4. Test the hypothesis of no difference in mean Classification Skill performance at the time of the posttest for groups of children attending different types of preschool ($\alpha = .05$). Use the scores for the specific 40 individuals randomly sampled from the "Sesame Street" data base.

001	037	055	089	126	147	190	224
003	039	061	091	129	148	194	228
008	041	063	105	130	152	200	230
013	047	065	107	136	154	214	232
021	048	079	113	142	164	220	237

REFERENCES

Ball, S.J., & Bogatz, G.A. (1970). *The first year of Sesame Street: An evaluation.* Princeton, NJ: Educational Testing Service.

Bettencourt, E.M., Gillett, M.H., Gall, M.D., & Hull, R.E. (1983). Effects of teacher enthusiasm training on student on-task behavior and achievement. *American Educational Research Journal, 20,* 435–450.

Bloom, B.S. (1968). "Learning for mastery," *Evaluation Comment, 1*(2).

Callenbach, C. (1973). The effects of instruction and practice in content-independent test-taking techniques upon the standardized reading test scores of selected second-grade students. *Journal of Educational Measurement, 10,* 25–30.

Campbell, D.T., & Stanley, J.C. (1966). *Experimental and quasi-experimental designs for research.* Skokie, IL: Rand McNally.

Carroll, J.B. (1963). A model of school learning. *Teachers College Record, 84,* 723–733.

Cohen, J. (1977). *Statistical power analysis for the behavioral sciences* (Rev. ed.). New York: Academic Press.

Cohen, J., & Cohen, P. (1983). *Applied multiple regression/correlation analysis for the behavioral sciences* (2nd ed.). Hillsdale, NJ: Lawrence Erlbaum Associates.

Coleman, J.S., Campbell, E.Q., Hobson, C.J., McPartland, J., Mood, A.M., Weinfeld, F.O., & York, R.L. (1966). *Equality of educational opportunity.* Washington, DC: U.S. Government Printing Office.

Cook, T.D., Appleton, H., Conner, R.F., Shaffer, A., Tamkin, G., & Weber, S.J. (1975). *"Sesame Street" revisited. Continuities in evaluation research.* New York: Russell Sage Foundation.

Deaton, W.L., Glasnapp, D.R., & Poggio, J.P. (1980). Effects of item characteristics on psychometric properties of forced choice scales. *Educational and Psychological Measurement, 40,* 599–610.

Gable, R.K., Roberts, A.D., & Owen, S.V. (1977). Affective and cognitive correlates of achievement. *Educational and Psychological Measurement, 37,* 977–986.

Gallup, G.H. (1977). Ninth annual Gallup Poll of the public's attitudes toward the public schools. *Phi Delta Kappan, 59,* 33–48.

Glass, G.V., Peckham, P.D., & Sanders, J.R. (1972). Consequences of failure to meet assumptions underlying the fixed effects analysis of variance and covariance. *Review of Educational Research, 42,* 237–288.

Halderson, J.S., & Glasnapp, D.R. (1972). Generalized rules for calculating the magnitude of an effect in factorial and repeated measures ANOVA designs. *American Educational Research Journal, 9,* 301–310.

Hays, W.L. (1981). *Statistics* (3rd ed.). New York: Holt, Rinehart and Winston.

Huff, D. (1954). *How to lie with statistics.* New York: Norton.

Karweit, N., & Slavin, R.E. (1981). Measurement and modeling choices in studies of time and learning. *American Educational Research Journal, 18,* 157–171.

Keppel, G. (1982). *Design and analysis: A researcher's handbook* (2nd ed.). Englewood Cliffs, NJ: Prentice-Hall.

Lindquist, E.F. (1953). *Design and analysis of experiments in psychology and education.* Boston: Houghton Mifflin.

Miskel, C., Glasnapp, D.R., & Hatley, R. (1975). A test of the inequity theory for satisfaction using educators' attitudes toward work motivation and work incentives. *Educational Administration Quarterly, 11,* 38–54.

National Assessment of Educational Progress. (1974). *NAEP Newsletter.* Denver: The Educational Commission of the States.

National Assessment of Educational Progress. (1976). *NAEP Newsletter.* Denver: The Educational Commission of the States.

Pearson, E.S., & Hartley, H.O. (Eds.) (1956). *Biometrika tables for statisticians.* (4th ed., Vol. 1) London: Cambridge University Press.

Pedhazur, E.J. (1982). *Multiple regression in behavioral research* (2nd ed.). New York: Holt, Rinehart and Winston.

Prawat, R.S. (1976). Mapping the affective domain in young adolescents. *Journal of Educational Psychology,* 68, 566–572.

RAND Corporation. (1955). *A million random digits with 100,000 normal deviates.* New York: Free Press.

Salkind, N.J. (1978). The development of norms for the Matching Familiar Figures Test. *Journal of Supplement Abstract Services, 8,* (Ms. no. 1718).

Salkind, N.J. & Wright, J.C. (1977). The development of reflection-impulsivity and cognitive efficiency: An integrated model. *Human Development, 20,* 377–387.

Schramm, W. (1976). The second harvest of two researcher-producing events: The Surgeon General's Inquiry and *Sesame Street. Proceedings of The National Academy of Education, 3,* 151–219.

Shepard, L.A., Smith, M.L., & Vojir, C.P. (1983). Characteristics of pupils identified as learning disabled. *American Educational Research Journal, 20,* 309–331.

Sherman, J., & Fennema, E. (1977). The study of mathematics by high school girls and boys: Related variables. *American Educational Research Journal, 14,* 159–168.

Stevens, S.S. (1951). Mathematics, measurement and psychophysics. In S.S. Stevens (Ed.), *Handbook of experimental psychology,* New York: John Wiley.

Streissguth, A.P., Barr, H.M., & Martin, D.C. (1983). Maternal alcohol use and neonatal habituation assessed with the Brazelton Scale. *Child Development,* 54, 1109–1118.

Tables of the Binomial Probability Distribution. (1949). Washington, DC: U.S. Government Printing Office, National Bureau of Standards, Applied Mathematics Series, 6.

Veale, J.R., & Foreman, D.I. (1983). Assessing cultural bias using foil response data: Cultural variation. *Journal of Educational Measurement, 20,* 249–258.

Wainer, H. (1984). How to display data badly. *The American Statistician, 38,* 137–147.

Walden, J.C. (1967). School board changes and superintendent turnover. *Administrator's Notebook, 15,* 1–4.

Weinraub, M., & Wolf, B.M. (1983). Effects of stress and social supports on mother-child interactions in single- and two-parent families. *Child Development, 54,* 1297–1311.

Willson, V.L. (1980). Research techniques in AERJ articles: 1969–1978. *Educational Researcher, 9* (6), 5–10.

Winer, B.J. (1971). Statistical principles in experimental design (2nd ed.). New York: McGraw-Hill.

Wood, P.H. (1982). The Nelson-Denny Reading Test as a predictor of college freshman grades. *Educational and Psychological Measurement, 42,* 575–583.

APPENDIX A

"Sesame Street" Data Base: Variable Description and Raw Data

DATA BASE DESCRIPTION

The "Sesame Street" data base contains raw scores on 26 variables for 240 individuals. The data are arranged such that each row represents the scores on the 26 variables for a single individual. The columns of scores constitute single variables and are labeled at the top of a column. The numbers in the first column are the individual identification numbers (001–240).

To provide more detailed information on each of the variables included, the following description is given along with the coding and scale score range where appropriate.

Variable Description

1. *Sampling Site.* The coding is as follows:

Individuals sampled from

- Site I were coded 1
- Site II were coded 2
- Site III were coded 3
- Site IV were coded 4
- Site V were coded 5

2. *Sex Identification.* Male = 1, Female = 2.
3. *Age* is in months.

4. *Viewing Category:*

- Code 1 = children who watched the show rarely (on the average of once a week) or never.
- Code 2 = children who watched the show on the average of two to three times a week.
- Code 3 = children who watched the show on the average of four or five times a week.
- Code 4 = children who watched the show on the average of more than five times a week.

5. *Setting* in which "Sesame Street" was viewed: home = 1, school = 2.

6. *Viewing Encouragement.* A treatment condition in which some children were encouraged to view "Sesame Street" (Code = 1) and others were not (Code = 2).

7–12. Pretest total scores on knowledge about

- *Body Parts*—Maximum possible score is 32. Content of items included (a) pointing to body parts, (b) naming body parts, (c) function of body parts (point), and (d) function of body parts (verbal).
- *Letters*—Maximum possible score is 58. Content of items included (a) recognizing letters, (b) naming capital letters, (c) naming lowercase letters, (d) matching letters in words, (e) recognizing letters in words, (f) initial sounds, and (g) reading words.
- *Forms*—Maximum possible score is 20. Content of items included (a) recognizing forms, and (b) naming forms.
- *Numbers*—Maximum possible score is 54. Content of items included (a) recognizing numbers, (b) naming numbers, (c) numerosity, (d) Counting, and (e) addition and subtraction.
- *Relational Terms*—Maximum possible score is 17. Content of items included (a) amount relationships, (b) size relationships, and (c) position relationships.
- *Classification Skills*—Maximum possible score is 24. Content of items included classification by (a) size, (b) form, (c) number, and (d) function.

13. *Frequency with which the mother reported reading to the child.* Category codes were 1 = never, 2 = less than once a week, 3 = once a week, 4 = several times a week, 5 = once a day, 6 = don't know.

14. *School attendance*—The category codes were 1 = does not attend school, 2 = attends kindergarten, 3 = attends nursery school, 4 = attends Headstart, 5 = attends day care center, 6 = other.

15–20. Posttest total scores on each of the six tests identified under variables 7–12:

- *Body Parts* posttest
- *Letters* posttest
- *Forms* posttest
- *Numbers* posttest
- *Relational Terms* posttest
- *Classification Skills* posttest

21. *Which Comes First Test.* There were 12 items on this test requiring the child to point to one of four pictures that was first or last in a sequence of events. This test was administered as a posttest.

22. *"Sesame Street" test.* This test consisted of 10 simple items used to assess the child's knowledge of "Sesame Street" as a television program. This test was administered as a posttest.

23. *Peabody Mental Age.* Mental age scores obtained from the administration of the *Peabody Picture Vocabulary Test* as a pretest measure of vocabulary maturity.

24. *Home Education Index.* This index was derived using responses to selected items from the parent pretest questionnaire following the procedure outlined by Cook et al. (1975). Items asked parents about their educational level and the educational level aspirations they had for their child.

25. *Home Stimulation—Affluence Index.* This is another index formulated from parental responses to selected items on the parent pretest questionnaire. The items this index comprises dealt with the number of home possessions, child's possessions, frequency of attending events outside the home, and amount of interaction between mother and child. Cook et al. considered such information indicative of home environmental richness, or affluence.

26. *Mother-Child Socialization Index.* This index was developed by the authors of the present text and includes a composite of responses to items from the parent pretest questionnaire. The items that form the index requested the mother to indicate the amount of interaction and communication she had with the child across different settings.

The data base appears on the following pages.

TABLE A.1
"Sesame Street" data base.

							Pretests								Posttests											
Child ID Number	Site	Sex	Age	Viewing Category	Setting	Viewing Encouragement	Body Parts	Letters	Forms	Numbers	Relational Terms	Classification Terms	Reading Freq.	School Attend.	Body Parts	Letters	Forms	Numbers	Relational Terms	Classification Skills	Comes First	Sesame St.	Peabody	Home Educ.	Home Stim.-Afflu.	Mother-Child Interaction
1	1	1	66	1	2	1	16	23	12	40	14	20	0	4	18	30	14	44	14	23	4	4	62	0	0	0
2	1	2	67	3	2	1	30	26	9	39	16	22	4	5	30	37	17	39	14	22	10	10	80	13	750	0
3	1	1	56	3	2	2	22	14	9	9	9	8	4	4	21	46	15	40	9	19	4	6	32	0	538	6
4	1	1	49	1	2	2	23	11	10	14	9	13	3	4	21	14	13	19	8	15	6	2	27	14	535	9
5	1	1	69	4	2	2	32	47	15	51	17	22	3	4	32	53	18	54	14	21	9	10	71	13	572	6
6	1	2	54	3	2	2	29	26	10	33	14	14	4	4	27	36	14	39	16	24	11	6	32	14	450	0
7	1	2	47	3	2	2	23	12	11	13	11	12	4	4	22	45	12	44	12	15	10	7	28	14	538	0
8	1	1	51	2	2	1	32	48	19	52	15	23	5	5	31	47	18	51	17	23	7	9	38	17	391	11
9	1	1	69	4	2	1	27	44	18	42	15	20	3	4	32	50	17	48	14	24	7	10	49	12	572	6
10	1	2	53	3	2	1	30	38	17	31	10	17	4	4	32	52	19	52	17	24	12	10	32	18	411	11
11	1	2	58	2	2	2	25	48	14	38	16	18	2	4	26	52	15	42	10	17	5	6	43	0	1105	7
12	1	2	58	4	2	2	21	25	13	29	16	21	3	4	27	29	15	40	13	19	7	8	58	11	610	6
13	1	2	49	1	2	2	28	8	9	13	8	12	6	5	20	16	9	18	10	13	4	3	39	15	0	0
14	1	1	64	2	2	1	26	11	15	21	10	15	5	4	26	28	15	35	16	14	6	8	43	18	467	10
15	1	2	58	2	2	1	23	15	9	16	9	11	4	4	28	21	10	22	10	17	6	10	56	0	575	0
16	1	1	49	3	2	1	25	12	17	24	12	18	4	4	28	45	14	45	13	21	8	10	37	20	640	0
17	1	1	57	2	2	1	25	15	13	26	10	18	4	4	25	24	16	28	8	18	3	6	43	11	708	0
18	1	1	45	4	2	1	16	12	8	11	6	3	4	5	25	16	11	17	9	9	4	9	29	19	655	0
19	1	1	45	3	2	1	25	16	12	23	10	13	5	4	32	46	18	35	14	19	7	10	45	16	570	0
20	1	1	60	3	2	2	19	19	8	23	14	10	2	4	28	50	12	38	12	13	2	5	51	0	894	7
21	1	2	65	4	2	1	29	24	14	41	10	23	4	5	29	48	20	51	15	24	11	10	55	15	393	0
22	1	1	44	4	1	1	25	15	17	22	11	16	2	1	32	42	19	45	15	19	4	10	49	17	422	8
23	1	2	38	3	1	1	20	9	2	7	8	9	2	1	22	23	17	19	15	14	4	8	31	0	0	0
24	1	1	35	4	1	1	11	6	8	16	8	9	5	1	22	27	16	20	14	15	7	9	40	19	465	6
25	1	2	42	2	1	1	15	7	8	11	12	7	1	1	14	13	7	21	12	10	5	4	48	0	0	0
26	1	2	50	2	1	1	26	14	10	36	13	17	4	1	25	18	14	42	13	18	5	6	35	15	0	6
27	1	1	61	4	2	2	28	42	16	40	16	11	2	1	24	27	15	20	14	14	9	9	62	12	602	9
28	1	2	34	4	1	1	17	13	7	10	5	11	4	1	21	17	13	13	10	15	5	9	42	15	455	0
29	1	1	60	3	1	2	23	13	9	23	11	14	4	1	28	20	18	45	14	21	5	10	58	11	0	5
30	1	2	39	2	1	1	11	5	5	5	5	1	4	1	27	15	13	9	8	12	3	4	29	16	583	7

TABLE A.1, *continued*

							Pretests								Posttests											
Child ID Number	Site	Sex	Age	Viewing Category	Setting	Viewing Encouragement	Body Parts	Letters	Forms	Numbers	Relational Terms	Classification Terms	Reading Freq.	School Attend.	Body Parts	Letters	Forms	Numbers	Relational Terms	Classification Skills	Comes First	Sesame St.	Peabody	Home Educ.	Home Stim.-Afflu.	Mother-Child Interaction
31	1	1	39	3	1	1	24	4	11	25	11	17	4	1	21	11	12	13	9	10	6	6	49	13	588	0
32	1	2	41	2	1	1	24	8	3	14	8	8	3	1	21	17	10	18	9	14	5	1	30	16	975	7
33	1	2	55	3	1	1	31	15	17	45	16	24	3	1	32	43	18	46	13	21	6	9	62	10	725	8
34	1	2	42	3	1	1	23	11	7	15	6	7	5	1	29	27	14	33	9	19	5	9	58	13	522	0
35	1	1	50	4	1	1	18	17	12	28	12	17	2	1	29	41	15	48	12	22	9	10	55	0	0	7
36	1	1	58	2	1	1	13	12	6	10	6	11	4	1	29	23	11	27	8	10	6	5	33	0	0	6
37	1	2	59	3	1	2	27	7	13	23	12	10	4	2	32	39	18	49	16	19	9	10	55	12	465	0
38	1	2	36	1	1	2	11	12	9	5	5	3	3	5	12	12	6	13	9	6	5	3	27	6	833	10
39	1	1	51	2	1	1	32	16	16	34	15	17	4	6	21	17	6	21	8	12	1	4	62	12	415	6
40	1	1	51	2	1	1	31	18	13	33	15	14	4	1	21	16	11	7	9	11	2	5	58	0	697	6
41	1	1	48	3	1	1	13	14	8	8	10	11	5	1	21	22	10	28	8	19	7	8	34	10	575	5
42	1	1	43	2	1	1	17	13	14	13	11	14	2	1	24	19	12	22	11	20	8	10	32	0	977	6
43	1	2	35	3	1	2	23	12	8	9	5	5	4	0	29	11	8	9	11	11	9	8	32	15	0	0
44	1	2	36	1	1	2	11	2	6	5	2	4	1	1	21	6	11	6	6	7	5	1	28	15	0	5
45	1	2	39	2	1	2	20	18	6	4	4	6	3	1	19	8	11	22	10	10	5	7	29	0	1025	7
46	1	1	45	4	1	2	14	13	9	16	9	12	6	1	29	48	17	48	14	19	2	10	35	11	0	7
47	1	1	58	3	1	2	30	38	15	45	14	18	4	1	32	48	19	46	14	23	6	10	67	21	435	0
48	1	2	38	3	1	1	13	10	7	8	5	7	1	4	26	36	14	20	10	16	8	7	29	14	675	5
49	1	2	57	4	1	1	26	15	11	22	10	15	4	1	24	20	18	28	12	18	5	10	35	13	510	0
50	1	1	49	3	1	2	26	35	10	47	13	17	4	1	26	13	7	12	11	14	5	5	67	0	586	6
51	1	1	55	1	2	2	24	11	10	18	8	10	5	4	20	10	14	23	10	10	5	2	39	14	1050	9
52	1	2	44	4	2	2	25	39	18	41	9	21	4	4	30	47	20	50	11	23	11	7	90	9	1055	0
53	1	1	56	1	2	2	13	11	6	15	10	11	5	4	15	13	9	13	10	6	4	3	46	13	640	9
54	1	2	48	2	2	2	17	11	9	14	8	9	5	4	26	32	15	27	11	11	5	6	39	11	0	0
55	1	1	50	2	2	2	16	10	8	9	7	6	3	4	21	15	17	17	12	15	4	9	34	13	508	6
56	1	1	52	1	2	2	16	15	6	13	11	13	4	4	19	14	14	18	7	16	5	4	38	11	583	7
57	1	2	51	2	2	1	24	14	10	20	10	17	4	4	28	21	17	36	14	17	9	8	34	0	628	0
58	1	2	58	1	2	2	25	17	17	23	14	15	4	4	25	16	19	28	9	20	0	0	37	0	533	0
59	1	2	48	3	2	2	13	10	10	13	9	7	3	4	27	15	14	23	11	12	4	5	33	9	495	7
60	1	1	54	4	2	2	16	13	10	10	9	7	5	4	19	20	14	19	12	11	6	7	36	10	488	7

TABLE A.1, *continued*

Child ID Number	Site	Sex	Age	Viewing Category	Setting	Viewing Encouragement	Pretests: Body Parts	Pretests: Letters	Pretests: Forms	Pretests: Numbers	Pretests: Relational Terms	Pretests: Classification Terms	Reading Freq.	School Attend.	Posttests: Body Parts	Posttests: Letters	Posttests: Forms	Posttests: Numbers	Posttests: Relational Terms	Posttests: Classification Skills	Comes First	Sesame St.	Peabody	Home Educ.	Home Stim.-Afflu.	Mother-Child Interaction
61	2	1	52	4	2	2	20	35	15	21	8	20	5	3	27	48	19	47	15	22	10	8	49	17	570	0
62	2	2	48	1	2	2	20	12	11	13	7	11	4	3	28	19	17	17	11	17	9	1	53	17	376	0
63	2	1	55	4	2	2	28	13	10	29	11	12	4	3	30	31	19	45	15	22	9	10	65	19	686	11
64	2	1	55	2	2	2	23	16	5	32	10	13	4	3	28	28	15	46	13	17	4	5	55	20	432	0
65	2	2	55	2	1	2	30	27	11	39	12	17	5	6	32	40	19	52	17	23	11	7	85	20	432	9
66	2	2	56	1	1	1	26	18	10	30	12	9	3	1	29	46	13	44	13	17	8	8	43	13	600	10
67	2	2	50	3	2	1	31	15	10	24	11	20	5	3	31	43	18	52	14	23	8	10	65	20	422	9
68	2	1	51	3	2	1	28	19	14	37	13	15	5	3	32	47	19	48	15	22	9	10	75	16	387	9
69	2	1	58	3	2	2	30	14	15	37	13	22	5	3	28	38	15	36	16	18	11	7	85	19	453	0
70	2	2	55	4	2	2	27	10	10	12	14	16	5	3	31	42	16	31	13	20	7	9	40	18	411	0
71	2	2	41	4	1	2	24	22	14	42	14	21	4	1	30	49	19	50	14	22	8	10	58	19	407	7
72	2	2	51	4	1	2	20	13	11	22	11	15	4	1	27	17	11	20	14	14	9	8	42	17	382	0
73	2	1	52	4	2	1	29	13	9	18	10	10	5	3	30	23	17	28	12	17	9	10	69	17	483	9
74	2	2	54	3	2	1	30	26	13	23	10	17	5	3	32	42	12	37	12	19	9	10	58	19	483	0
75	2	1	47	4	1	1	19	12	11	16	11	12	5	1	28	43	18	38	14	17	12	10	37	18	470	8
76	2	1	50	4	1	1	31	30	15	47	15	19	4	1	23	48	19	49	13	20	8	10	62	20	480	0
77	2	2	55	4	1	1	31	13	18	39	15	24	3	1	31	51	19	50	14	23	12	10	99	15	927	7
78	2	1	50	3	2	1	32	27	13	29	13	12	5	3	31	45	19	53	16	23	12	10	75	20	480	0
79	2	2	57	2	2	1	31	19	14	36	13	11	5	3	32	50	17	47	15	22	12	9	82	20	570	11
80	2	2	55	4	1	2	20	12	9	16	11	18	5	1	30	30	14	18	12	16	7	10	47	20	500	11
81	2	2	55	4	1	2	26	14	8	24	13	12	5	1	30	45	19	43	15	24	11	9	62	20	500	11
82	2	1	50	3	2	1	30	44	15	45	12	11	5	3	32	53	19	52	15	23	7	10	67	21	402	11
83	2	1	52	3	2	1	26	13	11	34	12	12	5	3	30	45	17	43	13	21	5	10	40	22	344	10
84	2	2	45	2	2	1	28	12	9	16	7	8	4	3	28	21	14	32	9	17	4	9	41	20	453	0
85	2	2	52	3	2	1	24	14	7	18	8	7	5	3	28	43	12	41	13	15	7	10	56	19	461	10
86	2	1	53	4	1	1	26	17	15	32	13	16	4	1	27	37	15	42	14	23	7	10	73	18	561	0
87	2	2	53	4	2	2	29	10	17	23	13	15	4	3	32	51	19	48	16	21	11	10	85	20	353	0
88	2	2	53	2	2	2	23	12	12	15	11	17	4	3	28	20	15	19	8	19	5	6	59	13	461	0
89	2	2	56	3	1	2	28	29	11	43	13	22	2	1	31	32	15	40	15	21	10	9	58	15	1016	7
90	2	2	54	4	2	2	32	46	15	48	13	21	5	3	32	51	19	50	16	23	12	10	92	17	353	0

TABLE A.1, *continued*

							Pretests								Posttests											
Child ID Number	Site	Sex	Age	Viewing Category	Setting	Viewing Encouragement	Body Parts	Letters	Forms	Numbers	Relational Terms	Classification Terms	Reading Freq.	School Attend.	Body Parts	Letters	Forms	Numbers	Relational Terms	Classification Skills	Comes First	Sesame St.	Peabody	Home Educ.	Home Stim.-Afflu.	Mother-Child Interaction
91	2	1	50	2	2	1	22	17	8	18	12	14	5	3	25	16	14	32	14	17	5	9	69	19	395	0
92	2	1	50	3	1	1	29	25	14	35	15	20	5	1	26	36	16	31	8	15	6	10	56	15	0	0
93	2	2	53	4	1	1	25	17	12	30	13	17	4	1	29	40	17	40	13	22	10	9	78	18	468	6
94	2	1	45	4	1	1	21	16	12	15	11	17	5	1	29	36	19	29	10	16	6	10	58	21	461	6
95	2	2	56	4	1	1	32	22	15	32	11	13	5	1	31	46	20	51	14	24	12	10	78	18	453	0
96	2	1	53	3	1	1	31	14	16	29	11	17	4	1	32	43	19	42	13	22	10	10	67	17	0	0
97	2	2	46	4	1	1	22	28	14	20	5	15	5	1	32	41	13	29	15	18	7	10	69	20	591	11
98	2	2	46	2	1	2	30	18	14	23	11	11	4	1	29	33	13	36	8	16	5	7	53	21	456	7
99	2	1	50	3	1	2	18	13	8	14	11	12	4	1	19	23	11	31	12	17	5	9	55	15	461	7
100	2	2	47	1	1	2	17	10	5	11	8	7	3	1	18	19	9	18	10	13	2	2	53	13	1140	0
101	2	1	56	2	2	1	27	11	15	22	13	14	4	3	30	47	20	45	15	24	10	10	67	19	650	0
102	2	2	46	3	1	1	27	13	13	22	12	13	3	1	28	36	17	46	11	20	7	10	62	18	570	0
103	2	2	45	4	1	1	21	23	14	27	13	20	5	1	31	48	19	44	11	24	9	10	59	18	597	0
104	2	2	46	4	1	1	19	17	4	19	10	13	4	1	28	36	16	29	10	15	4	10	46	20	395	0
105	2	2	47	3	1	1	31	9	14	24	11	16	5	1	32	42	17	42	11	20	11	10	58	16	552	10
106	2	1	52	2	2	1	26	16	12	30	14	16	4	3	27	29	20	41	15	21	7	10	92	19	577	0
107	2	2	52	4	2	1	24	15	12	30	14	14	4	3	31	45	18	49	13	23	12	10	48	17	422	0
108	2	1	56	2	1	1	21	22	12	25	12	15	4	1	29	37	14	46	14	18	5	9	65	17	633	5
109	2	1	48	4	1	1	28	22	13	19	11	8	5	1	32	48	18	43	15	19	9	10	67	19	397	9
110	2	1	49	3	1	1	25	16	13	18	15	15	4	1	27	48	13	45	15	18	6	9	59	19	510	0
111	2	2	55	4	2	1	32	8	13	23	11	17	3	1	29	35	18	36	11	19	6	10	39	17	777	5
112	2	2	45	3	1	1	22	15	12	20	9	14	4	1	25	21	15	21	9	16	5	10	47	17	452	0
113	2	1	45	3	1	1	32	16	14	30	11	17	5	1	25	26	17	39	13	19	7	9	51	20	607	10
114	2	2	48	4	1	1	31	6	8	13	7	13	4	1	29	32	15	23	7	16	1	10	43	18	491	0
115	2	1	58	1	1	2	19	14	12	23	11	10	4	1	28	15	10	34	11	7	2	4	65	19	406	0
116	3	1	55	2	1	1	20	16	7	14	9	9	3	1	21	11	13	20	13	13	6	3	39	13	0	5
117	3	2	48	1	1	2	20	15	5	13	8	7	4	1	21	14	3	17	10	7	4	0	33	12	658	7
118	3	2	52	3	1	1	14	6	3	9	4	8	3	1	18	19	20	18	12	20	11	7	34	12	633	7
119	3	2	58	1	1	1	20	11	5	25	9	7	6	1	26	16	10	23	12	12	3	4	35	0	0	9
120	3	1	50	3	2	2	13	12	5	11	10	9	5	4	21	18	10	19	10	9	4	5	32	0	1083	7

TABLE A.1, *continued*

Child ID Number	Site	Sex	Age	Viewing Category	Setting	Viewing Encouragement	Pretests: Body Parts	Pretests: Letters	Pretests: Forms	Pretests: Numbers	Pretests: Relational Terms	Pretests: Classification Terms	Reading Freq.	School Attend.	Posttests: Body Parts	Posttests: Letters	Posttests: Forms	Posttests: Numbers	Posttests: Relational Terms	Posttests: Classification Skills	Comes First	Sesame St.	Peabody	Home Educ.	Home Stim.-Afflu.	Mother-Child Interaction
121	3	1	58	3	2	2	22	19	11	35	10	17	5	5	23	44	18	46	13	19	2	7	44	17	527	0
122	3	2	49	4	2	2	14	13	7	5	5	7	0	5	17	16	12	19	12	15	1	8	31	0	539	0
123	3	1	56	4	2	2	24	17	9	16	11	10	4	4	29	35	17	40	14	19	7	10	44	10	1113	0
124	3	1	50	1	1	2	7	14	4	10	5	5	6	1	19	15	13	14	13	17	5	1	27	8	666	0
125	3	2	49	1	1	1	20	18	3	14	6	6	6	1	11	12	6	8	9	5	1	2	29	11	675	0
126	3	1	46	2	1	1	15	14	8	23	14	12	5	2	25	18	14	30	12	16	4	8	47	10	500	7
127	3	1	57	2	1	1	26	14	9	23	15	9	5	1	28	15	14	24	12	9	2	5	35	11	633	8
128	3	1	44	1	2	2	12	9	7	14	9	14	2	4	13	13	7	17	9	7	5	0	32	9	1130	4
129	3	2	41	2	2	2	16	14	9	10	11	9	5	4	22	17	9	23	8	13	6	2	36	12	985	4
130	3	2	58	4	2	1	17	9	11	28	8	13	4	5	29	13	12	29	12	15	1	10	38	16	591	10
131	3	2	60	4	2	1	31	19	11	27	11	16	3	5	31	31	17	38	13	22	4	10	42	13	523	6
132	3	2	40	2	1	1	12	14	3	17	6	6	6	1	16	13	6	17	10	9	4	3	27	0	1050	0
133	3	2	37	2	1	1	7	4	6	4	5	4	1	1	13	13	6	14	9	11	4	4	60	10	930	3
134	3	1	45	1	1	1	12	5	5	9	7	7	4	1	15	13	12	20	12	11	5	3	28	13	930	7
135	3	1	60	3	1	1	17	18	9	14	7	6	5	1	32	36	13	32	13	12	6	10	33	15	0	8
136	3	1	52	2	1	2	18	13	9	24	10	16	3	5	25	15	12	26	11	10	4	5	55	13	0	8
137	3	2	46	4	1	1	20	12	4	17	8	8	3	1	28	22	17	38	14	20	11	10	29	18	1030	0
138	3	2	60	4	1	1	23	16	9	25	11	14	3	1	29	26	17	38	16	22	12	10	46	15	1019	0
139	3	1	60	3	1	1	17	11	10	15	10	14	4	1	11	13	7	16	9	13	2	6	33	12	0	8
140	3	1	59	3	1	1	7	16	11	10	6	10	5	1	15	14	9	14	8	10	4	5	32	12	639	0
141	3	2	52	3	1	1	29	20	8	37	13	13	4	1	28	46	12	42	13	15	7	10	47	7	1150	4
142	3	1	60	3	1	1	29	13	12	17	12	16	5	1	29	25	17	32	13	19	7	8	90	11	411	0
143	3	2	56	2	1	1	21	12	12	17	9	14	4	1	23	26	16	34	10	20	4	5	61	9	452	0
144	3	2	54	2	2	2	18	23	9	14	12	16	5	4	27	42	8	37	11	15	6	3	36	9	500	10
145	3	2	61	3	1	1	13	12	8	16	7	11	1	1	28	15	15	18	7	15	6	7	35	15	1105	0
146	3	2	61	3	1	1	29	18	12	22	11	14	3	1	30	25	17	39	13	19	4	8	48	8	528	7
147	3	2	51	3	1	1	17	15	5	11	10	11	5	1	32	43	14	44	15	21	4	10	35	11	575	0
148	3	1	49	4	2	1	19	17	7	16	3	6	4	5	27	27	18	43	15	20	6	10	35	15	450	0
149	3	1	52	2	1	1	22	13	10	20	11	14	4	1	22	14	9	21	9	7	4	4	35	14	691	0
150	3	2	55	3	2	1	25	13	12	16	10	14	5	5	26	17	6	31	13	9	4	8	35	14	1240	9

TABLE A.1, ***continued***

Child ID Number	Site	Sex	Age	Viewing Category	Setting	Viewing Encouragement	Pretests: Body Parts	Pretests: Letters	Pretests: Forms	Pretests: Numbers	Pretests: Relational Terms	Pretests: Classification Terms	Reading Freq.	School Attend.	Posttests: Body Parts	Posttests: Letters	Posttests: Forms	Posttests: Numbers	Posttests: Relational Terms	Posttests: Classification Skills	Comes First	Sesame St.	Peabody	Home Educ.	Home Stim.-Afflu.	Mother-Child Interaction
151	3	2	60	4	2	1	28	10	10	22	12	15	5	4	28	15	15	30	11	17	2	10	45	11	350	8
152	3	2	43	1	2	2	14	9	5	15	5	6	4	5	16	16	12	14	10	14	8	3	33	19	402	0
153	3	1	55	3	2	1	14	7	9	15	9	12	4	5	18	15	16	22	11	16	3	10	42	7	0	0
154	3	2	52	4	2	1	18	11	9	15	8	12	5	4	23	23	15	40	13	18	6	10	32	13	566	8
155	3	2	56	1	2	1	26	24	13	25	9	10	6	3	17	7	3	13	6	5	4	3	40	10	0	12
156	3	1	56	4	1	1	24	11	11	28	14	17	6	1	27	14	15	40	12	21	4	9	42	11	766	8
157	3	2	47	2	1	1	20	19	9	25	12	8	3	1	26	24	13	35	11	17	4	8	46	14	1030	7
158	3	2	56	2	1	1	17	18	8	17	5	9	1	1	24	17	10	19	10	15	1	8	39	7	1233	4
159	3	2	52	3	1	1	28	15	13	27	9	15	4	2	31	16	16	22	12	18	2	7	69	16	0	10
160	3	2	51	4	1	1	23	14	11	23	8	11	4	1	31	37	17	42	14	18	5	10	36	13	940	0
161	3	1	51	1	1	1	7	13	6	11	7	6	3	1	12	8	14	22	10	16	7	3	35	0	691	0
162	3	2	53	3	1	1	15	15	8	18	8	11	3	1	29	32	12	28	10	14	6	10	34	18	583	7
163	3	1	50	4	1	1	26	11	14	23	10	11	4	5	30	22	16	40	14	17	4	10	32	10	733	0
164	3	2	59	4	1	1	16	10	8	21	9	12	0	4	25	22	14	31	10	18	5	10	38	0	608	0
165	3	1	53	3	1	1	14	12	7	9	9	5	4	5	22	28	9	30	9	10	4	9	32	14	655	0
166	3	1	55	3	1	1	15	10	7	9	6	11	4	4	24	20	14	27	9	9	1	7	34	0	422	0
167	3	1	57	1	1	1	6	13	2	8	7	7	3	1	18	6	4	0	1	4	2	0	35	17	1066	0
168	3	1	58	2	1	1	16	5	5	8	6	9	5	1	13	14	11	11	9	16	8	3	34	0	1155	0
169	3	1	44	3	1	1	10	12	4	9	10	11	4	1	13	15	3	8	3	5	1	5	28	15	1225	0
170	3	1	39	1	1	2	14	12	4	5	7	5	3	1	13	11	8	19	10	8	3	5	29	9	525	5
171	3	1	53	4	2	1	21	17	12	16	10	13	5	4	27	20	14	29	15	16	11	9	37	13	450	8
172	3	2	52	4	1	1	23	10	9	9	7	6	4	1	21	16	11	20	9	9	5	8	32	0	683	0
173	3	1	57	3	1	2	25	11	10	19	11	13	3	1	28	29	20	25	16	19	11	7	35	12	975	5
174	3	2	40	3	1	1	11	10	7	14	4	8	5	1	16	22	11	21	9	9	2	5	35	0	780	6
175	3	2	47	2	1	1	16	13	7	7	6	9	6	1	22	13	4	18	11	9	5	2	32	0	1110	8
176	3	1	51	2	1	1	25	19	11	24	12	8	4	1	26	20	15	24	11	13	5	8	47	15	415	5
177	3	1	48	2	1	1	11	7	4	14	3	13	4	1	11	12	8	27	11	10	1	1	35	15	465	0
178	3	2	49	1	1	2	15	16	6	9	4	7	4	1	20	16	7	17	10	5	6	1	35	11	0	6
179	3	1	50	2	1	1	12	8	5	17	8	10	4	1	18	19	12	13	12	17	8	5	30	11	525	7
180	4	2	53	1	2	2	10	13	4	13	7	8	4	3	19	16	9	16	7	11	4	0	35	10	480	0

TABLE A.1, *continued*

Child ID Number	Site	Sex	Age	Viewing Category	Setting	Viewing Encouragement	Pretests: Body Parts	Pretests: Letters	Pretests: Forms	Pretests: Numbers	Pretests: Relational Terms	Pretests: Classification Terms	Reading Freq.	School Attend.	Posttests: Body Parts	Posttests: Letters	Posttests: Forms	Posttests: Numbers	Posttests: Relational Terms	Posttests: Classification Skills	Comes First	Sesame St.	Peabody	Home Educ.	Home Stim.-Afflu.	Mother-Child Interaction
181	4	2	52	1	2	2	13	15	8	19	8	9	2	5	21	11	8	16	7	11	3	2	39	0	1155	8
182	4	1	51	1	2	2	19	12	9	17	8	12	4	4	27	16	12	27	11	16	4	0	39	11	950	0
183	4	1	52	1	2	2	20	16	12	22	11	17	4	4	25	19	14	26	11	15	5	1	36	17	467	0
184	4	1	46	1	2	2	13	3	3	1	4	4	6	5	24	11	10	13	11	13	2	2	27	10	408	0
185	4	2	51	1	2	2	21	19	12	25	13	14	4	4	24	15	11	25	8	14	1	0	45	0	488	0
186	4	2	47	1	2	2	19	12	13	27	8	11	4	4	24	14	15	21	7	13	4	0	28	15	583	0
187	4	2	51	3	2	1	25	13	12	21	12	16	4	4	31	16	15	25	11	18	7	10	40	14	0	0
188	4	2	54	1	2	1	8	20	5	8	7	6	6	3	14	13	11	11	8	10	5	4	47	18	808	9
189	4	2	54	2	2	1	12	4	9	4	7	6	6	5	17	13	10	12	9	8	1	8	36	18	688	9
190	4	1	57	2	2	1	24	11	10	28	12	11	4	3	30	24	18	26	14	16	6	10	39	21	680	0
191	4	1	53	2	2	1	17	12	8	9	5	11	5	5	26	13	11	20	10	15	4	6	39	13	493	0
192	4	2	50	2	2	1	20	16	8	18	9	13	3	3	28	25	15	15	9	12	5	4	43	11	450	8
193	4	2	57	1	2	2	28	23	16	33	14	11	5	5	26	25	16	42	14	11	7	4	69	0	0	0
194	4	2	58	1	2	2	31	30	12	44	14	17	3	4	32	43	16	44	11	13	8	8	69	19	513	0
195	4	2	58	1	2	2	28	29	9	33	14	8	4	5	29	44	9	44	15	10	8	5	38	0	966	0
196	4	2	53	1	2	2	19	19	14	24	11	16	1	4	21	13	9	31	10	16	1	3	39	16	1133	0
197	4	2	49	1	2	2	20	17	7	13	9	10	6	4	30	15	6	21	10	9	7	1	30	16	480	0
198	4	2	51	1	2	2	10	1	2	2	4	0	2	4	13	0	0	0	0	0	0	0	34	0	0	11
199	4	1	58	1	2	2	22	13	9	13	10	9	0	0	18	18	11	13	11	8	5	0	36	0	0	0
200	4	2	51	1	2	2	18	12	4	10	5	9	4	5	17	10	8	14	5	10	3	2	48	0	0	0
201	4	2	53	1	2	2	21	17	9	18	9	11	3	4	28	15	9	19	12	9	3	3	49	0	0	0
202	4	2	56	3	2	1	29	17	17	32	10	20	4	5	30	33	17	38	12	20	6	10	49	19	570	9
203	4	2	51	3	1	1	19	11	10	19	8	7	4	1	22	19	11	39	11	21	6	7	37	15	800	0
204	4	1	47	1	1	1	23	12	11	14	11	13	2	1	29	15	13	22	14	16	5	2	45	0	466	0
205	4	2	54	4	1	1	23	14	12	23	8	15	0	1	28	41	16	35	16	22	8	10	46	0	0	10
206	4	1	54	4	1	1	17	15	6	15	4	11	0	1	24	30	13	42	17	20	7	10	32	0	0	11
207	4	2	46	1	1	1	22	14	7	15	3	14	6	1	29	24	18	36	13	23	3	8	30	16	725	5
208	4	2	52	2	1	1	20	15	14	19	5	13	1	1	27	45	16	38	17	22	5	0	35	0	980	7
209	4	2	48	1	1	1	24	18	5	21	9	11	4	1	23	17	10	16	15	9	5	2	36	22	575	0
210	4	1	49	2	1	2	17	21	7	23	9	4	3	1	13	14	13	35	15	13	3	7	45	16	1175	7

TABLE A.1, *continued*

							Pretests								Posttests											
Child ID Number	Site	Sex	Age	Viewing Category	Setting	Viewing Encouragement	Body Parts	Letters	Forms	Numbers	Relational Terms	Classification Terms	Reading Freq.	School Attend.	Body Parts	Letters	Forms	Numbers	Relational Terms	Classification Skills	Comes First	Sesame St.	Peabody	Home Educ.	Home Stim.-Afflu.	Mother-Child Interaction
211	4	1	58	1	1	2	14	7	3	17	13	6	3	1	22	15	11	23	13	9	6	3	59	0	0	0
212	4	1	46	3	1	2	18	13	10	11	7	9	3	1	22	14	13	23	10	13	3	6	42	10	650	6
213	4	1	57	1	1	2	27	19	11	20	10	15	4	1	27	19	8	29	12	11	8	1	41	0	452	0
214	4	1	48	4	1	2	27	12	15	23	11	16	1	2	27	17	13	27	13	10	8	8	39	9	1150	0
215	4	2	52	2	1	2	23	8	9	16	12	8	1	1	20	16	13	23	10	12	9	8	94	0	1300	0
216	4	1	57	2	1	1	29	17	12	24	12	16	3	4	31	32	12	17	9	10	6	4	55	13	0	0
217	4	1	46	3	1	1	18	9	10	12	9	14	5	1	24	18	11	13	8	10	3	9	28	0	0	0
218	4	1	55	2	1	1	14	12	8	14	11	11	1	1	27	40	18	35	16	21	2	9	32	0	0	8
219	4	1	44	3	1	1	8	11	6	10	6	9	6	1	31	23	15	18	10	13	6	8	28	17	975	8
220	4	1	56	1	2	2	26	15	12	29	11	12	5	5	25	23	13	40	12	16	8	2	59	0	716	0
221	4	2	44	2	1	1	14	14	12	10	10	12	3	1	25	23	16	26	13	12	8	5	27	0	404	7
222	4	1	59	4	1	1	28	17	12	27	10	11	0	4	31	46	19	49	16	20	10	10	61	0	0	9
223	5	1	48	2	1	1	16	8	8	9	8	8	3	1	24	11	9	11	8	7	3	4	35	0	580	7
224	5	1	56	2	1	1	22	17	11	23	14	13	5	4	30	20	17	38	13	21	8	5	58	17	0	0
225	5	2	58	2	1	1	20	18	8	26	11	10	3	3	30	44	12	40	13	21	10	5	59	18	561	8
226	5	2	53	1	1	1	15	11	2	8	10	5	1	1	18	19	10	14	6	8	4	1	34	15	883	0
227	5	2	53	1	1	1	26	16	8	14	10	9	4	4	28	13	12	18	10	11	5	1	41	14	633	10
228	5	2	65	1	1	2	15	16	5	24	12	12	6	6	22	15	12	26	11	19	8	0	44	16	0	0
229	5	1	46	1	1	2	15	5	5	4	9	4	5	1	15	13	10	10	8	11	1	2	41	0	745	0
230	5	1	49	1	1	2	19	12	12	16	13	15	5	6	28	16	14	36	14	16	8	1	59	9	0	0
231	5	1	55	1	1	2	21	40	8	36	9	10	4	1	27	49	13	47	9	17	5	2	53	13	570	0
232	5	2	46	2	1	1	20	9	6	17	10	11	4	1	29	13	12	23	12	8	5	10	31	19	1205	8
233	5	2	58	4	1	1	30	55	19	52	15	23	5	3	31	54	19	54	15	23	11	10	78	20	591	10
234	5	1	47	4	1	1	18	13	9	20	8	11	4	1	28	34	17	33	11	18	7	10	43	18	483	0
235	5	1	53	4	1	1	26	25	14	36	13	13	5	3	30	44	19	43	15	23	9	9	90	20	1047	11
236	5	2	51	2	1	1	30	15	8	12	10	10	4	1	30	33	12	45	12	20	3	6	49	15	427	0
237	5	1	49	4	1	1	17	16	12	15	8	15	5	6	25	26	15	20	12	11	3	9	43	16	610	7
238	5	1	43	2	1	1	16	13	6	11	8	9	3	1	22	19	10	10	9	7	3	5	30	21	0	10
239	5	2	60	3	1	1	23	16	9	33	14	16	4	1	29	35	18	50	13	23	6	10	69	17	490	0
240	5	1	51	4	1	1	21	11	10	27	10	12	5	1	25	32	17	47	11	19	6	10	65	16	441	9

Table A.2 summarizes the "Sesame Street" data base.

TABLE A.2
Means, standard deviations, and sample sizes* for the "Sesame Street" variables.

Variable	*n*	Mean		Standard Deviation
1. Site	240		(nominal level)	
2. Sex	240		(nominal level)	
3. Age in months	240	51.53		6.28
4. Viewing category	240		(ordinal level)	
5. Setting	240		(nominal level)	
6. Viewing encouragement	240		(nominal level)	
PRETEST SCORES (7–12)				
7. Body Parts	240	21.40		6.39
8. Letters	240	15.92		8.51
9. Forms	240	9.91		3.72
10. Numbers	240	20.90		10.69
11. Relational Terms	240	9.94		3.07
12. Classificational Skills	239	12.29		4.60
13. Frequency of Reading	233		(ordinal level)	
14. School Attendance	238		(nominal level)	
POSTTEST SCORES (15–20)				
15. Body Parts	240	25.26		5.41
16. Letters	239	26.81		13.19
17. Forms	239	13.75		3.92
18. Numbers	238	30.43		12.60
19. Relational Terms	239	11.70		2.74
20. Classification Skills	239	15.80		5.06
21. Which Comes First	238	5.95		2.84
22. "Sesame Street" Test	228	7.05		2.99
23. Peabody Mental Age	240	46.78		15.99
24. Home Education Index	192	15.00		3.68
25. Home Stimulation-Affluence Index	200	647.28		239.82
26. Mother-Child Stimulation Index	123	7.69		1.95

*Some children had missing data for specific variables. The means and standard deviations were calculated using only information for children who had legitimate scores. As noted, the sample sizes are not the same for all variables.

APPENDIX B
Statistical Tables

TABLE B.1
Random numbers.

Row No.	Column No. (1)	(2)	(3)	(4)	(5)	(6)	(7)	(8)	(9)	(10)
01	47147	95884	04990	32755	99408	44077	36789	05150	74945	33529
02	54729	39291	25596	90156	85412	24759	55771	15888	33271	57233
03	45311	30677	57040	40773	97920	61734	93076	10040	85821	74950
04	57242	41316	48596	57867	65857	79837	91213	46229	95391	00067
05	32923	29542	84533	90641	62556	95644	35910	11043	08495	57183
06	43200	74748	44146	48844	91259	06173	67581	93604	81962	74620
07	88790	59106	93535	56299	54869	29391	96724	64264	75126	28413
08	82047	36348	37174	41845	22356	26620	99406	67449	52954	37300
09	79562	35490	87052	26187	21822	16953	78854	80032	58907	83763
10	51255	13429	04674	86598	81278	31536	97277	88406	16852	27255
11	12931	52107	92823	39994	74677	99471	54065	54980	96256	80823
12	25479	70823	39006	79294	93170	34469	59114	22708	93248	59289
13	55256	56255	77578	94944	47516	40515	45115	17476	20271	55838
14	86117	66860	94360	70106	49962	00061	24219	57296	76444	00458
15	17423	04477	38806	04944	35383	61780	89590	34020	46149	49792
16	61254	51523	88244	10567	28533	84001	35276	84621	07586	14088
17	12663	81554	56251	79368	26192	74790	12458	94377	08234	94962
18	55370	47939	48927	02034	77798	76874	66444	57951	68349	71552
19	28267	76310	03656	98338	71083	91424	01622	80685	80358	54595
20	70207	49158	65097	03018	63286	17034	68546	16351	69004	56700
21	28455	69176	68765	91828	29572	15431	95163	30939	98053	33475
22	81138	00771	57380	98656	88926	47550	97085	03839	13053	97026
23	43837	84225	58891	76514	30786	51106	89871	13084	35979	42100
24	24563	85476	46368	85265	24481	29504	83964	59735	22504	69875
25	82530	60460	79251	52107	87403	52840	28535	66137	43842	44246
26	00590	08549	44158	44684	95419	62062	87979	65596	02458	88721
27	71082	28785	76823	43859	60693	05722	78930	89541	48410	20330
28	91013	83774	50113	73411	82638	88788	21856	24677	73325	85629
29	82631	02955	74528	75115	53738	91846	07496	60883	91168	23526
30	63249	23246	02936	34085	80918	02431	84160	59779	95088	67047
31	52500	01406	15818	82812	02705	73353	39117	52919	73430	47760
32	89512	99718	64050	84846	70375	01969	33979	13858	28824	95786
33	99088	39796	84242	46073	59508	03203	01128	13205	66087	50948
34	78736	26329	44739	06177	90223	90238	05682	85209	35953	69589
35	64164	99380	91053	66522	54091	46301	38038	46711	23840	07275
36	39132	48801	32809	15694	22467	37172	84658	50580	42640	43623
37	50974	41385	19767	54567	63340	08727	20097	68250	89451	78545
38	33284	04590	70900	01300	76609	51678	60920	64760	86366	78888
39	94793	18540	70278	12104	54921	46897	13211	22913	55124	59022
40	36392	52878	82359	39020	63243	64233	48691	23356	47177	37360
41	91456	80324	19822	68560	72085	22143	57777	13120	88573	09530
42	72288	69167	52080	48039	08143	86370	48121	17762	45825	39928
43	18220	69968	38725	42390	37292	83453	25772	54255	03022	07732
44	43810	06189	78781	23697	58534	03120	04230	53553	15389	58439
45	52155	29705	10619	81987	82405	80608	23074	02431	57233	41879
46	06310	16389	87804	50245	91678	97604	10906	65281	71069	61275
47	42305	85246	44987	35597	65741	76572	57932	85747	54640	81504
48	22230	22518	02403	36233	10624	69212	89634	64297	09270	97242
49	36625	86811	55911	45418	71329	18649	51209	95421	83655	64030
50	37284	35767	62009	27871	27645	88439	82082	12143	48603	76568

Reprinted from pages 47, 176, 262, and 294 of *A Million Random Digits with 100,000 Normal Deviates* by The Rand Corporation (New York: The Free Press, 1955). Copyright 1955 and 1983 by The Rand Corporation. Used by permission.

TABLE B.1, *continued*

Row No.	(11)	(12)	(13)	(14)	(15)	(16)	(17)	(18)	(19)	(20)
	Column No.									
01	46968	57727	92862	64521	89917	14264	64582	27581	74914	84618
02	44191	53263	55877	73329	45003	13484	05729	04427	78220	76217
03	34398	06224	33816	72811	34792	63534	23099	40080	55316	32791
04	74978	03065	07713	00184	47458	97074	13049	05127	61346	38744
05	22146	16086	38337	11333	72081	61350	56485	12948	30934	56629
06	40039	64003	47246	00921	21180	00586	52854	07373	57683	63391
07	52337	64456	15477	57873	78198	70698	47480	26203	42053	25502
08	37592	04093	17822	41990	80873	27459	84244	97917	23758	34200
09	11025	55748	84743	62267	19201	59253	37137	55489	56732	33907
10	10337	92634	54713	38974	68482	31749	18810	55285	70057	31117
11	67592	82547	89116	56520	42239	64164	20308	98764	37933	69635
12	93507	52009	18543	46802	54872	30543	36982	99793	87083	09298
13	24432	37280	64342	72346	15361	06321	59697	60262	31624	76034
14	39916	31838	27107	88884	42320	09792	00049	45005	75827	08035
15	74621	96754	18289	23369	98967	43572	11682	50895	31791	17954
16	91033	56081	16619	72514	78212	20525	47263	78039	19064	10200
17	42576	67672	91036	34846	44013	35362	28805	95505	02161	31754
18	06366	30615	03723	80158	25376	64856	19564	12928	76497	43950
19	92709	67089	80871	42549	25946	27826	15413	94875	64696	10638
20	91088	85352	13048	22300	26364	74404	08848	00744	76515	24159
21	68859	77447	53900	59084	87482	50111	05451	14315	60403	18459
22	59541	30913	80422	96303	24641	24328	10016	56207	79830	55961
23	39183	26933	46581	93403	59289	73102	65474	77039	74173	02265
24	67433	10912	60195	76378	11801	09715	70048	98178	54848	78342
25	61753	71956	90753	90356	49927	29527	52874	71252	54624	32313
26	78109	11100	63196	37458	69035	13860	36535	67706	69038	99124
27	93237	15809	78080	30771	79322	51961	04403	31206	78919	78895
28	37554	88263	89925	91472	19172	25190	20036	49660	48019	54484
29	62131	17117	23292	51385	33350	76925	68579	25711	84440	13366
30	29899	74703	13208	62245	59986	45389	64948	15587	73815	84642
31	16948	58289	07173	02989	89809	83625	36530	24914	34035	20920
32	04931	05975	12988	46093	68168	76011	50465	65845	88725	04611
33	95714	36897	18851	71308	00507	75046	92452	69721	48222	30832
34	86053	91435	48681	83657	09624	02887	08747	99108	27124	33203
35	59306	01041	31006	96377	01899	46019	02708	87233	38882	06086
36	05453	54054	03989	67627	77848	00864	60443	45424	85208	57732
37	71851	77466	27851	95556	51364	89232	44404	71038	22524	22996
38	80203	28098	00409	25751	52831	45531	99732	34007	64544	49921
39	13507	80273	39668	20128	67517	56633	97475	84244	88092	85806
40	67926	54145	36779	64621	14395	63670	15744	36269	82307	72877
41	72567	34426	04628	11535	79269	95728	22259	48062	95489	82386
42	28868	56494	11587	84536	34459	13851	10683	68781	16773	01936
43	69574	08044	94608	29493	98014	85057	69607	76960	74220	57717
44	71200	33079	25741	77834	54450	47742	59757	86499	37031	80336
45	89985	59822	45124	98271	57332	86950	59150	92579	39428	41870
46	58748	28114	02010	57794	65577	03942	48568	43159	18704	17460
47	50547	38191	07812	62545	49612	28841	20001	55993	51606	56563
48	49337	29010	20652	84463	95867	57869	24416	58610	34103	20093
49	11850	14365	02856	95688	34296	43548	15701	17783	01348	29104
50	34224	64184	74270	25777	47937	20295	63757	46969	61343	19213

TABLE B.1, *continued*

Row No.	Column No. (21)	(22)	(23)	(24)	(25)	(26)	(27)	(28)	(29)	(30)
01	31692	51607	89056	74472	91284	20263	16039	94491	33767	73915
02	82997	58320	04852	52595	95514	56543	06636	61291	67504	57205
03	05043	40582	46051	60261	04996	82256	47375	87507	05112	88489
04	75781	38768	70475	00601	18378	32077	36523	30843	07057	78326
05	21033	15175	30741	45814	92222	16704	00197	51267	33224	40276
06	99092	60991	12571	71753	65214	33885	82939	50723	88987	69761
07	07204	93373	85112	29610	30375	64836	18459	08235	67650	72930
08	88859	97254	07771	21393	64657	42013	12753	03028	24224	24918
09	30497	91407	72900	15699	58653	38063	25072	48698	88083	48040
10	09726	18075	45852	54968	43743	82050	78412	79456	95032	10984
11	95330	01985	24128	60514	42539	91907	25694	37097	39566	24043
12	09760	32388	05601	49923	66126	54146	67213	52234	48381	89442
13	01534	81967	15337	95831	84643	40792	47562	95494	62087	18064
14	11234	59350	48368	57195	36287	03046	87136	36057	93913	70080
15	71056	48762	80221	59683	27504	21121	94711	11807	80882	48359
16	34208	05374	60304	43178	97247	24875	26259	67622	14657	80354
17	47132	62839	82198	92445	60650	76219	02772	48651	66449	89213
18	55685	93302	43019	45861	95493	16106	12783	37248	83533	25440
19	17803	18184	10510	27159	83008	20544	41665	99439	70606	28974
20	55045	17219	66737	59080	78489	12626	60661	53733	70062	14289
21	01923	33647	98442	59293	83318	33425	76412	87062	01295	11083
22	07202	76476	71888	54845	17468	41964	68694	59662	55905	26898
23	68825	68242	95750	11033	58634	78411	08523	19313	29327	47526
24	68525	06496	17446	41378	32368	82019	66101	56733	43308	82641
25	80819	33515	97373	43064	16221	99697	37951	07947	12935	49391
26	64200	96929	26044	49283	56545	67200	21325	85056	51345	06309
27	30156	29121	75874	42399	41121	90643	19585	06364	47203	19679
28	50467	14282	89098	66717	14753	73356	47781	34165	82842	00121
29	53764	83212	26675	64184	64455	29023	03181	13674	08838	83829
30	81727	35572	95469	36825	81882	95083	68323	14965	34166	32351
31	30807	55558	96026	97398	21723	86560	52617	07771	61886	48234
32	75104	23682	78756	72728	85940	57290	75507	78715	01426	02310
33	06180	62724	36835	80288	25075	32609	33312	21348	87710	55457
34	22098	34834	66117	36252	82717	50585	43639	79999	07414	84003
35	13173	64783	20984	11929	18849	26211	77375	49561	96747	67007
36	75273	36108	55265	15653	82270	99216	27805	60088	06056	97377
37	89849	65756	44454	04602	14292	74458	57777	35934	05160	26359
38	91108	43562	18883	16569	49599	73871	67101	12054	56492	15981
39	51843	01542	17881	12954	94913	39583	94969	61146	35907	72184
40	02644	23564	85464	62947	92571	89377	85004	84654	20465	86212
41	38608	83374	74032	62183	08740	05279	30455	31032	71512	16476
42	43164	28909	88624	14992	85359	10193	32491	14769	63694	92640
43	80933	52950	45646	36636	05085	28053	27596	54873	68476	65823
44	67690	96766	69250	19344	47855	43489	77479	62418	54079	40069
45	68579	17014	25362	15114	30982	27250	29052	71115	83369	46776
46	46353	39733	44677	50133	26623	15979	10651	04263	34087	67005
47	30039	09532	52215	09164	20930	88230	43403	63230	83525	93550
48	89200	92772	42195	91634	39272	46462	76835	27755	03151	75692
49	58118	57942	14807	68214	76093	47484	24468	91764	52907	16675
50	97230	33027	70166	43232	98802	70715	30216	35586	18909	79658

TABLE B.1, *continued*

Row No.	(31)	(32)	(33)	(34)	(35)	(36)	(37)	(38)	(39)	(40)
	Column No.									
01	79228	94510	57711	64366	89040	43278	69072	22003	89465	61483
02	48103	56760	82564	33649	35176	32278	51357	05489	47462	55931
03	70969	27677	99621	63065	73194	70462	19316	77945	45004	39895
04	69931	20237	75246	59124	12484	22012	79731	82435	56301	99752
05	37208	22741	41946	74109	03760	24094	40210	76617	52317	50643
06	60151	92327	85150	27728	64813	47667	66078	03628	95240	03808
07	46210	47674	53747	95354	67757	75477	26396	09592	96239	50854
08	55399	48142	12284	95298	56399	61358	87541	12998	79639	63633
09	23677	64950	97041	43088	80143	34294	91468	01066	90350	78891
10	41947	70066	90311	17133	11674	00826	75760	37586	33621	14199
11	16972	42181	87945	94104	95701	00743	75411	51930	54869	98991
12	74938	79042	38473	89672	45752	35715	89537	78155	09851	24983
13	78075	53671	81047	92759	94519	59473	91679	90536	41676	35230
14	76744	26190	21649	79753	21287	17698	39490	00533	34823	08134
15	82273	69293	23383	59365	18258	54530	47274	69686	55081	28731
16	30239	23081	09526	26055	87099	41372	55542	32754	87317	94638
17	41177	77163	38252	10349	49511	17540	61781	32769	51662	55606
18	07715	88600	69730	78912	19642	39764	47146	19472	84012	08887
19	16855	47454	98638	15189	87345	80509	33392	50866	17629	28208
20	27985	61979	02979	98092	41184	73815	57939	91057	04860	66667
21	77411	98433	42302	86602	26596	64175	64359	97570	64437	55592
22	19453	18731	01039	18933	92188	83767	56148	56261	79920	78514
23	03381	35119	30355	08287	00448	32800	24106	04054	70572	71063
24	11659	27315	09204	26213	57325	51470	56108	23141	16121	53925
25	35032	14283	20642	15311	36238	12079	67596	00017	51789	90737
26	32061	51250	39825	08554	88716	40945	68579	33784	62025	32535
27	81855	16888	24630	15077	47256	08529	54837	24161	95621	53483
28	48422	09247	43406	16093	01168	28523	31406	49360	99243	85090
29	86190	56195	31409	88248	52436	70161	98500	74702	99546	74570
30	90627	37048	50285	69189	97489	83007	31477	13908	97472	74448
31	60103	76739	57644	56746	63005	08804	47081	65928	65045	58629
32	09606	69465	16536	94055	86328	56533	16670	57295	26249	18524
33	62479	29610	03235	51050	15855	66828	08115	16166	32854	74206
34	40232	52840	02512	99258	09327	55073	86030	29933	00528	67359
35	10690	55550	81275	78369	33658	47000	89425	60573	81137	25474
36	73958	38949	99568	72713	22665	03244	17399	83950	66820	08704
37	56554	57926	41529	00619	51972	09442	60298	81066	28362	41165
38	35676	20333	77622	93718	57255	09780	26798	60083	58959	45691
39	01383	85677	96572	16401	31379	88519	41325	33938	36342	03327
40	29448	88487	05814	82402	42132	85708	89754	57495	57655	78644
41	56863	94737	68661	43498	33376	81659	07422	58435	24855	15523
42	20269	34456	48608	11787	86056	88290	17463	66628	03033	80771
43	06790	99803	86439	94235	48560	62912	82302	43198	97087	97104
44	73690	79726	06492	77431	49864	69775	46450	02122	09083	92746
45	76222	20006	98660	88690	01190	05588	76651	03461	11987	80756
46	18434	21893	80472	19499	80423	58643	27088	66458	78358	56606
47	20463	75133	41713	84279	56045	79079	20212	91560	60548	95128
48	27105	77095	72016	23683	01386	40381	74673	11811	36625	62958
49	47736	56338	07546	36084	73126	33364	78730	47282	76795	95719
50	60938	13970	90288	79457	50343	92054	12541	93216	58624	37392

TABLE B.2

Proportional areas of the unit normal distribution for given z score values.*

z Score Values: Only positive z values are given. Because the distribution shape is symmetrical, the proportional areas for negative z values ($-z$) are the same as for the positive values.

Column A: The proportional areas of the entire distribution between the mean ($z = 0$) and the positive or negative value of z.

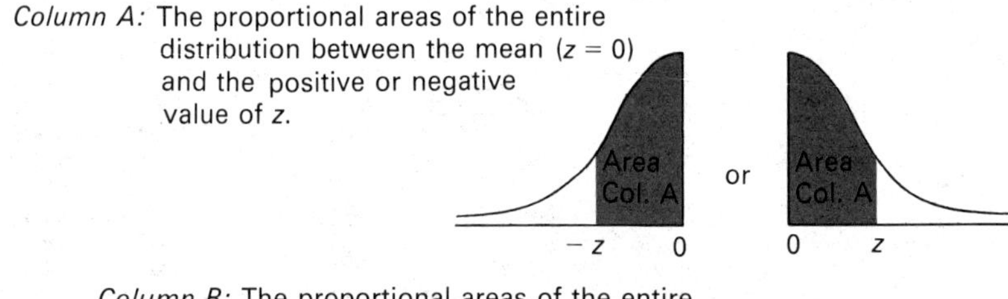

Column B: The proportional areas of the entire distribution beyond the given z value (to the left of $-z$ or to the right of z).

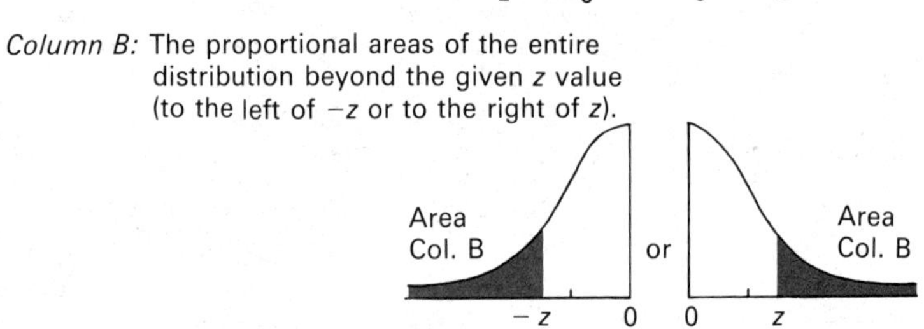

z or $-z$	A. Area Between Mean and z or $-z$	B. Area $> z$ or $< -z$	z or $-z$	A. Area Between Mean and z or $-z$	B. Area $> z$ or $< -z$
0.00	.0000	.5000	0.15	.0596	.4404
0.01	.0040	.4960	0.16	.0636	.4364
0.02	.0080	.4920	0.17	.0675	.4325
0.03	.0120	.4880	0.18	.0714	.4286
0.04	.0160	.4840	0.19	.0753	.4247
0.05	.0199	.4801	0.20	.0793	.4207
0.06	.0239	.4761	0.21	.0832	.4168
0.07	.0279	.4721	0.22	.0871	.4129
0.08	.0319	.4681	0.23	.0910	.4090
0.09	.0359	.4641	0.24	.0948	.4052
0.10	.0398	.4602	0.25	.0987	.4013
0.11	.0438	.4562	0.26	.1026	.3974
0.12	.0478	.4522	0.27	.1064	.3936
0.13	.0517	.4483	0.28	.1103	.3897
0.14	.0557	.4443	0.29	.1141	.3859

*Table values were generated from a Monroe 1776 desk calculator.

TABLE B.2, *continued*

z or $-z$	A. Area Between Mean and z or $-z$	B. Area $> z$ or $< -z$	z or $-z$	A. Area Between Mean and z or $-z$	B. Area $> z$ or $< -z$
0.30	.1179	.3821	0.65	.2422	.2578
0.31	.1217	.3783	0.66	.2454	.2546
0.32	.1255	.3745	0.67	.2486	.2514
0.33	.1293	.3707	0.68	.2517	.2483
0.34	.1331	.3669	0.69	.2549	.2451
0.35	.1368	.3632	0.70	.2580	.2420
0.36	.1406	.3594	0.71	.2611	.2389
0.37	.1443	.3557	0.72	.2642	.2358
0.38	.1480	.3520	0.73	.2673	.2327
0.39	.1517	.3483	0.74	.2704	.2296
0.40	.1554	.3446	0.75	.2734	.2266
0.41	.1591	.3409	0.76	.2764	.2236
0.42	628	.3372	0.77	.2794	.2206
0.43	.1664	.3336	0.78	.2823	.2177
0.44	.1700	.3300	0.79	.2852	.2148
0.45	.1736	.3264	0.80	.2881	.2119
0.46	.1772	.3228	0.81	.2910	.2090
0.47	.1808	.3192	0.82	.2939	.2061
0.48	.1844	.3156	0.83	.2967	.2033
0.49	.1879	.3121	0.84	.2995	.2005
0.50	.1915	.3085	0.85	.3023	.1977
0.51	.1950	.3050	0.86	.3051	.1949
0.52	.1985	.3015	0.87	.3078	.1922
0.53	.2019	.2981	0.88	.3106	.1894
0.54	.2054	.2946	0.89	.3133	.1867
0.55	.2088	.2912	0.90	.3159	.1841
0.56	.2123	.2877	0.91	.3186	.1814
0.57	.2157	.2843	0.92	.3212	.1788
0.58	.2190	.2810	0.93	.3238	.1762
0.59	.2224	.2776	0.94	.3264	.1736
0.60	.2257	.2743	0.95	.3289	.1711
0.61	.2291	.2709	0.96	.3315	.1685
0.62	.2324	.2676	0.97	.3340	.1660
0.63	.2357	.2643	0.98	.3365	.1635
0.64	.2389	.2611	0.99	.3389	.1611

TABLE B.2, ***continued***

z or $-z$	A. Area Between Mean and z or $-z$	B. Area $> z$ or $< -z$
1.00	.3413	.1587
1.01	.3438	.1562
1.02	.3461	.1539
1.03	.3485	.1515
1.04	.3508	.1492
1.05	.3531	.1469
1.06	.3554	.1446
1.07	.3577	.1423
1.08	.3599	.1401
1.09	.3621	.1379
1.10	.3643	.1357
1.11	.3665	.1335
1.12	.3686	.1314
1.13	.3708	.1292
1.14	.3729	.1271
1.15	.3749	.1251
1.16	.3770	.1230
1.17	.3790	.1210
1.18	.3810	.1190
1.19	.3830	.1170
1.20	.3849	.1151
1.21	.3869	.1131
1.22	.3888	.1112
1.23	.3907	.1093
1.24	.3925	.1075
1.25	.3944	.1056
1.26	.3962	.1038
1.27	.3980	.1020
1.28	.3997	.1003
1.29	.4015	.0985
1.30	.4032	.0968
1.31	.4049	.0951
1.32	.4066	.0934
1.33	.4082	.0918
1.34	.4099	.0901
1.35	.4115	.0885
1.36	.4131	.0869
1.37	.4147	.0853
1.38	.4162	.0838
1.39	.4177	.0823
1.40	.4192	.0808
1.41	.4207	.0793
1.42	.4222	.0778
1.43	.4236	.0764
1.44	.4251	.0749
1.45	.4265	.0735
1.46	.4279	.0721
1.47	.4292	.0708
1.48	.4306	.0694
1.49	.4319	.0681
1.50	.4332	.0668
1.51	.4345	.0655
1.52	.4357	.0643
1.53	.4370	.0630
1.54	.4382	.0618
1.55	.4394	.0606
1.56	.4406	.0594
1.57	.4418	.0582
1.58	.4429	.0571
1.59	.4441	.0559
1.60	.4452	.0548
1.61	.4463	.0537
1.62	.4474	.0526
1.63	.4484	.0516
1.64	.4495	.0505
1.645	.4500	.0500
1.65	.4505	.0495
1.66	.4515	.0485
1.67	.4525	.0475
1.68	.4535	.0465
1.69	.4545	.0455

TABLE B.2, *continued*

z or $-z$	A. Area Between Mean and z or $-z$	B. Area $> z$ or $< -z$	z or $-z$	A. Area Between Mean and z or $-z$	B. Area $> z$ or $< -z$
1.70	.4554	.0446	2.05	.4798	.0202
1.71	.4564	.0436	2.06	.4803	.0197
1.72	.4573	.0427	2.07	.4808	.0192
1.73	.4582	.0418	2.08	.4812	.0188
1.74	.4591	.0409	2.09	.4817	.0183
1.75	.4599	.0401	2.10	.4821	.0179
1.76	.4608	.0392	2.11	.4826	.0174
1.77	.4616	.0384	2.12	.4830	.0170
1.78	.4625	.0375	2.13	.4834	.0166
1.79	.4633	.0367	2.14	.4838	.0162
1.80	.4641	.0359	2.15	.4842	.0158
1.81	.4649	.0351	2.16	.4846	.0154
1.82	.4656	.0344	2.17	.4850	.0150
1.83	.4664	.0336	2.18	.4854	.0146
1.84	.4671	.0329	2.19	.4857	.0143
1.85	.4678	.0322	2.20	.4861	.0139
1.86	.4686	.0314	2.21	.4864	.0136
1.87	.4693	.0307	2.22	.4868	.0132
1.88	.4699	.0301	2.23	.4871	.0129
1.89	.4706	.0294	2.24	.4875	.0125
1.90	.4713	.0287	2.25	.4878	.0122
1.91	.4719	.0281	2.26	.4881	.0119
1.92	.4726	.0274	2.27	.4884	.0116
1.93	.4732	.0268	2.28	.4887	.0113
1.94	.4738	.0262	2.29	.4890	.0110
1.95	.4744	.0256	2.30	.4893	.0107
1.96	.4750	.0250	2.31	.4896	.0104
1.97	.4756	.0244	2.32	.4898	.0102
1.98	.4761	.0239	2.33	.4901	.0099
1.99	.4767	.0233	2.34	.4904	.0096
2.00	.4772	.0228	2.35	.4906	.0094
2.01	.4778	.0222	2.36	.4909	.0091
2.02	.4783	.0217	2.37	.4911	.0089
2.03	.4788	.0212	2.38	.4913	.0087
2.04	.4793	.0207	2.39	.4916	.0084

TABLE B.2, ***continued***

z or $-z$	A. Area Between Mean and z or $-z$	B. Area $> z$ or $< -z$	z or $-z$	A. Area Between Mean and z or $-z$	B. Area $> z$ or $< -z$
2.40	.4918	.0082	2.74	.4969	.0031
2.41	.4920	.0080	2.75	.4970	.0030
2.42	.4922	.0078	2.76	.4971	.0029
2.43	.4925	.0075	2.77	.4972	.0028
2.44	.4927	.0073	2.78	.4973	.0027
2.45	.4929	.0071	2.79	.4974	.0026
2.46	.4931	.0069	2.80	.4974	.0026
2.47	.4932	.0068	2.81	.4975	.0025
2.48	.4934	.0066	2.82	.4976	.0024
2.49	.4936	.0064	2.83	.4977	.0023
2.50	.4938	.0062	2.84	.4977	.0023
2.51	.4940	.0060	2.85	.4978	.0022
2.52	.4941	.0059	2.86	.4979	.0021
2.53	.4943	.0057	2.87	.4979	.0021
2.54	.4945	.0055	2.88	.4980	.0020
2.55	.4946	.0054	2.89	.4981	.0019
2.56	.4948	.0052	2.90	.4981	.0019
2.57	.4949	.0051	2.91	.4982	.0018
2.575	.4950	.0050	2.92	.4982	.0018
2.58	.4951	.0049	2.93	.4983	.0017
2.59	.4952	.0048	2.94	.4984	.0016
2.60	.4953	.0047	2.95	.4984	.0016
2.61	.4955	.0045	2.96	.4985	.0015
2.62	.4956	.0044	2.97	.4985	.0015
2.63	.4957	.0043	2.98	.4986	.0014
2.64	.4959	.0041	2.99	.4986	.0014
2.65	.4960	.0040	3.00	.4987	.0013
2.66	.4961	.0039	3.01	.4987	.0013
2.67	.4962	.0038	3.02	.4987	.0013
2.68	.4963	.0037	3.03	.4988	.0012
2.69	.4964	.0036	3.04	.4988	.0012
2.70	.4965	.0035	3.05	.4989	.0011
2.71	.4966	.0034	3.06	.4989	.0011
2.72	.4967	.0033	3.07	.4989	.0011
2.73	.4968	.0032	3.08	.4990	.0010

TABLE B.2, ***continued***

z or $-z$	A. Area Between Mean and z or $-z$	B. Area $> z$ or $< -z$	z or $-z$	A. Area Between Mean and z or $-z$	B. Area $> z$ or $< -z$
3.09	.4990	.0010	3.24	.4994	.0006
3.10	.4990	.0010	3.25	.4994	.0006
3.11	.4991	.0009	3.30	.4995	.0005
3.12	.4991	.0009	3.35	.4996	.0004
3.13	.4991	.0009	3.40	.4997	.0003
			3.50	.4998	.0002
3.14	.4992	.0008			
3.15	.4992	.0008	3.60	.4998	.0002
3.16	.4992	.0008	3.70	.4999	.0001
3.17	.4992	.0008			
3.18	.4993	.0007			
3.19	.4993	.0007			
3.20	.4993	.0007			
3.21	.4993	.0007			
3.22	.4994	.0006			
3.23	.4994	.0006			

TABLE B.3
Student's t distribution values at selected percentile levels.

Percentile point t values are given for proportional areas in the outer tails of the t distribution corresponding to potential level of significance values in both one- and two-tailed tests of a null hypothesis. The first column identifies the specific t distribution according to its degrees of freedom. The remaining columns give critical t values that partition the t distribution into proportional areas corresponding to selected significance levels. The areas given are those beyond a given t value. Because the t distribution is symmetrical, the area to the left of negative values of t is the same as to the right of positive t values.

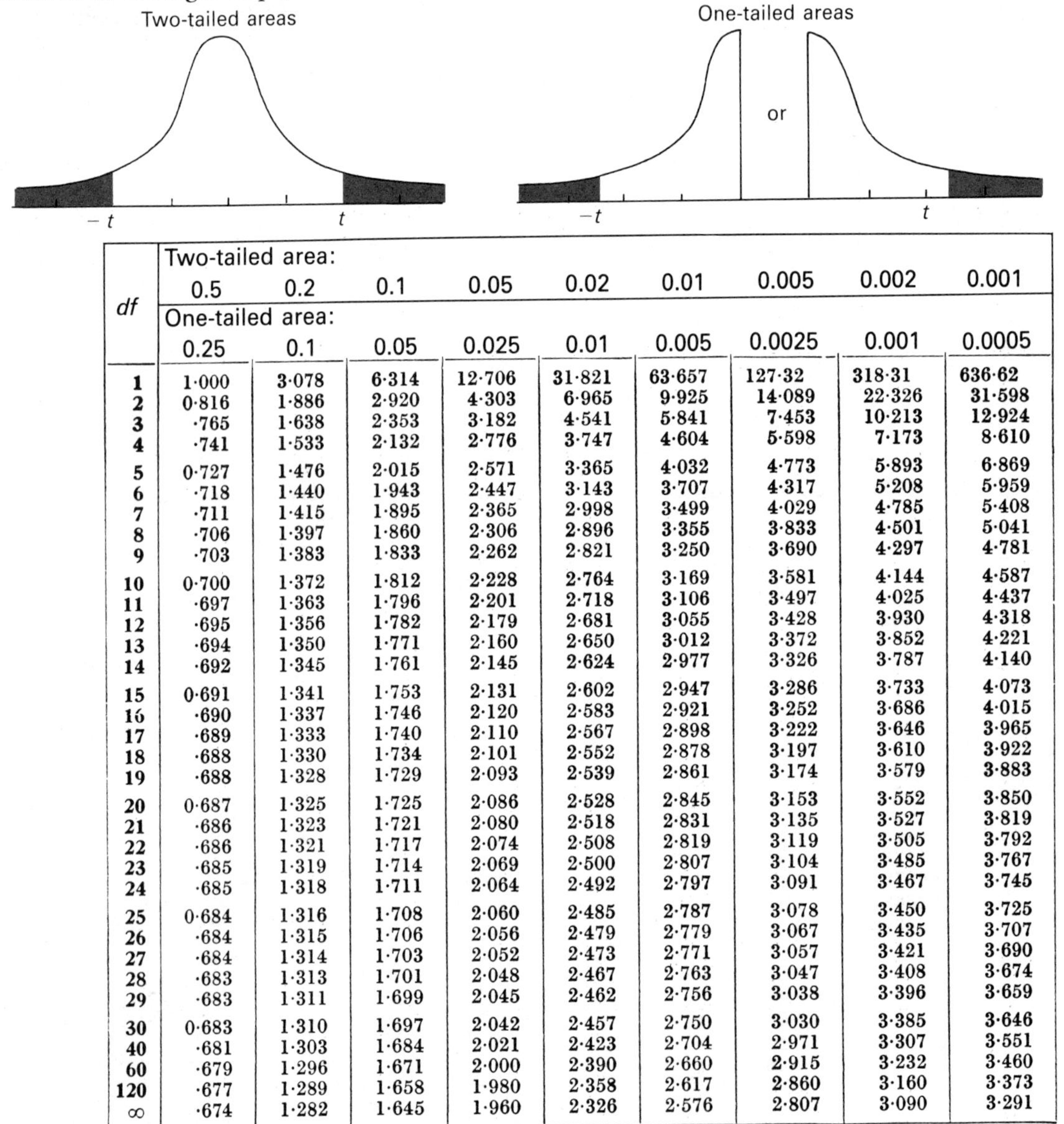

df	Two-tailed area: 0.5	0.2	0.1	0.05	0.02	0.01	0.005	0.002	0.001
	One-tailed area: 0.25	**0.1**	**0.05**	**0.025**	**0.01**	**0.005**	**0.0025**	**0.001**	**0.0005**
1	1·000	3·078	6·314	12·706	31·821	63·657	127·32	318·31	636·62
2	0·816	1·886	2·920	4·303	6·965	9·925	14·089	22·326	31·598
3	·765	1·638	2·353	3·182	4·541	5·841	7·453	10·213	12·924
4	·741	1·533	2·132	2·776	3·747	4·604	5·598	7·173	8·610
5	0·727	1·476	2·015	2·571	3·365	4·032	4·773	5·893	6·869
6	·718	1·440	1·943	2·447	3·143	3·707	4·317	5·208	5·959
7	·711	1·415	1·895	2·365	2·998	3·499	4·029	4·785	5·408
8	·706	1·397	1·860	2·306	2·896	3·355	3·833	4·501	5·041
9	·703	1·383	1·833	2·262	2·821	3·250	3·690	4·297	4·781
10	0·700	1·372	1·812	2·228	2·764	3·169	3·581	4·144	4·587
11	·697	1·363	1·796	2·201	2·718	3·106	3·497	4·025	4·437
12	·695	1·356	1·782	2·179	2·681	3·055	3·428	3·930	4·318
13	·694	1·350	1·771	2·160	2·650	3·012	3·372	3·852	4·221
14	·692	1·345	1·761	2·145	2·624	2·977	3·326	3·787	4·140
15	0·691	1·341	1·753	2·131	2·602	2·947	3·286	3·733	4·073
16	·690	1·337	1·746	2·120	2·583	2·921	3·252	3·686	4·015
17	·689	1·333	1·740	2·110	2·567	2·898	3·222	3·646	3·965
18	·688	1·330	1·734	2·101	2·552	2·878	3·197	3·610	3·922
19	·688	1·328	1·729	2·093	2·539	2·861	3·174	3·579	3·883
20	0·687	1·325	1·725	2·086	2·528	2·845	3·153	3·552	3·850
21	·686	1·323	1·721	2·080	2·518	2·831	3·135	3·527	3·819
22	·686	1·321	1·717	2·074	2·508	2·819	3·119	3·505	3·792
23	·685	1·319	1·714	2·069	2·500	2·807	3·104	3·485	3·767
24	·685	1·318	1·711	2·064	2·492	2·797	3·091	3·467	3·745
25	0·684	1·316	1·708	2·060	2·485	2·787	3·078	3·450	3·725
26	·684	1·315	1·706	2·056	2·479	2·779	3·067	3·435	3·707
27	·684	1·314	1·703	2·052	2·473	2·771	3·057	3·421	3·690
28	·683	1·313	1·701	2·048	2·467	2·763	3·047	3·408	3·674
29	·683	1·311	1·699	2·045	2·462	2·756	3·038	3·396	3·659
30	0·683	1·310	1·697	2·042	2·457	2·750	3·030	3·385	3·646
40	·681	1·303	1·684	2·021	2·423	2·704	2·971	3·307	3·551
60	·679	1·296	1·671	2·000	2·390	2·660	2·915	3·232	3·460
120	·677	1·289	1·658	1·980	2·358	2·617	2·860	3·160	3·373
∞	·674	1·282	1·645	1·960	2·326	2·576	2·807	3·090	3·291

Note: Modified from *Biometrika Tables for Statisticians,* 4th ed., Vol. 1, (p. 138) by E.S. Pearson and H.O. Hartley, eds., 1956, New York: Cambridge University Press. Copyright 1956 by the Biometrika Trustees. Used by permission.

TABLE B.4
Chi square (χ^2) distribution values at selected percentile levels.

Chi square values at points along the χ^2 value scale continuum correspond to selected areas of several chi square distributions. The degrees of freedom (df) defining a specific chi square distribution appear in column 1. The proportions heading each column give the proportional area of the entire χ^2 distribution to the right of the χ^2 value listed for a specific distribution.

df	Areas to the right of χ^2 value						
	0·995	**0·990**	**0·975**	**0·950**	**0·900**	**0·750**	**0·500**
1	392704.10^{-10}	157088.10^{-9}	982069.10^{-9}	393214.10^{-8}	0·0157908	0·1015308	0·454937
2	0·0100251	0·0201007	0·0506356	0·102587	0·210720	0·575364	1·38629
3	0·0717212	0·114832	0·215795	0·351846	0·584375	1·212534	2·36597
4	0·206990	0·297110	0·484419	0·710721	1·063623	1·92255	3·35670
5	0·411740	0·554300	0·831211	1·145476	1·61031	2·67460	4·35146
6	0·675727	0·872085	1·237347	1·63539	2·20413	3·45460	5·34812
7	0·989265	1·239043	1·68987	2·16735	2·83311	4·25485	6·34581
8	1·344419	1·646482	2·17973	2·73264	3·48954	5·07064	7·34412
9	1·734926	2·087912	2·70039	3·32511	4·16816	5·89883	8·34283
10	2·15585	2·55821	3·24697	3·94030	4·86518	6·73720	9·34182
11	2·60321	3·05347	3·81575	4·57481	5·57779	7·58412	10·3410
12	3·07382	3·57056	4·40379	5·22603	6·30380	8·43842	11·3403
13	3·56503	4·10691	5·00874	5·89186	7·04150	9·29906	12·3398
14	4·07468	4·66043	5·62872	6·57063	7·78953	10·1653	13·3393
15	4·60094	5·22935	6·26214	7·26094	8·54675	11·0365	14·3389
16	5·14224	5·81221	6·90766	7·96164	9·31223	11·9122	15·3385
17	5·69724	6·40776	7·56418	8·67176	10·0852	12·7919	16·3381
18	6·26481	7·01491	8·23075	9·39046	10·8649	13·6753	17·3379
19	6·84398	7·63273	8·90655	10·1170	11·6509	14·5620	18·3376
20	7·43386	8·26040	9·59083	10·8508	12·4426	15·4518	19·3374
21	8·03366	8·89720	10·28293	11·5913	13·2396	16·3444	20·3372
22	8·64272	9·54249	10·9823	12·3380	14·0415	17·2396	21·3370
23	9·26042	10·19567	11·6885	13·0905	14·8479	18·1373	22·3369
24	9·88623	10·8564	12·4011	13·8484	15·6587	19·0372	23·3367

Note: Reproduced from *Biometrika Tables for Statisticians,* 4th ed., Vol. 1, (pp. 130–131) by E.S. Pearson and H.O. Hartley, eds., 1956, New York: Cambridge University Press. Copyright 1956 by the Biometrika Trustees. Used by permission.

TABLE B.4, *continued*

df	Areas to the right of χ^2 value						
	0·995	0·990	0·975	0·950	0·900	0·750	0·500
25	10·5197	11·5240	13·1197	14·6114	16·4734	19·9393	24·3366
26	11·1603	12·1981	13·8439	15·3791	17·2919	20·8434	25·3364
27	11·8076	12·8786	14·5733	16·1513	18·1138	21·7494	26·3363
28	12·4613	13·5648	15·3079	16·9279	18·9392	22·6572	27·3363
29	13·1211	14·2565	16·0471	17·7083	19·7677	23·5666	28·3362
30	13·7867	14·9535	16·7908	18·4926	20·5992	24·4776	29·3360
40	20·7065	22·1643	24·4331	26·5093	29·0505	33·6603	39·3354
50	27·9907	29·7067	32·3574	34·7642	37·6886	42·9421	49·3349
60	35·5346	37·4848	40·4817	43·1879	46·4589	52·2938	59·3347
70	43·2752	45·4418	48·7576	51·7393	55·3290	61·6983	69·3344
80	51·1720	53·5400	57·1532	60·3915	64·2778	71·1445	79·3343
90	59·1963	61·7541	65·6466	69·1260	73·2912	80·6247	89·3342
100	67·3276	70·0648	74·2219	77·9295	82·3581	90·1332	99·3341

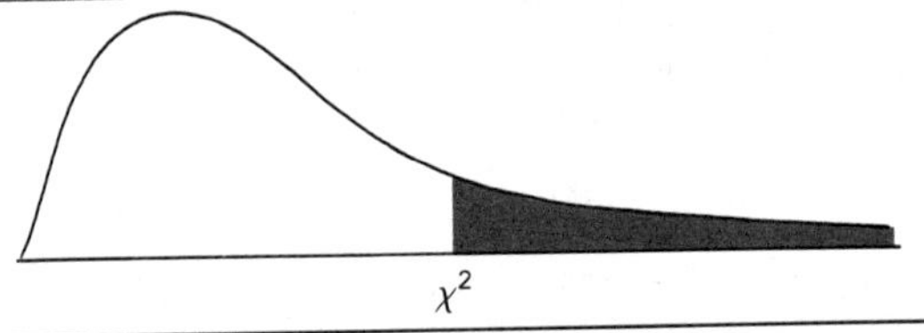

df	Areas to the right of χ^2 value						
	0·250	0·100	0·050	0·025	0·010	0·005	0·001
1	1·32330	2·70554	3·84146	5·02389	6·63490	7·87944	10·828
2	2·77259	4·60517	5·99147	7·37776	9·21034	10·5966	13·816
3	4·10835	6·25139	7·81473	9·34840	11·3449	12·8381	16·266
4	5·38527	7·77944	9·48773	11·1433	13·2767	14·8602	18·467
5	6·62568	9·23635	11·0705	12·8325	15·0863	16·7496	20·515
6	7·84080	10·6446	12·5916	14·4494	16·8119	18·5476	22·458
7	9·03715	12·0170	14·0671	16·0128	18·4753	20·2777	24·322
8	10·2188	13·3616	15·5073	17·5346	20·0902	21·9550	26·125
9	11·3887	14·6837	16·9190	19·0228	21·6660	23·5893	27·877
10	12·5489	15·9871	18·3070	20·4831	23·2093	25·1882	29·588
11	13·7007	17·2750	19·6751	21·9200	24·7250	26·7569	31·264
12	14·8454	18·5494	21·0261	23·3367	26·2170	28·2995	32·909
13	15·9839	19·8119	22·3621	24·7356	27·6883	29·8194	34·528
14	17·1170	21·0642	23·6848	26·1190	29·1413	31·3193	36·123

TABLE B.4, *continued*

df	Areas to the right of χ^2 value						
	0·250	**0·100**	**0·050**	**0·025**	**0·010**	**0·005**	**0·001**
15	18·2451	22·3072	24·9958	27·4884	30·5779	32·8013	37·697
16	19·3688	23·5418	26·2962	28·8454	31·9999	34·2672	39·252
17	20·4887	24·7690	27·5871	30·1910	33·4087	35·7185	40·790
18	21·6049	25·9894	28·8693	31·5264	34·8053	37·1564	42·312
19	22·7178	27·2036	30·1435	32·8523	36·1908	38·5822	43·820
20	23·8277	28·4120	31·4104	34·1696	37·5662	39·9968	45·315
21	24·9348	29·6151	32·6705	35·4789	38·9321	41·4010	46·797
22	26·0393	30·8133	33·9244	36·7807	40·2894	42·7956	48·268
23	27·1413	32·0069	35·1725	38·0757	41·6384	44·1813	49·728
24	28·2412	33·1963	36·4151	39·3641	42·9798	45·5585	51·179
25	29·3389	34·3816	37·6525	40·6465	44·3141	46·9278	52·620
26	30·4345	35·5631	38·8852	41·9232	45·6417	48·2899	54·052
27	31·5284	36·7412	40·1133	43·1944	46·9630	49·6449	55·476
28	32·6205	37·9159	41·3372	44·4607	48·2782	50·9933	56·892
29	33·7109	39·0875	42·5569	45·7222	49·5879	52·3356	58·302
30	34·7998	40·2560	43·7729	46·9792	50·8922	53·6720	59·703
40	45·6160	51·8050	55·7585	59·3417	63·6907	66·7659	73·402
50	56·3336	63·1671	67·5048	71·4202	76·1539	79·4900	86·661
60	66·9814	74·3970	79·0819	83·2976	88·3794	91·9517	99·607
70	77·5766	85·5271	90·5312	95·0231	100·425	104·215	112·317
80	88·1303	96·5782	101·879	106·629	112·329	116·321	124·839
90	98·6499	107·565	113·145	118·136	124·116	128·299	137·208
100	109·141	118·498	124·342	129·561	135·807	140·169	149·449

TABLE B.5

F Distribution values at selected percentile levels.

Upper 5% points

df_d \ df_n	1	2	3	4	5	6	7	8	9	10	12	15	20	24	30	40	60	120	∞
1	161·4	199·5	215·7	224·6	230·2	234·0	236·8	238·9	240·5	241·9	243·9	245·9	248·0	249·1	250·1	251·1	252·2	253·3	254·3
2	18·51	19·00	19·16	19·25	19·30	19·33	19·35	19·37	19·38	19·40	19·41	19·43	19·45	19·45	19·46	19·47	19·48	19·49	19·50
3	10·13	9·55	9·28	9·12	9·01	8·94	8·89	8·85	8·81	8·79	8·74	8·70	8·66	8·64	8·62	8·59	8·57	8·55	8·53
4	7·71	6·94	6·59	6·39	6·26	6·16	6·09	6·04	6·00	5·96	5·91	5·86	5·80	5·77	5·75	5·72	5·69	5·66	5·63
5	6·61	5·79	5·41	5·19	5·05	4·95	4·88	4·82	4·77	4·74	4·68	4·62	4·56	4·53	4·50	4·46	4·43	4·40	4·36
6	5·99	5·14	4·76	4·53	4·39	4·28	4·21	4·15	4·10	4·06	4·00	3·94	3·87	3·84	3·81	3·77	3·74	3·70	3·67
7	5·59	4·74	4·35	4·12	3·97	3·87	3·79	3·73	3·68	3·64	3·57	3·51	3·44	3·41	3·38	3·34	3·30	3·27	3·23
8	5·32	4·46	4·07	3·84	3·69	3·58	3·50	3·44	3·39	3·35	3·28	3·22	3·15	3·12	3·08	3·04	3·01	2·97	2·93
9	5·12	4·26	3·86	3·63	3·48	3·37	3·29	3·23	3·18	3·14	3·07	3·01	2·94	2·90	2·86	2·83	2·79	2·75	2·71
10	4·96	4·10	3·71	3·48	3·33	3·22	3·14	3·07	3·02	2·98	2·91	2·85	2·77	2·74	2·70	2·66	2·62	2·58	2·54
11	4·84	3·98	3·59	3·36	3·20	3·09	3·01	2·95	2·90	2·85	2·79	2·72	2·65	2·61	2·57	2·53	2·49	2·45	2·40
12	4·75	3·89	3·49	3·26	3·11	3·00	2·91	2·85	2·80	2·75	2·69	2·62	2·54	2·51	2·47	2·43	2·38	2·34	2·30
13	4·67	3·81	3·41	3·18	3·03	2·92	2·83	2·77	2·71	2·67	2·60	2·53	2·46	2·42	2·38	2·34	2·30	2·25	2·21
14	4·60	3·74	3·34	3·11	2·96	2·85	2·76	2·70	2·65	2·60	2·53	2·46	2·39	2·35	2·31	2·27	2·22	2·18	2·13
15	4·54	3·68	3·29	3·06	2·90	2·79	2·71	2·64	2·59	2·54	2·48	2·40	2·33	2·29	2·25	2·20	2·16	2·11	2·07
16	4·49	3·63	3·24	3·01	2·85	2·74	2·66	2·59	2·54	2·49	2·42	2·35	2·28	2·24	2·19	2·15	2·11	2·06	2·01
17	4·45	3·59	3·20	2·96	2·81	2·70	2·61	2·55	2·49	2·45	2·38	2·31	2·23	2·19	2·15	2·10	2·06	2·01	1·96
18	4·41	3·55	3·16	2·93	2·77	2·66	2·58	2·51	2·46	2·41	2·34	2·27	2·19	2·15	2·11	2·06	2·02	1·97	1·92
19	4·38	3·52	3·13	2·90	2·74	2·63	2·54	2·48	2·42	2·38	2·31	2·23	2·16	2·11	2·07	2·03	1·98	1·93	1·88
20	4·35	3·49	3·10	2·87	2·71	2·60	2·51	2·45	2·39	2·35	2·28	2·20	2·12	2·08	2·04	1·99	1·95	1·90	1·84
21	4·32	3·47	3·07	2·84	2·68	2·57	2·49	2·42	2·37	2·32	2·25	2·18	2·10	2·05	2·01	1·96	1·92	1·87	1·81
22	4·30	3·44	3·05	2·82	2·66	2·55	2·46	2·40	2·34	2·30	2·23	2·15	2·07	2·03	1·98	1·94	1·89	1·84	1·78
23	4·28	3·42	3·03	2·80	2·64	2·53	2·44	2·37	2·32	2·27	2·20	2·13	2·05	2·01	1·96	1·91	1·86	1·81	1·76
24	4·26	3·40	3·01	2·78	2·62	2·51	2·42	2·36	2·30	2·25	2·18	2·11	2·03	1·98	1·94	1·89	1·84	1·79	1·73
25	4·24	3·39	2·99	2·76	2·60	2·49	2·40	2·34	2·28	2·24	2·16	2·09	2·01	1·96	1·92	1·87	1·82	1·77	1·71
26	4·23	3·37	2·98	2·74	2·59	2·47	2·39	2·32	2·27	2·22	2·15	2·07	1·99	1·95	1·90	1·85	1·80	1·75	1·69
27	4·21	3·35	2·96	2·73	2·57	2·46	2·37	2·31	2·25	2·20	2·13	2·06	1·97	1·93	1·88	1·84	1·79	1·73	1·67
28	4·20	3·34	2·95	2·71	2·56	2·45	2·36	2·29	2·24	2·19	2·12	2·04	1·96	1·91	1·87	1·82	1·77	1·71	1·65
29	4·18	3·33	2·93	2·70	2·55	2·43	2·35	2·28	2·22	2·18	2·10	2·03	1·94	1·90	1·85	1·81	1·75	1·70	1·64
30	4·17	3·32	2·92	2·69	2·53	2·42	2·33	2·27	2·21	2·16	2·09	2·01	1·93	1·89	1·84	1·79	1·74	1·68	1·62
40	4·08	3·23	2·84	2·61	2·45	2·34	2·25	2·18	2·12	2·08	2·00	1·92	1·84	1·79	1·74	1·69	1·64	1·58	1·51
60	4·00	3·15	2·76	2·53	2·37	2·25	2·17	2·10	2·04	1·99	1·92	1·84	1·75	1·70	1·65	1·59	1·53	1·47	1·39
120	3·92	3·07	2·68	2·45	2·29	2·17	2·09	2·02	1·96	1·91	1·83	1·75	1·66	1·61	1·55	1·50	1·43	1·35	1·25
∞	3·84	3·00	2·60	2·37	2·21	2·10	2·01	1·94	1·88	1·83	1·75	1·67	1·57	1·52	1·46	1·39	1·32	1·22	1·00

Tabled values of *F* correspond to the 95th, 97.5th, and 99th percentile points found in the upper tails of specific *F* distributions. These values correspond to α levels of .05, .025, and .01, respectively. The denominator degrees of freedom (df_d) are given in the far left column and label each row. The numerator degrees of freedom (df_n) are given in the upper row and label each column. The intersection of the column for a specific df_n and row for a specific df_d identifies the critical *F* values for a specific *F* distribution with df_n and df_d degrees of freedom.

TABLE B.5, *continued*

Upper 2·5 % points

df_d \ df_n	1	2	3	4	5	6	7	8	9	10	12	15	20	24	30	40	60	120	∞
1	647·8	799·5	864·2	899·6	921·8	937·1	948·2	956·7	963·3	968·6	976·7	984·9	993·1	997·2	1001	1006	1010	1014	1018
2	38·51	39·00	39·17	39·25	39·30	39·33	39·36	39·37	39·39	39·40	39·41	39·43	39·45	39·46	39·46	39·47	39·48	39·49	39·50
3	17·44	16·04	15·44	15·10	14·88	14·73	14·62	14·54	14·47	14·42	14·34	14·25	14·17	14·12	14·08	14·04	13·99	13·95	13·90
4	12·22	10·65	9·98	9·60	9·36	9·20	9·07	8·98	8·90	8·84	8·75	8·66	8·56	8·51	8·46	8·41	8·36	8·31	8·26
5	10·01	8·43	7·76	7·39	7·15	6·98	6·85	6·76	6·68	6·62	6·52	6·43	6·33	6·28	6·23	6·18	6·12	6·07	6·02
6	8·81	7·26	6·60	6·23	5·99	5·82	5·70	5·60	5·52	5·46	5·37	5·27	5·17	5·12	5·07	5·01	4·96	4·90	4·85
7	8·07	6·54	5·89	5·52	5·29	5·12	4·99	4·90	4·82	4·76	4·67	4·57	4·47	4·42	4·36	4·31	4·25	4·20	4·14
8	7·57	6·06	5·42	5·05	4·82	4·65	4·53	4·43	4·36	4·30	4·20	4·10	4·00	3·95	3·89	3·84	3·78	3·73	3·67
9	7·21	5·71	5·08	4·72	4·48	4·32	4·20	4·10	4·03	3·96	3·87	3·77	3·67	3·61	3·56	3·51	3·45	3·39	3·33
10	6·94	5·46	4·83	4·47	4·24	4·07	3·95	3·85	3·78	3·72	3·62	3·52	3·42	3·37	3·31	3·26	3·20	3·14	3·08
11	6·72	5·26	4·63	4·28	4·04	3·88	3·76	3·66	3·59	3·53	3·43	3·33	3·23	3·17	3·12	3·06	3·00	2·94	2·88
12	6·55	5·10	4·47	4·12	3·89	3·73	3·61	3·51	3·44	3·37	3·28	3·18	3·07	3·02	2·96	2·91	2·85	2·79	2·72
13	6·41	4·97	4·35	4·00	3·77	3·60	3·48	3·39	3·31	3·25	3·15	3·05	2·95	2·89	2·84	2·78	2·72	2·66	2·60
14	6·30	4·86	4·24	3·89	3·66	3·50	3·38	3·29	3·21	3·15	3·05	2·95	2·84	2·79	2·73	2·67	2·61	2·55	2·49
15	6·20	4·77	4·15	3·80	3·58	3·41	3·29	3·20	3·12	3·06	2·96	2·86	2·76	2·70	2·64	2·59	2·52	2·46	2·40
16	6·12	4·69	4·08	3·73	3·50	3·34	3·22	3·12	3·05	2·99	2·89	2·79	2·68	2·63	2·57	2·51	2·45	2·38	2·32
17	6·04	4·62	4·01	3·66	3·44	3·28	3·16	3·06	2·98	2·92	2·82	2·72	2·62	2·56	2·50	2·44	2·38	2·32	2·25
18	5·98	4·56	3·95	3·61	3·38	3·22	3·10	3·01	2·93	2·87	2·77	2·67	2·56	2·50	2·44	2·38	2·32	2·26	2·19
19	5·92	4·51	3·90	3·56	3·33	3·17	3·05	2·96	2·88	2·82	2·72	2·62	2·51	2·45	2·39	2·33	2·27	2·20	2·13
20	5·87	4·46	3·86	3·51	3·29	3·13	3·01	2·91	2·84	2·77	2·68	2·57	2·46	2·41	2·35	2·29	2·22	2·16	2·09
21	5·83	4·42	3·82	3·48	3·25	3·09	2·97	2·87	2·80	2·73	2·64	2·53	2·42	2·37	2·31	2·25	2·18	2·11	2·04
22	5·79	4·38	3·78	3·44	3·22	3·05	2·93	2·84	2·76	2·70	2·60	2·50	2·39	2·33	2·27	2·21	2·14	2·08	2·00
23	5·75	4·35	3·75	3·41	3·18	3·02	2·90	2·81	2·73	2·67	2·57	2·47	2·36	2·30	2·24	2·18	2·11	2·04	1·97
24	5·72	4·32	3·72	3·38	3·15	2·99	2·87	2·78	2·70	2·64	2·54	2·44	2·33	2·27	2·21	2·15	2·08	2·01	1·94
25	5·69	4·29	3·69	3·35	3·13	2·97	2·85	2·75	2·68	2·61	2·51	2·41	2·30	2·24	2·18	2·12	2·05	1·98	1·91
26	5·66	4·27	3·67	3·33	3·10	2·94	2·82	2·73	2·65	2·59	2·49	2·39	2·28	2·22	2·16	2·09	2·03	1·95	1·88
27	5·63	4·24	3·65	3·31	3·08	2·92	2·80	2·71	2·63	2·57	2·47	2·36	2·25	2·19	2·13	2·07	2·00	1·93	1·85
28	5·61	4·22	3·63	3·29	3·06	2·90	2·78	2·69	2·61	2·55	2·45	2·34	2·23	2·17	2·11	2·05	1·98	1·91	1·83
29	5·59	4·20	3·61	3·27	3·04	2·88	2·76	2·67	2·59	2·53	2·43	2·32	2·21	2·15	2·09	2·03	1·96	1·89	1·81
30	5·57	4·18	3·59	3·25	3·03	2·87	2·75	2·65	2·57	2·51	2·41	2·31	2·20	2·14	2·07	2·01	1·94	1·87	1·79
40	5·42	4·05	3·46	3·13	2·90	2·74	2·62	2·53	2·45	2·39	2·29	2·18	2·07	2·01	1·94	1·88	1·80	1·72	1·64
60	5·29	3·93	3·34	3·01	2·79	2·63	2·51	2·41	2·33	2·27	2·17	2·06	1·94	1·88	1·82	1·74	1·67	1·58	1·48
120	5·15	3·80	3·23	2·89	2·67	2·52	2·39	2·30	2·22	2·16	2·05	1·94	1·82	1·76	1·69	1·61	1·53	1·43	1·31
∞	5·02	3·69	3·12	2·79	2·57	2·41	2·29	2·19	2·11	2·05	1·94	1·83	1·71	1·64	1·57	1·48	1·39	1·27	1·00

Note: Modified from *Biometrika Tables for Statisticians,* 4th ed., Vol. 1, (pp. 157–163) by E.S. Pearson and H.O. Hartley, eds., 1956, New York: Cambridge University Press. Copyright 1956 by the Biometrika Trustees. Used by permission.

TABLE B.5, *continued*

Upper 1% points

df_d \ df_n	1	2	3	4	5	6	7	8	9	10	12	15	20	24	30	40	60	120	∞
1	4052	4999·5	5403	5625	5764	5859	5928	5982	6022	6056	6106	6157	6209	6235	6261	6287	6313	6339	6366
2	98·50	99·00	99·17	99·25	99·30	99·33	99·36	99·37	99·39	99·40	99·42	99·43	99·45	99·46	99·47	99·47	99·48	99·49	99·50
3	34·12	30·82	29·46	28·71	28·24	27·91	27·67	27·49	27·35	27·23	27·05	26·87	26·69	26·60	26·50	26·41	26·32	26·22	26·13
4	21·20	18·00	16·69	15·98	15·52	15·21	14·98	14·80	14·66	14·55	14·37	14·20	14·02	13·93	13·84	13·75	13·65	13·56	13·46
5	16·26	13·27	12·06	11·39	10·97	10·67	10·46	10·29	10·16	10·05	9·89	9·72	9·55	9·47	9·38	9·29	9·20	9·11	9·02
6	13·75	10·92	9·78	9·15	8·75	8·47	8·26	8·10	7·98	7·87	7·72	7·56	7·40	7·31	7·23	7·14	7·06	6·97	6·88
7	12·25	9·55	8·45	7·85	7·46	7·19	6·99	6·84	6·72	6·62	6·47	6·31	6·16	6·07	5·99	5·91	5·82	5·74	5·65
8	11·26	8·65	7·59	7·01	6·63	6·37	6·18	6·03	5·91	5·81	5·67	5·52	5·36	5·28	5·20	5·12	5·03	4·95	4·86
9	10·56	8·02	6·99	6·42	6·06	5·80	5·61	5·47	5·35	5·26	5·11	4·96	4·81	4·73	4·65	4·57	4·48	4·40	4·31
10	10·04	7·56	6·55	5·99	5·64	5·39	5·20	5·06	4·94	4·85	4·71	4·56	4·41	4·33	4·25	4·17	4·08	4·00	3·91
11	9·65	7·21	6·22	5·67	5·32	5·07	4·89	4·74	4·63	4·54	4·40	4·25	4·10	4·02	3·94	3·86	3·78	3·69	3·60
12	9·33	6·93	5·95	5·41	5·06	4·82	4·64	4·50	4·39	4·30	4·16	4·01	3·86	3·78	3·70	3·62	3·54	3·45	3·36
13	9·07	6·70	5·74	5·21	4·86	4·62	4·44	4·30	4·19	4·10	3·96	3·82	3·66	3·59	3·51	3·43	3·34	3·25	3·17
14	8·86	6·51	5·56	5·04	4·69	4·46	4·28	4·14	4·03	3·94	3·80	3·66	3·51	3·43	3·35	3·27	3·18	3·09	3·00
15	8·68	6·36	5·42	4·89	4·56	4·32	4·14	4·00	3·89	3·80	3·67	3·52	3·37	3·29	3·21	3·13	3·05	2·96	2·87
16	8·53	6·23	5·29	4·77	4·44	4·20	4·03	3·89	3·78	3·69	3·55	3·41	3·26	3·18	3·10	3·02	2·93	2·84	2·75
17	8·40	6·11	5·18	4·67	4·34	4·10	3·93	3·79	3·68	3·59	3·46	3·31	3·16	3·08	3·00	2·92	2·83	2·75	2·65
18	8·29	6·01	5·09	4·58	4·25	4·01	3·84	3·71	3·60	3·51	3·37	3·23	3·08	3·00	2·92	2·84	2·75	2·66	2·57
19	8·18	5·93	5·01	4·50	4·17	3·94	3·77	3·63	3·52	3·43	3·30	3·15	3·00	2·92	2·84	2·76	2·67	2·58	2·49
20	8·10	5·85	4·94	4·43	4·10	3·87	3·70	3·56	3·46	3·37	3·23	3·09	2·94	2·86	2·78	2·69	2·61	2·52	2·42
21	8·02	5·78	4·87	4·37	4·04	3·81	3·64	3·51	3·40	3·31	3·17	3·03	2·88	2·80	2·72	2·64	2·55	2·46	2·36
22	7·95	5·72	4·82	4·31	3·99	3·76	3·59	3·45	3·35	3·26	3·12	2·98	2·83	2·75	2·67	2·58	2·50	2·40	2·31
23	7·88	5·66	4·76	4·26	3·94	3·71	3·54	3·41	3·30	3·21	3·07	2·93	2·78	2·70	2·62	2·54	2·45	2·35	2·26
24	7·82	5·61	4·72	4·22	3·90	3·67	3·50	3·36	3·26	3·17	3·03	2·89	2·74	2·66	2·58	2·49	2·40	2·31	2·21
25	7·77	5·57	4·68	4·18	3·85	3·63	3·46	3·32	3·22	3·13	2·99	2·85	2·70	2·62	2·54	2·45	2·36	2·27	2·17
26	7·72	5·53	4·64	4·14	3·82	3·59	3·42	3·29	3·18	3·09	2·96	2·81	2·66	2·58	2·50	2·42	2·33	2·23	2·13
27	7·68	5·49	4·60	4·11	3·78	3·56	3·39	3·26	3·15	3·06	2·93	2·78	2·63	2·55	2·47	2·38	2·29	2·20	2·10
28	7·64	5·45	4·57	4·07	3·75	3·53	3·36	3·23	3·12	3·03	2·90	2·75	2·60	2·52	2·44	2·35	2·26	2·17	2·06
29	7·60	5·42	4·54	4·04	3·73	3·50	3·33	3·20	3·09	3·00	2·87	2·73	2·57	2·49	2·41	2·33	2·23	2·14	2·03
30	7·56	5·39	4·51	4·02	3·70	3·47	3·30	3·17	3·07	2·98	2·84	2·70	2·55	2·47	2·39	2·30	2·21	2·11	2·01
40	7·31	5·18	4·31	3·83	3·51	3·29	3·12	2·99	2·89	2·80	2·66	2·52	2·37	2·29	2·20	2·11	2·02	1·92	1·80
60	7·08	4·98	4·13	3·65	3·34	3·12	2·95	2·82	2·72	2·63	2·50	2·35	2·20	2·12	2·03	1·94	1·84	1·73	1·60
120	6·85	4·79	3·95	3·48	3·17	2·96	2·79	2·66	2·56	2·47	2·34	2·19	2·03	1·95	1·86	1·76	1·66	1·53	1·38
∞	6·63	4·61	3·78	3·32	3·02	2·80	2·64	2·51	2·41	2·32	2·18	2·04	1·88	1·79	1·70	1·59	1·47	1·32	1·00

TABLE B.6

Fisher's Z transformation values corresponding to given values of r_{xy}.

r	r (3rd decimal) ·000	·002	·004	·006	·008
·00	·0000	·0020	·0040	·0060	·0080
1	·0100	·0120	·0140	·0160	·0180
2	·0200	·0220	·0240	·0260	·0280
3	·0300	·0320	·0340	·0360	·0380
4	·0400	·0420	·0440	·0460	·0480
·05	·0500	·0520	·0541	·0561	·0581
6	·0601	·0621	·0641	·0661	·0681
7	·0701	·0721	·0741	·0761	·0782
8	·0802	·0822	·0842	·0862	·0882
9	·0902	·0923	·0943	·0963	·0983
·10	·1003	·1024	·1044	·1064	·1084
1	·1104	·1125	·1145	·1165	·1186
2	·1206	·1226	·1246	·1267	·1287
3	·1307	·1328	·1348	·1368	·1389
4	·1409	·1430	·1450	·1471	·1491
·15	·1511	·1532	·1552	·1573	·1593
6	·1614	·1634	·1655	·1676	·1696
7	·1717	·1737	·1758	·1779	·1799
8	·1820	·1841	·1861	·1882	·1903
9	·1923	·1944	·1965	·1986	·2007
·20	·2027	·2048	·2069	·2090	·2111
1	·2132	·2153	·2174	·2195	·2216
2	·2237	·2258	·2279	·2300	·2321
3	·2342	·2363	·2384	·2405	·2427
4	·2448	·2469	·2490	·2512	·2533
·25	·2554	·2575	·2597	·2618	·2640
6	·2661	·2683	·2704	·2726	·2747
7	·2769	·2790	·2812	·2833	·2855
8	·2877	·2899	·2920	·2942	·2964
9	·2986	·3008	·3029	·3051	·3073
·30	·3095	·3117	·3139	·3161	·3183
1	·3205	·3228	·3250	·3272	·3294
2	·3316	·3339	·3361	·3383	·3406
3	·3428	·3451	·3473	·3496	·3518
4	·3541	·3564	·3586	·3609	·3632
·35	·3654	·3677	·3700	·3723	·3746
6	·3769	·3792	·3815	·3838	·3861
7	·3884	·3907	·3931	·3954	·3977
8	·4001	·4024	·4047	·4071	·4094
9	·4118	·4142	·4165	·4189	·4213
·40	·4236	·4260	·4284	·4308	·4332
1	·4356	·4380	·4404	·4428	·4453
2	·4477	·4501	·4526	·4550	·4574
3	·4599	·4624	·4648	·4673	·4698
4	·4722	·4747	·4772	·4797	·4822
·45	·4847	·4872	·4897	·4922	·4948
6	·4973	·4999	·5024	·5049	·5075
7	·5101	·5126	·5152	·5178	·5204
8	·5230	·5256	·5282	·5308	·5334
9	·5361	·5387	·5413	·5440	·5466
r	·000	·002	·004	·006	·008
	r (3rd decimal)				

r	r (3rd decimal) ·000	·002	·004	·006	·008
·50	·5493	·5520	·5547	·5573	·5600
1	·5627	·5654	·5682	·5709	·5736
2	·5763	·5791	·5818	·5846	·5874
3	·5901	·5929	·5957	·5985	·6013
4	·6042	·6070	·6098	·6127	·6155
·55	·6184	·6213	·6241	·6270	·6299
6	·6328	·6358	·6387	·6416	·6446
7	·6475	·6505	·6535	·6565	·6595
8	·6625	·6655	·6685	·6716	·6746
9	·6777	·6807	·6838	·6869	·6900
·60	·6931	·6963	·6994	·7026	·7057
1	·7089	·7121	·7153	·7185	·7218
2	·7250	·7283	·7315	·7348	·7381
3	·7414	·7447	·7481	·7514	·7548
4	·7582	·7616	·7650	·7684	·7718
·65	·7753	·7788	·7823	·7858	·7893
6	·7928	·7964	·7999	·8035	·8071
7	·8107	·8144	·8180	·8217	·8254
8	·8291	·8328	·8366	·8404	·8441
9	·8480	·8518	·8556	·8595	·8634
·70	·8673	·8712	·8752	·8792	·8832
1	·8872	·8912	·8953	·8994	·9035
2	·9076	·9118	·9160	·9202	·9245
3	·9287	·9330	·9373	·9417	·9461
4	·9505	·9549	·9594	·9639	·9684
·75	0·973	0·978	0·982	0·987	0·991
6	0·996	1·001	1·006	1·011	1·015
7	1·020	1·025	1·030	1·035	1·040
8	1·045	1·050	1·056	1·061	1·066
9	1·071	1·077	1·082	1·088	1·093
·80	1·099	1·104	1·110	1·116	1·121
1	1·127	1·133	1·139	1·145	1·151
2	1·157	1·163	1·169	1·175	1·182
3	1·188	1·195	1·201	1·208	1·214
4	1·221	1·228	1·235	1·242	1·249
·85	1·256	1·263	1·271	1·278	1·286
6	1·293	1·301	1·309	1·317	1·325
7	1 333	1·341	1·350	1·358	1·367
8	1·376	1·385	1·394	1·403	1·412
9	1·422	1·432	1·442	1·452	1·462
·90	1·472	1·483	1·494	1·505	1·516
1	1·528	1·539	1·551	1·564	1·576
2	1·589	1·602	1·616	1·630	1·644
3	1·658	1·673	1·689	1·705	1·721
4	1·738	1·756	1·774	1·792	1·812
·95	1·832	1·853	1·874	1·897	1·921
6	1·946	1·972	2·000	2·029	2·060
7	2·092	2·127	2·165	2·205	2·249
8	2·298	2·351	2·410	2·477	2·555
9	2·647	2·759	2·903	3·106	3·453
r	·000	·002	·004	·006	·008
	r (3rd decimal)				

Note: Abridged from *Biometrika Tables for Statisticians,* 4th ed., Vol. 1, (p. 139) by E.S. Pearson and H.O. Hartley, eds., 1956, New York: Cambridge University Press. Copyright 1956 by the Biometrika Trustees. Used by permission.

APPENDIX C
Summation Notation Rules

Early in the text's presentation of statistical procedures, summation notation was introduced as an efficient system for symbolically expressing necessary summing operations. Summation expressions introduce and define topics conceptually and provide equivalent computational formulas. The text omits the algebraic steps demonstrating a proof or development of one statistical expression from another. A few simple rules of summation can help demonstrate the algebraic development of several statistical expressions presented.

Rule 1: If some constant value a is summed over n observations, the resulting sum equals n times a, or

$$\sum_{i=1}^{n} a = na$$

As an example suppose the mean for a group of five observations equals 20. We have substituted the mean as a score for each observation and want to find the sum.

$$\sum_{i=1}^{n} a = \sum_{i=1}^{5} 20 = (20 + 20 + 20 + 20 + 20) = 100$$

or

$$\sum_{i=1}^{5} 20 = 5(20) = 100$$

Summation of the mean as a constant occurs frequently in the development of many statistical expressions. In general terms,

$$\sum_{i=1}^{n} \bar{X} = n\bar{X}$$

However, $n\bar{X} = n(\Sigma X/n) = \Sigma X$. Using the summation rule is just another way to demonstrate that n times the group mean will give you the sum of the raw scores ($n\bar{X} = \Sigma X$).

Rule 2: The summation sign may be distributed across an expression's terms if the expression to be summed is itself a sum or difference, that is:

$$\Sigma(X + Y) = \Sigma X + \Sigma Y \quad \text{and}$$
$$\Sigma(X - Y) = \Sigma X - \Sigma Y$$

This rule applies to an expression involving any number of terms.

Both rules 1 and 2 were applied in Chapter 6 to demonstrate that the sum of the deviations of raw scores from the group mean equals zero. This simple algebraic proof is reproduced here:

$$\begin{aligned} \Sigma x_i &= \Sigma(X_i - \bar{X}) \\ &= \Sigma X_i - \Sigma \bar{X} && \textbf{(rule 2)} \\ &= \Sigma X_i - n\bar{X} && \textbf{(rule 1)} \\ &= \Sigma X_i - \not{n}\left(\frac{\Sigma X_i}{\not{n}}\right) && \textbf{(substitution)} \\ &= \Sigma X_i - \Sigma X_i \\ &= 0 \end{aligned}$$

Rule 3: The sum of the product of a constant a and a variable X_i equals the constant times the summation of the variable, or $\Sigma aX = a\Sigma X$. This rule and rules 1 and 2 were applied in Chapter 6 to derive the raw score formula (6.5) for computing the sum of squared deviations.

$$\begin{aligned} \Sigma x_i^2 = \Sigma(X_i - \bar{X})^2 &= \Sigma(X_i^2 - 2X_i\bar{X} + \bar{X}^2) \\ &= \Sigma X_i^2 - 2\bar{X}\Sigma X_i + n\bar{X}^2 \\ & && \textbf{(rules 1, 2, and 3)} \\ &= \Sigma X_i^2 - 2\left(\frac{\Sigma X_i}{n}\right)\Sigma X_i + n\left(\frac{\Sigma X_i}{n}\right)^2 \\ & && \textbf{(substitution)} \\ &= \Sigma X_i^2 - 2\frac{(\Sigma X_i)^2}{n} + n\left(\frac{X_i}{n^2}\right)^2 \\ &= \Sigma X_i^2 - 2\frac{(\Sigma X_i)^2}{n} + \frac{(\Sigma X_i)^2}{n} \\ \Sigma x_i^2 &= \Sigma X_i^2 - \frac{(\Sigma X_i)^2}{n} \end{aligned}$$

Rule 4: The sum of the product (or quotient) of two variables does *not* equal the product (or quotient) of the two sums, that is,

$$\Sigma X_i Y_i \neq (\Sigma X_i)(\Sigma Y_i) \qquad \text{and} \qquad \sum \frac{(X_i)}{Y_i} \neq \frac{(\Sigma X_i)}{(\Sigma Y_i)}$$

The product (or quotient) of two variables must be treated as one term in the summation. Rule 4 is applied in Chapters 8 and 9. Specifically, the sum of the deviation cross-products is an important quantity in correlation and regression. Development and application of the computational formula presented as formula 8.4 applies to rules 1, 2, 3, and 4.

$$\begin{aligned}
\Sigma xy &= \Sigma(X - \overline{X})(Y - \overline{Y}) \\
&= \Sigma(XY - X\overline{Y} - Y\overline{X} + \overline{X}\overline{Y}) \\
&= \Sigma XY - \Sigma X\overline{Y} - \Sigma Y\overline{X} + \Sigma\overline{X}\overline{Y} && \textbf{(rule 2)} \\
&= \Sigma XY - \overline{Y}\Sigma X - \overline{X}\Sigma Y + \Sigma\overline{X}\overline{Y} && \textbf{(rule 3)} \\
&= \Sigma XY - \overline{Y}\Sigma X - \overline{X}\Sigma Y + n\overline{X}\overline{Y} && \textbf{(rule 1)} \\
&= \Sigma XY - \left(\frac{\Sigma Y}{n}\right)\Sigma X - \left(\frac{\Sigma X}{n}\right)\Sigma Y + \cancel{n}\left(\frac{\Sigma X}{\cancel{n}}\right)\left(\frac{\Sigma Y}{n}\right) \\
& && \textbf{(substitution)} \\
&= \Sigma XY - \frac{(\Sigma X)(\Sigma Y)}{n} && \text{Cancel similar terms. Apply Rule 4.}
\end{aligned}$$

When computing Σxy using the raw score sums, rule 4 must be applied.

Applications of these summation rules are used for more algebraically complex derivations resulting in statistical expressions. For example, Chapter 16 uses double summation symbols. The four rules given still hold for the more complex notational systems.

APPENDIX D
Answers to Problem Sets A and C

CHAPTER TWO

Problem Set A

1. Each of the potential populations listed as sources from which sampling could occur are restrictive in some way. Different segments of the total community population would be excluded, depending on the source used in sampling. No information would be collected from these excluded groups, thereby affecting sample estimates of total community population attitudes. The primary characteristic that makes each source inappropriate for surveying attitudes toward high school graduation requirements is as follows:

   - Excludes persons on the basis of education and economic characteristics.
   - Excludes over 80% of the community population in many cases.
   - Excludes parents whose children have graduated, parents of preschool age children, potential parents, and nonparents who have a vested interest in the local school system and the product it produces.
   - Excludes persons on economic, employment, education, and age characteristics.
   - Excludes individuals primarily on occupational characteristics as yellow pages predominantly list community businesses.

   A better source than any of these may be the white page telephone listings. This source would be slightly restrictive on economic conditions but would be more comprehensive as a community population source than any of the preceding sources.

2. From the information given, it is impossible to describe the population. You would need information on other variables for the persons not sampled and those not responding to demonstrate that they do not differ on important characteristics related to salary level. If the large urban district and/or the 22% not responding had salaries systematically higher or lower than the state population of teachers, the sample data used to estimate the population of average salaries for state teachers would be biased.

3. a. Number of yearly subscriptions for each magazine (ratio scale).
   b. Number of magazines of each type sold over the counter per year (ratio scale).
   c. Conduct a popularity survey asking a sample of individuals to rank order the magazines (ordinal scale), or rate the magazines on a six-point scale from like-very-much to dislike-very-much (ordinal scale which probably would be treated as quasi-interval data), or have respondents name their favorite magazines (nominal).

4. a. Ratio.
   b. Nominal: The first digit rank orders the rooms by height above the ground, but the entire number (e.g., 102, 108) merely classifies the room into a floor-level category.
   c. Ratio
   d. Ordinal (quasi-interval)
   e. Ratio
   f. Ordinal
   g. Potentially ratio if based on the number of persons watching per set number of T.V. sets.
   h. Ratio
   i. Ratio
   j. Interval (Quasi-interval)
   k. Ordinal
   l. Nominal, classifies by state and county
   m. Nominal, classifies by position played
   n. Interval

Problem Set C

1.

	Score on Variable		
Individual	Site Location	Sex	Forms Pretest
2	1	2	9
3	1	1	9
13	1	2	9
15	1	2	9
29	1	1	9
38	1	2	9
46	1	1	9
54	1	2	9
	Constant	Variable	Constant

2. Nominal classifies each individual into a separate category.
3. All belong to a category of preschool-aged children from three to five years old.
4. It is a parameter, because it was assumed that the data base represented a population.

CHAPTER THREE

Problem Set A

1. a. Two primary decisions would need to be made about which elements to include as part of the population of elements: (1) Are only words in the main body of the material to be included or are chapter headings and subheadings, table descriptions and labels, figure descriptions, and so on to be included? and (2) Are symbols and numbers, which can be translated into words, to be included?

   b. The advantages and disadvantages of each of the sampling units depend somewhat on the sampling method and the number of units to be sampled. The worst sampling unit would be the chapter because there are a limited number from which to sample and the lengths of the chapters vary considerably. There would be greater potential for bias in the estimate of the number of words using a subset of the chapters as the sample.

   If a large number of units are sampled (10%–15% of the total) at random or on a stratified basis, the page, the line, and the paragraph should provide reasonable estimates. In this example, however, the more similar the sampling units in number of words, the better the element as the sampling unit. In this respect the line would be best, followed by the paragraph and then the page. The number of words is less consistent from page to page than from paragraph to paragraph because of the tables and figures.

2. a. It would be better to start at a random point initially and then select every 10th name. The arrangement by zip code and alphabetical order would not seem to bias the sample selected in this case in terms of the number endorsing an increase in annual dues.

   b. Because names are arranged by zip code, this sampling procedure will result in a sample from a limited number of locations in the state and from the same geographical area of the state.

   c. Each zip code area is represented by one person in the sample and the last names of the sample elements will all start with letters in the middle of the alphabet. Locations would not be represented in proportion to their size in the population.

   d. Persons attending annual meetings tend to be more interested in and committed to the profession and would more likely endorse an increase than those not attending.

   e. An economic bias might affect endorsement responses in this sample assuming that these persons are new to the profession and may not be making as much money as older members.

3. a. Type of physician specialty would be one stratification variable. Even within specialty areas, the incomes of physicians would differ substantially depending on other factors, such as location of practice in the country or type of client served.

b. Examples of stratification variables include age, socioeconomic status, occupation, or sex.
c. Stratification variables might include aptitude, high-school GPA, sex, type of home community, or parent's occupation.
d. Relationship of writer to the applicant might be one stratification variable that would be related to the objectivity of the recommendation.

4. a. Seven cases would be selected. Cases with IDs 71 and 64 repeat and would be used only once. The IDs selected are 64, 21, 16, 12, 71, 45, and 67.
b. The single digits in any table of random numbers are grouped into sets only for convenience in reading the table. They really exist as single random digits. For this problem, we need groups of five random digits. These groupings could be formed by rows or by columns once a random starting point in the table is identified. Assume the starting point is the fourth row, first column. One plan might result in the following pattern:

8106 9816 7430 etc.

81069 81674

Another plan is

8106 9816 7430

81069

10698

06981 etc.

One could move down columns:

8106 9 816 → 81069
1918 8 512 → 19188
1756 1 371 → 17561

Another possible plan would be to use single digits from five rows:

8108 9816
1918 8512
1756 1371
9933 6445
0291 3271

81190
19792
01539

etc.

These are only a few plans to draw random samples with five-digit ID numbers from the example table. Many other possibilities exist.

c. and d.

Plan 1: Assign ID numbers to each individual from 01 to 30. Starting at a random point in the table, assign the first individual selected to group 1, the second randomly selected individual to group 2, and so forth. Once the random assignment of individuals to groups is completed, it would be best to assign the groups to specific treatment conditions at random. For example, start at the third row, second column, first two digits in the set (5 and 8), and move through the table in the following manner:

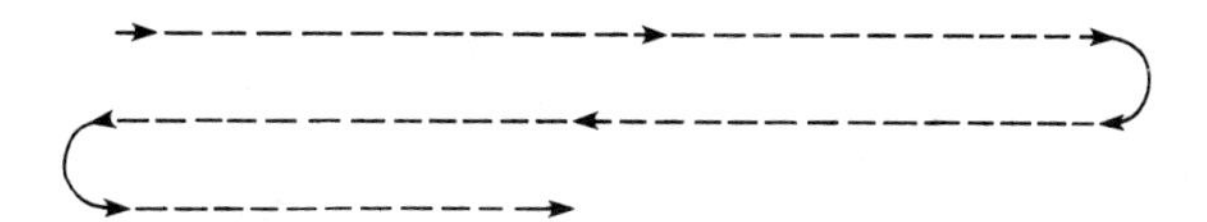

and taking consecutive two-digit sets of numbers results in the following assignment:

Group 1	Group 2	Group 3
21	07	29
30	16	08
19	18	12
05	10	01
28	24	13
17	15	27
02	06	*

*The table has been exhausted using this plan. One would change plans or groupings of digits until all IDs had been selected and assigned.

Plan 2: Select at random a starting point in the table and one of the cases in the list of 30. For each case in the list, select the number 1, 2, or 3 that occurs first in the table of random numbers and assign the case to that group. Continue randomly assigning individuals until 10 cases have been assigned to all groups. For example, start in row 6, in column 4 with the first single-digit number moving through the table using single-digit columns in the following manner:

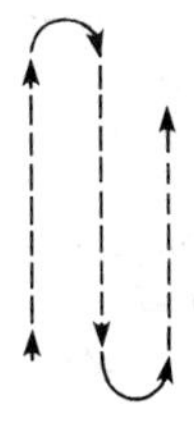

Start with case 14 as the first to be assigned, then case 15, and so on. The assignment of cases to groups would result in the following:

Group 1	Group 2	Group 3
20	14	17
21	15	19
24	16	25
26	18	01
27	22	02
29	23	04
03	28	(Move to the first column, first digit, last row)
05	30	08
06	09	11
07	10	12
(This group is complete. Only select 2s & 3s from the table.)	(This group is complete. Remaining cases go in group 3.)	13

Problem Set C

1. Because the cases are listed at random within each sampling-site location, a systematic sampling plan would be appropriate. One could select a starting point at random within each site and then select every *i*th case, depending on the sample size needed. Simple random sampling would also work, but would be less efficient. Inappropriate sampling plans would be any plans that predetermined the numbers of males and females to be selected or resulted in a biased inclusion or exclusion of males or females from the samples selected.

2. a. The sampling units were children three to five years of age from advantaged and disadvantaged environments.
   b. Sampling was conducted to obtain samples of children representative of each of the following target populations:
      (1) three- to five-year-old disadvantaged children from innercity areas
      (2) four-year-old advantaged suburban children
      (3) advantaged and disadvantaged rural children
      (4) disadvantaged, Spanish-speaking children
   c. While random sampling and assignment were included in the sampling process, the procedure might best be characterized as one of purposive sampling.

3. The simple random sampling plan we employed is described as follows:
   a. Start at the 16th row and 6th column of the five-digit sets of numbers.
   b. Use the second, third, and fourth random digits in the five-digit set to identify the three-digit ID to be selected.
   c. Move to the right through the table going across pages on row 16, then back on row 17, then to row 18, and so on, using all columns of the table before moving to another row.

The IDs for the children sampled and their Letters pretest scores are as follows:

ID	Letters Pretest	ID	Letters Pretest	ID	Letters Pretest
103	23	023	9	117	15
052	39	137	12	065	27
020	19	166	10	219	11
030	5	178	16	175	13
035	17	034	11	216	17

For this second sample, none of the same children were selected. The scores in the present sample ranged from a low of 5 to a high of 39. This range is much greater than for the sample in Section 3.4 (low = 10, high = 29). Except for the range, the scores for each sample appear comparable. Eleven of the 15 scores are in the 10 to 20 range for each sample.

4. The number of children in each viewing category are 54, 60, 64, and 62, respectively. The proportions of the data base each represents is

$$\frac{54}{240} = .225; \frac{60}{240} = .250; \frac{64}{240} = .267; \frac{62}{240} = .258$$

Multiplying these proportions times 80 gives us the number to be selected from each category in our proportional stratified sample.

Category 1: (.225)(80) = 18
Category 2: (.250)(80) = 20
Category 3: (.267)(80) = 21.36 or 21
Category 4: (.258)(80) = 20.64 or 21

CHAPTER FOUR
Problem Set A

1.

Score Interval	f	cf	Exact Limits	Midpoint
39–40	2	45	38.5–40.5	39.5
37–38	7	43	36.5–38.5	37.5
35–36	11	36	34.5–36.5	35.5
33–34	8	25	32.5–34.5	33.5
31–32	5	17	30.5–32.5	31.5
29–30	4	12	28.5–30.5	29.5
27–28	4	8	26.5–28.5	27.5
25–26	1	4	24.5–26.5	25.5
23–24	2	3	22.5–24.5	23.5
21–22	1	1	20.5–22.5	21.5

2. a.

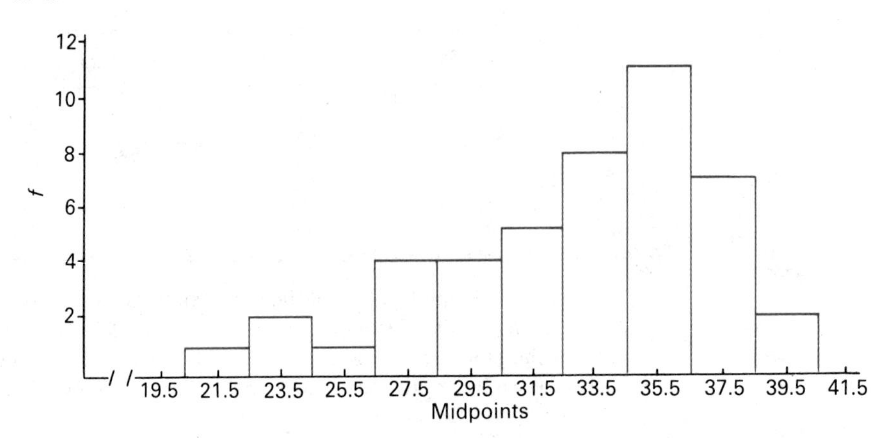

2. b.

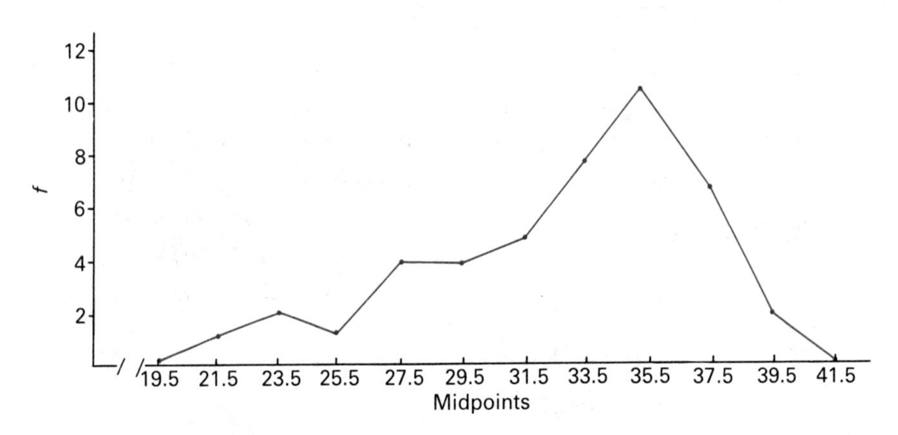

3.

Intervals	Midpoints	Exact Limits
a. 85–87	86	84.5–87.5
82–84	83	81.5–84.5
b. 2.5–2.9	2.7	2.45–2.95
2.0–2.4	2.2	1.95–2.45
c. 19,000–24,000	21,500	18,500–24,500
13,000–18,000	15,500	12,500–18,500
d. 6–11	8.5	5.5–11.5
0–5	2.5	−.5–5.5
e. 131.50–131.74	131.62	131.495–131.745
131.25–131.49	131.37	131.245–131.495

4. a.

	Site I		Site II	
Category	f	$\%f$	f	$\%f$
6	0	.000	7	.146
5	2	.083	12	.250
4	7	.292	3	.063
3	5	.208	5	.104
2	7	.292	11	.229
1	3	.125	10	.208

b.

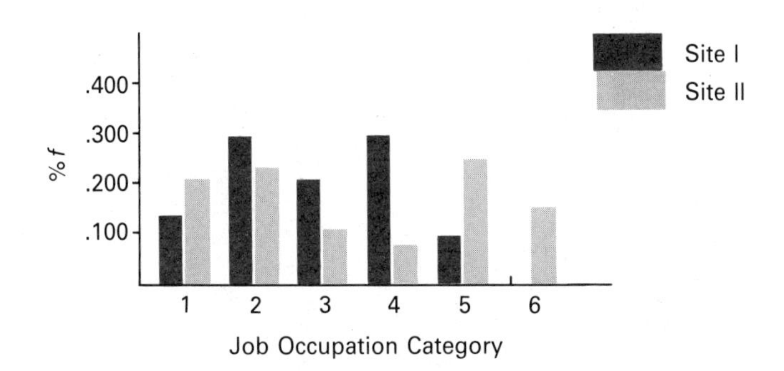

c. The greatest discrepancy occurs in category 4 with Site I having proportionately more blue-collar workers. This, combined with categories 2 and 3, indicates that Site I is a more industrial manufacturing community, whereas Site II is more diverse with a greater proportion of professional/upper management types of persons plus a farming community.

5.

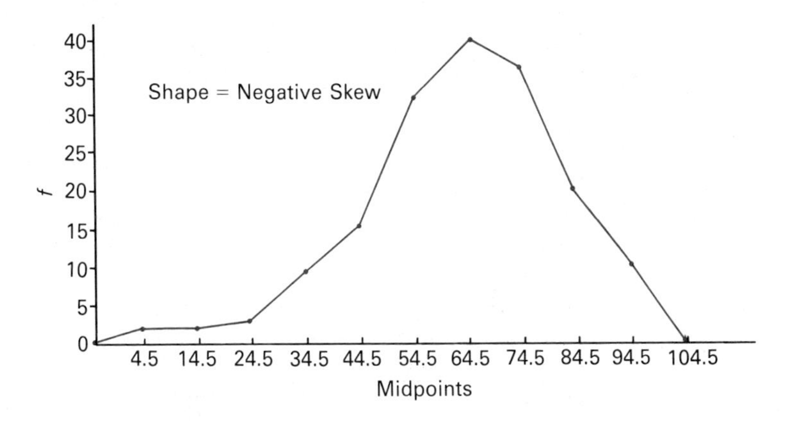

6.

Interval	Homework Assigned %f	No Homework Assigned %f
90–99	20	5
80–89	33	11
70–79	28	22
60–69	10	23
50–59	5	20
40–49	2	9
30–39	1	6
20–29	0	2
10–19	0	1
0–9	1	1

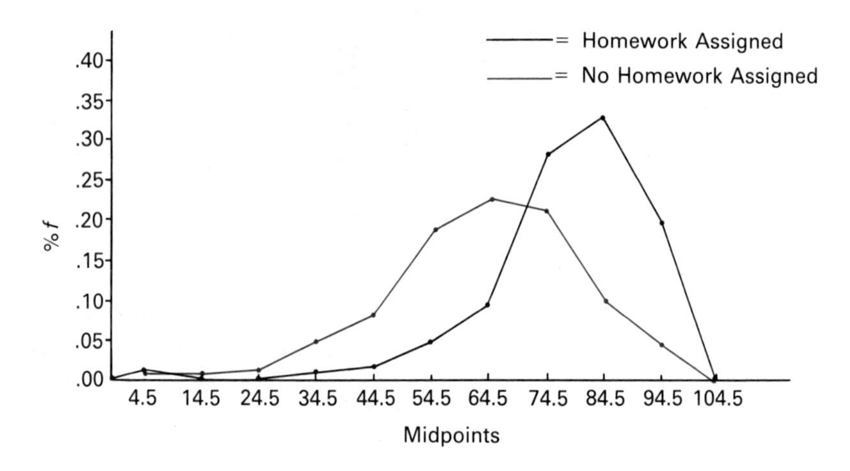

The group that was assigned homework had more negatively skewed scores than the group that had no assigned homework.

7. a. District: $P_{50} = 41.28$
   Region: $P_{50} = 36.92$
   b. District: $PR_{23} = 6.88$ or 7
   Region: $PR_{23} = 14.47$ or 14
   c. Region: $P_{25} = 28.89$
   $P_{75} = 44.47$
   d. District: $PR_{37} = 37.25$ or 37
   Region: $PR_{37} = 50.28$ or 50

The student's score of 37 is at or above 37.25% of all scores in the district and at or above 50.28% of all scores in the region.

Problem Set C

1.

Interval	Pretest Scores f	cf	%f	Posttest Scores f	cf	%f
53–56	1	240	0	2	100	2
49–52	0	239	0	9	98	9
45–48	4	239	2	17	89	17
41–44	3	235	1	10	72	10
37–40	4	232	2	6	62	6
33–36	2	228	1	5	56	5
29–32	4	226	2	6	51	6
25–28	9	222	4	5	45	5
21–24	11	213	5	9	40	9
17–20	37	202	15	12	31	12
13–16	79	165	33	13	19	13
9–12	62	86	26	4	6	4
5– 8	18	24	7	2	2	2
1– 4	6	6	2	0	0	0

2.

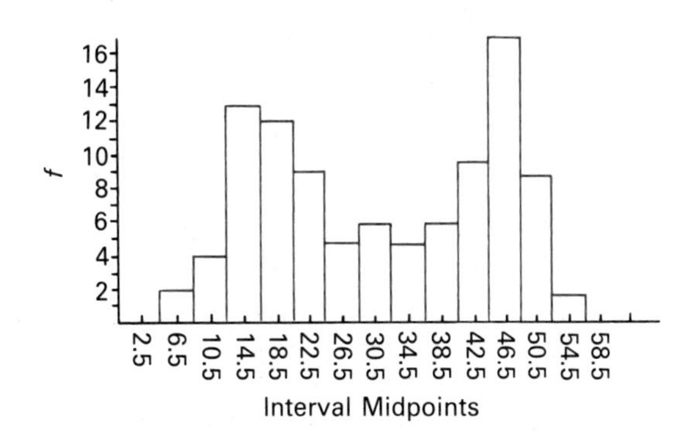

3.

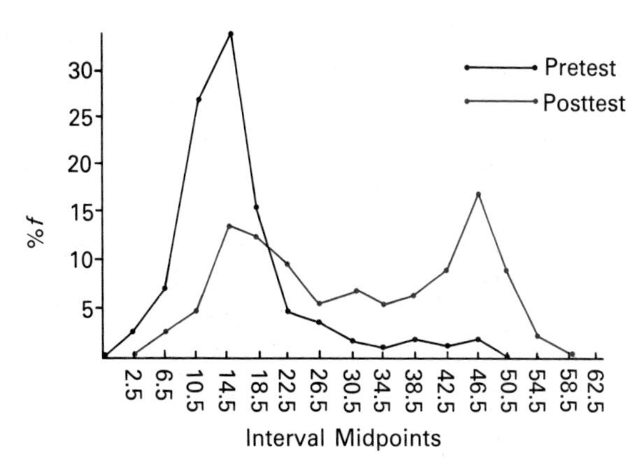

4. Pretest P_{50} = 14.22 Posttest P_{50} = 31.83
Pretest P_{75} = 18.12 Posttest P_{75} = 45.21
Posttest third-quartile interval (45.21 − 31.83 = 13.38) is much larger than the pretest interval (18.12 − 14.22 = 3.90)

5. Pretest: PR_{20} = 82
Posttest: PR_{20} = 29.5
Pretest scores should be lower than posttest scores. For the pretest, 82% had scores lower than 20, while only 29.5% had scores lower than 20 on the posttest.

CHAPTER FIVE

Problem Set A

1. a. $\overline{X} = 14.33$
   b. Median = P_{50} = 12.5
   Using the frequency distribution, the median is 12.5. Actually, any point between 12.5 and 14.5 satisfies the definition.
   c. Mode = 20

2. a. $\overline{X} = 13$
   b. Median = P_{50} = 12.0
   c. Mode = 12

3. a. $\overline{X} = 20.14$
   b. Median = P_{50} = 15.0
   c. Mode = 20

4. $(8 - 14.33) + (11 - 14.33) + (12 - 14.33) + (15 - 14.33) + (20 - 14.33) + (20 - 14.33) = 0$

5. a. Mean = median = mode = unimodal and symmetrical
   b. Mean > median > mode = positive skew
   c. Mean < median < mode = negative skew
   d. Mean = median, two modes = bimodal

6. Median, it would be little affected by the few lower scores.

7. $\overline{X}_{\text{total}} = 22$

8. The effective weights must sum to zero: $\Sigma(X - \overline{X}) = 0$.
   $(3 - 4) + (7 - 1) + (2 - 4) + (? - 4) = 0$
   $? = 4$

9. The mean, median, and mode would all increase by 30 points. For example, assume a set of four scores (3, 7, 2, 4) with $\overline{X} = 4$. If 30 points are added to each score, the set is (33, 37, 32, 34). The mean for this last set is

$$\Sigma X = 136, X = \frac{136}{4} = 34$$

The mean is increased by 30 points.

Problem Set C

1. a.

	Viewing Category 1		Viewing Category 4	
	Pre	Post	Pre	Post
ΣX	292	332	443	542
$\overline{X}$	9.73	11.07	14.77	18.07
Mode	11	16	11	17
Median	10.17	11.0	14.83	18.5

b. The pretest score distribution for Viewing Category 1 appears to have a slight negative skew (mean < median < mode) whereas this is not true for Viewing Category 4 (mean = median > mode). The means and medians for the two groups' posttest scores are approximately the same, indicating no noticeable skewness. Based on the pre- and posttest mean and median values, it would appear that children in Viewing Category 4 learned more than children in Category 1. This is less evident if the modes are used as the measure of central tendency. All the mode values are in line with the mean and median values except for the posttest mode for Viewing Category 1, which is much higher and distorts the interpretation.

2. $\overline{X}_{\text{Cat. 1 and 2}} = 42.92$
$\overline{X}_{\text{Cat. 3 and 4}} = 49.78$
$\overline{X}_{\text{total}} = 46.52$

CHAPTER SIX

Problem Set A

1.

	Data Set			
	1	2	3	4
Score Interval	f	f	f	f
10	1	1	4	6
9				
8		2		
7				
6	8	4	2	3
5				
4		2		
3				
2	1	1	4	1

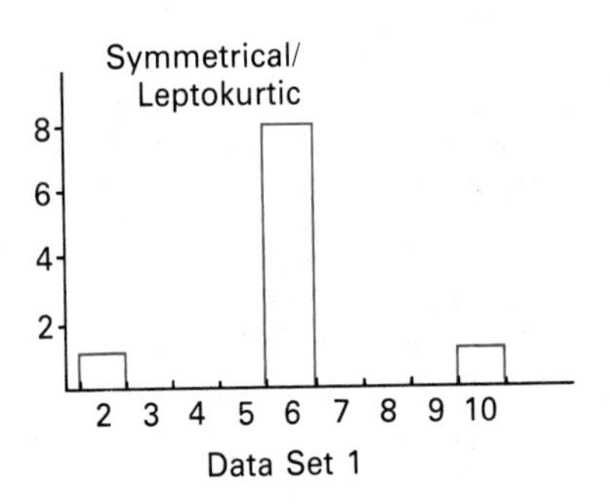

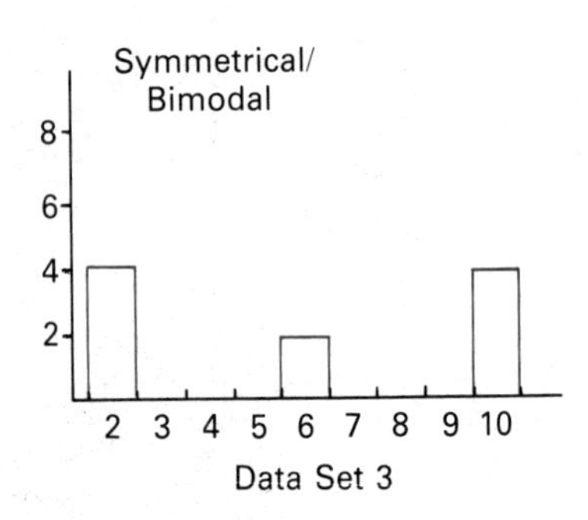

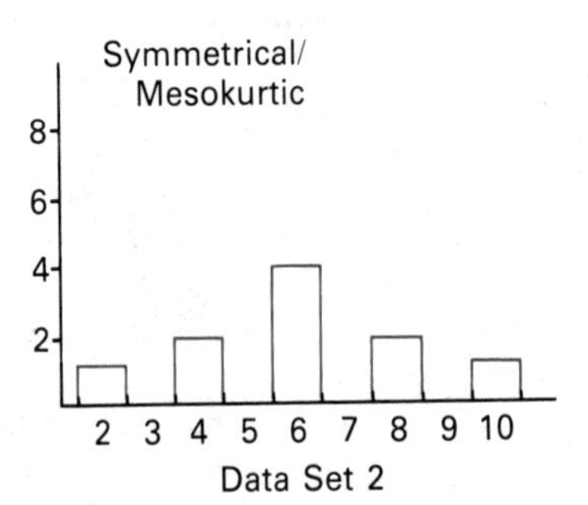

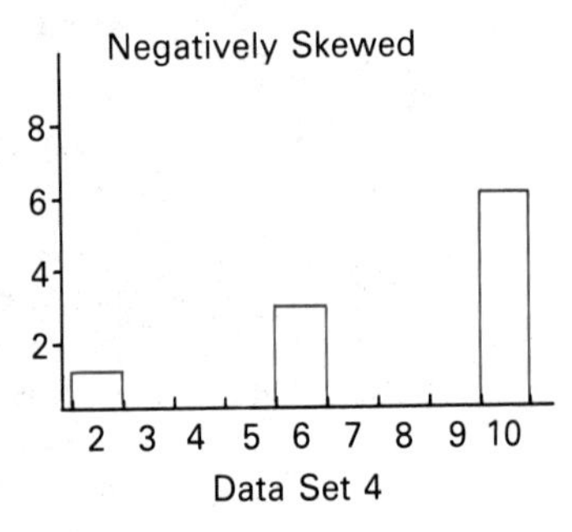

2. Data Set 2: $P_{75} = 7.75$
$P_{50} = 6.0$
$P_{25} = 4.25$
$(P_{75} - P_{50}) = (P_{50} - P_{25}) = 1.75$
Distribution is symmetrical

Data Set 4: $P_{75} = 9.75$
$P_{50} = 9.67$
$P_{25} = 6.00$
$(P_{75} - P_{50}) < (P_{50} - P_{25})$
$.08 < 3.67$
Distribution is negatively skewed

For both data sets 1 and 3, $P_{75} - P_{50}$ should equal $P_{50} - P_{25}$. The distributions are symmetrical.

3.

	Data Set			
	1	2	3	4
ΣX	60	60	60	80
$\overline{X}$	6	6	6	8
Mdn	6	6	6	9.67
Mode	6	6	2, 10	10

The mean, median, and mode are equal for data sets 1 and 2, indicating a symmetrical distribution. The mean equals the median for data set 3, but there are two modes equidistant from $\overline{X}$ and *Mdn,* indicating a bimodal, symmetrical distribution. For data set 4, $\overline{X} < Mdn <$ Mode, indicating a negatively skewed distribution.

4. Data set 1 should be least variable and data set 3 should be most variable. It is difficult to tell whether data set 2 or 4 is more variable, but both should have variability index values between those for sets 1 and 3.

5. For data set 2, $Q = 1.75$. For data set 4, $Q = 1.875$
Data set 4 is slightly more variable based on the calculated values of Q.

6. Deviation Formula: $\Sigma(X - \overline{X})^2 = \Sigma x^2$.
For data set 1, 32; for data set 2, 48; for data set 3, 128; for data set 4, 72

7. For example, if the score chosen is 7:
$(10 - 7)^2 + (10 - 7)^2 + (10 - 7)^2 + (10 - 7)^2 + (10 - 7)^2 + (10 - 7)^2 + (6 - 7)^2 + (6 - 7)^2 + (6 - 7)^2 + (2 - 7)^2 = 82$
The sum of the squared deviations around the value of 7 is 82, which is greater than the value of 72 calculated in problem 6 using the mean.

8. For data set 1, $s^2 = 3.56$, $s = 1.89$; data set 2, $s^2 = 5.33$, $s = 2.31$; data set 3, $s^2 = 14.22$, $s = 3.77$; data set 4, $s^2 = 8.00$, $s = 2.83$.

9. For a population, Σx^2 is divided by n rather than $n - 1$ to compute the variance. Assuming a population,

$$\sigma^2 = 3.2$$

$$\sigma = 1.79$$

The calculated standard deviation assuming a population data set is 1.79, whereas 1.89 was the calculated value assuming sample data.

10. $\overline{X} = 9$, $s^2 = 3.56$, $s = 1.89$
Adding 3 to each value in the data set increased the mean by 3 units, but the variance and standard deviation were unchanged. When multiplying each score by 3, the predicted values using the formula in Section 6.4 are $\overline{X}_{\text{new}} = 6(3) = 18$, $s^2_{\text{new}} = 3.56(9) = 32.04$, and $s_{\text{new}} = 1.89(3) = 5.67$
For the new data set (18, 18, 6, 18, 18, 18, 30, 18, 18, 18),

$$\overline{X} = \frac{180}{10} = 18$$

$$\Sigma x^2 = 3528 - \frac{180^2}{10} = 288$$

$$s^2 = \frac{288}{9} = 32$$

$$s = \sqrt{32} = 5.66$$

Problem Set C

1. *Viewing Category 1*
a. Range = 15
b. $Q = 3.50$
c. $\Sigma x^2 = 513.87$
d. $s^2 = 17.72$
e. $s = 4.21$

Viewing Category 4
a. Range = 15
b. $Q = 2.75$
c. $\Sigma x^2 = 527.87$
d. $s^2 = 18.20$
e. $s = 4.27$

2. In Problem Set C, Chapter 5, the means and medians for the two sets of posttest data were approximately equal, indicating no identifiable skewness in the distribution.

Another method to identify skewness is to compare $P_{75} - P_{50}$ to $P_{50} - P_{25}$. For each distribution, P_{75} and P_{25} were computed in problem 1 to calculate the semi-interquartile range (Q). The posttest score medians for Viewing Categories 1 and 4, respectively, were 11.0 and 18.5. Comparing $P_{75} - P_{50}$ to $P_{50} - P_{25}$ for each group,

Viewing Category					
1			4		
$P_{75} - P_{50}$	?	$P_{50} - P_{25}$	$P_{75} - P_{50}$	?	$P_{50} - P_{25}$
14.75 − 11.0	?	11.0 − 7.75	21.5 − 18.5	?	18.5 − 16.0
3.75	>	3.25	3.0	>	2.5

Similar to the mean and median comparisons, these differences are not great and indicate a slight positive skew, if any. Based on the calculated indices of variability, the range and standard deviations indicate no difference in the variability of scores for the two groups. The values of Q indicate that the scores in the middle 50% of the distribution are slightly more variable for Viewing Category 1 data than for Category 4's group scores.

CHAPTER SEVEN

Problem Set A

1. a. .0668
b. .0107
c. .1255
d. .7580
e. .9115
f. .4844
g. .0775
h. .0857
i. .0500
j. .6528

2. a. .0668
b. .0062
c. .4060
d. .1151
e. .8641

3. a. approximately 421
b. approximately 467
c. approximately 146
d. approximately 379

4. a. 37
b. 36

c. 33.92 and 38.08
d. 29.42 and 41.12

5. $P_{20} = z$ score of $-.84$; $P_{63} = z$ score of .33; $P_{50} = z$ score of .00

6. a. .7734
b. Approximately 25

7. 26.68

8. 125; 121

Problem Set C

I.

	Pretest *z* score				Posttest *z* score			
Individual	BP	L	N	CS	BP	L	N	CS
035	−.53	.13	.66	1.02	.69	1.08	1.39	1.23
148	−.38	.13	−.46	−1.37	.32	.01	1.00	.83
179	−1.47	−.93	−.37	−.50	−1.34	−.59	−1.38	.24

Individual 35 made relative gains in all areas with the largest improvement in BP and the smallest in CS. Individual 148 made comparable gains on three of the four variables, the exception being on letters. Large gains were demonstrated in Numbers and Classification Skill scores. Individual 179 had a sizable loss for Numbers and slight gains on the other three variables. The largest increase was made for 148 on CS.

II. 45; 53

CHAPTER EIGHT

Problem Set A

1. a. 6.0
b. 44
c. 2.97
d. 6.0
e. 32
f. 2.53
g. 12
h. 2.40
i.–1. .32

2. The change in Y scores for individuals C and F result in paired z_x and z_y scores that are more similar than in problem 1. The scores covary more closely in this latter data set; therefore, r_{xy} should increase ($r_{xy} = .85$).

3. For the new data set, $r_{xy} = .88$. While the scores are matched in terms of their rank order from lowest to highest, each paired X and Y score does not have the identical z score equivalencies. The scatter plot demonstrates that the points do not form a straight line.

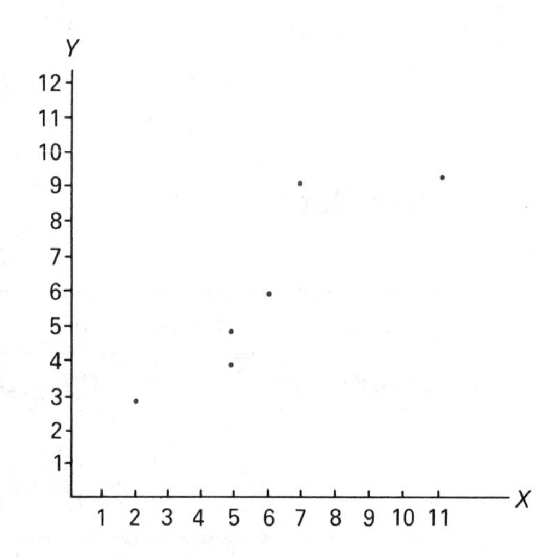

4. $r_{xy} = -.85$. Distribution shapes affect the value of r_{xy}. If the distribution were perfectly symmetrical for X and Y, the magnitude of the relationship would be the same for this problem as in problem 3, but r_{xy} would be opposite in sign.

5. $\overline{Y} = 16$; $s_y = 2.53$; $s_{xy} = 2.40$; $r_{xy} = .32$. The mean for Y is increased by 10 score points, but adding a constant does not affect (change) the values of s_y, s_{xy}, or r_{xy}.

6. $\overline{Y} = 24$; $s_y^2 = 102.4$; $s_y = 10.12$; $s_{xy} = 9.60$; $r_{xy} = .32$
Multiplying each Y score by the constant value of 4 has the following effects: (a) the mean is changed by the product of the constant or 4 times—$\overline{Y}_{\text{new}} = (4)\overline{Y}_{\text{old}} = 4(6) = 24$; (b) the variance is changed by the product of the constant squared or 4^2 times—$s^2_{\text{new}} = (4^2)s^2_{\text{old}} = 16(6.40) = 102.4$; (c) the standard deviation is changed by the product of the constant or 4 times—$s_{\text{new}} = (4)s_{\text{old}} = 4(2.53) = 10.12$; (d) the covariance is changed by the product of the constant or 4 times—$s_{xy\,\text{new}} = (4)s_{xy\,\text{old}} = 4(2.40) = 9.60$; and (e) the correlation coefficient value does not change because both the numerator covariance term s_{xy} and the denominator s_y were each increased 4 times, thereby not differentially affecting r_{xy}.

Problem Set C

Three persons (40, 119, and 166) have missing data on the Home Education Index and would need to be eliminated from the analysis. The data for the remaining 27 individuals are shown here.

Student ID	X	Y	Student ID	X	Y	Student ID	X	Y
7	12	14	91	17	19	182	12	11
12	25	11	96	14	17	188	20	18
19	16	16	109	22	19	189	4	18
21	24	15	126	14	10	194	30	19
30	5	16	146	18	8	202	17	19
56	15	11	150	13	14	226	11	15
67	15	20	153	7	7	230	12	9
71	22	19	160	14	13	234	13	18
82	44	21	163	11	10	238	13	21

1. Letters Pretest

Intervals	f
43–45	1
40–42	0
37–39	0
34–36	0
31–33	0
28–30	1
25–27	1
22–24	3
19–21	1
16–18	4
13–15	8
10–12	5
7–9	1
4–6	2

HEI

Intervals	f
21–22	2
19–20	6
17–18	4
15–16	4
13–14	3
11–12	3
9–10	3
7–8	2

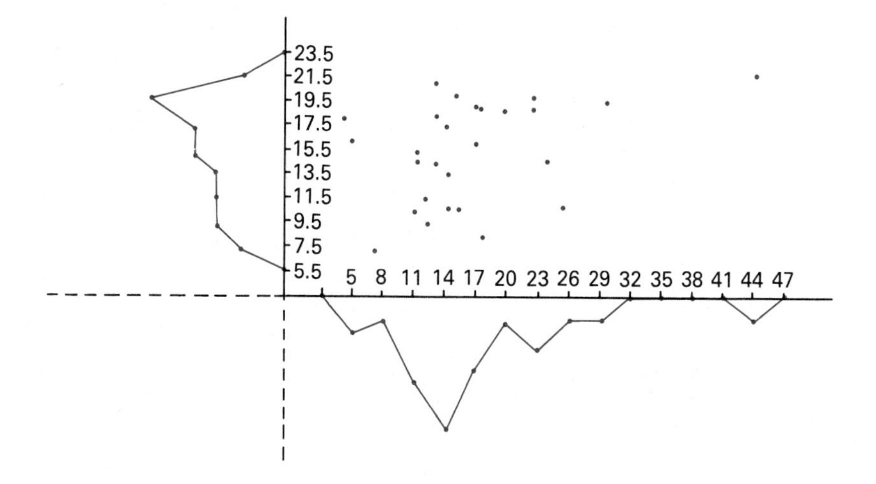

The Letters pretest distribution is positively skewed and the HEI distribution appears to be negatively skewed. Because the distributions are skewed in opposite directions, we would expect the size of r_{xy} to be restricted. The relationship appears to be positive but very low. Without the individual who had a score of 44 on X and 21 on Y, the scatter plot would be circular.

2. $\overline{X} = 16.30$; $s_x = 8.09$; $\overline{Y} = 15.11$; $s_y = 4.22$.

3. $s_{xy} = 12.31$, thus the relationship is positive, or direct.

4. $r_{xy} = .36$
The relationship is low moderate in size. If the data point (44, 21) for individual 82 were removed, $r_{xy} = .24$.

CHAPTER NINE

Problem Set A

1. Criterion variable = teacher-made test scores, predictor variable = ability.

2. Criterion variable = activity level, predictor variable = sugar per unit weight.

3. $Y' = .7X + 1.2$; $b = .7$, indicating that for every change in X of one unit, Y' should change .7 of a unit. When X changes from 8 to 9, Y' should change .7 of a unit. To illustrate, when $X = 8$, $Y' = (.7)(8) + 1.2 = 6.8$. When $X = 9$, $Y = (.7)(9) + 1.2 = 7.5$ The difference or change in $Y' = 7.5 - 6.8 = .7$ When X changes from 5 to a value of 14, a change of 9 units occurs. Thus, Y' should change .7 units for each of the 9 units change in X or $(.7)(9) = 6.3$ units total. To illustrate, when $X = 5$, $Y' = (.7)(5) + 1.2 = 4.7$. When $X = 14$, $Y' = (.7)(14) + 1.2 = 11$. The difference or change in $Y' = 11 - 4.7 = 6.3$

4. Regression equation: $Y' = 1.5X - 1$
For values of $X = 0, 1, 4, 6, 10$, $Y' = -1, .5, 5, 8, 14$. If $Y = 12$ for $X = 10$, error $= Y - Y' = 12 - 14 = -2$. If $Y = 15$ for $X = 10$, error $= 15 - 14 = 1$.

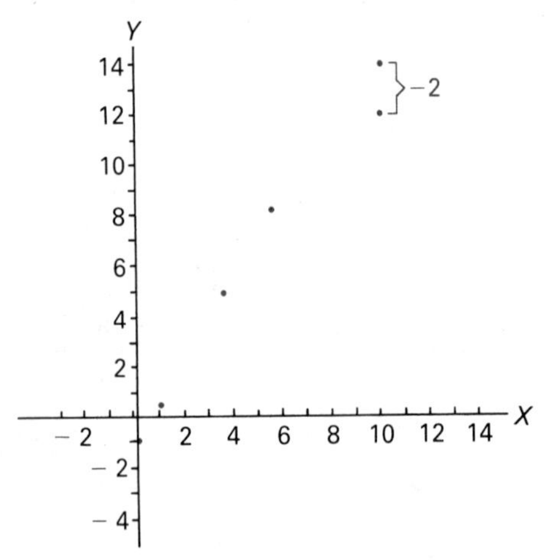

5. a. $Y' = -2X + 2$
For $X = 0$, $Y' = 2$
For $X = 5$, $Y' = -8$

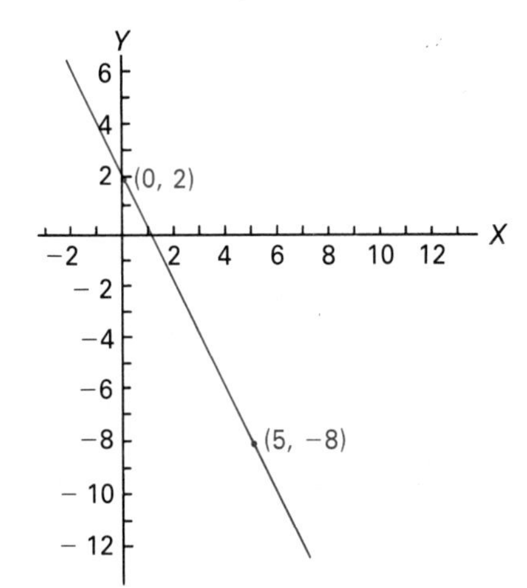

b. $Y' = -.8X - 2.5$
For $X = 0$, $Y' = -2.5$
For $X = 5$, $Y' = -6.5$

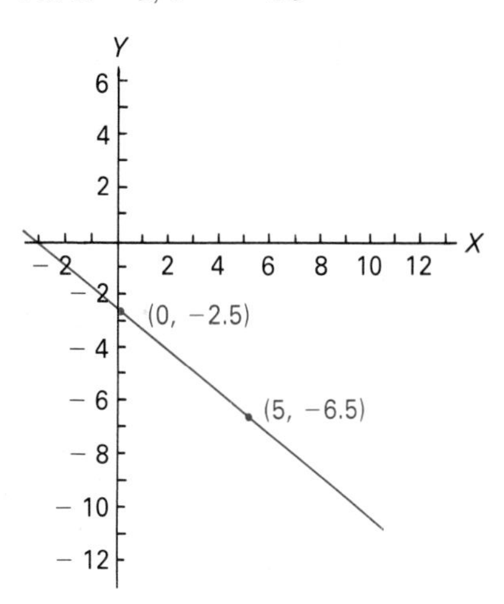

6. a. The value of the regression coefficient, b, indicates the change in Y for every unit change in X ($b = 1.00$).
b. $Y' = 1.00X + 1.5$.
c. To plot the regression line, select any two points that satisfy the equation. Draw the line through these two points, for example, when $X = 0$, $Y' = 1.5$. When $X = 5$, $Y' = 6.5$.
d. The error scores for each individual, respectively, are −2.5, .5, −.5, −.5, .5, 2.5, −.5, and .5. The mean, variance, and standard deviation of the error scores are 0, 2.00, and 1.41, respectively.
e. $r_{xy} = .88$; $s^2_{y \cdot x} = 1.98$; $s_{y \cdot x} = 1.41$
f. .7744
g. $s^2_{y'} = 6.79$; $r_{xy}(s^2_y) = (.88)^2(8.79) = 6.81$

7. $Y' = -.40X + 45.2$

8. Since $b = r_{xy}(s_y/s_x)$, increasing s_y will increase the absolute value of b. Decreasing s_y will decrease the absolute value of b. The opposite effect occurs for increases or decreases in s_x.
In problem 7, when $r_{xy} = -.80$, $s_x = 12$, and $s_y = 6$, then $b = -.40$
If s_y is decreased to 3, $b = -.20$
If s_y is increased to 9, $b = -.60$
If s_x is decreased to 9, $b = -.53$
If s_x is increased to 15, $b = -.32$

9. a. $Y' = .933X + 14.68$
b. $Y' = 54.8$
c. $s_{y \cdot x} = 5.71$
d. .4168

e. .1635
f. Variance of Y' scores = 31.36
variance of error scores = 32.60

Problem Set C

1. $b = .42$; thus, the change in Numbers Pretest score predicted for every 10 units change in PMA scores is 10(.42) = 4.2

2. $Y' = .42 \text{ PMA} + 1.25$
For PMA = 40, $Y' = 18.05$; $s_{y \cdot x} = 8.36$

3. The individuals with PMA scores of 40 are 24, 70, 83, 155, and 187. Their respective error scores are −2.05, −6.05, 15.95, 6.95, and 2.95.

4. When predicting Letters posttest scores, $Y_1' = 4.96X + 14.30$. When predicting Relational Terms posttest scores, $Y_2' = 1.11X + 9.24$.

5. Letters posttest (Y_1) would show the greatest change when moving from one quartile to the next ($b = 4.96$ vs. $b = 1.11$).
The regression coefficient when predicting Y_1 is greater primarily because of the greater variability in Y_1 scores ($s_{y_1} = 13.80$) in comparison to Y_2 scores ($s_{y_2} = 2.67$).

6. Knowledge of viewing category is most beneficial in accounting for variability in the Relational Terms Posttest because the correlation between it and VQ is greater: $r_{xy_1} = .50$, $r^2_{xy_1} = .25$; $r_{xy_2} = .58$, $r^2_{xy_2} = .34$

CHAPTER TEN

Problem Set A

1. a. .12
b. .05
c. .21
d. .46
e. .84
f. .01
g. .0021.

2. There are eight possible different outcomes for the three trials: (H, H, H), (H, H, T), (H, T, H), (H, T, T), (T, H, H), (T, H, T), (T, T, H), and (T, T, T).
a. $p(\text{T, T, T}) = .125$
b. $p(\text{two heads, one tail}) = .375$
c. $p(\text{H, H, T}) = .125$
d. $p(\text{at least one head}) = .875$

3. a. $C\binom{6}{2} = 15$
b. $C\binom{15}{8} = 6435$

c. $P\binom{6}{2} = 30$
d. $P\binom{15}{8} = 259{,}459{,}920$
e. $C\binom{500}{4} = 2{,}573{,}031{,}125$
f. $P\binom{500}{4} = 30{,}876{,}373{,}500$

4. a. .10
b. .0048
c. .0002

5. a. $p(r = 3) = .0146$
b. $p(r = 0 \quad \text{or} \quad r = 1) = .8857$
c. $p(r \geq 4) = .0013$

6. a. $p(r = 0) = C\binom{10}{0}(.4)^0(.6)^{10} = .0060$
b. $p(r \geq 1) = 1 - p(r = 0) = 1 - .0060 = .9940$

7. Seven or more successes ($r \geq 7$) would meet the probability criterion.

8. No, the normal approximation should only be used if np or nq, whichever is smaller, is greater than or equal to five. The value $nq = 10(.6) = 6$ meets this criterion, but $np = 10(.4) = 4$ does not.

9. Mean $= 20(.40) = 8$
Variance $= 20(.40)(.60) = 4.8$
Standard deviation $= 4.8^{1/2} = 2.19$
a. $p(r \geq 10) = .2483$
b. $p(r = 3) = .0142$
c. $p(r \leq 6) = .1271$
d. $p(4 \leq r \leq 8) = .3889$

Problem Set C

1. a. $p(X = 10) = .317$
b. $p(X \leq 4) = .225$
c. $p(X > 7) = .525$
d. $p(X = 5, 6, \text{ or } 7) = .200$

2. a. .521
b. .479
c. .214
d. .508
e. .324
f. .009
g. .244

3. $p(\text{success}) = .214$; $p(r \geq 5) = .0020$.

4. $p(\text{success}) = .214$; use the normal approximation to the binomial.
a. $p(r = 10) = p(-.41 \leq z_r \leq -.07) = .1312$
b. $p(r \geq 40) = p(z_r \geq 9.93) = .0000$
c. $p(r < 5) = p(z_r < -2.14) = .0162$

CHAPTER ELEVEN

Problem Set A

1. a. .2119
b. .0808
c. .0228
d. .0000

2. a. .0228
b. .2514
c. .3446

3. a. .1587
b. .8413
c. .0456
d. .8610

4. a. $z = -8.00$
b. $z = -4.67$
c. $z = -3.50$
d. $z = -3.33$
While none of the probabilities are large, that for population d would be the largest.

5. Approximately 34

6. $C(9.648 \leq \mu \leq 14.352) = .95$; $C(10.026 \leq \mu \leq 13.974) = .90$

7. $C(120.38 \leq \mu \leq 124.42) = .80$

8. $C(66.395 \leq \mu \leq 73.605) = .99$

9. Over repeated samplings of size 25 from a population with $\sigma = 7$, 99% of the confidence interval limits identified will contain the value of μ within their boundaries.

Problem Set C

1. a. .1635
b. .3410
c. .1112

2. a. $f = 16$; proportion = .16 (vs. .1635 expected)
b. $f = 36$; proportion = .36 (vs. .3410 expected)
c. $f = 12$; proportion = .12 (vs. .1112 expected)

3. For the random samples we selected, the following means were computed. The mean for $n = 5$ appears to be unusually discrepant from $\mu = 30.43$. For $n = 5$, $\overline{X} = 20.2$; for $n = 18$, $\overline{X} = 28.57$.

4. For $n = 5$, $\sigma_{\overline{X}} = 5.64$ and $p(\overline{X} \leq 20.2) = .0351$. For $n = 18$, $\sigma_{\overline{X}} = 2.97$ and $p(\overline{X} \leq 28.57) = .2643$.
The sample $\overline{X}$ for $n = 5$ has a low probability of occurring.

5. For $n = 5$, $C(10.92 \leq \mu \leq 29.48) = .90$
For $n = 18$, $C(23.68 \leq \mu \leq 33.46) = .90$
The value of $\mu = 30.43$ does not lie within the limits of the confidence interval for $n = 5$.

CHAPTER TWELVE

Problem Set A

1. The following are not appropriate expressions:
a. Null hypotheses are made about population parameters, not sample statistical values, (e.g., H: $\mu = 40$ would be appropriate).
c. Null hypotheses are statements of equality or no difference.
e. Appropriate if x and y represent measurement at two different times for the same group. If x and y represent two different variables, the hypothesis is not legitimate.
f. Statement of inequality is not appropriate.
h. Statement must be about population parameters.
i. Same as *h*.

2. c. H_0: $\mu = 540$
f. H_0: $\mu_x = \mu_y$ where x and y represent different groups or measurement time, but not different variables.

3. a. $z_{obs} = -1.73$
$z_{crit} = \pm 1.645$
reject H_0
a Type I error is made
b. $z_{obs} = -1.73$
$z_{crit} = -1.645$
reject H_0
correct decision is made
c. $z_{obs} = -1.73$
$z_{crit} = \pm 1.96$
do not reject H_0
correct decision is made
d. $z_{obs} = -1.73$
$z_{crit} = -2.33$
do not reject H_0
a Type II error is made
e. $z_{obs} = 1.80$
$z_{crit} = \pm 1.96$
do not reject H_0
correct decision is made
f. $z_{obs} = 2.77$
$z_{crit} = 2.33$
reject H_0
correct decision is made

g. $z_{obs} = -7.00$
   $z_{crit} = \pm 2.575$
   reject H_0
   a Type I error is made
h. $z_{obs} = 2.5$
   $z_{crit} = 1.645$
   reject H_0
   correct decision is made

4. Note: A Type I error can exist and therefore be affected only when H_0 is true. When H_0 is false, only a Type II error can be affected.

	Probability of Making	
	Type I Error	*Type II Error*
a.	decreased	—
b.	not affected	—
c.	—	increased
d.	not affected	—
e.	—	decreased
f.	—	increased
g.	increased	—
h.	not affected	—
i.	—	increased
j.	—	decreased

5. Power is the complement to a Type II error and has meaning only when H_0 is false. Power is affected in the opposite direction of the Type II error probability (e.g., in c, power will decrease).

6. case A, power = .7642; case B, power = .2266; case C, power = .6591

7. a. .6179
   b. .3336
   c. .1131
   d. .4721
   e. .1611

8. a. The true population mean, $\mu_T = 105$, is farther from the null hypothesized population mean value in problem 7 than in problem 6. This increased discrepancy between μ_T and μ_0 increases power. The greater the discrepancy between μ_T and μ_0 when all other factors are held constant, the greater is the power when testing H_0. In problem 6 with $\mu_T = 103$, the power was .2266. In problem 7 with $\mu_T = 105$, power was .6179.
   b. A decrease in the value of the population variance, σ, decreases the value of the standard error of the mean ($\sigma_{\bar{x}} = \sigma / n^{1/2}$) and thus increases the accuracy of the estimate of the true population mean, making the test of H_o more powerful.
   Power in problem 6 when $\sigma = 12$ is .2266.
   Power in problem 7 when $\sigma = 10$ is .3336.
   c. Decreasing sample size increases $\sigma_{\bar{x}}$, the standard error, and decreases power. When $n = 40$ (problem 6), power = .2266. When $n = 20$ (problem 7), power = .1131.

d. Increasing the value of α (larger probability of a Type I error) increases power. When $\alpha = .01$, power $= .2266$. When $\alpha = .05$, power $= .4721$.
e. A nondirectional test of H_0 is less powerful than a directional test of H_0 because the critical values are larger.
For a directional test of H_0, power $= .2266$.
For a nondirectional test of H_0, power $= .1611$.

9. 34; 61

10. 22; 11

Problem Set C

1. $z_{obs} = -.28$
$z_{crit} = \pm 1.645$;
do not reject H_o

2. *Body Parts*
$\overline{X} = 28.2$
$z_{obs} = 3.31$
$z_{crit} = 1.645$
reject H_0
Relational Terms
$\overline{X} = 11.5$
$z_{obs} = 1.71$
$z_{crit} = \pm 1.96$
do not reject H_0
Classification Skills
$\overline{X} = 18.8$
$z_{obs} = 3.69$
$z_{crit} = 2.33$
reject H_0
The evidence would suggest that more frequent viewing as defined by Viewing Category 3 results in statistically higher mean performance in knowledge of Body Parts and Classification Skills, but not in Relational Terms.

3. The steps involved in the solution to this problem are
a. Draw a random sample of size 12 from those individuals who are in Viewing Category 4.
b. Record their scores on the Body Parts posttest.
c. Calculate the sample mean.
d. Test H_0: $\mu_{BP} = 22.54$ using the information in problem 2. Substitute your sample mean into the following formula:

$$z_{obs} = \frac{\overline{X} - 22.54}{\frac{5.4}{\sqrt{12}}}$$

The critical value is 1.645.

4. Power $= .7794$

CHAPTER THIRTEEN

Problem Set A

1. $t_{crit} = \pm 2.069$
2. $z_{crit} = \pm 1.96$
3. $z_{crit} = -2.33$
4. $t_{crit} = 2.624$
5. $t_{crit} = -2.492$
6. $t_{crit} = \pm 2.797$
7. $t_{crit} = 2.492$
8. $t_{crit} = \pm 2.11$
9. $t_{crit} = -1.678$
10. $t_{crit} = \pm 2.015$
11. $t_{obs} = -5.88$
    $t_{crit} = \pm 2.131$; reject H_0
12. Given $\sigma^2 = 2$
    $z_{obs} = -3.39$
    $z_{crit} = -2.33$; reject H_0
13. $t_{obs} = 2.01$
    $t_{crit} = 1.73$; reject H_0
14. $t_{obs} = -1.87$
    $t_{crit} = \pm 2.074$; do not reject H_0
15. $t_{obs} = -3.49$
    $t_{crit} = -1.771$; reject H_0
16. $s_{\bar{x}} = 2.367$
    $t_{obs} = -.97$
    $t_{crit} = \pm 2.262$; do not reject H_0
17. $z_{obs} = -4.10$
    $z_{crit} = \pm 2.58$
    Conclusion: The mean of 38 is considered to be too discrepant to have occurred as a result of random sampling.
18. Power = .450
19. H_0: $\mu_{Exp} = \mu_{Cont}$
    H_A: $\mu_{Exp} > \mu_{Cont}$
    $\bar{X} = 41.13$
    $\bar{Y} = 31.80$

$t_{obs} = .92$
$t_{crit} = 1.70$
Do not reject H_0. The data were *not* sufficient to conclude that reports made aloud were influenced more by the presence of a stooge than reports written down.

20. $\bar{D} = 2.65$
$t_{obs} = 2.34$
$t_{crit} = \pm 2.86$
Do not reject H_0. The sample data were *not* sufficient to conclude that there is a difference in mean reaction time when responding to visual versus auditory stimuli.

Problem Set C

1. $z_{obs} = -.82$
$z_{crit} = \pm 1.645$
Do not reject H_0. The sample mean $\bar{X} = 5.125$ is a likely occurring event.

2. $t_{obs} = .615$; $t_{crit} = \pm 2.365$ with $df = 7$
The sample mean of 10.625 is a likely occurring event.

3. $\bar{X}_1 = 17.7$
$\bar{X}_4 = 27.9$
$t_{obs} = -2.11$
$t_{crit} = \pm 2.101$
Reject H_0. The mean Numbers pretest score was significantly lower at the .05 level of significance for the children who viewed less.

4. $\bar{Y}_1 = 23.6$
$\bar{Y}_4 = 39.3$
$t_{obs} = -2.86$
$t_{crit} = \pm 2.101$
Reject H_0. Those viewing "Sesame Street" less had a significantly lower mean Numbers posttest score at the .05 level of significance.

5. $\bar{D} = 5.9$
$t_{obs} = 3.73$
$t_{crit} = 1.833$
Reject H_0. The posttest mean was significantly higher than the pretest mean.

6. $\bar{D} = 11.4$
$t_{obs} = 5.51$
$t_{crit} = 1.833$
Reject H_0. The posttest mean was significantly higher than the pretest mean.

7. Children from Viewing Category 4 had significantly higher mean scores than those in Viewing Category 1 at the time of both the pretest and posttest. Both groups had significantly higher mean posttest scores than pretest scores, indicating that learning had taken place in both groups. The hypothesis *not* tested was a comparison of the two groups on the change in mean scores from pre- to posttest.

CHAPTER FOURTEEN

Problem Set A

1. a. $H_0: \rho_{xy} = .00$
$H_A: \rho_{xy} \neq .00$
t test
$t_{crit} = \pm 2.048$

b. $H_0: P_{73} = P_{83}$
$H_A: P_{73} > P_{83}$
normal curve
$z_{crit} = 1.645$

c. $H_0: \sigma^2_{pre} = \sigma^2_{post}$
$H_A: \sigma^2_{pre} \neq \sigma^2_{post}$
dependent groups
t test
$t_{crit} = \pm 2.069$; $df = 23$

d. $H_0: P = .43$
$H_A: P > .43$
normal curve
$z_{crit} = 1.645$

e. $H_0: \sigma^2 = a$
$H_A: \sigma^2 \neq a$
chi square
$\chi^2_{crit} = 129.56$ and 74.22

f. $H_0: \sigma^2_{males} = \sigma^2_{fem}$
$H_A: \sigma^2_{males} \neq \sigma^2_{fem}$

$$F_{obs} = \frac{\sigma^2_m}{\sigma^2_{fem}}$$

${}_{.975}F_{(30,15)} = 2.64$
${}_{.025}F_{(30,15)} = .43$

2. a. $t_{obs} = -3.73$; $t_{crit,.10} = \pm 1.697$
$t_{crit,.05} = \pm 2.042$
$t_{crit,.01} = \pm 2.75$
Reject H_0 at all three α levels

b. $\chi^2_{obs} = 74.97$

	$\alpha = .10$	$\alpha = .05$	$\alpha = .01$
Upper χ^2_{crit}	84.81	89.16	98.09
Lower χ^2_{crit}	47.47	44.62	39.40

do not reject

c. $z_{obs} = 1.74$; $z_{crit,.10} = 1.28$
$z_{crit,.05} = 1.645$
$z_{crit,.01} = 2.33$
Reject H_0 at the .10 and .05 levels, but not at the .01 level.

d. $F_{obs} = .505$

	$\alpha = .10$	$\alpha = .05$	$\alpha = .01$
Upper F_{crit}	2.74	3.37	no table
Lower F_{crit}	.44	.38	no table

Do not reject at any α level.

3. H_0: $\rho_{13} = \rho_{23}$
   H_A: $\rho_{13} < \rho_{23}$
   where
   1 = grades in Jones'
   2 = grades in Smith's
   3 = GPA
   $t_{obs} = -2.38$
   $t_{crit} = \pm 2.04$ at $\alpha = .05$
   Reject H_0; conclude that Professor Smith's grades are significantly better predictors of GPA than are Professor Jones' grades.

4. $z_{obs} = -.30$
   $z_{crit} = \pm 1.96$
   do not reject

5. $F_{obs} = 1.26$; $F_{crit} = 2.20$ and .43. Do not reject H_0.

6. $z_{obs} = -6.13$
   $z_{crit} = -2.33$
   Reject H_0; the return rate value of 77% is significantly below the expected rate of 90%.

7. $\chi^2_{obs} = 49.20$
   Since lower χ^2_{crit} = approximately 57.15, reject H_0

Problem Set C

The answers given here are specific to the random samples we selected. Your answers may differ, but the process should be the same.

1. For our sample, the following sums were obtained for scores on Body Parts (X) and Letters (Y) posttests.
   $\Sigma X = 533$
   $\Sigma X^2 = 14807$
   $\Sigma Y = 574$
   $\Sigma Y^2 = 20478$
   $\Sigma XY = 16213$;
   The computed value of r_{xy} was .59.
   $t_{obs} = 3.11$
   $t_{crit} = \pm 2.101$ with $df = 18$
   reject H_0

2. For our sample,
   $\Sigma X = 175$
   $\Sigma X^2 = 2207$

$\Sigma Y = 244$
$\Sigma Y^2 = 4412$
$\Sigma XY = 2979$
$r_{xy} = .49$
Testing H_0: $\rho_{xy} = .42$
$z_{obs} = .30$
$z_{crit} = \pm 2.575$
do not reject H_0

3.

Group 1 Random Sample	Numbers		Group 2 Random Sample	Numbers	
Viewed at Home	Pre (X)	Post (Y)	Viewed at School	Pre (X)	Post (Y)
044	5	6	054	14	27
100	11	18	088	15	19
116	14	20	154	15	40
203	19	39	195	33	44
230	16	36	188	8	11
143	17	34	130	28	29
177	14	27	057	20	36
225	26	40	051	18	23
092	35	31	129	10	23
080	16	18	073	18	28
			007	13	44
			155	25	13

Group 1: $\Sigma X = 173$, $\Sigma Y = 269$, $\Sigma X^2 = 3601$, $\Sigma Y^2 = 8347$, $\Sigma XY = 5194$

Group 2: $\Sigma X = 217$, $\Sigma Y = 337$, $\Sigma X^2 = 4525$, $\Sigma Y^2 = 10831$, $\Sigma XY = 6380$

$r_{xy_1} = .657$
$r_{xy_2} = .315$
H_0: $\rho_{xy_1} = \rho_{xy_2}$
H_A: $\rho_{xy_1} \neq \rho_{xy_2}$
$z_{obs} = .92$
$z_{crit} = \pm 1.96$
do not reject H_0

4. $s_1^2 = 67.57$; $s_2^2 = 54.63$; $F_{obs} = 1.24$
upper $F_{crit}(9, 11) = 3.59$
do not reject H_0

5. For our data, $s_y^2 = 31.64$

$$\chi^2_{obs} = \frac{(15 - 1)(31.64)}{25.6} = 17.3$$

Upper $\chi^2_{crit} = 23.68$ with $df = 14$
do not reject H_0

6. The number of children in the data base from Site III is 64, therefore $p = 64/240 = .267$
$z_{obs} = 3.04$
$z_{crit} = \pm 1.96$
Reject H_0, since the proportion of children from Site III in the text data base is significantly higher than that in the total evaluation sample. A proportion of .267 would not be a likely outcome if sampling had been random.

7. For our sample, the frequency of males in group 1 was 5 ($f_1 = 5$) and in group 2, 11 ($f_2 = 11$).
$z_{obs} = -1.29$
$z_{crit} = -1.645$
Do not reject H_0. There is insufficient evidence to indicate that the proportions of males in the two reading frequency groups are different.

CHAPTER FIFTEEN

Problem Set A

1. a. $df = 1$; $\chi^2_{crit} = 3.84$
b. $df = 5$; $\chi^2_{crit} = 15.09$
c. $df = 12$; $\chi^2_{crit} = 26.22$
d. $df = 10$; $\chi^2_{crit} = 15.99$
e. $df = 4$; $\chi^2_{crit} = 9.49$

2. $f_e = 33.75$
$\chi^2_{obs} = 3.22$; $\chi^2_{crit} = 11.34$
Do not reject; the data are not sufficient to support a difference.

3. $\chi^2_{obs} = 4.34$; $\chi^2_{crit} = 11.34$
Does not deviate significantly from expected distribution.

4. $\chi^2_{obs} = .031$ without Yates' correction. With Yates' correction, $\chi^2_{obs} = .004$.
$\chi^2_{crit} = 3.84$ at $\alpha = .05$
do not reject

5. a. $f_{ei} = 30$
$\chi^2_{obs} = 9.27$
Reject H_0
$\chi_{crit} = 7.81$

b.

	Incorrect	*Correct*	
f_{oi}	77	43	$\chi^2_{obs} = 7.52$
f_{ei}	90	30	$\chi^2_{crit} = 6.63$

reject H_0

6. $\chi^2_{obs} = 5.08$
$\chi^2_{crit} = 18.31$
do not reject

7. $\chi^2_{obs} = 24.77$
$\chi^2_{crit} = 21.03$
reject H_0
$C = .299$

8. $\chi^2_{obs} = 4.47$
$\chi^2_{crit} = 9.49$
do not reject

Problem Set C

1.

ID#	Sex	Age	Read. Freq.	ID#	Sex	Age	Read. Freq.	ID#	Sex	Age	Read. Freq.
075	1	47	5	095	2	56	5	225	2	58	3
056	1	52	4	177	1	48	4	223	1	48	3
085	2	52	5	081	2	55	5	186	2	47	4
238	1	43	4	182	1	51	4	131	2	60	3
072	2	51	4	030	2	39	4	142	1	60	5
093	2	53	4	202	2	56	4	044	2	36	1
156	1	56	6	153	1	55	4	057	2	51	4
197	2	49	6	042	1	43	2	134	1	45	4
087	2	53	4	031	1	39	4	062	2	48	4
200	2	51	4	236	2	51	4	230	1	49	5
013	2	49	6	061	1	52	5	051	1	55	5
045	2	39	3	108	1	56	4	130	2	58	4
185	2	51	4	229	1	46	5	002	2	67	4
121	1	58	5	034	2	42	5	077	2	55	3
132	2	40	6	109	1	48	5	221	2	44	3
224	1	56	5	163	1	50	4	160	2	51	4
233	2	58	5	063	1	55	4	113	1	45	5
198	2	51	2	092	1	50	5	129	2	41	5
220	1	56	5	106	1	52	4	073	1	52	5
167	1	57	3	024	1	35	5	005	1	69	3

Age Categories	f_o	f_e
42 or below	8	15
43–48	12	15
49–54	20	15
55 or above	20	15

$\chi^2_{obs} = 7.21$
$\chi^2_{crit} = 7.81$ with $df = 3$
Do not reject H_0. The obtained frequencies do not differ significantly from those expected based on an equal distribution ($f_e = 15$) in each of the categories.

2.

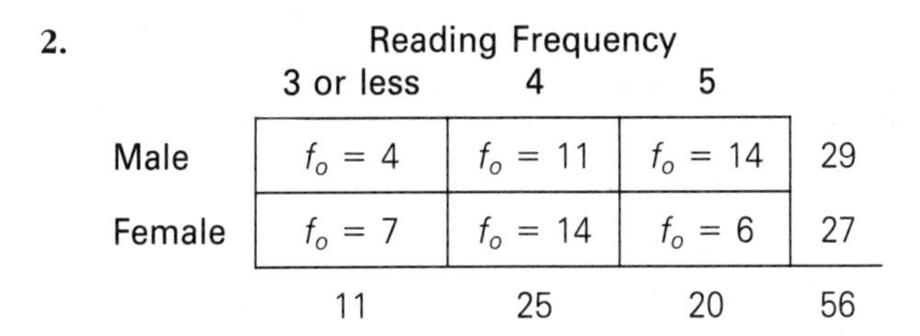

	Reading Frequency			
	3 or less	4	5	
Male	$f_o = 4$	$f_o = 11$	$f_o = 14$	29
Female	$f_o = 7$	$f_o = 14$	$f_o = 6$	27
	11	25	20	56

Note: Four individuals had Reading Frequency codes of 6 and are not included.

$\chi^2_{\text{obs}} = 4.32$

$\chi^2_{\text{crit}} = 4.61$ with $df = 2$

Do not reject H_0. It would appear that sex of child is not related significantly to the reported frequency of reading to the child. Based on the pattern of observed frequencies, it appears that males are not read to with a greater frequency than are females.

3. *Observed Frequencies (f_o) (School Attendance Categories)*

Viewing Setting	1	2	3	4	5	6	
Home	53	0	0	1	1	2	57
School	3	0	23	26	5	0	57
	56	0	23	27	6	2	114

School Attendance categories 2, 5, and 6 are to be eliminated for the test of H_0 as $fe < 5$. The observed frequency pattern is unusual, but not unexpected. Most children who viewed at home were those not attending any school-like setting. Those viewing at school primarily attended nursery school (School Attendance Category 1), preschool (Category 3), or Headstart (Category 4). Note that because an equal number of children are viewing at home and at school in this sample (row observed frequencies), the expected frequencies across these two categories are equal within each school attendance category.

Using only the data from School Attendance categories 1, 3, and 4, the observed chi square value is computed as follows:

$\chi^2_{\text{obs}} = 90.78$

$\chi^2_{\text{crit}} = 5.99$ with $df = 2$

Reject H_0. Not unexpected, where one viewed "Sesame Street" is dependent on school attendance classification. The hypothesis that the two variables are independent was rejected at the .05 level of significance.

4. Observed frequencies for a random sample of 80:

		Viewing Category				
		1	2	3	4	
setting	Home	10	14	11	15	50
	School	10	5	8	7	30
		20	19	19	22	80

$\chi^2_{obs} = 2.81$
$\chi^2_{crit} = 7.81$ with $df = 3$
Do not reject H_0. The pattern of amount of reported viewing did not differ significantly depending on whether viewing occurred in the home or at school.

CHAPTER SIXTEEN

Problem Set A

1. $df_B = 5$; $df_W = 30$; $df_T = 35$
2. $df_B = 12$; $df_W = 130$; $df_T = 142$
3. $df_B = 6$; $df_W = 61$; $df_T = 67$
4. $df_W = df_{W\,\text{group 1}} + \cdots + df_{W\,\text{group 7}} = (n_{.1} - 1) + (n_{.2} - 1) + \cdots + (n_{.7} - 1) = 5 + 10 + 10 + 8 + 14 + 7 + 7 = 61$
5. ${}_{.95}F_{(5,30)} = 2.53$; ${}_{.99}F_{(5,30)} = 3.70$
   ${}_{.95}F_{(12,130)} = 1.83$; ${}_{.99}F_{(12,130)} = 2.34$
   ${}_{.95}F_{(6,61)} = 2.25$; ${}_{.99}F_{(6,61)} = 3.12$
6.

	Treatment Group			
	1	2	3	
$\Sigma X_{.j}$	22	15	20	$\Sigma X_{..} = 57$
$\Sigma X^2_{.j}$	116	81	102	$\Sigma X^2_{..} = 299$
$\overline{X}_{.j}$	4.4	3.0	4.0	$\overline{G} = 3.8$

Source	*df*	*SS*	*MS*	F_{obs}
Between Groups	2	5.2	2.60	.40
Within Groups	12	77.2	6.43	
TOTAL	14	82.4		

${}_{.95}F_{(2,12)} = 3.89$
do not reject H_0

7.

	Treatment Group			
	1	2	3	
$\Sigma X_{.j}$	32	20	40	$\Sigma X_{..} = 92$
$\Sigma X^2_{.j}$	224	116	342	$\Sigma X^2_{..} = 682$
$\overline{X}_{.j}$	6.4	4.0	8.0	$\overline{G} = 6.13$

Source	*df*	*SS*	*MS*	F_{obs}
Between Groups	2	40.53	20.27	3.15
Within Groups	12	77.20	6.43	
TOTAL	14	117.73		

Since ${}_{.95}F_{(2,12)} = 3.89$, the treatment effects added to the data still were not sufficiently large to lead us to reject H_0 given the sample sizes. However, note the increase in SS_B, MS_B, and F_{obs} caused just by the addition of constant *differential* treatment effects to each of the groups. Note that SS_W and MS_W did not change in value from problem 6.

8. $\omega^2 = .22$

9. $F_{obs} = 2.24$
$F'_{crit} = 7.78$
do not reject H_0

10. $F_{obs} = 1.00$
$F'_{crit} = 7.78$
do not reject H_0

11. $F_{obs} = 6.22$
$F'_{crit} = 7.78$
do not reject H_0

12. H_0: $F_{obs} = 4.06$
$F'_{crit} = 7.78$
do not reject H_0

13.

	Contrast Weights		
	$\overline{X}_1$	$\overline{X}_2$	$\overline{X}_3$
C_1: H_0: $\mu_1 = \mu_2$	1	−1	0
C_2: H_0: $\mu_1 = \mu_3$	1	0	−1
C_3: H_0: $\mu_2 = \mu_3$	0	1	−1
C_4: H_0: $\dfrac{\mu_1 + \mu_2}{2} = \mu_3$	1	1	−2

	C_1 vs. C_2	C_1 vs. C_3	C_1 vs. C_4	C_2 vs. C_3	C_2 vs. C_4	C_3 vs. C_4
$\Sigma w_{ij} w_{i'j}$	1	−1	0	1	3	3

The $df_B = 2$, indicating that a set of orthogonal contrasts will contain at most two contrasts. The only pair of contrasts for which $\Sigma w_{ij} w_{i'j} = 0$ are C_1 and C_4; therefore, the subhypotheses defined by C_1 and C_4 define a mutually orthogonal set of contrasts.

$SS_{C_1} = 14.40$
$SS_{C_4} = 26.13$
$SS_B = 40.53$ (From the ANOVA Summary Table, problem 7)
$\Sigma SS_{C_i} = SS_{C_1} + SS_{C_4}$
$= 14.40 + 26.13 = 40.53$

Problem Set C

1. The following random samples of six individuals were selected from each of the four amount of viewing categories.

Viewing Categories			
1	2	3	4
201	091	068	111
100	232	169	076
051	158	110	148
185	054	135	107
066	084	059	237
117	127	187	035

The Forms pretest and posttest scores for each are as follows:

Viewing Categories							
1		2		3		4	
Pre	Post	Pre	Post	Pre	Post	Pre	Post
9	9	8	14	14	19	13	18
5	9	6	12	4	3	15	19
10	14	8	10	13	13	7	18
12	11	9	15	9	13	12	18
10	13	9	14	10	14	12	15
5	3	9	14	12	15	12	15

										Pre	*Post*
$\Sigma X_{.j}$	51	59	49	79	62	77	71	103	$\Sigma X_{..} =$	233	318
$\Sigma X_{.j}^2$	475	657	407	1057	706	1129	875	1783	$\Sigma X_{..}^2 =$	2463	4626
$\bar{X}_{.j}$	8.50	9.83	8.17	13.17	10.33	12.83	11.83	17.17	$\bar{G} =$	9.71	13.25
s_j^2	8.30	15.37	1.37	3.37	13.07	28.17	6.97	2.97			

For the Forms Pretest Hypothesis:

Source	*df*	*SS*	*MS*	*F*	F_{crit}
Viewing Categories	3	52.46	17.49	2.35	3.10
Within Groups	20	148.50	7.43		
TOTAL	23	200.96			

Do *not* reject the null hypothesis of no difference in mean Forms pretest scores across the viewing categories. The four group means do not differ sufficiently beyond that expected under a null hypothesis of equality.

For the Forms posttest hypothesis:

Source	df	SS	MS	F	F_{crit}
Viewing Categories	3	163.17	54.39	4.36*	3.10
Within Groups	20	249.33	12.47		
TOTAL	23	412.50			

*Significant at the .05 level

2. $H_0: \frac{\mu_1 + \mu_2}{2} = \frac{\mu_3 + \mu_4}{2}$

$F_{obs} = 5.90$
$F'_{crit} = 9.30$
do not reject H_0

3.

	Weights for Viewing Categories 1	2	3	4
C_1	1	1	−1	−1
C_2	1	−1	0	0
C_3	0	0	1	−1

The three orthogonal subhypotheses tested by these comparisons are

$H_{01}: \frac{\mu_1 + \mu_2}{2} = \frac{\mu_3 + \mu_4}{2}$ $SS_{c_1} = 73.50$ $F_{obs} = 5.90$
$H_{02}: \mu_1 = \mu_2$ $SS_{c_2} = 33.47$ $F_{obs} = 2.68$
$H_{03}: \mu_3 = \mu_4$ $SS_{c_3} = 56.51$ $F_{obs} = 4.53$
$\Sigma SS_{c_j} = 73.50 + 33.47 + 56.51 = 163.48$
Within rounding error, 163.48 equals SS_B (= 163.17).

4. The 40 individuals here are classified according to the type of preschool attended. The ID is given in parentheses along with Classification Skill posttest scores for each child.

		Preschool Types			
None (1)	Kindergarten (2)	Nursery (3)	Headstart (4)	Day Care (5)	Other (6)
(041) 19	(037) 19	(061) 22	(001) 23	(008) 23	(039) 12
(047) 23	(126) 16	(063) 22	(003) 19	(013) 13	(065) 23
(089) 21	(214) 10	(079) 22	(048) 16	(021) 24	(228) 19
(105) 20		(091) 17	(055) 15	(130) 15	(230) 16
(113) 19		(107) 23	(129) 13	(136) 10	(237) 11
(142) 19		(190) 16	(154) 18	(148) 20	
(147) 21			(164) 18	(152) 14	
(232) 8			(194) 13	(200) 10	
			(224) 21	(220) 16	

$\Sigma X_{.j}$	150	45	122	156	145	81	
$\Sigma X_{.j}^2$	2958	717	2526	2798	2551	1411	
$n_{.j}$	8	3	6	9	9	5	$\Sigma X_{..} = 699$
$\overline{X}_{.j}$	18.75	15.00	20.33	17.33	16.11	16.20	$\Sigma X_{..}^2 = 12961$
$S_{.j}^2$	20.79	21.00	9.07	11.75	26.86	24.70	$\overline{G} = 17.475$

Source	*df*	*SS*	*MS*	*F*	F_{crit}
Preschool Type	5	105.45	21.09	1.12	2.49
Within Group	34	640.52	18.84		
Total	39	745.97			

F of 1.12 is not statistically significant.

INDEX